Deformation and Evolution of Life in Crystalline Materials

An Integrated Creep-Fatigue Theory

Xijia Wu
National Research Council Canada
Ottawa, Ontario, Canada

CRC Press is an imprint of the
Taylor & Francis Group, an **informa** business
A SCIENCE PUBLISHERS BOOK

Cover credit:
Photograph on the cover provided by Dr. Dongyi Seo, National Research Council Canada. The copyright belongs to Government of Canada.

CRC Press
Taylor & Francis Group
6000 Broken Sound Parkway NW, Suite 300
Boca Raton, FL 33487-2742

First issued in paperback 2020

ISBN-13: 978-1-138-29673-2 (hbk)
ISBN-13: 978-0-367-78002-9 (pbk)

Library of Congress Cataloging-in-Publication Data

Names: Wu, Xijia, 1958- author.
Title: Deformation and evolution of life in crystalline materials : an integrated creep-fatigue theory / Xijia Wu (National Research Council Canada, Ottawa, Ontario, Canada).
Description: Boca Raton, FL : CRC Press, 2019. | "A science publishers book." | Includes bibliographical references and index.
Identifiers: LCCN 2019001686 | ISBN 9781138296732 (hardback)
Subjects: LCSH: Crystallography. | Crystal growth. | Crystals. | Crystals--Plastic properties. | Dislocations in crystals. | Minerals.
Classification: LCC QD905.2 .W8 2019 | DDC 530.4/11--dc23
LC record available at https://lccn.loc.gov/2019001686

Visit the Taylor & Francis Web site at
http://www.taylorandfrancis.com

and the CRC Press Web site at
http://www.crcpress.com

Preface

Ever since fatigue and creep were first studied more than one hundred years ago, life prediction for materials in service has been mostly an experimental science. The phenomena that mask the underlying physics and mechanisms are so complicated, especially at high temperatures, that it requires a multi-disciplinary understanding of materials science, solid mechanics, and fracture mechanics. This book attempts to give a holistic treatment of all the material deformation behaviours observed from mechanical testing to date, using a mechanism-delineated approach—the *integrated creep-fatigue theory (ICFT)*—for the first time in the literature.

To obtain this holistic view, we start with the understanding of basic material physical nature—crystalline structures and imperfections (dislocations and vacancies), and the kinetics of their movements in a microstructure strengthened by mechanisms involving interactions of dislocations with various point defects (solute atoms), line defects (dislocation themselves), planar defects (grain boundaries and interfacial discontinuities) and volume defects (precipitates and inclusions), in Chapter 1. Because of the nature of crystalline materials, the theory of anisotropic elasticity is introduced with the Stroh formalism. Indeed, crystals only deform elastically, except the movements of crystalline defects leading to permanent shape change, i.e. inelastic deformation. In Chapter 2, the fundamental deformation mechanisms involving various types of dislocation and vacancy movements are summarized, and their constitutive equations are given to describe both the transient and steady states. In Chapter 3, physical damage processes are discussed in relation to the responsible deformation mechanisms. It classifies damage into either intragranular damage such as dislocation pile-ups and persistent slip band, which tend to cause surface or subsurface crack nucleation; or intergranular damage such as creep cavities and microcracks, which are regarded as *internally distributed damage*. The integrated creep-fatigue theory framework is introduced in Chapter 4, where constitutive laws are formulated based on physical decomposition of mechanism strains, and a holistic damage accumulation equation is derived, considering nucleation and propagation of surface/subsurface cracks in coalescence with internally distributed damage. The process can also be assisted by environmental effects such as oxidation. In the next four chapters, Chapter 5 to Chapter 8, the ICFT is consistently applied to creep, low-cycle fatigue (LCF), thermomechanical fatigue (TMF) and high-cycle fatigue (HCF) processes and life prediction. Detailed mechanism-delineated treatments are given to the above deformation and damage processes, which include: for example, a mechanism-based true-stress model of the three-stage creep behaviour and long-term creep life prediction with oxidation effect, a dislocation-based mechanical fatigue life model and its extension to high-temperature LCF and TMF, and HCF with foreign object damage, dwell damage and prior creep damage. Such descriptions are not provided in any other books. To describe crack nucleation and microscopic crack growth, Chapter 9 presents a mathematical theory to treat continuously distributed dislocation pile-ups in two basic forms—the Zener-Stroh-Koehler (ZSK) type and Bilby-Cottrel-Swinden (BCS) type—in anisotropic crystalline materials with due considerations for dislocation-microstructure interactions. The short crack growth phenomenon is shown as an example. In Chapter 10, the processes of macroscopic crack growth, or phenomenologically long crack growth, under fatigue and creep conditions are treated with the consideration of the average effect of the aforementioned dislocation mechanisms in the crack-tip plastic/creep zone. The emphasis is on explanation of the relationships between the crack growth rate and the crack-tip field controlling parameters such as the stress intensity factor, and the microstructure and environment dependence of such relationships. The crack closure concept is challenged with an energy approach for the K-similitude, and the existence of fatigue crack growth threshold is interrogated. The above deformation and crack growth processes are actually integral parts of the holistic structural integrity process (HOLSIP) that are divided into phases of 1) crack nucleation, 2) small crack growth, 3) long crack growth, and 4) unstable fracture. In line with material development for advanced gas turbine engines, Chapter 11 treats single crystal Ni-base superalloys with consideration of anisotropy in creep, fatigue and fatigue crack growth. In Chapter 12, microstructural evolution and failure mechanisms in thermal barrier coatings are discussed, and a crack number density theory description is given to the crack evolution process. In Chapter 13, constitutive and life prediction models are developed for tensile, creep and fatigue behaviors of ceramics matrix composites, which emerge as advanced gas turbine materials to endure the hottest operating conditions. Last but not the least, Chapter 14 reviews the current and trending component lifing philosophies and provides a few case studies of component analyses using ICFT. Particularly, a new paradigm—the holistic structural integrity process concept—is introduced, which is better suited to the industrial trend of prognosis health management schemes.

The ICFT is developed from the author's work at the National Research Council Canada, in collaboration with colleagues, as governmental research for various industrial clients. The background phenomena and theories have been well documented in numerous publications in the literature. The author can only give a brief summarization with a few references, given the limited space of the book. The readers are encouraged to consult other books on special topics such as theories of elasticity, plasticity and fracture mechanics for further details.

Finally, the author is grateful to Bill Wallace, David Simpson and Jerzy Komorowski, who supported the author's research endeavour when they acted as Director Generals of NRC Aerospace (research institute/portfolio/center) and now are retired. The author also would like to thank several colleagues, Drs. Rick Kearsey, Scott Yandt, and Zhong Zhang, who provided constructive inputs into the book. Part of the analysis work of Chapter 5 was performed by Dr. Xiaozhou Zhang, as part of his Ph.D. thesis under the supervision of the author and Prof. R. Liu from Carleton University.

Xijia Wu

Contents

CHAPTER

1

Crystal Structure and Dislocation Kinetics

1.1 Introduction

About 13.5 billion years ago, matter, energy, time and space came into being in what is known as the Big Bang (from *Sapiens—A Brief History of Humankind* by Y.N. Harari 2014). About 4.5 billion years ago, planet Earth formed. About 10,000 years ago, when humans on planet Earth started to use copper, the time is called the Copper Age. About 5,000 years ago, people started to use bronze, an alloy of copper and tin, and the time is called the Bronze Age. About 3,000 years ago, people started to use iron, and the time is called the Iron Age. Only over the last 300 years, people started to make more complicated alloys, from steels to superalloys, more recently.

Modern civilization is sustained by the use of advanced engineering materials ranging from semiconductors to nickel-base superalloys. The building blocks of all these materials are crystals. When an engineering device, be it as small as a cell phone or as big as a gas turbine engine, enters into service, the clock of its life starts to tick. Design engineers are constantly striving for a better design of high efficiency and durable products. End-users always want to maximize the performance with low-cost product life cycle management. Both need advanced life prediction, which requires the engineers to understand the material behaviour and damage mechanisms operating under service load-environment (static, cyclic, thermal and fluid dynamic) conditions.

Physical insights into the material behaviour begin with understanding of the basic crystalline structures, their defects and the kinetics of defect motion. The interactions between motion of defects and microstructure result in various deformation and strengthening mechanisms. Material designers often utilize these mechanisms to strengthen materials for various applications. For description of deformation of a crystalline solid, one first needs to understand solid mechanics concerning stress and strain distribution within the solid. Usually, stress is generated via external stimuli, and strain is the material's response. Then, particularly, the stress-strain relationship in defect motion needs to be established at the microstructural level. Because of the nature of crystalline solids, anisotropic mechanical response should be considered. Therefore, for generality, the Stroh formalism is briefly introduced to describe the stress distribution around crystalline defects. However, for simplification, most discussions are still given in terms of "isotropic" material properties, especially for homogeneous polycrystalline materials. Interested readers can extend the "isotropic" treatments to "anisotropic" treatments, should advanced designs require such consideration. With understanding of the concepts elucidated in this chapter, the readers shall proceed to indulge in physical deformation and damage mechanisms, the integrated creep-fatigue theory and its application to various deformation processes in the following chapters.

1.2 Crystal Structures and Imperfections

1.2.1 Basic Structures

A crystalline solid contains numerous crystallites (a.k.a. grains). A perfect crystal structure can be constructed by repeated packing of unit cells. A unit cell consists of a group of atoms arranged in a specific configuration such as face-centered cubic (f.c.c.), body-centered cubic (b.c.c.), and hexagonal close-packed (h.c.p.), as shown in Figure 1.1. By this nature, the positions of all the atoms in the crystal are known, once the configuration of the unit cell is defined. The minimum amount of information needed to specify a crystal structure is thus the unit cell type (e.g. f.c.c., b.c.c., h.c.p., etc.), the cell parameters (a, c) and the positions of the atoms in the unit cell.

The response of a crystal to an external stimulus, such as a tensile stress, electric field and so on, is usually dependent on the direction of the applied stimulus. It is therefore important to specify crystallographic directions and planes in an unambiguous fashion. Directions are generally written in terms of Miller Indices as $[u\ v\ w]$, and $[\bar{u}\ \bar{v}\ \bar{w}]$ indicates the opposite direction of $[u\ v\ w]$. The index number is usually a fraction of the unit cell edge. As with Miller indices, it is sometimes convenient to group together all directions that are identical by virtue of the symmetry of the structure. These are represented by the notation $<u\ v\ w>$. For example, in a cubic crystal, <100> represents the six directions [100], $[\bar{1}00]$, [010], $[0\bar{1}0]$, [001], $[00\bar{1}]$. Similarly, the facets of a crystal or internal planes through a crystal structure are also specified in terms of Miller Indices, often written as $(h\ k\ l)$. The family of such planes is noted as $\{h\ k\ l\}$, by virtue of symmetry. Examples of crystallographic directions and planes are given in Figure 1.2 for cubic and hexagonal crystals. Note that the indices, h, k and l represent not just one plane but the set of all parallel planes.

In deformation, certain crystallographic planes and directions are of particular importance, particularly those close-packed with the shortest

Burgers vector (distance to a nearby atom on the same plane). By the nature of crystalline lattice, exchanging bonds between neighbouring atoms has to overcome the Peierls-Nabarro force (Hirth and Lothe 1992):

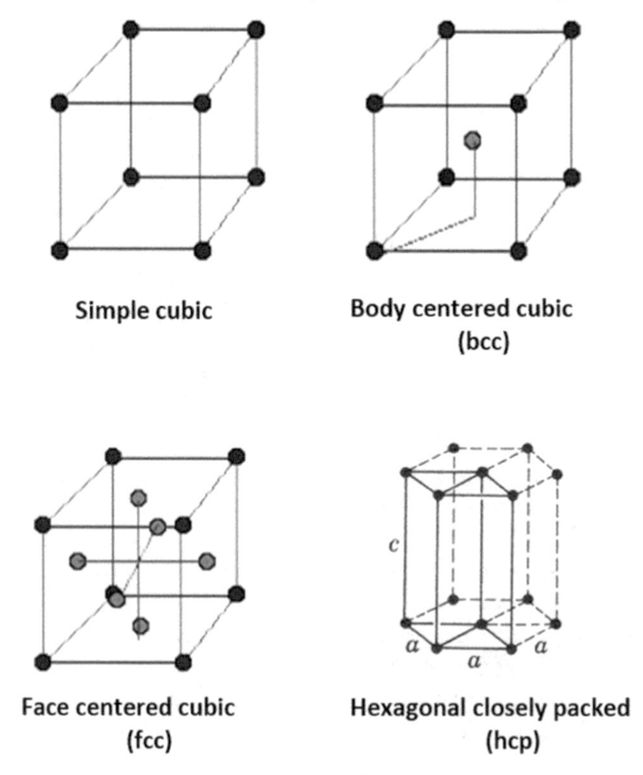

Figure 1.1. Crystal unit cells.

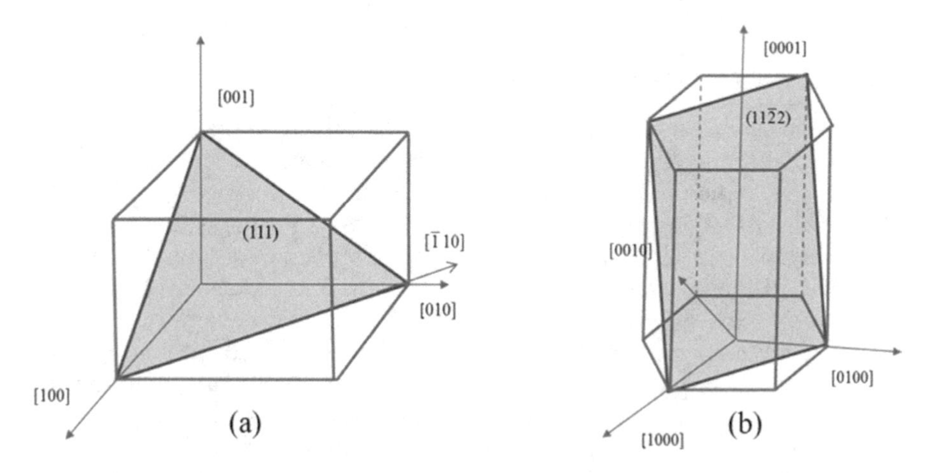

Figure 1.2. Miller indices for crystallographic directions and planes in a) cubic and b) hexagonal crystals.

$$\tau_{p-n} \propto \mu \exp\left(-\frac{a}{(1-\nu)b}\right) \tag{1.1}$$

where μ is the shear modulus, a is distance between similar planes, ν is Poisson's ratio and b is Burgers vector.

Therefore, relative shearing displacements are easy to occur between those close-packed planes and in close-packed directions with maximum a/b ratio, which results in crystallographic slip, leading to plastic deformation. In an f.c.c. crystal, the {111} planes and <110> are more closely-packed than other planes and directions, where slip tends to occur more favourably. The $\{111\} < 0\bar{1}0 >$ slip systems are called octahedral slip systems, and there are 12 such slip systems in an f.c.c. crystal. In b.c.c. crystals, there are no truly close-packed planes, but thermal energy can activate many slip systems such as $\{110\} < \bar{1}11 >$(12), $\{211\} < \bar{1}11 >$ (12), $\{321\} < \bar{1}11 >$ (24), so there are total 48 possible slip systems. In an h.c.p. crystal, the possible slip systems are $\{0001\} < 11\bar{2}0 >$(3), $\{10\bar{1}0\} < 11\bar{2}0 >$ (3), $\{10\bar{1}1\} < 11\bar{2}0 >$(6), so there are 12 in total.

In real materials of any type of crystalline structure, defects always exist, which are imperfections relative to the atomic arrangement of parent crystal. Crystal defects can modify the properties of metals, but they do not always play a "bad" role as it may seem. It has been known that the actual strength of a metal is far less than its theoretical strength of perfect atomic bonding because of the existence of linear defects (dislocations). On the other hand, metals become ductile when dislocations are free to move. This property allows metals to be made into various shapes and forms. Without this property of metals, humans may still live in the Stone Age. Therefore, it is important to have an idea of the types and forms of crystalline defects and the roles they play on the material properties, in order to understand the behaviour of solids.

1.2.2 Point Defects

A point defect is the simplest defect in a crystal, which occurs as a "mistake" at a single atomic site. There are four types of point defects in a crystal: 1) *vacancies*, which are missing atoms at normally occupied positions; 2) *interstitials*, which occur by incorporating atoms at positions not normally occupied; 3) *substitutionals*, where foreign atoms replace the atoms at normally occupied positions in the parent lattice, as schematically shown in Figure 1.3, and 4) *antisite defects*, which may exist in an ordered alloy or compound when atoms of different type exchange positions. Intersitial/substutional defects are also called impurity defects. When quite large numbers of impurity atoms enter a crystal, without changing the crystal structure, the resultant phase is referred to as a solid solution. Solid solutioning is often utilized deliberately as an alloying mechanism to strengthen materials. On the other hand, point defects may move by thermal activation via diffusion within the

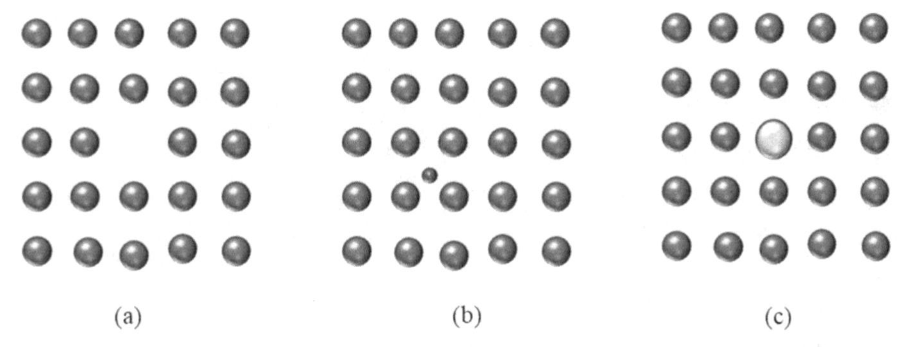

Figure 1.3. Point defects in a crystal: (a) a vacancy and (b) an interstitial, and (c) substitutional.

crystal. Such movement can change the material's response under stress, especially at elevated temperatures, as a viable deformation mechanism as discussed in Chapter 2.

1.2.3 Linear Defects

Metals can be deformed easily and retain the new shape via a process called plastic deformation, whereas ceramic solids often fracture without appreciable shape change. Mechanical properties and behaviours of metals and alloys can be largely attributed to the presence and movements of linear defects called dislocations.

Linear defects in crystals can be thought of as combinations of two fundamental types: i) edge dislocations, and ii) screw dislocations, as shown in Figure 1.4. An edge dislocation consists of an extra half-plane of atoms inserted into the crystal (Figure 1.4a). The edge dislocation is represented by $\perp$. The dislocation line runs normal to the plane of the drawing. The plane on which the dislocation line lies is the slip plane. The edge dislocation moves in the direction parallel to its Burgers vector. A screw dislocation, which is shown in Figure 1.4b, can be visualized as a crystal being cut part way through and slipped by one lattice vector. Unlike edge dislocations, a screw dislocation is not necessarily confined on one slip plane. It has an infinite number of possible slip planes since the dislocation line is parallel to the Burgers vector. Thus, the movement of a screw dislocation takes place by cross slip. Screw dislocations have greater mobility and more alternative ways to cross slip over to another plane, if impedance by some obstacles is present ahead on its original slip plane.

As one can see, the presence of a dislocation distorts nearby planes of atoms. The Burgers vector **b** marks the direction and the magnitude of the local distortion. When the crystal is pushed by a shear force in the Burgers vector direction, atoms may exchange mutual bonding such that the dislocation effectively moves along the intersection plane, which results in dislocation glide. When a row of vacancies are attracted to the middle inserted

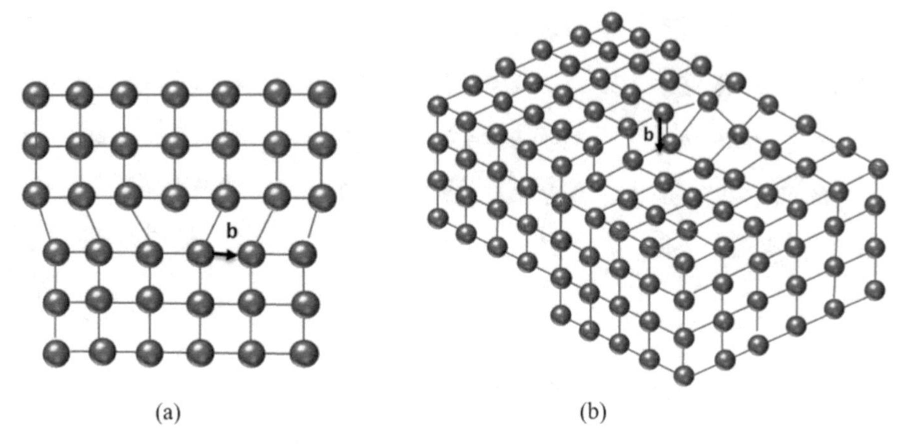

Figure 1.4. (a) An edge dislocation, and (b) a screw dislocation. Arrows represent the Burgers vector.

row, replacing the atoms at the bottom line, then the dislocation climbs up one lattice step, which is called dislocation climb. Crystal deformation often occurs by a combination of both glide and climb. The fundamental theory and observations of dislocations have long been established (Orowan 1934, Taylor 1934, Polanyi 1934, Orowan 1940, Hirth and Lothe 1992).

1.2.4 Planar Defects

Planar defects are two-dimensional surface defects. The most important planar defects in a crystalline solid are grain boundaries, which are the boundaries between crystallites in a polycrystalline array, as schematically shown in Figure 1.5. Since grain boundaries are the boundaries between the same types of crystals but oriented differently, grain boundaries can be considered as agglomerates of vacancies and dislocations. Another type of planar defect is inter-phase boundary between different phases within a material.

Grain boundaries are high-energy areas, much of which is surface energy. At low temperatures, grain boundaries can be obstacles to impede the movement of intragranular dislocations because of the interruption of slip planes by neighbour grains with different orientations. However, at elevated temperatures, grain boundaries can be both sources and sinks of vacancies, which either facilitate

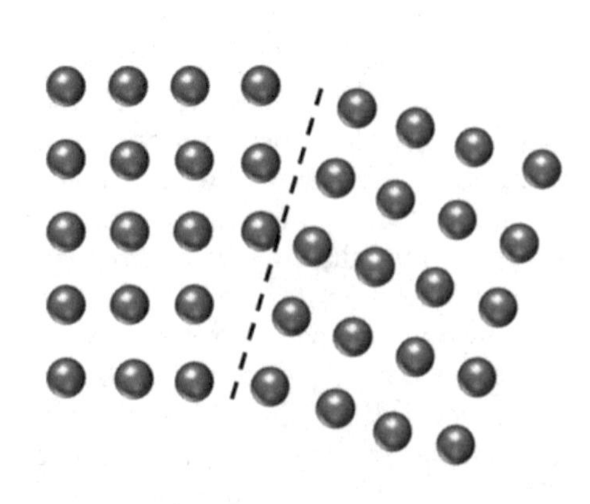

Figure 1.5. Schematic of grain boundary.

vacancy flow leading to grain boundary dislocation climb or condensation of vacancies into cavities. Therefore, grain boundaries are invariably weaker than the crystal matrix at high temperatures. If the temperature is high enough to allow for extensive atom diffusion, given enough time, atomic rearrangement at grain boundaries will occur, leading to lower-energy configurations. This may result in grain growth with annihilation of nearby small grains.

1.2.5 Volume Defects

Precipitates and inclusions are volume defects in a solid material, which occupy regions of the parent matrix, as impurity phases. Precipitates form on slow cooling from the solid solution state, due to the solubility limits of solute atoms in the solvent material. The precipitation process can result in second phase particles with new crystal structure within the parental matrix. Precipitates have important effects on the mechanical, electronic and optical properties of materials. Precipitation hardening is an important process used to strengthen alloys. In this technique, precipitates are induced to form in the alloy matrix by carefully controlled heat treatments. The precipitates may interfere with dislocation movement and have an effect to harden the alloy. However, it may be noted that not all precipitates provide the hardening effect, such as sigma phases in steels and superalloys.

One prominent example of precipitation is the γ' precipitates in Ni-base superalloy, as shown in Figure 1.6. The γ' phase is an intermetallic compound based on $Ni_3(Al,Ti)$. It is a coherently precipitating phase (i.e. the crystal planes of the precipitate are in registry with the γ matrix) with an ordered $L1_2$ (f.c.c.) crystal structure. The close match in matrix/precipitate

Figure 1.6. The γ' precipitates in γ matrix (etched out) of Ni-base superalloy (image provided by Rick Kearsey, National Research Council Canada).

lattice parameter (~0-1%) combined with the chemical compatibility allows γ' to precipitate homogeneously throughout the matrix and have long-time stability. Interestingly, the flow stress of γ' increases with increasing temperature, up to about 650°C. In addition, γ' is quite ductile and thus imparts strength to the matrix without lowering the fracture toughness of the alloy. In some advanced single crystal Ni-base alloys, the volume fraction of the γ' precipitate is around 70%. There are many factors that contribute to the hardening imparted by γ' precipitates, which include γ' anti-phase energy, γ' strength, coherency strains, volume fraction of γ', and γ' particle size.

1.3 Stress and Strain in Solids

Stress is the most common stimulus of a material, since most materials are subjected to mechanical and thermal loads in their working environments. Strain is the material's response to that stimulation. In engineering, stress and strain are defined in a continuum sense. For polycrystalline materials, because the dimensions of the structural body are far greater than the sizes of their microstructural features, the materials can be considered as homogeneous and isotropic. But, for single crystal components, because of the atomic arrangement in the crystal orientation, the material is generally anisotropic. Therefore, for generality, it is important to understand the 3D nature of stress and strain.

1.3.1 Definition of Stress and Strain

Stress—Stress is defined as the internal force acting on an infinitesimal area of an internal plane of the solid. In a Cartesian coordinate system, there are three orthogonal planes cutting through a material point in 3D space, and on each plane the internal force can be projected into three directions. Thus, totally, there are nine (9) stress components, as shown in Figure 1.7; but by symmetry, it reduces to 6 independent components. Therefore, stress is a second-order tensor, by nature. The stress tensor can be written in a matrix form:

$$[\sigma] = \begin{bmatrix} \sigma_{xx} & \sigma_{xy} & \sigma_{xz} \\ \sigma_{yx} & \sigma_{yy} & \sigma_{yz} \\ \sigma_{zx} & \sigma_{zy} & \sigma_{zz} \end{bmatrix} \tag{1.2}$$

The first subscript index indicates the plane on which the force acts and the second indicates the force direction. The stress component is positive if both the force direction and the plane normal are all in the same (positive or negative) direction; the stress component is negative if one of them is in the opposite direction of the other.

Strain—The definition of point strain is derived from the point movement in a 3D solid in the Lagrangian coordinate system, where under small-scale deformation conditions the displacement can be written as:

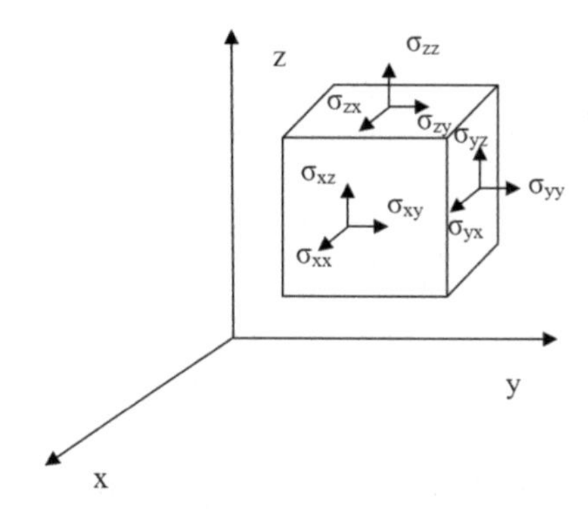

Figure 1.7. The stress tensor components.

$$u = u_0 + \begin{bmatrix} \frac{\partial u}{\partial x} & \frac{\partial u}{\partial y} & \frac{\partial u}{\partial z} \\ \frac{\partial v}{\partial x} & \frac{\partial v}{\partial y} & \frac{\partial v}{\partial z} \\ \frac{\partial w}{\partial x} & \frac{\partial w}{\partial y} & \frac{\partial w}{\partial z} \end{bmatrix} \begin{pmatrix} \Delta x \\ \Delta y \\ \Delta z \end{pmatrix} = u_0 + [J]\Delta x \tag{1.3}$$

The displacement derivative matrix J (a.k.a. Jacobian matrix) can be further decomposed into a symmetrical and an asymmetrical part, as

$$[J] = [\varepsilon] + [\Omega] \tag{1.4}$$

The symmetrical matrix $[\varepsilon]$ is regarded as the strain tensor, as given by

$$[\varepsilon] = \begin{bmatrix} \frac{\partial u}{\partial x} & \frac{1}{2}\left(\frac{\partial u}{\partial y} + \frac{\partial v}{\partial x}\right) & \frac{1}{2}\left(\frac{\partial u}{\partial z} + \frac{\partial w}{\partial x}\right) \\ \frac{1}{2}\left(\frac{\partial u}{\partial y} + \frac{\partial v}{\partial x}\right) & \frac{\partial v}{\partial y} & \frac{1}{2}\left(\frac{\partial v}{\partial z} + \frac{\partial w}{\partial y}\right) \\ \frac{1}{2}\left(\frac{\partial u}{\partial z} + \frac{\partial w}{\partial x}\right) & \frac{1}{2}\left(\frac{\partial v}{\partial z} + \frac{\partial w}{\partial y}\right) & \frac{\partial w}{\partial z} \end{bmatrix} \tag{1.5}$$

The asymmetrical matrix $[\Omega]$ is the rotation matrix, as given by

$$[\Omega] = \begin{bmatrix} 0 & \frac{1}{2}\left(\frac{\partial u}{\partial y} - \frac{\partial v}{\partial x}\right) & \frac{1}{2}\left(\frac{\partial u}{\partial z} - \frac{\partial w}{\partial x}\right) \\ \frac{1}{2}\left(\frac{\partial v}{\partial x} - \frac{\partial u}{\partial y}\right) & 0 & \frac{1}{2}\left(\frac{\partial v}{\partial z} - \frac{\partial w}{\partial y}\right) \\ \frac{1}{2}\left(\frac{\partial w}{\partial x} - \frac{\partial u}{\partial z}\right) & \frac{1}{2}\left(\frac{\partial w}{\partial y} - \frac{\partial v}{\partial z}\right) & 0 \end{bmatrix} \tag{1.6}$$

The effects of [ε] and [Ω] are illustrated by the example of an infinitesimal cube in the *x-z* plane, as shown in Figure 1.8. The strain tensor [ε] represents the volume and shape change (in the case of pure shear as shown in Figure 1.8, there is no volume change), and [Ω] represents rigid body rotation.

The strain tensor is written in matrix form as:

$$[\varepsilon] = \begin{bmatrix} \varepsilon_{xx} & \varepsilon_{xy} & \varepsilon_{xz} \\ \varepsilon_{yx} & \varepsilon_{yy} & \varepsilon_{yz} \\ \varepsilon_{zx} & \varepsilon_{zy} & \varepsilon_{zz} \end{bmatrix} \tag{1.7}$$

By its definition, the strain tensor [ε] is a symmetrical second-order tensor.

In engineering, it is convenient to define the engineering shear strains as: $\gamma_{xy} = 2\varepsilon_{xy}$, $\gamma_{yz} = 2\varepsilon_{yz}$, $\gamma_{xz} = 2\varepsilon_{xz}$, because they characterize the total shape change on the respective plane.

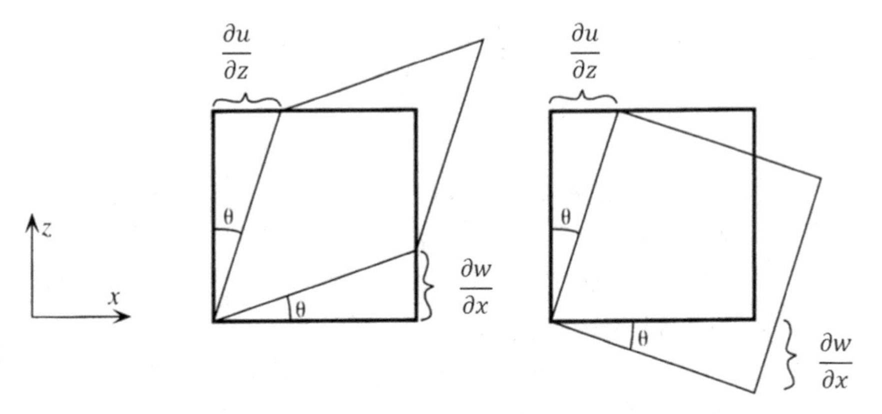

Figure 1.8. Deformation of an infinitesimal cube by off-diagonal components in the x-z plane.

1.3.2 Transformation of Stress and Strain

The stress and strain tensors together represent the material's deformation state. Its physical nature does not change with the choice of reference coordinates. Generally, stress and strain are described with a particular reference coordinate system $\{x, y, z\}$. For example, to describe stress-strain and properties of a cubic crystal, it is more convenient to use the crystallographic cube axes as the reference coordinate system. However, in describing an engineering problem in application, a different reference coordinate, $\{x', y', z'\}$, may be chosen for convenience. For example, a material engineer conducting a uniaxial tension test may designate the loading direction as x', irrespective of the crystal orientation of the sample tested. Therefore, it is necessary to find the relationship between the stress and strain tensors in one reference coordinate system with respective to another.

The coordinate transformation relationship between the old and the new coordinate systems can be generally expressed as:

$$\mathbf{e}_i = \beta_{ki}\mathbf{e}'_k \tag{1.8}$$

where $\mathbf{e}_i$ represents the unit vector of the coordinate axis, and β_{ki} is the cosine of the angle between the $\mathbf{e}_i$ and $\mathbf{e}'_k$ directions (i and k = 1, 2, 3). Here and henceforth, the index convention is kept as $x = 1$, $y = 2$, and $z = 3$, and the Einstein summation convention applies.

Then, the tensor expressed in the old coordinate system will be transformed into the new system, following

$$\sigma_{ij}\mathbf{e}_i\mathbf{e}_j = \sigma_{ij}(\beta_{ki}\mathbf{e}'_k)(\beta_{lj}\mathbf{e}'_l) = (\sigma_{ij}\beta_{ki}\beta_{lj})\mathbf{e}'_k\mathbf{e}'_l \tag{1.9}$$

By the invariance of tensor properties, the new stress tensor is:

$$\sigma'_{kl} = \sigma_{ij}\beta_{ki}\beta_{lj} \tag{1.10a}$$

The β matrix is composed of directional cosines between the old and the new coordinate system, as given in Table 1.1.

Table 1.1. Directional cosines

	x	y	z
x'	$\beta_{11} = \cos(x, x')$	$\beta_{12} = \cos(y, x')$	$\beta_{13} = \cos(z, x')$
y'	$\beta_{21} = \cos(x, y')$	$\beta_{22} = \cos(y, y')$	$\beta_{23} = \cos(z, y')$
z'	$\beta_{31} = \cos(x, z')$	$\beta_{32} = \cos(y, z')$	$\beta_{33} = \cos(z, z')$

Then, Eq. (1.10a) can be written in terms of matrix operations as:

$$[\sigma'] = [\beta][\sigma][\beta]^T \tag{1.10b}$$

By the transformation of Eq. (1.10), no matter what coordinate system the observer may choose, the physical stress state in the solid remains the same.

Similar to stress transformation, strain transformation from the old coordinate system $\{x, y, z\}$ to a new coordinate system $\{x', y', z'\}$ is given by the following relationship:

$$\varepsilon'_{kl} = \varepsilon_{ij}\beta_{ki}\beta_{lj} \tag{1.11a}$$

Or, in the matrix form:

$$[\varepsilon'] = [\beta][\varepsilon][\beta]^T \tag{1.11b}$$

1.3.3 Stress/Strain Invariants

The second-order tensor possesses certain invariant properties, which also applies to stress and strain. To reveal the invariants of stress tensor, we first look for the principal planes on which there are no shear stresses but normal principal stresses. Generally, the stress resolved on a given plane with the normal v is $\sigma_{ij}v_j$. What we are looking for here is to find such a plane on which the stress becomes pv_j (p is a scalar value representing the principal stress), i.e. only the normal component is present. Letting them to be equal in the tensor sense, we have:

$$(\sigma_{ij} - \delta_{ij}p)v_j = 0 \tag{1.12}$$

where δ_{ij} is the Kronecker delta function ($\delta_{ij} = 1$, when $i = j$; $\delta_{ij} = 0$, when $i \neq j$)

By the matrix theory, in order to solve for the principal stress and the corresponding plane normal v_j, the determinant of Eq. (1.12) must be zero:

$$\det\left|\sigma_{ij} - \delta_{ij}p\right| = \begin{vmatrix} \sigma_{xx} - p & \sigma_{xy} & \sigma_{xz} \\ \sigma_{xy} & \sigma_{yy} - p & \sigma_{yz} \\ \sigma_{xz} & \sigma_{yz} & \sigma_{zz} - p \end{vmatrix} = p^3 - I_1p^2 + I_2p - I_3 = 0 \tag{1.13a}$$

where

$$I_1 = \sigma_{xx} + \sigma_{yy} + \sigma_{zz} \tag{1.13b}$$

$$I_2 = \sigma_{xx}\sigma_{yy} + \sigma_{yy}\sigma_{zz} + \sigma_{zz}\sigma_{xx} - \sigma_{xy}^2 - \sigma_{yz}^2 - \sigma_{zx}^2 \tag{1.13c}$$

$$I_3 = \sigma_{xx}\sigma_{yy}\sigma_{zz} + 2\sigma_{xy}\sigma_{yz}\sigma_{xz} - \sigma_{zz}\sigma_{xy}^2 - \sigma_{xx}\sigma_{yz}^2 - \sigma_{yy}\sigma_{zx}^2 \tag{1.13d}$$

Eq. (1.13) is the eigenvalue equation for the stress state, which has three real roots, σ_1, σ_2, and σ_3 (ranked in a descending order), called the principal stresses. The quantities I_1, I_2, and I_3 are the first, second and third stress tensor invariants. It can be proven by solid mechanics that these quantities do not change with the choice of coordinate system, and so do the principal stresses. After solving for σ_i ($i = 1, 2, 3$), the values are substituted into Eq. (1.12), then the directional vector can be obtained to define the corresponding principal stress direction, v_i ($i = 1, 2, 3$). The three principal directions are orthogonal. Therefore, in the new coordinate system represented by v_i ($i = 1, 2, 3$), the stress tensor has only three diagonal components. Also, by the invariant principle, Eq. (1.13b) can be written as

$$I_1 = \sigma_{xx} + \sigma_{yy} + \sigma_{zz} = \sigma_1 + \sigma_2 + \sigma_3$$

Similarly, the strain tensor also has three principal values, which can be obtained by solving the following equations:

$$\det\left|\varepsilon_{ij} - \delta_{ij}\lambda\right| = \begin{vmatrix} \varepsilon_{xx} - \lambda & \varepsilon_{xy} & \varepsilon_{xz} \\ \varepsilon_{xy} & \varepsilon_{yy} - \lambda & \varepsilon_{yz} \\ \varepsilon_{xz} & \varepsilon_{yz} & \varepsilon_{zz} - \lambda \end{vmatrix} = \lambda^3 - \Theta_1\lambda^2 + \Theta_2\lambda - \Theta_3 = 0 \tag{1.14a}$$

where

$$\Theta_1 = \theta = \varepsilon_{xx} + \varepsilon_{yy} + \varepsilon_{zz} \tag{1.14b}$$

$$\Theta_2 = \varepsilon_{xx}\varepsilon_{yy} + \varepsilon_{yy}\varepsilon_{zz} + \varepsilon_{zz}\varepsilon_{xx} - \varepsilon_{xy}^2 - \varepsilon_{yz}^2 - \varepsilon_{zx}^2 \tag{1.14c}$$

$$\Theta_3 = \varepsilon_{xx}\varepsilon_{yy}\varepsilon_{zz} + 2\varepsilon_{xy}\varepsilon_{yz}\varepsilon_{xz} - \varepsilon_{zz}\varepsilon_{xy}^2 - \varepsilon_{xx}\varepsilon_{yz}^2 - \varepsilon_{yy}\varepsilon_{zx}^2 \tag{1.14d}$$

Eq. (1.14) is the eigenvalue equation for the strain state, which has three real roots, ε_1, ε_2, and ε_3 (usually ranked in a descending order by value), called the principal strains. The quantities Θ_1, Θ_2, and Θ_3 are the first, second and third strain tensor invariants. It can be proven that these quantities do not change with the choice of coordinates, and so do the principal strains. Also, the following relation holds for the strain tensor:

$$\Theta_1 = \varepsilon_{xx} + \varepsilon_{yy} + \varepsilon_{zz} = \varepsilon_1 + \varepsilon_2 + \varepsilon_3$$

It represents the volume change. It should be noted that Θ_1 or θ for plastic and creep strains are always zero, following the rule of incompressible flow. So volume change can only be induced by stress, elastically, except by phase transformation or other chemical reactions.

1.3.4 The Deviatoric Stress and Strain

A special stress state is the hydrostatic stress state, where $\sigma_{xx} = \sigma_{yy} = \sigma_{zz} = \sigma_0$. In this case, the material is under expansion or contraction equally in all directions under pressure, causing no shape changes. For example, when a ball is submerged in deep water, it is in the hydrostatic stress state. For an arbitrary stress state, if we extract the hydrostatic state represented by the average of normal stresses: $\sigma_m = (\sigma_{xx} + \sigma_{yy} + \sigma_{zz})/3$, we obtain the so-called deviatoric stress:

$$s_{ij} = \sigma_{ij} - \delta_{ij}\sigma_m \tag{1.15}$$

It represents the stress state responsible for changing the shape of the solid. The first invariant of the deviatoric stress tensor is $J_1 = 0$ (no volume change), and the second invariant of the deviatoric stress tensor is obtained, following the same procedure as in the previous section, as

$$J_2 = \frac{1}{3}I_1^2 - I_2 = \frac{1}{6}\left[(\sigma_{xx} - \sigma_{yy})^2 + (\sigma_{yy} - \sigma_{zz})^2 + (\sigma_{zz} - \sigma_{xx})^2\right] + \sigma_{xy}^2 + \sigma_{yz}^2 + \sigma_{zx}^2 \tag{1.16}$$

The second invariant of the deviatoric stress tensor happens to be proportional to the square of the shear stress on the octahedral plane ($1/\sqrt{3}$, $1/\sqrt{3}$, $1/\sqrt{3}$), as it can be proven as follows.

On the octahedral plane, the total traction force is calculated by

$$T_i = \sigma_{ij}\nu_j = \frac{1}{\sqrt{3}}(\sigma_{xx} + \sigma_{xy} + \sigma_{xz})e_1 + \frac{1}{\sqrt{3}}(\sigma_{xy} + \sigma_{yy} + \sigma_{yz})e_2 + \frac{1}{\sqrt{3}}(\sigma_{xx} + \sigma_{xy} + \sigma_{xz})e_3 \tag{1.17}$$

The magnitude of the normal stress on the octahedral plane can be calculated as

$$\sigma_{oct} = \nu_i \sigma_{ij} \nu_j = \frac{1}{3}(\sigma_{xx} + \sigma_{yy} + \sigma_{xz} + 2\sigma_{xy} + 2\sigma_{xz} + 2\sigma_{yz}) \tag{1.18}$$

Then, the magnitude of the octahedral shear stress can be calculated as

$$\begin{aligned} \tau &= \sqrt{|T|^2 - \sigma_{oct}^2} \\ &= \sqrt{\frac{1}{9}\left[(\sigma_{xx} - \sigma_{yy})^2 + (\sigma_{yy} - \sigma_{zz})^2 + (\sigma_{zz} - \sigma_{xx})^2\right] + \frac{2}{3}\left[\sigma_{xy}^2 + \sigma_{yz}^2 + \sigma_{zx}^2\right]} \\ &= \sqrt{\frac{2}{3} J_2} \end{aligned} \tag{1.19}$$

The octahedral stress τ is also called the von Mises stress (von Mises 1913), which is often used to define the yield surface in the theory of plasticity. For a generalized material, when the von Mises stress exceeds certain value, permanent shape change would occur. Therefore, J_2 is an important stress-state quantity as it is the driving force for distortion.

Similarly, if one separates the total strain into volumetric strain and distortion strain, one can define the deviatoric strain as

$$e_{ij} = \varepsilon_{ij} - \frac{1}{3}\delta_{ij}\theta \tag{1.20}$$

which is called the deviatoric strain tensor.

The first invariant of the deviatoric strain tensor Θ'_1 is zero, which means no volume change in this stress-strain state. The *octahedral shear strain* (also called the *equivalent strain*) is related to the second invariant of the deviatoric strain tensor Θ'_2, as:

$$\gamma = \frac{2}{\sqrt{3}}\sqrt{\Theta'_2} = \frac{\sqrt{2}}{3}\sqrt{\left[(\varepsilon_{xx} - \varepsilon_{yy})^2 + (\varepsilon_{yy} - \varepsilon_{zz})^2 + (\varepsilon_{zz} - \varepsilon_{xx})^2\right] + 6\left[\varepsilon_{xy}^2 + \varepsilon_{yz}^2 + \varepsilon_{zx}^2\right]} \tag{1.21}$$

Hence, γ is a deformation quantity that characterizes the shape change of a solid. For plastic deformation, the constitutive relationship is often expressed in terms of the von Mises stress τ and the equivalent strain γ.

1.4 Crystal Elasticity and Stroh Formalism

Ideally, crystalline solids only deform elastically, except the moving of dislocations within the body. Elasticity refers to the material's ability to return to their original shape when the external forces are removed. Elastic deformation of crystalline solids involves stretching of atomic bonds without change in the atomic configuration. Almost all crystalline materials exhibit a linear portion of stress-strain relationship when deformation is small, which is called *linear elasticity*.

The elasticity of materials is described by the generalized Hooke's law, which can be written in the tensor form as:

$$\sigma_{ij} = C_{ijkl}\varepsilon_{kl} \tag{1.22a}$$

or

$$\varepsilon_{ij} = S_{ijkl}\sigma_{kl} \tag{1.22b}$$

where C_{ijkl} are the fourth-order tensor of elastic stiffness, and S_{ijkl} are the fourth-order tensor of elastic compliance.

The relationship (1.22) is for generally anisotropic materials, e.g. single crystal materials. For polycrystalline materials where the grain size is far less than the dimension of the solid body, the material can be regarded as isotropic, because the average property is practically the same along any direction containing numerous grains. Then, by consideration of the tensor symmetry and isotropy on rotation of material reference axes, Eq. (1.22a) can be reduced to:

$$\sigma_{ij} = 2\mu\varepsilon_{ij} + \lambda\theta\delta_{ij} \tag{1.22c}$$

where μ is the shear modulus, and λ is the Lame constant. These two constants are related to the elastic modulus, E, and Poisson's ratio, ν, as given by

$$\mu = \frac{E}{2(1+\nu)}, \quad \lambda = \frac{\nu E}{(1+\nu)(1-2\nu)} \tag{1.23}$$

The elastic modulus E is often measured as the slope of the linear portion of the stress-strain curve, and the Poisson's ratio is defined as the ratio of lateral to longitudinal deformation in uniaxial tension.

Anisotropic elastic properties are usually given with respect to the coordinates of crystalline symmetry, e.g. cube axes of cubic crystals. To deal with various mechanical problems of crystalline solids, one often needs to choose a different coordinate system with respect to which the loading and boundary conditions are known. For example, a Ni-base single crystal turbine blade is usually directionally solidified along [001] axis in the main centrifugal loading direction, but its secondary axis is often not controlled, i.e. not necessarily in alignment with [100] or [010] direction. In the new coordinate system, using Eq. (1.10) and (1.11), we can obtain the transformed stress-strain relation as

$$\sigma'_{ij} = \beta_{ik}\beta_{jl}\beta_{pr}\beta_{qs}C_{klpq}\varepsilon'_{rs} \tag{1.24a}$$

with the transformed tensor of elasticity defined as

$$C'_{ijrs} = \beta_{ik}\beta_{jl}\beta_{pr}\beta_{qs}C_{klpq} \tag{1.24b}$$

As discussed above, elastic deformation does not involve change of atomic configuration of the crystalline solid, but plastic deformation does via movements of dislocations. For the purpose of this book, it is useful to understand the stress fields generated by dislocations. The stress and displacement fields of dislocations in a homogeneous anisotropic elastic body have been elegantly formulated by Stroh (1957, 1958), based on Eshelby-Read-Shockley's theory (1953), which is called the Stroh formalism and has been widely used to deal with anisotropic elastic problems of various point and line defects (Ting 1996, Asundi and Deng 1995, Wu et al. 2001, Wu 2005). In Stroh formalism, the elementary field potential of a unit line dislocation without the line tension force is given by

$$\mathbf{h}(z) = \frac{1}{2\pi i} < \ln z > \mathbf{L}^T \mathbf{b} \tag{1.25}$$

where **b** is the Burgers vector, and **L** is an anisotropic material matrix satisfying the equilibrium condition. The function $<f(z)>$ denotes a 3×3 diagonal matrix with $f(z_1)$, $f(z_2)$, and $f(z_3)$ at each diagonal position, $z_\alpha = x_1 + p_\alpha x_2$ is a complex variable with the origin of the coordinates set at the dislocation core, and p_α is the complex variable eigenvalue with a positive imaginary part (Im $(p_\alpha) > 0$, $\alpha = 1, 2, 3$), $i = \sqrt{-1}$ is the imaginary number. The coordinate axes represent Burgers vector directions such that x_1 is in the Burgers vector direction of an edge glide dislocation, x_2 is in the Burgers vector direction of an edge climb dislocation, and x_3 is in the Burgers vector direction of a screw dislocation, respectively.

By the Stroh formalism, the displacement vector **u**, and the stress vectors, **s** and **t**, are expressed in terms of the complex field potential vector **h**(z), as

$$\mathbf{u} = \{u_1, u_2, u_3\}^T = \mathbf{A}\mathbf{h}(z) + \overline{\mathbf{A}}\overline{\mathbf{h}(z)} \tag{1.26a}$$

$$\mathbf{s} = \{\sigma_{11}, \sigma_{12}, \sigma_{13}\}^T = -\mathbf{L}\mathbf{P}\mathbf{h}'(z) + \overline{\mathbf{L}\mathbf{P}}\overline{\mathbf{h}'(z)} \tag{1.26b}$$

$$\mathbf{t} = \{\sigma_{21}, \sigma_{22}, \sigma_{23}\}^T = \mathbf{L}\mathbf{h}'(z) + \overline{\mathbf{L}}\overline{\mathbf{h}'(z)} \tag{1.26c}$$

where $A_{i\alpha}$ is a matrix of equilibrium eigenvectors, $L_{i\alpha} = [C_{i2k1} + p_\alpha C_{i2k2}]A_{k\alpha}$, and **P** is a diagonal matrix of the three complex eigenvalues $p_\alpha(\alpha = 1, 2, 3)$; $\overline{\mathbf{A}}$, $\overline{\mathbf{L}}$ and $\overline{\mathbf{P}}$ are their conjugates, respectively.

It can be shown that the displacement and stress functions given in Eq. (1.26) satisfy the equilibrium condition:

$$\sigma_{ij,j} = C_{ijkl} \cdot \frac{1}{2}(u_{k,l} + u_{l,k})_{,j} = C_{ijkl} u_{k,lj} = 0 \tag{1.27}$$

where the comma denotes differentiation with respect to the indexed coordinates.

With the displacement function defined as given by Eq. (1.26), Eq. (1.27) turns into an algebraic equation, as

$$(C_{i1k1} + p_\alpha C_{i1k2} + p_\alpha C_{i2k1} + p_\alpha^2 C_{i2k2})A_{k\alpha} = 0, \alpha = 1, 2, 3 \tag{1.28}$$

The eigenvalue equation for p_α (α = 1, 2, 3) is obtained by solving the sextic equation:

$$\det\left| C_{i1k1} + pC_{i1k2} + pC_{i2k1} + p^2C_{i2k2} \right| = 0 \tag{1.29}$$

For each eigenvalue, its unit eigenvector is obtained by solving Eq. (1.28). Then, the matrix **L** is constructed with these eigenvalues and eigenvectors, satisfying the equilibrium condition, Eq. (1.28). The above solution procedure can easily be implemented in a computer code solving for the algebraic equation, Eq. (1.29).

We now define the matrix **F** as

$$\mathbf{F} = -2i\mathbf{L}\mathbf{L}^T \tag{1.30}$$

Then, we find that the stress component in the Burgers vector direction of an edge glide dislocation (x_1-axis) on the gliding plane is

$$\tau_{21} = t_1(x_1) = \frac{F_{1j}b_j}{2\pi x_1} \tag{1.31}$$

To appreciate the significance of orientation effect, a numerical example is given here for dislocations on a (111) plane in single crystal nickel. The material's elastic compliance coefficients are given with respect to the cubic reference coordinates (Nye 1957):

$$S_{11} = S_{22} = S_{33} = 0.00799\ (\text{GPa})^{-1};$$

$$S_{12} = S_{23} = S_{31} = -0.00312\ (\text{GPa})^{-1};$$

$$S_{44} = S_{55} = S_{66} = 0.00844\ (\text{GPa})^{-1}.$$

Choosing the Cartesian coordinates as: x—$[1\bar{1}0]$, y—[111] and z—$[11\bar{2}]$, the matrix **F** can be solved, following the above transformation and solution procedures, as:

$$F = \begin{bmatrix} 263 & 0 & 503 \\ 0 & 125 & 0 \\ 503 & 0 & 1680 \end{bmatrix} (\text{GPa})$$

Apparently, due to anisotropic coupling, the existence of screw dislocations can affect the stress distribution of edge dislocations, and vice versa. For isotropic materials, **F** reduces to a diagonal matrix with $F_{11} = F_{22} = \mu/(1-\nu)$ and $F_{33} = \mu$, where μ is the shear modulus and ν is the Poisson's ratio. Without anisotropic elastic coupling, Eq. (1.31) reduces to the stress solution for isotropic materials, as given by Hirth and Lothe (1992).

1.5 The Absolute Rate Theory

Inelastic deformation involves atomic configuration change, which inevitably starts from crystal imperfections and generally proceeds as a rate process. Accumulation of physical damage can be associated with inelastic deformation, which limits the useful life of the material. In this section, we shall first understand how elementary rate processes proceed at the atomistic level. Their play at the macroscopic level will be discussed in later chapters.

Imagine materials are giant molecules containing millions of atoms. If the atoms were perfectly arranged, the entire solid would consist of repeated arrangements of unit cells, and deformation under stress would be perfectly elastic as the unit cells restore their original shape and volume when the stress is removed. This way, the material can live up to its ultimate theoretical strength forever, if chemical effects from environment are absent. The reason that materials fail below their ideal strengths is because of the existence of crystalline defects such as vacancies, dislocations and grain boundaries (agglomerates of dislocations and vacancies between differently orientated crystals). These defects exist naturally through the manufacturing process, and they evolve, under either static or cyclic loading, into larger flaws such as voids or cracks that can eventually cause the material to fracture.

When dislocations and vacancies move from one lattice position to another, the atomic configuration of the material undergoes permanent changes, resulting in plastic deformation. The change of atomic configurations can be described by the absolute rate theory, which is briefly introduced below.

When a material configuration undergoes transition from one equilibrium state to another, it usually needs to overcome some energy barriers. These energy barriers appear as "humps" on the potential energy surface over the configuration space. For a system of N atoms, the configuration space has dimensions of 3(N-2) degrees of freedom. An exact description of the atomistic potential energy surface for a material requires computational quantum mechanics involving billions of atoms, which is a formidable task, even given the computational power today. However, its basic principle can be illustrated by a schematic of the potential energy surface for a linear three-atom system (A-B-C), as shown in Figure 1.9. In this system, one axis represents the interaction distance between atoms A and B, and the other axis represents the interaction distance between B and C. The transition process can be envisaged as a person walking on a mountainous terrain: he initially stands in one valley (a stable configuration), and he needs to find a path that is usually across a saddle point to arrive at a nearby valley to rest. He must spend the least effort or energy. The saddle point represents the transition state or the activated state. The possibility of success to reach that state, of course, depends on the relative height of the saddle point and the path length, but it can also be energized by the thermal energy it preserves for itself. By the principles of statistical quantum mechanics, the absolute rate of atomistic transition or reaction is (Krausz and Eyring 1975):

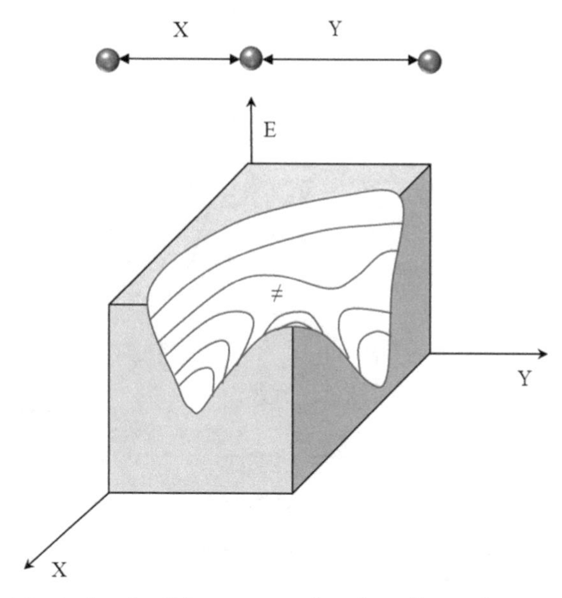

Figure 1.9. A typical potential energy surface for a linear three-atom system.

$$\kappa = \frac{kT}{h}\exp\left(-\frac{\Delta G^{\neq}}{kT}\right) \tag{1.32}$$

where k is the Boltzmann constant, h is Planck's constant, T is the absolute temperature, and $\Delta G^{\neq}$ is the activation energy.

Therefore, for one particular reaction, there always exists a two dimensional path, as shown in Figure 1.10, that represents the most favourable activation process. In practice, if one knows the reactants and the product, one can carefully conduct an experiment to measure the rate of reaction or process, and thus determine the activation energy. The apparent activation energy is equal to the Boltzmann constant times the slope measured from the Arrhenius plot (the logarithm of reaction rate vs. the inverse of absolute temperature), as inferred from Eq. (1.32).

Rigorously speaking, the absolute rate theory is only valid when the activation complexes are in equilibrium with the stable configurations. This would be approximately true when the activation energy is relatively large compared to the thermal energy of the atoms involved ~kT. Some quantum mechanics and non-equilibrium statistical mechanics analyses have shown that the equilibrium assumption is acceptable when (Krausz and Eyring 1975)

$$\frac{\Delta G^{\neq}}{kT} \geq 5 \cdot$$

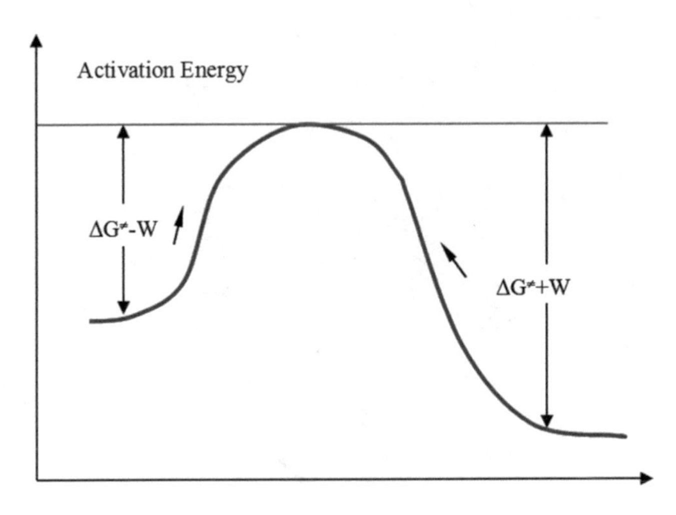

Figure 1.10. The schematic of activation energy barrier with both forward and backward steps.

Usually, in a deformation process, the apparent activation energy is an evolutionary state variable that depends on the current deformation state as characterized by the applied stress, τ, the plastic strain, γ_p, and temperature, T, i.e.

$$\Delta G^{\neq} = \Delta G^{\neq}(\tau, \gamma_p, T) \tag{1.33}$$

By the first-order Taylor expansion, the Gibbs free energy of the activation system may be expressed in terms of the state-variables (τ, γ_p, T) as

$$\Delta G_f^{\neq} = \Delta G_0^{\neq} + \frac{\partial \Delta G^{\neq}}{\partial \tau}\tau + \frac{\partial \Delta G^{\neq}}{\partial \gamma_p}\gamma_p + \frac{\partial \Delta G^{\neq}}{\partial T}T = \Delta G_0^{\neq} - V(\tau - H\gamma_p) - \Delta S T \tag{1.34}$$

where

$V = -\left(\frac{\partial \Delta G^{\neq}}{\partial \tau}\right)$ is the activation volume,

$H = \frac{1}{V}\left(\frac{\partial \Delta G^{\neq}}{\partial \gamma_p}\right)$ is the work hardening coefficient, and

$\Delta S = -\left(\frac{\partial \Delta G^{\neq}}{\partial T}\right)$ is the entropy change of the system.

Usually, we consider deformation within a stable structure, such that $\Delta S = 0$.

Statistically, thermal activation over an energy barrier can take place in both directions. In the activation system as represented by Figure 1.10, forward activation can be promoted by mechanical work, which effectively reduces the height of the energy barrier, such that

$$\Delta G_f^{\neq} = \Delta G_0^{\neq} - V\tau_{eff} \tag{1.35a}$$

where τ_{eff} is the effective shear stress acting on the activation system. On the other hand, backward steps may also occur against the mechanical work, such that

$$\Delta G_b^{\neq} = \Delta G_0^{\neq} + V\tau_{eff} \tag{1.35b}$$

Then the net dislocation velocity passing over a single energy barrier can be expressed as (Krausz and Eyring 1975):

$$\begin{aligned} v &= l\left(\kappa_f - \kappa_b\right) = l\frac{kT}{h}\left[\exp\left(-\frac{\Delta G_f^{\neq}}{kT}\right) - \exp\left(-\frac{\Delta G_b^{\neq}}{kT}\right)\right] \\ &= 2l\frac{kT}{h}\exp\left(-\frac{\Delta G_0^{\neq}}{kT}\right)\sinh\left(\frac{V\tau_{eff}}{kT}\right) \end{aligned} \tag{1.36}$$

where l is the activation step.

Dislocation kinetics has been extensively discussed by Krausz and Eyring (1975) on the basis of thermal activation. An example is shown in Figure 1.11 for dislocation velocity in Ge. Good agreement is found between the description of Eq. (1.36) and the experiment. The thermo-dynamics of slip has also been discussed by Kocks et al. (1975). These works have laid the foundation for the physics-based constitutive equations of rate-dependent deformation processes.

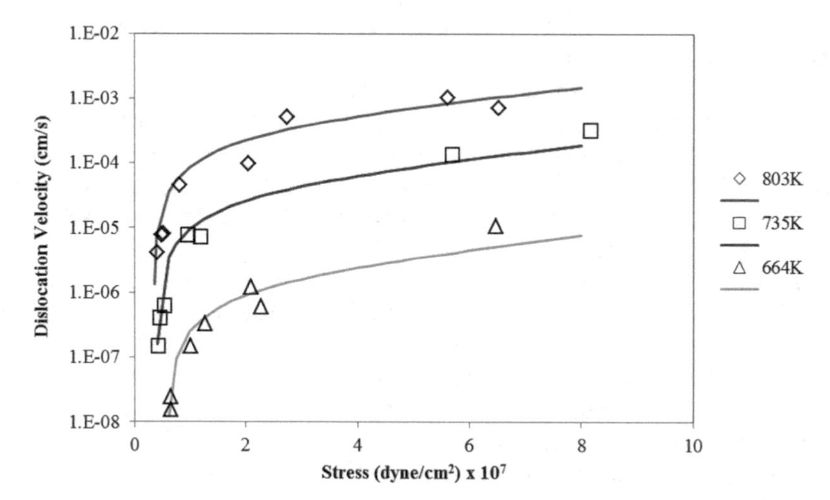

Figure 1.11. The stress and temperature dependence of dislocation velocity in Ge, after Krausz (1968). The symbols represent the observed values and the lines represent Eq. (1.36).

1.6 The Kinetics of Dislocation Mobility

1.6.1 Dislocation Velocity and Strain Rate

Inelastic deformation in crystalline materials mostly occurs by dislocation movement, which results from changing the relative positions with the surrounding atoms, breaking the previous bonds and establishing new bonds under stress at temperature. It is in essence a thermally activated process similar to chemical reactions resulting in a new atomic configuration.

Orowan (1940) proposed a relationship between plastic shear strain rate $\dot{\gamma}$ and dislocation mobility, as

$$\dot{\gamma}_p = \alpha b \rho_m v \tag{1.37}$$

where α (~1) is a geometrical factor; b is the Burgers vector; ρ_m is the mobile dislocation density; and v is the dislocation velocity, which can be expressed in the form of Eq. (1.36), such that

$$\dot{\gamma}_p = 2\dot{\gamma}_0 exp\left(-\frac{\Delta G_0^{\neq}}{kT}\right)\sinh\left(\frac{V\tau_{eff}}{kT}\right) \tag{1.38}$$

where $\dot{\gamma}_0$ is the pre-exponential strain rate factor. Henceforth, without further specification, scalar symbols (σ, τ) and (ε, γ) represent either uniaxial or equivalent stress and strain, respectively.

1.6.2 Dislocation Pile-up

In real materials at the microstructure level, dislocation density can change with deformation and the movement of dislocations often encounters obstacles such as second phase particles or inclusions. These obstacles impede local dislocation motion via Orowan looping or precipitate cutting, effectively providing strengthening mechanisms for the material. Hence, Eq. (1.38) is understood to be effectively the average strain rate for engineering materials. Nevertheless, it is of interest to look at the local dislocation pile-up behaviour, as the constitutive behaviour of crystalline materials arises from interactions of moving dislocations with the surrounding microstructure including grain boundaries, precipitates, and inclusions etc.

When dislocations approach an obstacle one by one, they form a dislocation pile-up in front of the obstacle. In the kinetic process, a dynamic equilibrium may exist for the pile-up, where some dislocations are pushed to the pile-up by force and others leave by climbing through the lattice. Then, dislocation pile-up accumulation rate would be equal to the number of dislocations arriving by glide, subtracting the number of dislocations climb out per unit time. The dynamic process of dislocation pile-up is schematically illustrated in Figure 1.12.

Let ρ be the dislocation density, v be the average dislocation glide velocity and κ be the rate of climb. The number of new dislocations arriving

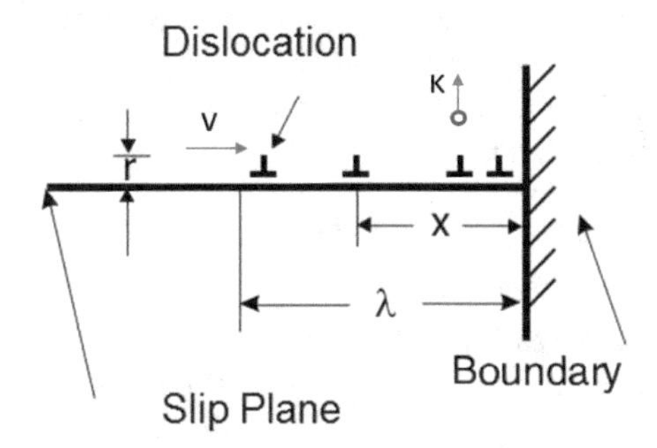

Figure 1.12. A schematic of dislocation pile-up against an obstacle.

per unit time is given by $\rho v r$, where r is the slip-band width, and the number of dislocations leaving is equal to $n\kappa$. Hence, for a single slip system, the change in the number of pile-up dislocations in a unit time is given by

$$\frac{dn}{dt} = \rho v r - \kappa n \tag{1.39}$$

According to Eq. (1.37), $\dot{\gamma}_p = b\rho v$ (r ≈ b), Eq. (1.39) can be written as

$$\frac{dn}{dt} = \dot{\gamma}_p - \kappa n \tag{1.40}$$

At constant strain rate, the solution of Eq. (1.40) can be obtained as:

$$n = \frac{\dot{\gamma}_p}{\kappa}\left(1 - e^{-\kappa t}\right) \tag{1.41}$$

Equation (1.40) basically describes the competition of dislocation glide and climb in the dislocation pile-up process. Generally, the number of pile-up dislocations increases with deformation. However, as the temperature increases, vacancy flow will increase, allowing more dislocations to jump out of the pileup. Thus, at the steady-state, the number of dislocations in a pile-up tends to be stabilized at a constant level that is equal to the ratio of the inelastic strain rate to the vacancy flow rate, as implied by Eq. (1.41).

1.6.3 The Back Stress

Gliding dislocations may pile up against an obstacle such as grain boundary or particle, as schematically shown in Figure 1.12. The existing dislocation pile-up may exert a back stress opposing to the next incoming dislocation. A treatment for dislocation pile-up along grain boundaries in the presence of grain boundary precipitates has been given (Wu and Koul 1995). Here, it is generalized for both intragranular and intergranular cases. For simplicity, the distance between the dislocation source (e.g. a Frank-Reed source) and the obstacle is λ, which may be about half the grain size or the average precipitate spacing, depending on the microstructure. Suppose that, at time *t*, there are *n* dislocations in the pile-up. The back stress that pile-up dislocations together exert on a moving dislocation located at a distance x away is equal to n times the force of unit dislocation, as (note that isotropic material properties are assumed here for mathematical simplicity; if anisotropic material properties have to be considered, one may use Eq. (1.31))

$$\tau_b(x) = \frac{n\mu b}{2\pi(1-\nu)x} \tag{1.42}$$

where ν is the Poisson's ratio, and μ is the shear modulus.

The average back stress opposing dislocation glide, τ_{ig}, can be defined as the mechanical work to move the mobile dislocation against the dislocation pile-up divided by the distance λ that the dislocation travels, and can be expressed as:

$$\tau_{ig} = \frac{1}{\lambda}\int_{nb}^{\lambda} \tau_b(x)\,dx \tag{1.43}$$

Substituting Eq. (1.42) into Eq. (1.43), the integration leads to:

$$\tau_{ig} = \frac{n\mu b}{2\pi(1-\nu)\lambda}\ln\left(\frac{\lambda}{nb}\right) \tag{1.44}$$

Differentiating Eq. (1.44) with respect to time, and using Eq. (1.40), we obtain

$$\frac{d\tau_{ig}}{dt} = \frac{\mu b}{2\pi(1-\nu)\lambda}\left[\ln\left(\frac{\lambda}{nb}\right)-1\right]\left(\dot{\gamma}_p - \kappa n\right) \tag{1.45}$$

Consider that usually $\ln(\lambda/nb) >> 1$, Eq. (1.45) can be simplified as

$$\dot{\tau}_{ig} = H\dot{\gamma}_p - \kappa\tau_{ig} \tag{1.46}$$

Equation (1.46) expresses the evolution of back stress τ_{ig}. It also formulates the strain hardening-recovery mechanism. The term $H = [\mu b/2\pi(1-\nu)\lambda]\ln(\lambda/nb)$ is the work-hardening coefficient, which can be practically simplified as a material constant, since the logarithmic variation with n is usually small when n is large. The term $\kappa\tau_{ig}$, on the other hand, represents the recovery rate controlled by dislocation climb. By definition, the introduction of a back stress will reduce the effective stress on moving dislocations, as given by

$$\tau_{eff} = \tau - \tau_{ig} \tag{1.47}$$

With the back stress that reduces the effect of the applied stress, Eq. (1.47) can be substituted into Eq. (1.38) to describe the resulting strain rate with work-hardening and recovery.

1.6.4 Dislocation Climb in The Presence of Precipitates

When dislocations pile up between precipitates, vacancy migration must overcome the local stress field. Suppose an edge dislocation is lying on a gliding plane in the presence of precipitates of size r and spacing λ, as shown in Figure 1.13 (Wu and Koul 1995). Using Volterra's solution, the hydrostatic pressure in the upper half plane is given by (Nabarro 1967)

$$p(\rho,\theta)=\frac{1}{3}(\sigma_x+\sigma_y+\sigma_z)=-\frac{(1+\nu)\mu b}{3\pi(1-\nu)}\rho^{-1}\sin\theta \tag{1.48}$$

where ρ is the radial distance from the vacancy to the dislocation core, and θ is the pitching angle.

When a vacancy is forced to migrate from the position (ρ, θ) to the dislocation core, to allow the dislocation to climb, the preferred path for vacancy migration is the path that expends the least energy, i.e. along the arc S, as indicated in Figure 1.13. It can easily be proven that the minimum energy required is

$$\begin{aligned} W_{\min} &= -\int_S p(\rho,\theta)b^2 ds=\int_0^{\theta}\frac{(1+\nu)\mu b^3}{3\pi(1-\nu)}\rho^{-1}\sin\theta(\rho d\theta) \\ &= \frac{(1+\nu)\mu b^3}{3\pi(1-\nu)}(1-\cos\theta) \end{aligned} \tag{1.49}$$

Assume that dislocations are perfect sinks of vacancies and a dislocation climbs a Burgers vector when it absorbs a row of vacancies from the upper half plane (in the two-dimensional plane, Figure 1.13). Then, the minimum work needed to transport a vacancy to the dislocation core is

$$W_{\min}=\frac{(1+\nu)\mu b^3}{3\pi(1-\nu)}(1-\cos\theta_{\min})=\frac{(1+\nu)\mu b^3}{3\pi(1-\nu)}\left(1-\frac{\lambda}{\sqrt{\lambda^2+r^2}}\right) \tag{1.50}$$

Thus, additional work, $\Delta U = W_{\min}$, must be compensated by additional thermal energy such that vacancies can migrate to cause dislocation climb over the precipitates. The total activation energy for dislocation climb is hence equal to the activation energy for self-diffusion, U_d, plus additional energy as given by equation (1.50):

$$U_c=U_d+\Delta U=U_d+\frac{(1+\nu)\mu b^3}{3\pi(1-\nu)}\left(1-\frac{\lambda}{\sqrt{\lambda^2+r^2}}\right) \tag{1.51}$$

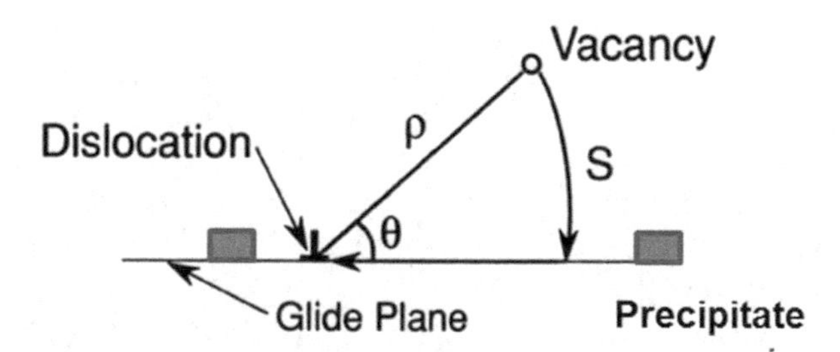

Figure 1.13. Vacancy migration in the stress field of an edge dislocation.

Hence, the diffusion coefficient should be modified

$$D = b^2 \nu_0 \exp\left(-\frac{U_c}{kT}\right) = b^2 \nu_0 \exp\left(-\frac{U_d + \Delta U}{kT}\right) \tag{1.52}$$

where ν_0 is the atomic frequency.

On the other hand, according to Rösler and Arzt (1988a, b), when a dislocation climbs along the particle/matrix interface, the dislocation-particle interaction will produce a back stress τ_{ic} which reduces the effective stress, acting on a climbing dislocation, to $\tau - \tau_{ic}$. This back stress, τ_{ic}, is independent of time as given by

$$\sigma_{ic} = \frac{\alpha \mu b}{\lambda} \tag{1.53}$$

where α is a factor less than unity, the value of which depends on the degree of coherency of the precipitate particle.

Hence, the vacancy diffusion rate constant can be modified as:

$$\kappa = \frac{D \mu b}{kT} \frac{\sigma - \sigma_{ic}}{\mu} \tag{1.54}$$

The above activation energy and back stress formulations will be used for creep deformation analysis in the later chapters.

1.7 Strengthening Mechanisms

Having understood how dislocations may move and be obstructed, we are now in a position to examine the ways by which metals may be strengthened. In fact, in material design and manufacturing, point defects, linear defects, planar defects and volume defects are all utilized to strengthen metals, making them into complex engineering alloys. The stress conditions for such strengthening mechanisms are discussed elsewhere (Herzberg 1996). The critical stress thus defined corresponds to σ_0 in Eq. (1.38), i.e. the minimum lattice resistance to plastic deformation.

1.7.1 Solid Solution Strengthening

Impurity atoms can be added into solid solution, which impose lattice strains on surrounding host atoms. The lattice strain field interacts with moving dislocations thus resulting in restriction of dislocation movement. This is one of the most common reasons to make alloys, which have higher strength than pure metals. For example, carbon is added into iron (Fe) to make steel, which is stronger and harder than pure iron. Some large atoms such as molybdenum (Mo) and tungsten (W) are often added to cobalt (Co) and nickel (Ni)-based superalloys to impede dislocation motion and diffusion, thus to increase their hardness and wear resistance and high-temperature creep resistance. Generally, substitutional solute atoms produce dilatational strains, whereas interstitial solute atoms tend to produce distortion strains. The effect of solute strengthening can be generalized as

$$\Delta\tau \propto \mu\varepsilon_{misfit}^{f} c^{g} \tag{1.55}$$

where ε_{misfit} is the lattice misfit strain, c is the concentration of solute in atomic fraction, and f and g are power-exponents.

Another role that solute atoms may play at high temperature is to lower the diffusion rate, thus enhancing the creep resistance of the material. For example, in Fe-2.3%W alloy, when tungsten was added to α-iron with no precipitates, the creep rate was reduced by three orders of magnitude (Maruyama et al. 2001).

1.7.2 Work Hardening

Work hardening or strain hardening refers to the phenomenon that a metal becomes stronger after being cold worked to a certain amount of strain. In essence, strain hardening arises from dislocation-dislocation interactions. As plastic deformation proceeds, dislocations from different sources and on different slip systems may become entangled with each other, forming a dislocation structure with walls that further reduce the mobility of other free-moving dislocations.

A necessary condition of work hardening is dislocation multiplication. Typically, dislocation density in an annealed state is in the order of 10^4-10^5/cm^2, and it can increase to 10^{11}-10^{12}/cm^2 in a cold-worked state (Hertzberg 1996). A widely accepted mechanism of dislocation multiplication is based on the Frank-Reed source. In this model, a segment of dislocation line is pinned by foreign atoms, particles or interaction with other dislocations, as shown in Figure 1.14. When a shear stress is applied, the dislocation line will bow out continuously, as shown in Figure 1.14 (a-c), until a critical point (c). The shear stress required to bow the dislocation is given by

$$\tau = \frac{\mu b}{2\pi(1-\nu)l} \tag{1.56}$$

where l is the distance between the pinning points. After this point, instability occurs in a way that the dislocation line loops around the pinning points, as shown in Figure 1.14 (c). Eventually, the loop pinches off when the two bent regions corresponding to screw dislocations of the opposite signs meet, as shown in Figure 1.14 (d). Afterwards, the loop and cusp straighten, leaving the same segment AB as before but with an additional dislocation loop having the same Burgers vector as the original segment, as in Figure 1.14 (e). With continued application of stress, this source can generate more and more dislocation loops, repeatedly. However, due to the back stress exerted by pile-ups of the outer loops against some unyielding obstacles, the source will eventually be shutdown.

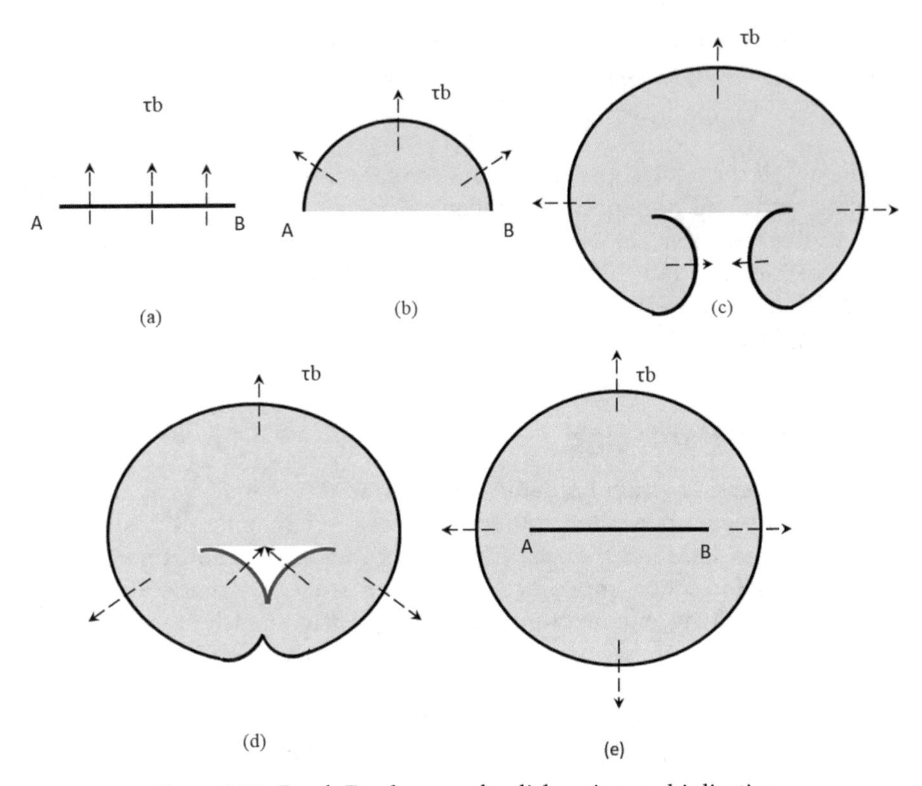

Figure 1.14. Frank-Reed source for dislocation multiplication (re-draw after Reed 1953).

Suppose that in a static pile-up situation, the mean free dislocation length l becomes a Frank-Read source, and the dislocation density ρ is proportional to l^{-2}, then the incremental stress, relative to a dislocation free state, is given by

$$\Delta\tau \propto \mu b\sqrt{\rho} \tag{1.57}$$

Cold working may generate a very high dislocation density (~10^{11} to $10^{13}/cm^2$) with high strain energy. At high temperatures, recrystallization of those cold-worked grains may occur to release the stored energy such that the material returns back to a low-energy state. This often happens during hot working of metals.

1.7.3 Grain Boundary Hardening

Planar defects such as grain boundaries can also be present as obstacles to impede dislocation motion. This occurs when dislocations generated from a Frank-Reed source in one grain come to meet another grain in different orientation and thus become immobile. Suppose there are n dislocations in a static pile-up over a length L (~ $d/2$, where d is the grain size), then the resolved shear stress at the Frank-Reed source to hold these pile-up dislocations is given by

$$\tau_s = \frac{n\mu b}{\pi(1-\nu)d} \tag{1.58}$$

The stress acting on the leading pile-up dislocation is found to be n times greater than τ_s. When this stress exceeds a critical value τ_c, the blocked dislocations are able to glide, passing through the grain boundary. Hence

$$\tau_c = n\tau_s = \frac{\pi(1-\nu)d\tau_s^2}{\mu b} \tag{1.59}$$

Since the resolved shear stress is equal to the applied stress less the frictional stress, i.e. $\tau_s = \tau - \tau_i$, Eq. (1.59) can be rearranged into (Hertzberg 1996)

$$\tau = \tau_i + \sqrt{\frac{\mu b \tau_c}{\pi(1-\nu)d}} \tag{1.60}$$

Eq. (1.60) is the theoretical basis for the well-known Hall-Petch relationship:

$$\tau_0 = \tau_i + kd^{-1/2} \tag{1.61}$$

where, τ_0 is the total lattice resistance to plastic deformation, τ_i is the lattice frictional stress, and k is a material constant. In a pure metal, τ_i corresponds to the Peierls-Nabarro stress defined by Eq. (1.1), but in alloys it may include contributions from other intragranular strengthening mechanisms.

Eq. (1.61) implies that no extensive dislocation glide beyond one grain can occur below this stress level, and hence, the Hall-Petch relationship is often used to describe the yield strength of a polycrystalline material. It infers that a fine-grained material is stronger than its coarse-grained counterpart, which has been observed to be true for numerous metals and alloys, indeed. The relationship, however, breaks down for nano-grained materials. At nanoscale, the material's yield strength decreases with decreasing grain size, which is referred to as the inverse Hall-Patch relationship (Masumura et al. 1998). This behaviour will be discussed in section 2.6, considering deformation mechanisms in nanocrystalline materials.

1.7.4 Precipitation Hardening

By alloy design, precipitation hardening can be achieved through solution treatment and aging of the alloy. When the solute concentration exceeds the limits of the solubility of the matrix phase, usually at temperatures below the solvus, nucleation and growth of a second phase occurs during cooling and subsequent aging heat treatment. Because nucleation and growth of precipitates are competing to draw solute atoms of a fixed amount, it needs the right aging process to achieve precipitates with the size and volume fraction to yield the maximum strength. For example, a new heat treatment process called retrogression and re-aging (RRA) has been developed to tailor the mechanical strength and corrosion resistance of 7000 series (Al-Zn-Mg-Cu) aluminum alloys through controlling the phase fractions of Guinier-Preston (GP) zones, η' and η phases (Wallace et al. 1981, Wu et al. 2001, 2005), where the intragranular GP zones impart the strength and η phase formed along the grain boundaries increases the corrosion resistance of 7000 aluminum to exfoliation and stress-corrosion cracking.

As discussed in section 1.2.4, precipitates are volume defects within the parent crystal. The presence of such second phase particles often causes lattice distortion, which impede the movement of dislocations in the crystal lattice. Other factors such as the structure of the second phase and the nature of the particle-matrix interface also influence dislocation motion in a way either to allow dislocations to cut through or by-pass. Since dislocations are often the dominant carriers of plasticity, the presence of precipitates serves to harden the material. Dislocation-particle interactions may occur in three ways:

Misfit—Difference in lattice parameter between the host and precipitate phase produces a misfit strain that alters the stress field around the precipitate. The presence of second phase precipitates also changes the local stiffness of the crystal. Dislocations are repulsed by regions of higher stiffness. Conversely, if the precipitate causes the material to be locally more compliant, then the dislocation will be attracted to that region. The contribution of misfit hardening can be expressed as (Glieter and Hornbogen 1967/68):

$$\Delta\tau \propto \mu\varepsilon_{misfit}^{3/2}\left(rV_f\right)^{1/2} \tag{1.62}$$

where ε_{misfit} is the lattice misfit strain, r is the precipitate size, and V_f is volume fraction.

At the first thought, it seems that the larger misfit strain is the better. But, other considerations have to be given in alloy design to limit the misfit strain. For example, for Ni-base superalloys, misfit strains are limited so as to maintain the coherency of large precipitates and to minimize Ostwald ripening, thus enhancing the creep resistance.

Precipitate-Cutting—If the precipitate particle maintains coherency with the host matrix, dislocation motion can cut through the particle. This creates new interphase and anti-phase boundary (APB) areas. For example, when a single dislocation passes through an ordered Ni_3(Ti, Al) (γ') precipitate in a Ni-base superalloy, it destroys the periodicity of the superlattice, the passage of a second identical dislocation on the same slip plane reorders the precipitate, and an anti-phase boundary is thus formed between these two superlattice dislocations, as shown in Figure 1.15. This precipitate cutting process contributes markedly to the alloy strengthening. Glieter and

Hornbogen (1967/68, Ardell and Huang 1988) reported the strengthening contribution associated with this mechanism to be of the form:

$$\Delta\tau \propto \gamma_{APB}^{3/2}\left(\frac{rV_f}{\mu}\right)^{1/2} \tag{1.63}$$

where γ_{APB} is the APB energy, r is the precipitate size, and V_f is volume fraction.

Precipitate-Looping—If the misfit strain is large, the precipitates are incoherent, or when the average particle separation is above a critical value, dislocations are unable to cut through the precipitates; instead, they loop round individual precipitates, as shown in Figure 1.16 (a). This process is called Orowan looping. A research group in Cornell University has simulated this type of dislocation-precipitate interaction using atomistic theory (Singh et al. 2011). The stress necessary for dislocations to loop around a particle is given by

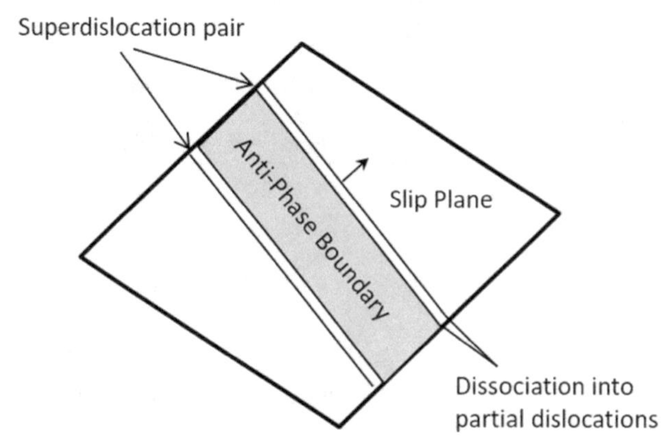

Figure 1.15. Superdislocation pairs creating anti-phase boundary.

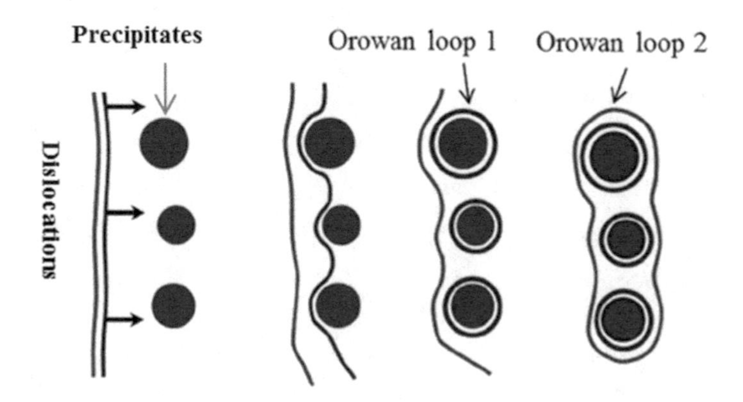

Figure 1.16. Dislocations looping around particles.

$$\Delta\tau = \frac{\mu b}{\pi(1-\nu)l} \tag{1.64}$$

which is the well-known Orowan-looping mechanism.

By the same token, grain boundary precipitates are often employed to strengthen grain boundaries against creep at high temperatures, which will be discussed in later chapters.

1.8 Summary

The aforementioned strengthening mechanisms can be summarized in Table 1.2 with mechanism parameters listed. These parameters may serve as the "genome" of a material, by which its behaviour is imparted. Given the above strengthening mechanisms, one question is immediately asked: for a complex engineering alloy with a microstructure containing solute atoms and second phase particles, which mechanisms controls the mechanical properties of the material? There is no straightforward answer to this question. Certainly,

Table 1.2. Strengthening mechanisms and parameters

Mechanism	Parameters	Illustration
Elasticity	μ ν	Shear modulus Poisson's ratio
Solute Strengthening $\Delta\tau \propto \mu\varepsilon_{misfit}^{f} c^{g}$	ε_{misfit} c f g	Lattice misfit strain Concentration of solute Power to misfit Power to concentration
Work Hardening $\Delta\tau = \mu b\sqrt{\rho}$	ρ	Dislocation density
Precipitate Hardening – Misfit $\Delta\tau = \mu\varepsilon_{misfit}^{3/2}\left(rV_f\right)^{1/2}$	ε_{misfit} r V_f	Misfit strain Precipitate size Volume fraction
Precipitate Hardening – APB $\Delta\tau = \gamma_{APB}^{3/2}\left(\frac{rV_f}{\mu}\right)^{1/2}$	γ_{APB}	APB energy
Precipitate Hardening – Loop $\Delta\tau = \frac{\mu b}{\pi(1-\nu)l}$	l	Precipitate spacing
Grain Boundary Strengthening $\tau_0 = \tau_i + kd^{-1/2}$	d τ_i k	Grain size Intragranular resistance G.B. strength constant

the above contributions are not all linearly added up. But, they can be classified into intragranular strengthening mechanisms and grain boundary strengthening mechanisms. Intragranular strengthening mechanisms such as solid solution strengthening and precipitate strengthening contribute directly to the frictional resistance of the lattice, such as the term τ_i in Eq. (1.60) and (1.61), and the critical stress τ_c for breaking into the next grain, which effectively change the Hall-Petch constant in Eq. (1.61). Grain boundary hardening mechanisms arise due to the presence of grain boundary precipitates and "hard" orientations, which impede dislocation motion along the grain boundaries or across the grain boundaries into neighbouring grains. If the loading rate is high, the strengthening mechanism that produces the maximum resistance will control the material yielding behaviour, before other low-stress-activated mechanisms could generate massive deformation. If the loading rate is slow, the low-stress-activated mechanisms can be operative to generate enough plastic deformation to cause material yielding, especially at elevated temperatures. As deformation proceeds, dislocation pile-up and entanglement may still continue to harden the material. Thus, various deformation mechanisms can be operative under different temperature conditions, which will be discussed in the next chapter.

CHAPTER

2

Deformation Mechanisms

As discussed in Chapter 1, many strengthening mechanisms are employed in alloy design to limit material deformation. A crystalline solid only undergoes elastic deformation without configuration change below the lattice resistance, and hence the stress-strain behavior is linear. Above this limit, dislocations can move along certain slip planes by shear forces and thermal activation, which often result in complicated non-linear deformation behaviors as observed in various mechanical tests of materials. The simplest mechanical test is monotonic tensile test, where the material is stretched at a constant loading rate under either stress or strain control and it exhibits a nonlinear stress-strain behaviour. Another simple test is creep test under constant load, often conducted at high temperature, where the material exhibits time-dependent deformation. The loading profile that a material may experience in service is more complicated, often comprised of cyclic and dwell loads at variable amplitudes and at high temperatures. Then, the material may exhibit unsymmetrical hysteresis and ratcheting behaviours.

Understanding the material behaviour is required in structural component design for engineering applications. Therefore, it is critical to understand the deformation mechanisms operating in the applied stress and temperature ranges, which govern the different material behaviours. Fundamental deformation mechanisms in metallic materials have been well understood from the earlier work mostly in the 1950 - 1980s and summarized into Ashby's deformation mechanism maps (Ashby 1972, Frost and Ashby 1982). This chapter gives a comprehensive review of those deformation mechanisms. Most importantly, extended treatments are given to elucidate their manifestation in materials' transient as well as steady-state behaviours under simple loading conditions such as tensile and creep loadings. These basic constitutive equations form the basis of a mechanism-based constitutive law framework, which will be used to describe various deformation processes in the later chapters.

2.1 Rate-Independent Plasticity

In Chapter 1, we have discussed the propensity of crystallographic slip and slip systems in f.c.c., b.c.c. and h.c.p. crystals. Slip generally proceeds in close-packed directions on close-packed planes within a crystallite. The consequence of block-like slip produces crystal offsets (slip steps) at the crystal boundaries, resulting in a permanent shape change, i.e. plastic deformation. In a polycrystalline material, plastic deformation is a result of collective slips of edge dislocations and cross-slips of screw dislocations in numerous grains. At homologous temperatures (the ratio of the absolute temperature to the melting temperature in Kelvin) below 0.3, plasticity manifests in a rate-independent manner, which can be described by a Ramberg-Osgood type equation:

$$\varepsilon_p = \left(\frac{\sigma - \sigma_0}{K}\right)^n \tag{2.1}$$

where σ_0 is the lattice resistance, and K is a plastic strength parameter (or drag stress), n is the stress sensitivity exponent. It should be noted here that Eq. (2.1) is merely to represent the non-linear stress-strain relationship of plasticity. In the finite element method (FEM) package, e.g. Abaqus and MSC. Marc, the incremental theory of plasticity obeying a power-law relationship can be implemented with kinematic and isotropic hardening options.

Rearranging Eq. (2.1), we have

$$\sigma_{ys} = \sigma_0 + \sqrt[n]{\varepsilon_p}K \tag{2.2}$$

Consider a typical tensile behaviour, as shown in Figure 2.1, at first the deformation process remains elastic up to a certain point (A). Beyond this point the material yields, but further load increase is usually required to maintain the plastic flow; this phenomenon is known as strain-hardening or work-hardening. If at some point (B), the specimen is unloaded, the stress will return to zero along the path BC, in parallel to the original elastic line, recovering the elastic deformation, but a permanent elongation offset is left at point C. This offset is called plastic deformation. If the material is re-loaded at this point, the stress-strain curve will retrace the unloading path CB until it again reaches the new plastic state. Further increases in stress will cause the curve to follow BD. However, if deformation continued from point C into compression (reverse in the previous loading direction), the material would yield at a new stress level E, which is often different in magnitude from the yield stress of the first loading.

In engineering, the yield strength of a material is usually defined at 0.2% offset, which can be evaluated from Eq. (2.2) or measured experimentally on the first loading curve. As discussed above, the apparent yield strength would change after deformation passing the initial yield point (A) upon reversing the load, apparently due to the effect of prior deformation history. If $\overline{CE} = \overline{BC}$, the material exhibits *isotropic hardening*. If $\left(\overline{CE} + \overline{BC}\right) = 2\overline{OA}$, the material exhibits *kinematic hardening*, which means that the center of tension/compression symmetry has been shifted by an amount of $\left(\overline{BC} - \overline{CE}\right)/2$

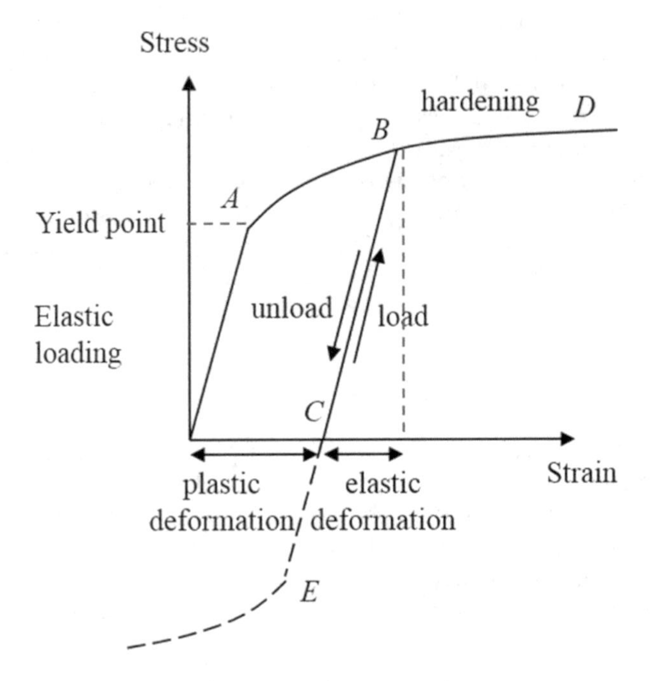

Figure 2.1. A typical stress-strain curve.

without changing the radius of the yield surface by the previous loading history. In this case, $\overline{CE} < \overline{OA} < \overline{BC}$, and the phenomenon is called the *Bauschinger effect*, which is a manifestation of kinematic hardening caused by dislocation pile-up formation.

Owing to its simplicity, Eq. (2.1) is widely used to describe materials' tensile behaviour. However, at high temperatures, materials' tensile behaviours are often rate-dependent, which invalidate n and K as "material constants", because they would change with the test condition. Therefore, for mechanism characterization, Eq. (2.1) should be strictly used as the constitutive equation of rate-independent plasticity. Descriptions of Eq. (2.1) for rate-independent cyclic stress-strain behaviours will be discussed in Chapter 6.

2.2 Rate-Dependent Plasticity by Dislocation Glide

Rate-dependent plastic deformation arises due to thermally activated dislocation glide, which proceeds by shearing or looping around second-phase precipitates (see section 1.7.4). Kocks et al. (1975) proposed an exponential function to describe dislocation glide as limited by various types of obstacles:

$$\dot{\gamma}_p = \dot{\gamma}_0 \exp\left(-\frac{\Delta F}{kT}\left[1-\left(\frac{\tau}{\hat{\tau}}\right)^p\right]^q\right) \tag{2.3}$$

where $\hat{\tau}$ is the athermal flow strength; ΔF represents the obstacle strength (it is in the order of 2μb³ for strong obstacles such as large precipitates or dispersions, 0.2-1 μb³ for medium obstacles such as forest dislocations, <0.2 μb³ for weak obstacles such as lattice resistance by solutes); p and q are constants accounting for the shape of obstacles, usually $0 \le p \le 1$, $1 \le q \le 2$. For pure metals, Frost and Ashby assumed $\hat{\tau} = \mu b\sqrt{\rho}$ to describe work hardening due to forest dislocations (Frost and Ashby 1982). Eq. (2.3) has been used in combination with a dislocation density evolution model to describe the stress-strain behaviour during large deformation processes (Arsenlis and Parks 2002). Dislocation networks would induce isotropic hardening.

For small deformation which is mostly concerned in lifing practices, we can simply assume a *constant structure* for the material. In this sense, dislocation glide can be generally regarded as a thermally activated process over a constant single energy barrier. Then, the plastic strain rate can be expressed as (Wu and Krausz 1994):

$$\dot{\varepsilon}_g = 2\dot{\varepsilon}_0 \exp\left(-\frac{\Delta G_0^{\neq}}{kT}\right)\sinh\left(\frac{V\left(\sigma - H\varepsilon_g - \sigma_0\right)}{kT}\right) \tag{2.4}$$

Here, we use ε_g to signify plastic strain generated by rate-dependent intragranular dislocation glide (IDG). As discussed in Chapter 1, in the context of Eq. (1.38), Eq. (2.4) represents the kinetics of dislocation glide with both forward and backward activation steps. In a dislocation-glide-dominated process, due to the fast strain-loading rate, which is much greater than the dislocation climb rate ($\dot{\varepsilon}_g >> \kappa$) at the temperature, and because the strain hardening coefficient is usually much larger than the back stress (i.e., $H >> \sigma_{ig}$), the recovery term $\kappa\sigma_{ig}$ in Eq. (1.42) is negligible and hence linear strain hardening shall proceed in the presence of strong strengthening particles. When the material is under fast loading/unloading conditions such as during engine start-up and shutdown, we can assume the deformation process is controlled by dislocation glide within a *constant microstructure*. Then, the constant-strain-rate deformation process can be solve analytically as follows.

First, Eq. (2.4) can be re-written in a short form as:

$$\dot{\varepsilon}_g = 2A_g \sinh\Phi \tag{2.5a}$$

with

$$\dot{\Phi} = \frac{V\left(\dot{\sigma} - H\dot{\varepsilon}_g\right)}{kT} \tag{2.5b}$$

and

$$A_g = A_{g0} \exp\left(-\frac{\Delta G_0^{\neq}}{kT}\right) \tag{2.5c}$$

Here, Φ represents the mechanical energy over thermal energy per atom for the activation step.

Under constant-strain-rate loading conditions, as usually imposed in a tensile test, the total deformation satisfies:

$$\dot{\varepsilon} = \frac{\dot{\sigma}}{E} + \dot{\varepsilon}_g = \text{const.} \tag{2.6a}$$

or

$$\dot{\sigma} = E(\dot{\varepsilon} - \dot{\varepsilon}_g) \tag{2.6b}$$

Substituting Eq. (2.5a) and Eq. (2.6) into Eq. (2.5b), we obtain a first-order differential equation for the deformation energy, Φ, as given by

$$\dot{\Phi} = \frac{EV}{kT}\left[\dot{\varepsilon} - 2A_g\left(1 + \frac{H}{E}\right)\sinh\Phi\right] \tag{2.7}$$

Eq. (2.7) can be solved to obtain

$$\left(\frac{e^{-\Phi} - a}{e^{-\Phi} + b}\right) = \left(\frac{1-a}{\chi + b}\right)\exp\left[-\frac{VE\dot{\varepsilon}(t - t_0)\sqrt{1+\chi^2}}{kT}\right] \tag{2.8}$$

where

$$\chi = \frac{2A_g}{\dot{\varepsilon}}\left(1 + \frac{H}{E}\right),\ a = \frac{\sqrt{1+\chi^2} - 1}{\chi}, b = \sqrt{1+\chi^2} + 1$$

The initial time of plastic deformation is defined by

$$\Phi_0 = \left(\frac{V(\sigma - H\varepsilon_g - \sigma_0}{kT}\right) = 0 \tag{2.9}$$

where ε_g is the plastic strain accumulated from prior deformation history ($\varepsilon_g = 0$ for pristine materials), and σ_0 represents the initial lattice resistance to dislocation glide. Since deformation below σ_0 is purely elastic, i.e. $\sigma = E\dot{\varepsilon}t$, then $t_0 = \sigma_0/(E\dot{\varepsilon})$. Once the stress exceeds the initial lattice resistance in the material, i.e., $\sigma > \sigma_0$, plastic deformation commences.

From Eq. (2.8), we can obtain the stress-strain response as follows:

$$\sigma - H\varepsilon_p - \sigma_0 = -\frac{kT}{V}\ln\left(\frac{a + \omega b}{1 - \omega\chi}\right) \tag{2.10a}$$

where, $\omega(\varepsilon)$ is the response function defined by

$$\omega(\varepsilon) = \left(\frac{1-a}{\chi+b}\right)\exp\left\{-\frac{V(E\varepsilon-\sigma_0)\sqrt{1+\chi^2}}{kT}\right\} \qquad \left(\varepsilon > \frac{\sigma_0}{E}\right) \qquad (2.10b)$$

Eq. (2.10) basically describes the accumulation of plastic strain by a linear strain-hardening mechanism. Therefore, it describes a dislocation glide dominated deformation process where dislocation climb controlled recovery can be neglected.

An example is shown here for IN738LC with γ'-strengthening. Figure 2.2 shows the description of Eq. (2.10) for the tensile behaviours of IN738LC at 750-950 °C in comparison with the test data (Wu et al. 2002). The material exhibits a significant temperature-dependence as well as strain-rate dependence at temperatures above 750 °C. The parameters for this material model are given in Table 2.1.

Table 2.1. Model parameter values of the for IN738LC

Activation volume, V (cm^3)	**3.977×10^{-22}**		
Pre-exponential, A_{g0} (s^{-1})	**0.7**		
Activation Energy, $\Delta G_0^{\neq}$ (J)	**2.38 × 10^{-19}**		
	750 °C	850 °C	950 °C
Modulus of Elasticity, E (GPa)	175.5	151.4	137.0
Lattice resistance, σ_0 (GPa)	540	285	110
Work Hardening, H (MPa)	15	13.7	12.5

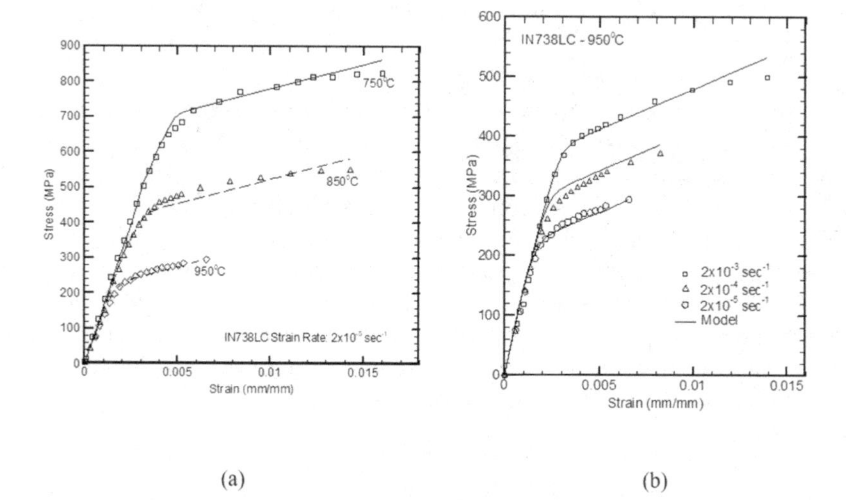

Figure 2.2. Stress-strain responses of IN738LC a) at strain rate of 2×10^{-5} at 750-950 °C, and b) at various strain rates at 950 °C.

In the above model, the parameters: E, H, V, σ_0, A_0 and $\Delta G_0^{\neq}$, are all physically-defined. In particular, the thermal activation parameters, V, A_0 and $\Delta G_0^{\neq}$ represent a "constant structure" at the nano-scale, which depends on composition of the material. On the other hand, microstructural effects arising from grain size and precipitates at the micron scale can be incorporated into the parameters H and σ_0 through incorporation of the strengthening mechanisms as discussed in section 1.7.

2.3 Power-Law Creep by Dislocation Climb-plus-Glide

At high temperatures, when vacancy flow is abundant via diffusion, dislocations acquire another degree of freedom: they can climb as well as glide. As discussed in section 1.6, while a gliding dislocation is being held up by an obstacle in the glide plane, climb may release it, allowing it to continue to glide on other easy-glide planes. Even though glide is almost exclusively responsible for producing the shape-changing strain, this high-temperature deformation process is apparently controlled by dislocation climb when the stress is not high enough to drive dislocations overcoming the obstacles by either cutting or looping around the obstacles. This is the reason why creep deformation can proceed well below the yield strength.

Weertman (1955) first developed a creep model based on the climb-plus-glide mechanism, which yields a power-law relationship with an exponent of 4.5. It has been found through experiments on many metals and alloys that the creep power-exponent mostly falls in the range of 3 to 8 (Mukherjee et al. 1969, Frost and Ashby 1982). In general, the phenomenon is called power-law creep, as described by the Norton equation (1929):

$$\dot{\varepsilon}_{ci} = A_c \sigma^m \tag{2.11a}$$

where A_c is an Arrhenius-type constant, and m is the power-law exponent. An example of power-law creep is shown in Figure 2.3 for ductile cast iron, more will be shown in Chapter 5.

During creep, dislocations may multiply from Frank-Reed sources. Johnston (1962) first observed that creep rate accelerated with increase of dislocation density. He proposed a linear dislocation multiplication model. Later, Dyson and Maclean (1983), and Dyson and Gibbons (1987) attributed the tertiary creep behaviour more favorably to dislocation multiplication than particle coarsening. Accordingly, the intragranular dislocation climb-controlled creep rate can be expressed as

$$\dot{\varepsilon}_c = (1 + M\varepsilon_c)\dot{\varepsilon}_{ci} \tag{2.11b}$$

where M is the dislocation multiplication factor and $\dot{\varepsilon}_{ci}$ is the initial climb-controlled creep rate.

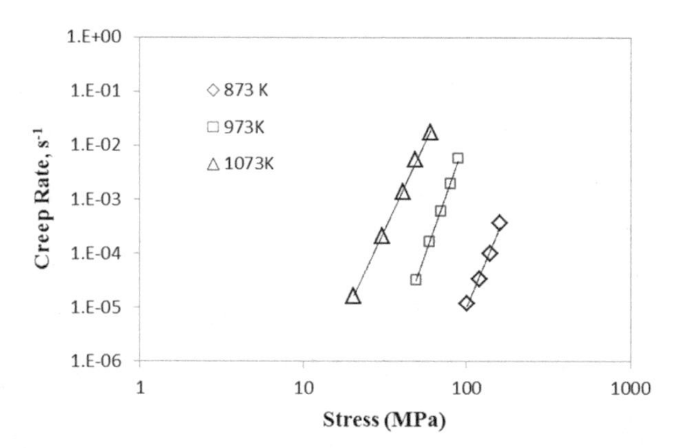

Figure 2.3. Creep rates in ductile cast iron (data from Wescast Industries Inc.)

Eq. (2.11) can be integrated into the form:

$$\varepsilon_c = \frac{1}{M}\left[e^{M\dot{\varepsilon}_{ci}t} - 1\right] \tag{2.12}$$

It is noted that when the initial creep rate is low such that $M\dot{\varepsilon}_{ci}t << 1$, the creep strain exhibits a nearly linear relationship with time.

At very high stresses, the stress dependence of creep rate becomes increasingly stronger than the power law, which is called power-law breakdown (Frost and Ashby 1982). Garofalo (1963) proposed a hyperbolic sine power relationship $\dot{\varepsilon}_{ci} = \sinh^m(\alpha\sigma)$ to cover both regions: it reduces to power-law at low stresses but an exponential function at high stresses. Actually, at high temperatures when the back stress achieves a balance between strain hardening and climb-controlled recovery in Eq. (1.46), the strain rate Eq. (1.38) predicts an exponential stress dependence of creep rate at high stresses, similar to the Garofalo equation. This indicates that, going from power-law to exponential stress dependence, creep deformation has undergone a transition from climb-controlled to glide-dominated mechanism. In engineering analysis, the Norton equation is often used to describe both regions, but the latter has a distinctively high power. For example, the creep rate of modified 9Cr1Mo steel is shown in Figure 2.4 (Zhang et al. 2017a). In this 9Cr1Mo steel, the low-stress region (normalized with its tensile strength) exhibits a power exponent of 6, the intermediate-stress region has a power exponent about 16, and the high-stress region appears to have a much higher exponent about 29. Power of 6 corresponds to intragranular dislocation climb (IDC)-controlled creep regime; the high power of 29 corresponds to the intragranular dislocation glide (IDG)-controlled process. In between, grain boundary sliding (GBS) occurs, which will be discussed in the next section.

From the deformation kinetics point of view, Wu and Krausz (1994) have demonstrated the equivalence of Eq. (1.38) and Eq. (2.4) with the hyperbolic sine power function. Chaboche (2008) has also compared the power-law, hyperbolic sine and hyperbolic sine power functions and shown that they

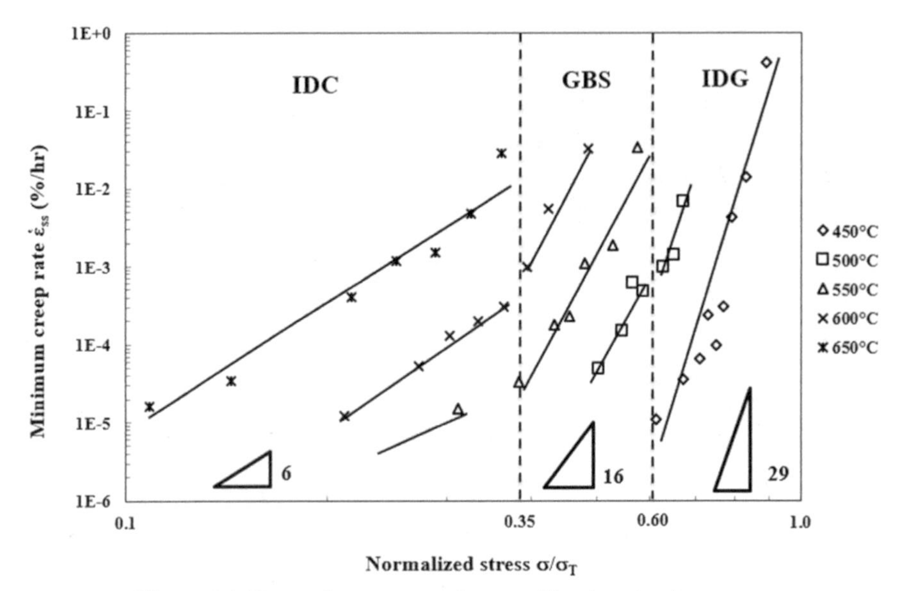

Figure 2.4. Power-law exponents according to rate-stress map (after Zhang et al. 2017).

are mathematically equivalent to describe the strain rate ranging from 10^{-8} to 10^{0} s^{-1} (within an acceptable error band), which covers the stress-dependence transition of the climb-plus-glide mechanism. Because of the common practice of using Norton's power-law, it is necessary to derive the link between the power-law exponent and the dislocation kinetics-based equation (1.38), as

$$m = \frac{\partial \log \dot{\varepsilon}_{ci}}{\partial \log \sigma} = \frac{V\sigma}{kT} \coth\left(\frac{V\sigma_{eff}}{kT}\right) \tag{2.13}$$

Power-law creep arising from dislocation climb-plus-glide can be influenced by alloying. Alloying elements are either present in the solid solution matrix or clustered to form second phase particles. Alloying elements in solid solution may change lattice parameters, stacking faulty energy, and diffusivity, etc., which in turn influence the creep behaviour. Second-phase particles or precipitates can impede dislocation motion and thus reduce the creep rate. While the power-law exponent for pure metals and their solid solutions appear to fall in the range of 3 to 5, the power-law exponent for alloys tends to be higher. Numerous creep studies have shown that, for most engineering alloys, the power-law creep has an exponent m = 3 ~ 8 (Frost and Ashby 1982). Therefore, alloying has been an effective way used by material engineers to enhance the material's creep resistance. For example, in Fe-2.3%W alloy, when tungsten was added to α-iron with no precipitates, the creep rate was reduced by three orders of magnitude (Maruyama, et al. 2001). Because the alloying effect is more specific to particular material chemistry, an extensive discussion of this subject is beyond the scope of this book.

2.4 Grain Boundary Sliding

In a polycrystalline material, grain boundaries are often the weakest link at high temperatures, because abundant crystalline defects (dislocations and vacancies) are agglomerated there, which can be mobilized by thermal activation at high temperatures. Grain boundary sliding (GBS) occurs by dislocation climb plus glide along grain boundaries (Langdon 1970, Wu and Koul 1995, 1997, Wadsworth et al. 2002). GBS generally needs to be accommodated at triple grain junctions, causing severe localized deformation in the impeding grain. When such accommodation is not perfect, due to disruption of slip plane and incompatible grain orientation, GBS would create triple junction wedge cracks. In addition, it can also lead to microcrack formation at grain boundary precipitates. GBS as a vital damage mechanism will be discussed in the next chapter. Here, we mainly consider GBS as an independent deformation mechanism. A general treatment of GBS in the presence of grain boundary precipitates along planar and wavy grain boundaries has been given by Wu and Koul (1995, 1997).

Grain boundary precipitates act as obstacles to both dislocation glide and climb along grain boundaries (Wu and Koul 1995), as shown schematically in Figure 2.5. Dislocations piling up in front of a grain boundary precipitate exert a back stress opposing dislocation glide, and the precipitate itself has a resistance to climb along the precipitate/matrix interface. With the above dislocation-precipitate interaction mechanism in consideration, Langdon's GBS equation for clean and planar boundaries can be further modified as follows.

According to Langdon (1970), the rate of dislocation climb along the boundary is given by

$$\dot{S} = Nb\kappa \tag{2.14}$$

where N is the number of grain boundary dislocations per unit length, κ is the climb rate given by Eq. (1.54).

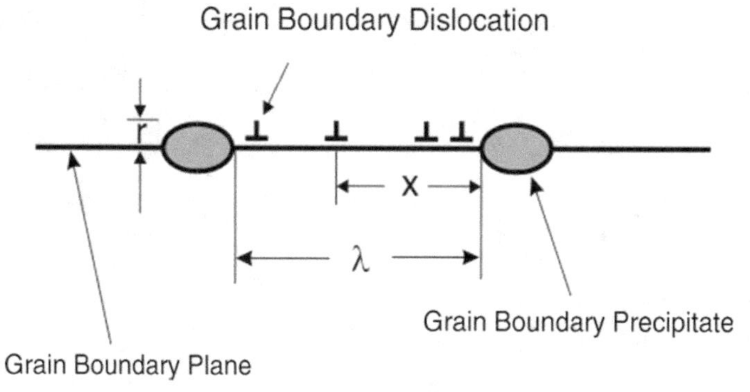

Figure 2.5. A schematic of grain boundary dislocation pile-up against a grain boundary precipitate.

The number of dislocations per unit length in a grain boundary can be related to the effective shear stress, $\tau - \tau_{ig}$ (in Langdon's model for clean boundaries, $\tau_{ig} = 0$), as

$$N = \frac{2\pi(1-\nu)(\tau - \tau_{ig})}{\mu b} \tag{2.15}$$

The shear strain rate produced by dislocation climb and glide processes along a grain boundary plane is given by

$$\dot{\gamma}_{gbs} = MAb\dot{S} \tag{2.16}$$

where M ($M \sim 6/\pi d^3$) is the number of grain boundaries per unit volume, d is the grain size, and A is the total area swept out by dislocations moving along the grain boundary. Langdon assumed $A \sim \pi d^2$ for clean grain boundaries. But, in the presence of grain boundary precipitates, as shown in Figure 2.6, A should be modified according to the grain boundary precipitate distribution morphology. For a discrete distribution, A should be proportional to $(\lambda + r)d$. For an almost continuous particle network, A is restrained to the area proportional to $\pi(\lambda + r)^2$.

Combining Eq. (2.14–2.16), with the consideration of κ as given by Eq. (1.54), we obtain

$$\dot{\gamma}_{gbs} = 12\pi\sqrt{3}(1-\nu)\left\{\frac{b}{d}, \frac{(\lambda+r)b}{\pi d^2}, \frac{(\lambda+r)^2 b}{d^3}\right\}\frac{D\mu b}{kT}\frac{(\tau-\tau_{ig})(\tau-\tau_{ic})}{\mu^2} \tag{2.17a}$$

which can be written in an abbreviated form

$$\dot{\gamma}_{gbs} = A_0\frac{D\mu b}{kT}\left(\frac{b}{d}\right)^q\left(\frac{\lambda+r}{b}\right)^{q-1}\frac{(\tau-\tau_{ig})(\tau-\tau_{ic})}{\mu^2} \tag{2.17b}$$

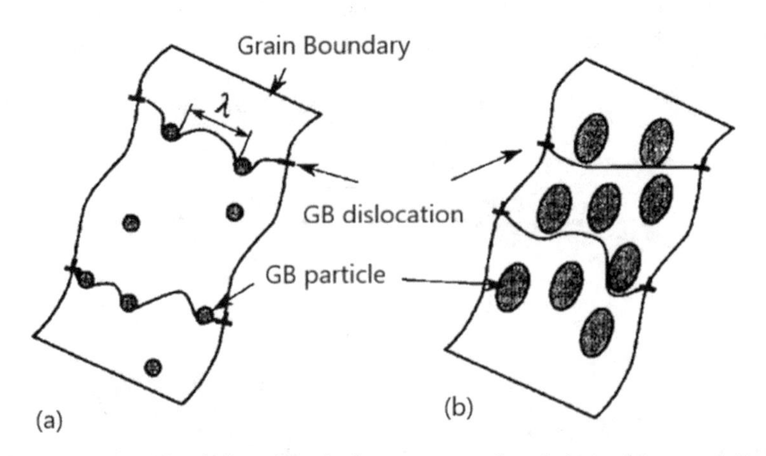

Figure 2.6. Area swept by dislocations in the presence of grain boundary precipitates, a) discrete distribution and b) almost continuous distribution.

where A_0 is a non-dimensional constant, q is the grain boundary precipitate distribution index which takes values of 1 (without particles), 2 (discrete particles) and 3 (continuous network of particles).

In microstructural design for damage tolerant materials, modification of grain boundary morphology generating wavy grain boundaries has been found to be an effective way to enhance creep resistance. For many Ni-base superalloys such as Alloy 718, IN738LC, Nimonic 115, Mar-M-247, and MERL 76, this could be achieved through special heat treatments (Koul and Gessinger 1983, Koul et al. 1988, Chang et al. 1994, Chang et al. 1996). The GBS model considering wavy grain boundaries provides guidance for microstructure design to improve material's creep properties (Wu and Koul 1997). In this section, two grain boundary wavy forms are considered: i) triangular wave and ii) sinusoidal wave with $h/2$ as the wave amplitude and λ as the wave length, as shown in Figure 2.7.

Raj and Ashby (1971) and Raj (1975) proposed a diffusion-controlled GBS model to describe anelastic transient creep. In their model, it is assumed that the shear stresses along the grain boundaries are relaxed to zero and that only the normal stress acts across the boundaries to drive the diffusion flow. However, they used the vector rule, instead of the tensor rule, to obtain the resolved normal stress on the grain boundary plane, which leads to an erroneous equation that predicts the sliding rate approaching infinity for perfectly planar boundaries, as $h/\lambda \to 0$.

To derive a GBS equation for wavy boundaries, the applied stress must be first resolved on the grain boundary plane using the stress-strain tensor transformation relationship as given in section 1.3. Between the two coordinate systems as shown in Figure 2.7, we can obtain the tangential shear stress on the triangular wave boundary as (Wu and Koul 1997):

$$\tau_\theta = \tau_a \cos 2\theta = \tau_a\left(\frac{2}{1+\tan^2\theta} - 1\right) = \tau_a\left(\frac{2}{1+\dfrac{4h^2}{\lambda^2}} - 1\right) \tag{2.18}$$

Similarly, suppose that the sinusoidal wave boundary is described by

$$x = \frac{h}{2}\cos\frac{2\pi y}{\lambda} \tag{2.19}$$

The tangential shear stress on the wave boundary can be obtained as (Wu and Koul 1997):

$$\tau_\theta = \tau_a \cos 2\theta = \tau_a\left(\frac{2}{1+\tan^2\theta} - 1\right) = \tau_a\left[\frac{2}{1+\left(\dfrac{\pi h}{\lambda}\right)^2 \sin^2\dfrac{2\pi y}{\lambda}} - 1\right] \tag{2.20}$$

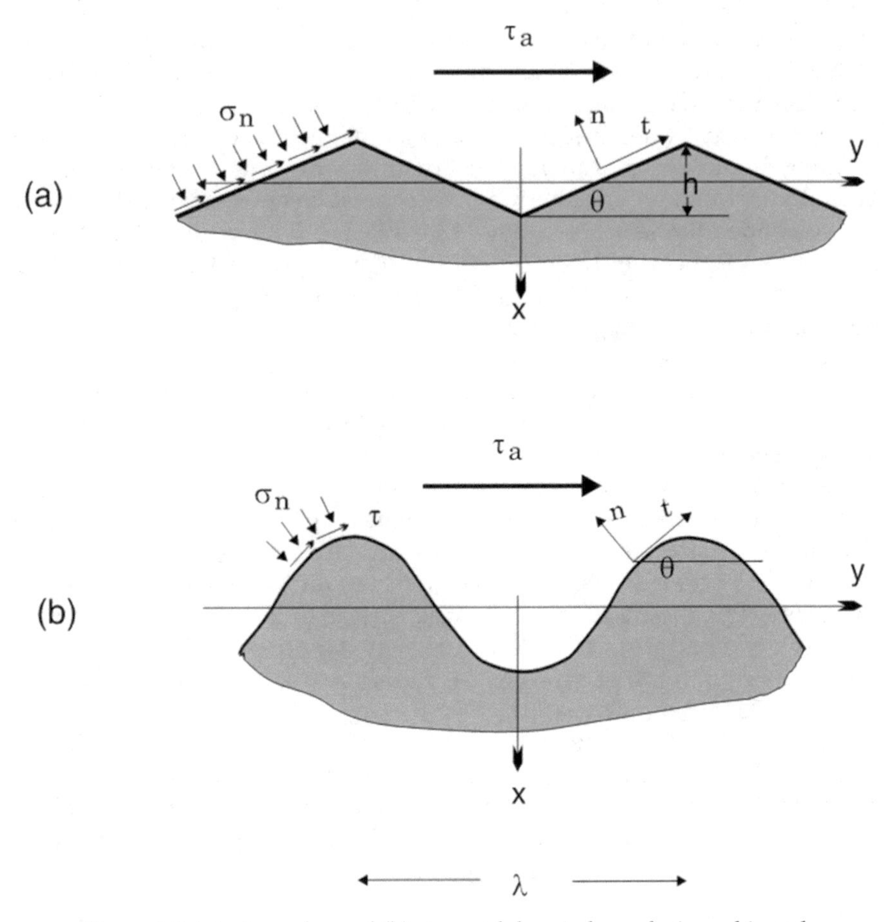

Figure 2.7. (a) triangular and (b) sinusoidal grain boundaries subjected to a remote shear stress.

In the local coordinate system (**n – t**), Eq. (2.17) holds true (the laws of physics do not change with coordinates) such that $\dot{\gamma}_\theta \propto \tau_\theta^2 \propto \tau_a^2 \cos^2 2\theta$. On the other hand, by strain transformation,

$$\dot{\gamma}(x, y) = \frac{\dot{\gamma}_\theta}{\cos 2\theta} = \dot{\gamma}_{\text{planar}} \cos 2\theta \tag{2.21}$$

where $\dot{\gamma}_{\text{planar}}$ is the GBS rate on planar grain boundaries as described by Eq. (2.17).

As a typical representation of the bulk material, averaging the local GBS rate, Eq. (2.21), over the wave length λ, we obtain

$$\dot{\gamma}_{\text{wavy}} = \frac{1}{\lambda} \int_{-\lambda/2}^{\lambda/2} \dot{\gamma}(x, y)\, dy = \phi \dot{\gamma}_{\text{planar}} \tag{2.22}$$

where

$$\phi = \begin{cases} \dfrac{2}{1+\left(\dfrac{2h}{\lambda}\right)^2} - 1 & \text{for triangular boundaries} \\ \dfrac{2}{\sqrt{1+\left(\dfrac{\pi h}{\lambda}\right)^2}} - 1 & \text{for sinusoidal boundaries} \end{cases} \tag{2.23}$$

With all the grain boundary microstructural features being in consideration, we can rewrite the GBS rate equation as:

$$\dot{\gamma}_{gbs} = \phi A_0 \frac{D\mu b}{kT}\left(\frac{b}{d}\right)^q \left(\frac{\lambda + r}{b}\right)^{q-1} \frac{(\tau - \tau_{ig})(\tau - \tau_{ic})}{\mu^2} \tag{2.24}$$

To solve for GBS strain, we first substitute Eq. (2.24) into Eq. (1.46):

$$\begin{aligned} \frac{d\tau_{ig}}{dt} &= H\phi A_1(\tau - \tau_{ig})(\tau - \tau_{ic}) - \tau_{ig}\kappa \\ &= H\phi A_1(\tau - \tau_{ic})\left[\tau - \left(1 + \frac{\kappa}{H\phi A_1(\tau - \tau_{ic})}\right)\tau_{ig}\right] \\ &= \eta\phi(\tau - \beta\tau_{ig}) \end{aligned} \tag{2.25}$$

where

$$A_1 = A_0 \frac{Db}{\mu kT}\left(\frac{b}{d}\right)^q \left(\frac{\lambda + r}{b}\right)^{q-1}$$

$$\eta = A_0 \frac{HDb}{\mu kT}\left(\frac{b}{d}\right)^q \left(\frac{\lambda + r}{b}\right)^{q-1} (\tau - \tau_{ic}) = \frac{\kappa}{\beta - 1}$$

$$\beta = 1 + \left(\frac{\phi A_0 H}{\mu}\right)^{-1} \left(\frac{b}{d}\right)^{-q} \left(\frac{\lambda + r}{b}\right)^{1-q}$$

The solution of the differential equation (2.25) takes the form

$$\tau_{ig} = \frac{\tau}{\beta}\left[1 - \exp(-\eta\phi\beta t)\right] \tag{2.26}$$

with a steady-state ($t \to \infty$) value of $\tau_{i\infty} = \tau/\beta$.

Then, substituting Eq. (2.26) into Eq. (2.24), we obtain the GBS creep strain by integration as

$$\gamma_{gbs} = \gamma_0 + \dot{\gamma}_{ss} t + \frac{\tau}{H\beta^2}\left[1 - e^{-\phi\eta\beta t}\right] \tag{2.27}$$

where ε_0 is the initial strain, and the steady-state GBS rate is given by

$$\dot{\gamma}_{ss} = \phi A_0 \frac{D\mu b}{kT}\left(\frac{b}{d}\right)^q \left(\frac{\lambda + r}{b}\right)^{q-1} \frac{(\tau - \tau_{ic})}{\mu} \frac{\tau}{\mu}\left(1 - \frac{1}{\beta}\right) \tag{2.28a}$$

or, in an abbreviated form:

$$\dot{\gamma}_{ss} = \phi A_{gbs}\left(\frac{b}{d}\right)^q \left(\frac{\lambda + r}{b}\right)^{q-1} \left(\frac{\tau}{\mu}\right)^p \tag{2.28b}$$

where p is the apparent stress-exponent.

For engineering convenience, Eq. (2.27) can be transformed from the shear mode (τ, γ, μ) into uniaxial tensile mode (σ, ε, E), as:

$$\varepsilon_{gbs} = \varepsilon_0 + \dot{\varepsilon}_{ss} t + \frac{\sigma}{H\beta^2}\left[1 - e^{-\frac{\beta^2 H \dot{\varepsilon}_{ss} t}{\sigma(\beta - 1)}}\right] \tag{2.29}$$

Eq. (2.29) describes the transient creep in terms of three parameters: $\dot{\varepsilon}_{ss}$, H and β. It shows that GBS is a transient creep phenomenon comprising primary and steady-state creep. It is also inferred that transient creep is mostly influenced by the grain boundary morphology and precipitate distribution, as oppose to intragranular microstructural features.

According to Eq. (2.29), the true steady-state creep occurs at a rate as described by Eq. (2.28) only after an infinitely long time. Practically, a quasi-steady-state may be observed relatively soon after commencing of creep, since the primary stage saturates exponentially with time. In real engineering materials, the quasi-steady-state creep may not last to the end, because tertiary creep can be onset, due to damage accumulation such as dislocation multiplication, precipitate coarsening and cavity nucleation and growth both in the grain interior and along grain boundaries. When these processes eventually become dominant, creep rate accelerates. Here, we focus on GBS related creep phenomena up to the occurrence of the minimum creep rate.

According to Eq. (2.29), the primary strain can be defined as:

$$\varepsilon_{tr}^p = \frac{\sigma}{\beta^2 H} \tag{2.30}$$

It states that the primary strain is proportional to the applied stress σ divided by the work hardening coefficient H. This relationship may provide guidance for grain boundary engineering to lower the primary strain in alloy applications. Practically, the primary strain can be attained ~99 percent when the exponential term reaches a value of 4.6, which defines the primary time as follows.

$$t_{tr}^p = 4.6\frac{(\beta - 1)\sigma}{\beta^2 H \dot{\varepsilon}_{ss}} \tag{2.31}$$

According to Eq. (2.30), the maximum primary strain normalized by the elastic strain ($\varepsilon_0 = \sigma/E$) is independent of stress and temperature, since H is proportional to E. Also, according to Eq. (2.31), the primary time is inversely proportional to the steady-state creep rate. These predictions are consistent with observations for numerous materials as reviewed by Malakondaiah et al. (1988).

2.4.1 Effects of Grain Boundary Precipitates

The present model predicts that the GBS strain rate ideally has a stress-dependence to the 2nd power. Because of the presence of grain boundary precipitates, which introduces a constant back stress σ_{ic}, the apparent stress dependence may change to a high exponent, as implied by Eq. (2.28). This scenario can be illustrated with creep data on Nimonic 115 (Furrillo et al. 1979). Furrillo et al. observed a stress exponent of 2 for $\dot{\varepsilon}_m$ without grain boundary carbides and a power exponent of 14.58 with grain boundary carbides. Here, we use $\dot{\varepsilon}_m$ to represent the steady-state GBS creep rate $\dot{\varepsilon}_{ss}$ when intragranular deformation is negligible. Upon plotting Furrillo et al.'s creep data for carbide free and carbide containing microstructures of Nimonic 115 against the effective stresses on a log-log scale, the data are well represented by two parallel lines with a slope of unity for both cases, Figure 2.8 (Wu and Koul 1995).

With regards to the temperature dependence of Nimonic 115 creep data, Furrillo et al. found that at high stresses (517 ~ 568 MPa) and moderate

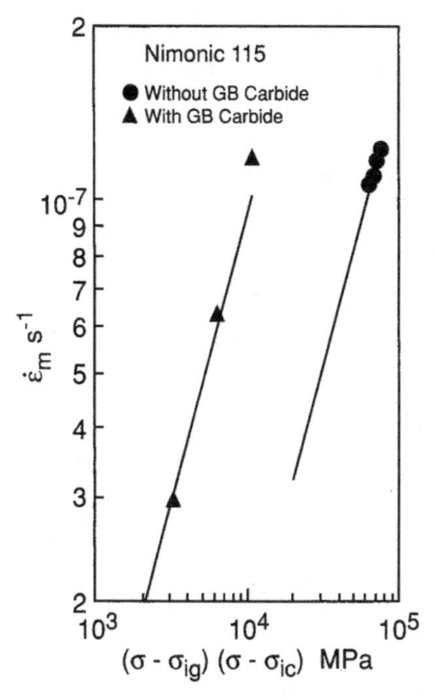

Figure 2.8. The minimum creep rate vs. effective stress in Nimonic 115 (Wu and Koul 1995). Test data are taken from Furrillo et al. (1979).

temperatures (746 ~ 788 °C) the deformation was controlled by volume diffusion with an activation energy of 334 KJ/mol without grain boundary carbides whereas the activation energy increased to 390 KJ/mol with grain boundary carbides. As discussed in section 1.6, an increase in activation energy is expected to occur in the presence of carbides because additional thermal energy is needed for vacancies to diffuse along grain boundaries such that dislocations can climb over the carbides. The experimental activation energies and the additional activation energies in the presence of grain boundary precipitates, as evaluated using Eq. (1.50) for Nimonic 115, are given in Table 2.2 (Wu and Koul 1995). The baseline activation energies correspond to the precipitate-free state. The calculated total activation energy for creep in Nimonic 115 with grain boundary carbide is 380.1 (334.4 + 45.7) kJ/mol, which agrees well with Furrillo et al.'s observation of 390 kJ/mol.

Castillo et al. (1988) generated low stress (90 MPa) creep data on new and service exposed IN738LC turbine blade materials over a range of temperatures varying between 899 °C and 988 °C. The new IN738LC material contained discrete $M_{23}C_6$ grain boundary precipitates ($q = 2$), whereas during long time service exposure, the precipitates grew into an almost continuous distribution ($q = 3$). The temperature dependence of IN738LC creep rate is shown in Figure 2.9. According to Castillo et al. (1988), the apparent activation energy is 230 kJ/mol below a transition temperature of 950 °C and the activation energy is 480 kJ/mol above this transition temperature. In the case of the service exposed IN738LC, the apparent activation energy increases to 250 kJ/mol below 950 °C and to 500 kJ/mol above 950 °C. The calculated total activation energy for the service-exposed IN738LC is 256.5 kJ/mol below 950°C and 506.5 kJ/mol above 950 °C, based on the

Table 2.2. Activation energies for Nimonic 115 and IN738LC Ni-base superalloy materials

Material	Carbides size and spacing (μm)	Baseline U_d kJ/mol.	Theoretical ΔU, KJ/mol.	Experimental U_c, kJ/mol.
Nimonic 115				
	$r = 0, \lambda = \infty$	334.4	0	334.4
	$r = 2.5, \lambda = 2$	334.4	45.7	390.4
IN738LC				
New	$r = 0.4, \lambda = 1.5$	222 (L.T.)	8.4	230 (L.T.)
		472 (H.T.)	8.4	480 (H.T.)
Service-exposed	$r = 0.7, \lambda = 0.7$	222 (L.T.)	34.5	250 (L.T.)
		472 (H.T.)	34.5	500 (H.T.)

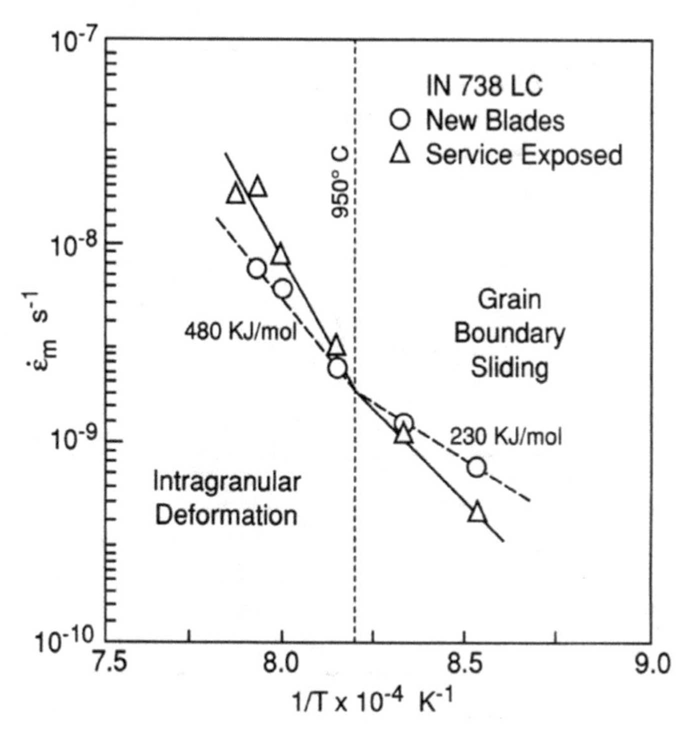

Figure 2.9. The temperature dependence of creep rate in new and service-exposed IN738LC (Castillo et al. 1988).

microstructural values of the grain boundary carbide size and distribution as given in Table 2.2. The theoretical evaluations are in very good agreement with the experimental observation.

The above analyses indicate that the GBS model in the context of Eq. (2.28) adequately accounts for the role of grain boundary carbides on temperature dependence of dislocation climb in complex engineering alloys.

2.4.2 Effect of Wavy Grain Boundary

Using special heat treatments, some Ni-base superalloys can be made with wavy grain boundaries. For example, Inconel 718 can be modified to have triangular grain boundaries with needle-like δ phase precipitates; and IN738LC can be modified to have sinusoidal grain boundaries. The creep behaviours of these two alloys with planar and wavy grain boundaries are shown in Figure 2.10 and Figure 2.11 for Alloy 718 and IN738LC, respectively. Creep curves predicted from Eq. (2.29) are also shown with the best-fit model parameters for the planar grain boundary case, the measured grain boundary serration profiles (wave length λ and amplitude h) and the calculated grain boundary wavy factor, ϕ, as given in Table 2.3. It is noted that $\dot{\varepsilon}_m$ for serrated grain boundaries is equal to $\dot{\varepsilon}_m$ for the planar grain

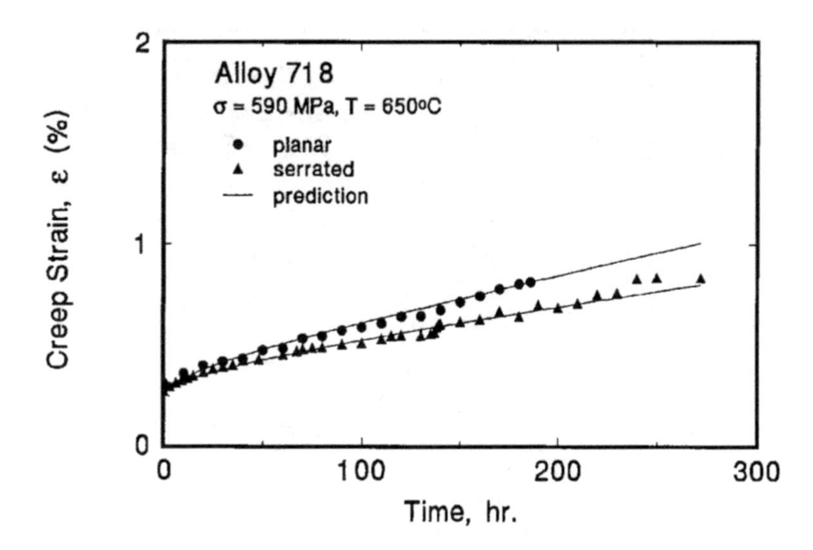

Figure 2.10. Creep curves of Alloy 718 with planar and triangular wave grain boundaries, as predicted by Eq. (2.29) (Wu and Koul 1997).

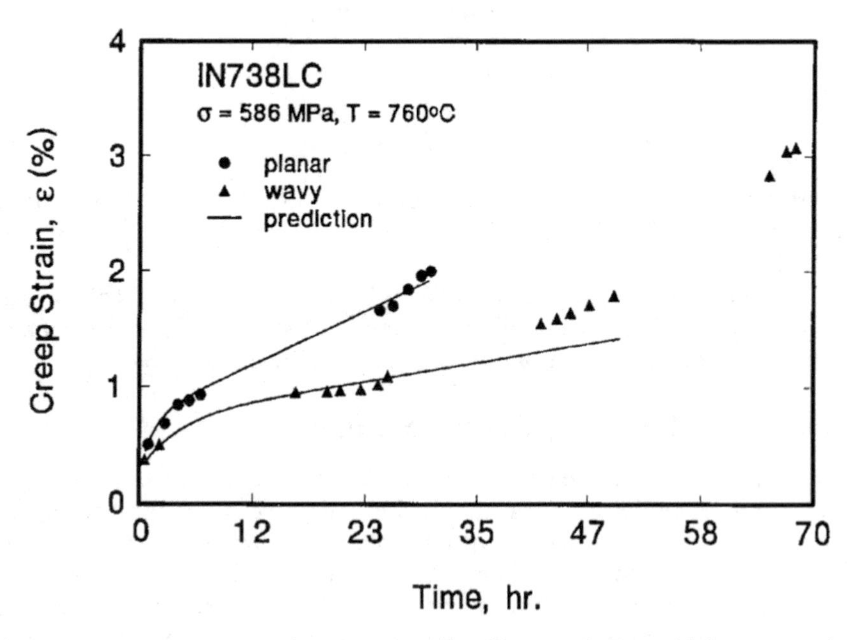

Figure 2.11. Creep curves of IN738LC with planar and sinusoidal wave grain boundaries, as predicted by Eq. (2.29) (Wu and Koul 1997).

boundary times the wavy factor, ϕ, while the parameters H and β remain the same for the same material. The theoretical curves are in good agreement with the experimental data. The above analyses demonstrate that GBS is indeed a controlling mechanism in the creep phenomena and Eq. (2.29) is suitable for describing the transient creep responses of complex engineering alloys.

Table 2.3. Microstructure and creep curve parameters

Material & test condition	Grain boundary	H (GPa)	β	ϕ	$\dot{\varepsilon}_m$ (hr^{-1})
Alloy 718 @	(planar)				
σ = 590MPa	$h = 0$	204.6	1.78	1.0	0.00233
T = 650 °C	λ = d				
	(triangular)				
	h = 2 μm	204.6	1.78	0.72	0.00168
	λ = 10 μm				
IN738LC @	(planar)				
σ = 586MPa	$h = 0$	106.5	1.13	1.0	0.0397
T = 760 °C	λ = d				
	(sinusoidal)				
	h = 5 μm	106.5	1.13	0.38	0.0152
	λ = 15 μm				

Here, it should be noted that Eq. (2.29) mathematically extends the transient regime up to infinite time. However, tertiary creep mechanisms such as grain boundary cavitation will intervene at some time and cause intergranular fracture. The material damage processes including grain boundary cavitation will be discussed in the next chapter.

2.5 Diffusion Creep

Diffusional matter flux may be induced by stress in materials at high temperatures, resulting in shape changes. The phenomena have been commonly classified as Nabarro-Herring creep via lattice diffusion and Coble creep via grain boundary diffusion. The rate equation for diffusional flow is given by (Frost and Ashby 1982):

$$\dot{\gamma}_d = \left(D_v + \frac{\pi\delta}{d} D_{gb} \right) \frac{42\Omega\sigma}{kTd^2} \tag{2.32}$$

where D_v and D_{gb} are lattice and boundary diffusion coefficients, Ω is the atomic volume and δ is the effective thickness of the boundary.

Diffusion creep may be manifested at very low stresses in pure metals. In complex engineering alloys, because of solid solution strengthening and grain boundary precipitate strengthening, diffusional matter flow is limited and rarely observed under normal application conditions (Wadsworth et al. 2002). Therefore, it will be ignored in the later treatment for creep in this book.

2.6 Deformation Mechanisms in Nanocrystalline Materials

Nanocrystalline (nc) materials are characterized by grain sizes <100 nm. With such small grain size, the atomic volume fractions of grain boundaries and crystallites are in the same order. Owing to this characteristic, the electronic, optical, magnetic, and mechanical properties of nc materials can be quite different from their conventional counterparts. For mechanical and structural applications, the deformation mechanisms in nc materials must be first understood, in order to design components of nc materials for appropriate applications.

Nanocrystalline metals can be fabricated through mechanical milling processes. During mechanical milling, the material was severely deformed to produce ultrafine grains. In this process, turbulent shear processes are operative by crystal rotation with disclinations, as observed by Murayama, et al. (2002). Based on such observations, Ovid'ko (2002) proposed a deformation mechanism of grain rotation. Since the nc materials are composed of nanometre-scale crystallites (grains) divided by interfaces (grain boundaries), and each crystallite is so small that a large fraction of its atoms (up to 50%) are located at interfaces, crystal lattice rotation may occur between dipoles of grain boundary disclinations. A disclination is a line defect characterized by rotation of the crystalline lattice around its line. Motion of a disclination dipole along grain boundaries causes plastic flow accompanied by crystal lattice rotation behind the disclinations. The disclination dipole motion has been suggested to be intensive in nanocrystalline materials, where the volume fraction of grain boundaries is high and disclinations are close to each other.

In addition to the rotational deformation, other deformation mechanisms such as grain boundary sliding and stress-induced mass transfer (diffusion) can occur effectively in nanocrystalline materials. These mechanisms are conducted by grain boundaries and compete effectively with conventional dislocation slip (shear) in crystallites, the latter being the dominant deformation mechanism in conventional materials. In this regard, Mohamed and Yang (2010) analysed deformation behaviours of ncNi and nc Cu using the Coble creep rate equation and various GBS rate equations. They concluded that dislocation-accommodated GBS is most likely the controlling deformation mechanism, as shown in Figure 2.12, whereas the stress dependence of the pure diffusion mechanism is too low to match the observed creep rate. The dislocation-accommodated GBS rate equation was proposed by Mohamed and Chauhan (2006), as:

$$\dot{\gamma}_{gbs} = \left(\frac{b}{d}\right)^q \frac{D_{gb}}{b^2}\left[e^{\frac{V\tau}{kT}} - 1\right] \tag{2.33}$$

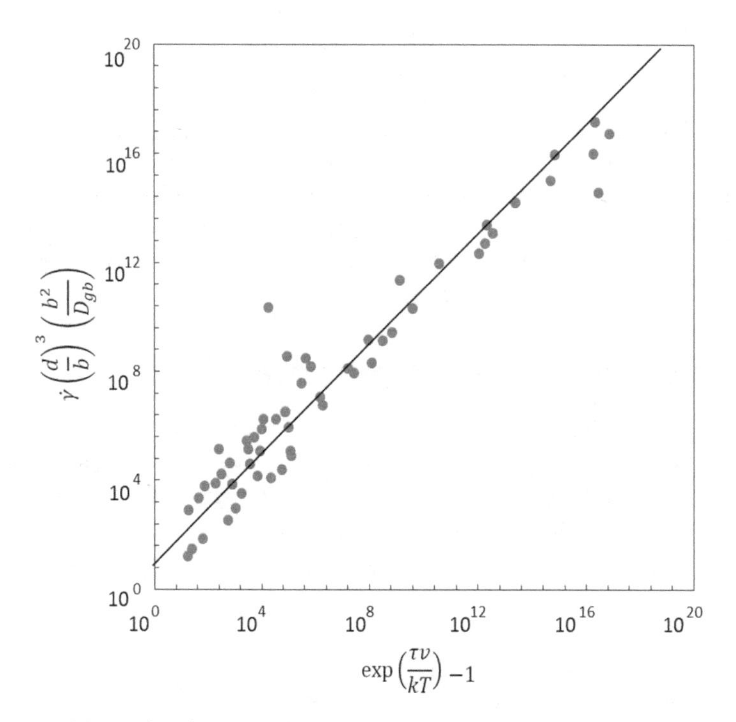

Figure 2.12. Normalized strain rates against exp (Vτ/kT) - 1 on a logarithmic scale for nc Ni and nc Cu, using the dislocation-accommodated GBS model (Mohamed and Yang 2010).

where D_{gb} is the grain boundary diffusion coefficient, V is the activation volume, q ($q = 3$) is the index of grain size dependence, k is the Boltzmann constant, τ is the applied stress, and T is the absolute temperature in Kelvin degree. Note that Eq. (2.33) is like "power-law breakdown" to Eq. (2.28) for GBS, as Eq. (2.4) to Eq. (2.11) for intragranular deformation.

The argument of GBS-dominated mechanism is supported by two experimental observations: first, observations on nc Ni showed no evidence of extensive dislocation debris as deformation in large-grained metals (Legros et al. 2000); second, the in-situ X-ray diffraction observation on *nc Ni* tested in tension at room temperature revealed no irreversible peak broadening (Farkas et al. 2005). These observations indicate that dislocations are not stored in the crystallites of nc materials as a result of deformation. Mohamed and Yang (2010) further rationalized the inverse Hall-Petch relationship for nc-materials based on dislocation-accommodated GBS mechanism.

Substituting Eq. (2.33) into Eq. (2.6), when the GBS rate approaches the loading strain rate, i.e., $\dot{\gamma}_{gbs} = \dot{\gamma}$, material yields by the GBS mechanism, and the flow stress can be solved, as

$$\tau = -\frac{kT}{V}\ln\left[\left(\frac{b}{d}\right)^q \frac{D_{gb}}{\dot{\gamma}^2}\right] \tag{2.34}$$

Eq. (2.34) shows that the yield strength of nc materials increases with the grain size in a logarithmic manner, when GBS is dominant. However, according to the strongest grain assumption (Argon and Yip 2006), when the grain size is sufficiently large such that $\dot{\gamma}_{gbs}$ becomes so low that slip within the grain is responsible for yielding of the material, the material's yield strength resumes back to the Hall-Petch relationship. In the transition of controlling mechanisms, a peak yield point would be exhibited in the relation of yield stress vs. the square root of the grain size, as shown in Figure 2.13.

2.7 Deformation Mechanism Map

As discussed in the previous sections, crystalline solids may deform plastically by a number of alternative, often competing mechanisms that involve dislocation glide and climb, grain boundary sliding (GBS) and diffusion, which may be summarized into the following categories:

1. Collapse at the ideal strength—(flow when the ideal shear strength is exceeded)
2. Low-temperature plasticity by dislocation glide.

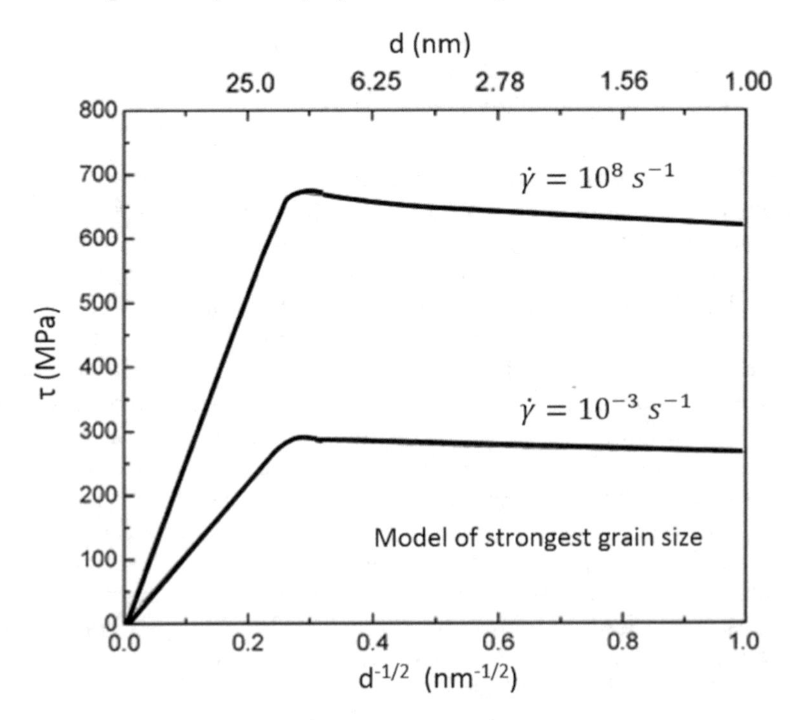

Figure 2.13. Stress as a function of the inverse square root of the grain size for Cu. Curve represents the model for two different strain rates, 10^{-3} and 10^{8} s^{-1} (Mohamed and Yang 2010).

3. Power-law creep by dislocation glide-plus-climb.
4. Diffusional flow via lattice diffusion (Nabarro-Herring creep) or grain boundary diffusion (Coble creep).
5. Grain boundary sliding (GBS) by dislocation glide-plus-climb along grain boundaries.

Ashby (1972) first proposed the concept of *deformation mechanism map*, which depicts the dominance of the first four category deformation mechanisms in the full range of normalized shear stress τ/μ and homologous temperature T/T_m. Frost and Ashby (1982) constructed deformation mechanism maps for most pure metals including f.c.c, b.c.c. and h.c.p. metals. Figure 2.14 shows the deformation-mechanism map of pure nickel, after Frost and Ashby (1982), as an example. The boarders on the map are marked by equal contributions from the two sides. Later on, Lüthy et al. (1979) modified the deformation mechanism maps for Al and Ni with consideration of GBS. Castillo et al. (1988) suggested that GBS should be included for Ni-base superalloy IN738LC. Still, deformation mechanism maps for complex engineering alloys are far from developed. The work by Langdon (1970), Lüthy et al. (1979), and Wu and Koul (1995) have shown that grain boundary sliding is an important deformation mechanism inducing transient creep behaviour. GBS is also considered to be responsible for

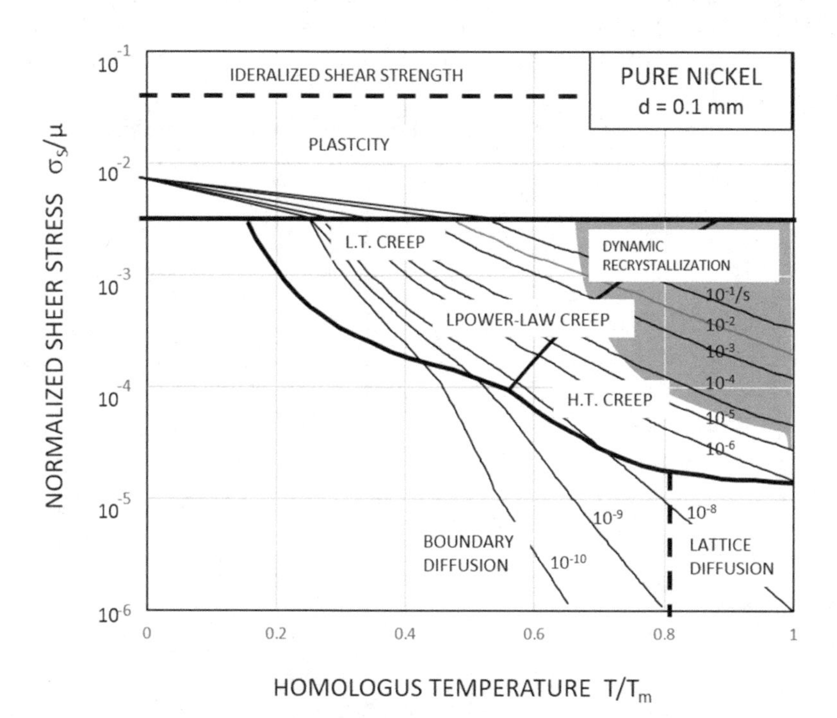

Figure 2.14. Deformation-mechanism map for pure nickel, after Frost and Ashby (1982).

damage accumulation via crack nucleation and propagation along grain boundaries, leading to intergranular fracture (Xu et al. 1999). Recently, a deformation mechanism map was constructed for Waspaloy, as shown in Figure 2.15 with consideration of GBS (Wu et al. 2012). In this map, the stress is normalized by the ultimate tensile strength (UTS) of the material for engineering convenience. Bano et al. (2014) also constructed a deformation mechanism map for 1.23Cr-1.2Mo-0.26V steel, using the neuro network fitting method, as shown in Figure 2.16. Such deformation-mechanism maps should be constructed with mechanism-based constitutive equations for more engineering alloys, in order to assist material design and selection for appropriate engineering applications. It is one of the major purposes of this book to use a mechanism-based approach constructing the constitutive law and life prediction models for complex engineering alloys.

For design engineers, the deformation mechanism maps show the dominance of particular deformation mechanisms operating under service conditions for the intended application of materials. It provides guidance for material testing to generate the design data. Long-term creep life prediction is often made by extrapolation of short-term creep tests conducted either at higher stress or higher temperatures than the service condition. Using empirical relationships such as the Larson-Miller parameter method or Monkman-Grant relationship cannot guarantee that the extrapolation does not go beyond the mechanism region from which the short-term data are collected. Without understanding whether the long-term operating

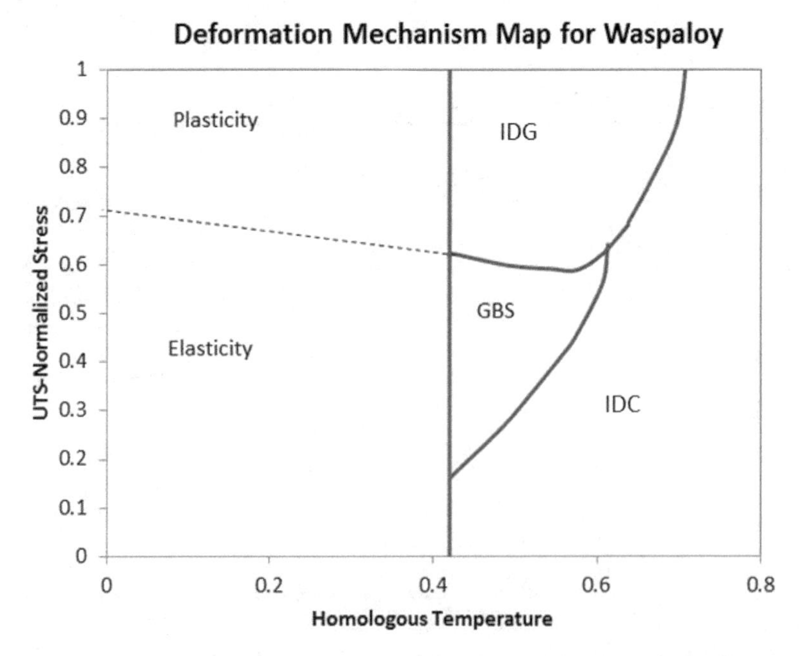

Figure 2.15. Deformation mechanism map of Waspaloy (Wu et al. 2012).

mechanism is the same as the short-term mechanism or not, such extrapolation often lacks a physical basis and the results can be erroneous.

In this book, the aforementioned fundamental deformation mechanisms are the vehicles carrying the consistency of analytical predictions from short-term tests to long-term life prediction throughout the following chapters. This will render a physics-based approach to material behaviour modeling and life prediction.

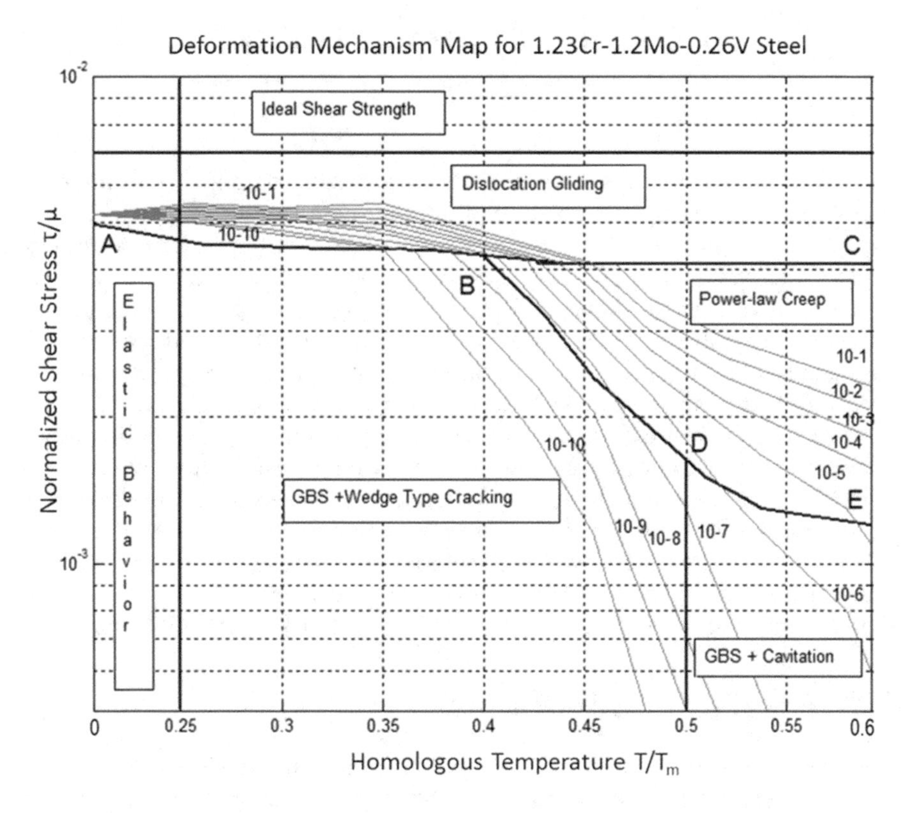

Figure 2.16. Deformation mechanism map of 1.23Cr-1.2Mo-0.26V steel, after Bano et al. (2014).

CHAPTER

3

Physics of Material Damage

In real materials under service conditions, physical damage accumulation may occur in multiple forms by different mechanisms associated with material deformation. To understand material life evolution, it is important to understand what types of physical damage may occur, where and how each type of damage accumulates and progresses. For example, fatigue damage is accumulated during cyclic deformation, which can lead to nucleation and propagation of sharp cracks at material surface or interfaces, whereas cavities accumulate during creep, which appear as round voids, mostly along grain boundaries. Therefore, damage accumulation controls the failure mode, transgranular or intergranular. On the other hand, due to loss of internal load-bearing area, damage accumulation can adversely affect the material behaviour. For example, creep cavitation can significantly increase the creep rate and reduce the ductility of the material by intergranular fracture. From material design and application point of view, it is important to know the material failure mode and its life under the service-loading conditions.

3.1 Persistent Slip Bands

Materials deform plastically via crystallographic slip, as slip steps are the basic carriers of plastic deformation. During the fatigue process under cyclic loading, dislocations multiply and accumulate within the material. These dislocations form unique structures with minimized energy to sustain the material stability during deformation. As a consequence, slip strain is localized in the form of persistent slip bands (PSBs) (Thompson et al. 1956). PSB formation consists of mutually blocked edge dislocations, which commences at $\gamma_p \sim 10^{-4}$ in a single crystallite. In pure metals, as it proceeds along slip planes in the grain, it can lead to formation of parallel wall (ladder) structures (Mughrabi 1980). PSB divides the material into a lamellar microstructure with periodic arrays of dislocation ladders or walls and channels. Slip in the channels between the walls occurs mainly by the glide of screw dislocations. At high γ_p, dislocations may be further arranged into labyrinth or cell structures. For example, the slip band/wall structure of a low-alloy steel fatigued at room temperature is shown in Figure 3.1.

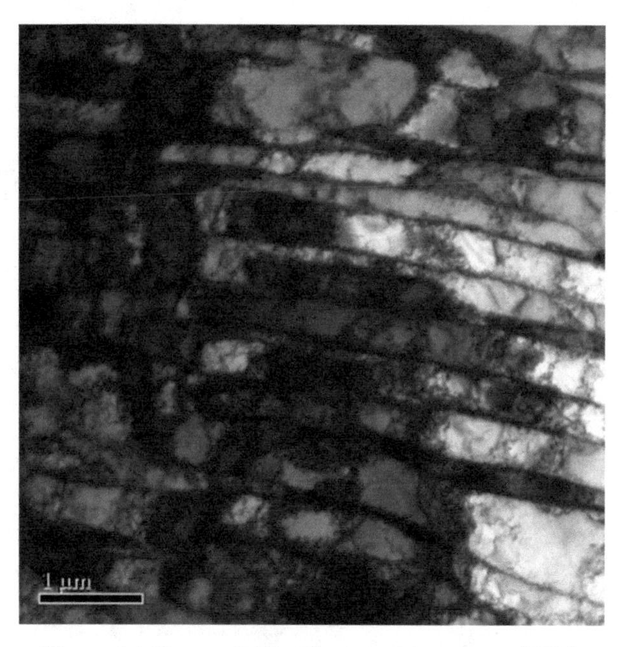

Figure 3.1. Transmission electron microscopy (TEM) micrograph of slip bands/walls.

More examples can be seen in other books (e.g., Suresh 1998). While previous observations were mainly made on single phase materials, PSB formation in complex engineering alloys was also observed. For example, Petrenec et al. (2005) observed PSB formation in Ni-based superalloys as dislocations cut through the matrix and γ' precipitates in a planar slip manner, forming multiple thin parallel bands.

Under monotonic loading, slip lines emerge at the material surface merely to form a staircase of slip steps, as schematically shown in Figure 3.2 (a). However, during cyclic deformation, due to slip irreversibility, those slip steps will not be totally reversed. The accumulation of irreversible slip steps on the surface can lead to a gradual roughening of the surface, with sites of local stress concentrations to nucleate cracks. Slip irreversibility may occur by the following micro-mechanisms (Mughrabi 2009):

1. cross-slip of screw dislocations on different slip planes upon reversing the load;
2. mutual annihilation of unlike screw and edge dislocations, which terminates the glide path;
3. slip plane asymmetry in the case of body centred-cubic (bcc) metals; and
4. random to-and-fro glide of dislocations.

In addition, annihilation of edge dislocations can occur preferentially in the regions of high edge dislocation density, i.e., in the walls, which leaves either vacancy or interstitial edge dipoles. Extrusions can be regarded as interstitial dipoles left at the surface, while intrusions consists of vacancy dipoles (Essmann et al. 1981), as schematically shown in Figure 3.2(b). Therefore, PSB is considered to be the precursor of fatigue crack nucleation. Crack nucleation at PSB has been observed by many researchers (Ma and Laird 1989, Polák et al. 2017). Figure 3.2 (c) shows the surface fatigue crack nucleation process in a polycrystalline copper. In addition to surface crack

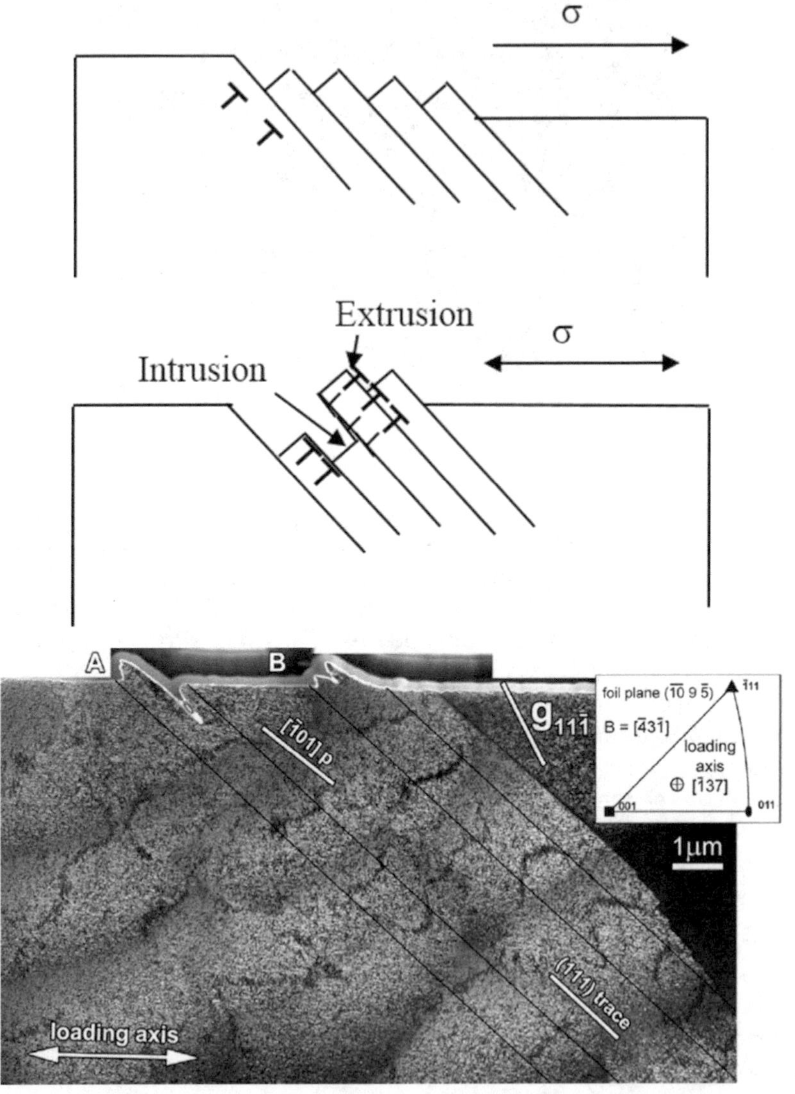

Figure 3.2. Schematic of slip steps at surface during (a) static, and (b) fatigue loading; and c) STEM bright field image of the dislocation structure in polycrystalline copper cycled with a plastic strain amplitude 7.5×10^{-4} for 5000 cycles, image after Polák et al. (2017). PSMs A and B are marked by black straight lines parallel to the trace of (111) plane.

nucleation, fatigue cracks may nucleate at other discontinuity sites such as pores, inclusion-matrix interfaces and/or grain and twin boundaries in a complex engineering alloy (Mughrabi 2009). Mathematical treatments of crack nucleation from dislocation pile-ups will be given in Chapter 6 for isotropic materials and Chapter 9 for generally anisotropic materials.

3.2 Grain Boundary Cavitation

Metals with good tensile ductility can fail with a relatively low elongation after creep for a prolonged period in the homologous temperature range of 0.3 to 0.9. This limited creep ductility, which can be as little as 1%, is related to either cracking at grain boundary precipitates, triple junctions or formation of cavities (or voids) on grain boundaries. Some types of creep damage are shown in Figure 3.3. By the nature of their generation, they are therefore called *internally-distributed damage* (IDD), which can be promoted by different deformation mechanisms, as discussed below.

3.2.1 Cavitation by Matter Diffusion

Creep cavitation can be promoted by matter diffusion or vacancy condensation along grain boundaries. Hull and Rimmer (1959) first developed a basic cavitation model, which expresses the rate of spherical cavity growth as:

$$\frac{dr}{dt} = \frac{D_b \delta_z (\sigma - P)\Omega}{rakT} \tag{3.1}$$

where r is the cavity radius, D_b is the grain boundary diffusivity, δ_z is the effective grain boundary thickness, Ω is the atomic volume, a is the inter-cavity spacing, σ is the applied normal stress, and P is the hydrostatic pressure. Cavities thus formed by vacancy condensation usually have a spherical shape and hence are called r-type cavities.

From Eq. (3.1), one expects that cavitation will occur mostly along grain boundaries that are perpendicular to the applied stress, and the material will fail when the cavities grow to the size of half inter-cavity spacing ($a/2$), as:

$$t_r = \frac{kTa^3}{D_b \delta_z (\sigma - P)\Omega} \tag{3.2}$$

Eq. (3.2) predicts a stress-dependence inversely to the power of 1, which is far below the observed stress dependence in rupture time inversely to the power of 3 to 14 ($t_r \propto \sigma^{-n}$) for most metals and alloys. Needleman and Rice (1980) considered the condition of constrained cavity growth. At high growth rates, the atoms diffusing away from the cavities cannot be distributed uniformly along the grain boundary. It requires the matrix creep straining to accommodate the matter. Based on deformation accommodation, the constrained cavity growth model predicts that

$$\dot{\varepsilon}_{ss} t_r = \text{const} \tag{3.3}$$

Actually, this relationship agrees with the Monkman-Grant relationship (1956) observed from short-term creep tests. However, the exponent of power-law creep and that of the creep life are often not the same.

All the above cavitation models assumed a constant density of cavities, i.e., nucleation of all cavities occur upon application of stress. Raj and Ashby (1975) considered continuous nucleation and growth of cavities. By the classical nucleation theory, cavity nucleation is controlled by thermally activation over the nucleation barrier. In this case, the cavity nucleation rate is given by

$$\dot{\rho}_c = \frac{4\pi\gamma_s D_{gb}\delta_z}{\sigma\Omega^{\frac{4}{3}}}(\rho_{\max} - \rho_c)\exp\left(\frac{\Omega\sigma - 4\gamma_s^3 F_v(\psi)/\sigma^2}{kT}\right) \tag{3.4}$$

where ρ_c is the cavity density, γ_s is the surface energy, ψ is the cavity dihedral angle, $F(\psi)$ is a cavity geometrical function.

Then, the total cavitation area can be calculated by the following compound integral (Raj and Ashby 1975):

$$A(t) = \rho_c \pi r_c^2 + 2\int_0^t\int_\tau^t r(t-\eta)\dot{r}(t-\eta)\dot{\rho}_c(\tau)\,d\eta\,d\tau \tag{3.5}$$

where r_c is the critical nucleation radius, and $r(t)$ is the cavity radius at time t.

Direct observations of cavitation are often made using metallographic techniques ranging from optical and scanning electron microscopy (SEM) measurements on polished surfaces to transmission electron microscopy (TEM) measurements on thin foils. The optical techniques have low resolutions and difficulties inherent in preparing a surface without grossly altering the cavities. The TEM studies usually have had difficulty finding any cavities due to the limited material volume sampled in each TEM specimen. Another metallographic technique is to fracture a specimen along its grain boundaries at low temperature after it has crept at elevated temperature, and to measure the cavitation using SEM. However, the low-temperature tensile fracture may select the most heavily cavitated boundaries which may not be representative of the overall cavitation. Indirect observation techniques such as small angle neutron scattering (SANS) and x-ray topography or x-radiography have also been used to measure cavitation (Fuller, Jr. et al. 1984). Given the nature of microstructural heterogeneity and without in-situ techniques, it is always difficult to separate cavity nucleation and growth from post-mortem observations.

3.2.2 Grain Boundary Cavitation/Cracking by GBS

Creep cavitation is not always found to be abundant along grain boundaries normal to the applied stress. As early as up to 1970s, many researchers had noticed that heterogeneous cavity nucleation could occur by shear along the grain boundaries or bi-crystal boundaries, as reviewed by Perry (1974). Such creep-stress condition excluded the vacancy condensation mechanism but rather favored GBS. In this case, the number of voids generated was found to be proportional to the amount of GBS regardless of temperature, which ruled out the vacancy mechanism, because otherwise the temperature effect would be apparent. The mechanism of void generation by sliding can generally be attributed to GBS intersecting grain boundary ledges or jogs, or any grain boundary discontinuities such as precipitates. GBS can be accommodated by sliding and/or migration along other boundaries at triple junctions. The dislocation pile-up at triple junctions can give rise to the localized deformation zone along the boundary, causing the so-called triple-point fold, and at the same time, causing wedge-type cracks (Miura et al. 1988), as shown in Figure 3.3 (c), where the shifting of the grid lines across the grain boundaries clearly shows evidence of GBS. Figure 3.3 (d) shows crack nucleation at grain boundary carbides in Waspaloy. On the other hand, the presence of grain boundary particles plays a role to reduce GBS (Wu and Koul 1995), as discussed in section 2.4, and hence reduce the onset of cavitation. This was indeed observed in early studies on Ni-base superalloysNimonic 80A (Weaver 1959-60).

Wrought materials that have experienced large compressive deformation during forming will possess more grain boundary ledges and thus are prone to grain boundary cavitation during creep. Cast materials, which have smooth grain boundaries, usually do not have small cavities; instead, triple-junction wedge cracking (also called w-type cavities) tends to occur in these materials, an example is shown in Figure 3.3 (c). Greenwood (1969) concluded that, by analyzing all the data at the time, the number of cavities depends linearly on creep strain. Figure 3.4 shows the cavity density as function of creep strain in Nimonic 80A (Dyson 1983). Dyson showed that tensile creep specimens that were cold worked at ambient temperature appeared to have a predisposition for creep cavitation. His observation supported the ledge-opening mechanism of void nucleation where the grain boundary ledges produced by intersecting slip lines during compression were opened up by GBS in tension at elevated temperatures. This fact would not appear to include vacancy accumulation, and it must be attributed to GBS or dislocation pile-ups at grain boundaries. This observation can be represented by

$$\rho_c = \frac{1}{\lambda_c^{\,2}} = C\varepsilon_v \tag{3.6}$$

where λ_c is the cavity spacing, and C is a microstructure-dependent constant.

Kassner and Hayes (2003) summarized the aforementioned cavitation mechanisms, as represented in Figure 3.5: a) GBS leading to cavity growth from ledges and triple junction points; b) cavity nucleation from vacancy condensation at a high stress region; c) cavity nucleation from a Zener–Stroh dislocation pileup mechanism; and d) formation of a cavity from a

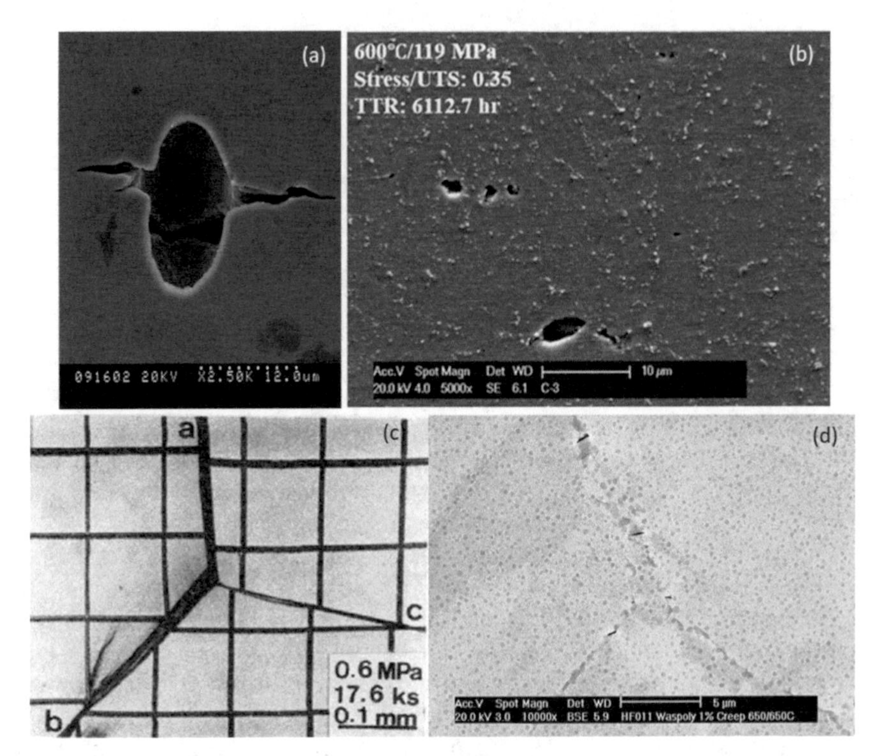

Figure 3.3. (a) Crack nucleation at an elongated pore in CMSX-10 Ni-base single crystal during creep; (b) creep voids formed at grain boundaries in a 9Cr-1Mo steel under stress of 119 MPa at 600°C; (c) triple-junction crack due to GBS in aluminum (Miura et al., 1988); and (d) grain boundary microcracks nucleated at grain boundary precipitates in Waspaloy crept to 1% strain at 650°C under 650 MPa.

grain boundary particle. In all the above mechanisms, with regards to cavity growth, we may simply formulate that the grain boundary cavity/crack size, l, is linearly proportional to the creep strain times the grain size, as

$$l = \varepsilon_v d \tag{3.7}$$

where ε_v signifies the viscous creep strain, i.e., strain produced by time-dependent mechanisms. Apparently, intergranular fracture occurs when l reaches a critical value l_c, which corresponds to the loss of load-bearing area to break the grain boundary cohesive strength. When GBS dominates, Eq. (3.7) reflects directly the experimental findings summarized in this section. In other general cases, it also corresponds to Eq. (3.4), if void formation occurs by a combination of vacancy condensation and matter diffusion accommodated by creep strain. With Eq. (3.6) and Eq. (3.7), one can evaluate the total cavitation area by Eq. (3.5) under GBS operating conditions.

During low cycle fatigue, alternating plasticity may produce slip lines intersecting grain boundaries, which may act effectively as ledges to generate cavities. In later chapters, we will discuss the interaction of internally

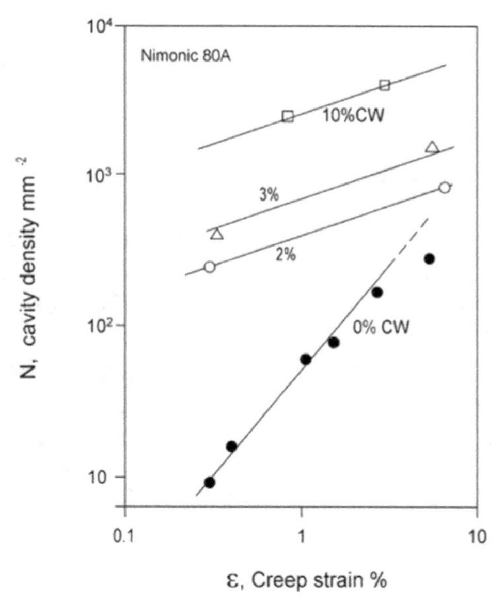

Figure 3.4. Variation of the cavity concentration versus creep strain in Nimonic 80A for annealed and pre-strained (cold-worked) alloy (adapted from Dyson, 1983).

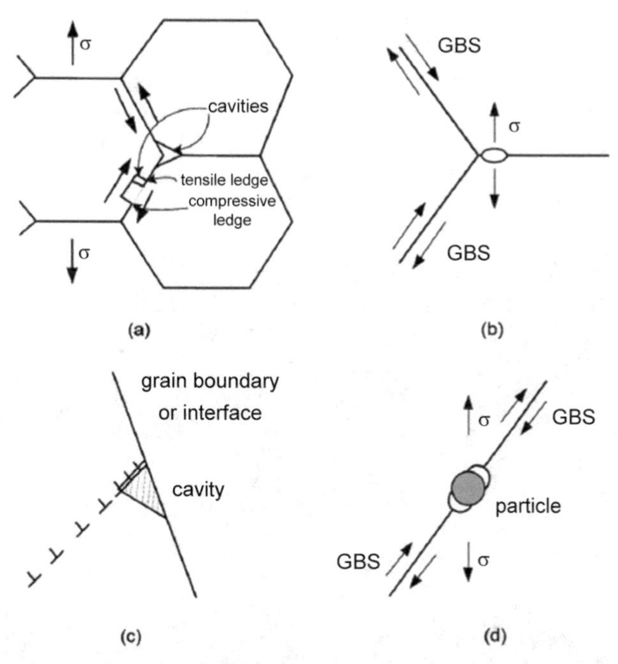

Figure 3.5. Cavity nucleation mechanisms: (a) sliding leading to cavitation from ledges (and triple points); (b) cavity nucleation from vacancy condensation at a high stress region; (c) cavity nucleation from a Zener–Stroh mechanism; (d) the formation of a cavity from a particle-obstacle in conjunction with the mechanisms described in (a-c); after Kassner and Hayes (2003).

distributed damage as promoted by time-dependent mechanisms with fatigue crack propagation during low cycle fatigue and thermomechanical fatigue.

3.3 Environmental Attack

Environment has a profound influence on the life of materials and structures. It generally accelerates both crack nucleation and propagation. Aircraft frequently flying in humid air containing salt can suffer from corrosion. The tragic accident of Aloha Flight 243 (1988) was due to wide-spread corrosion-fatigue damage hidden beneath the rivets. Gas turbine components also suffer from oxidation and hot corrosion and need to be protected by coatings. Corrosion is a broad science involving chemical environment interaction with specific materials and microstructures. It is beyond the scope of this book to discuss the chemical reactions leading to corrosion but focusing on some physical and mechanical aspects that may lead to crack nucleation, e.g. oxidation and corrosion pitting.

3.3.1 Oxidation

Oxidation is a form of surface degradation by reaction of metals with oxygen. It can interact with deformation processes such as fatigue and creep at high temperature, which may assist both crack nucleation and propagation. Figure 3.6 shows a fatigue crack originating from an oxide spike in a single crystal Ni-base superalloy CMSX-2 at 1100 °C. The precipitate free zone formed below the surface oxide layer was due to aluminum depletion from the γ' (Ni_3Al) phase.

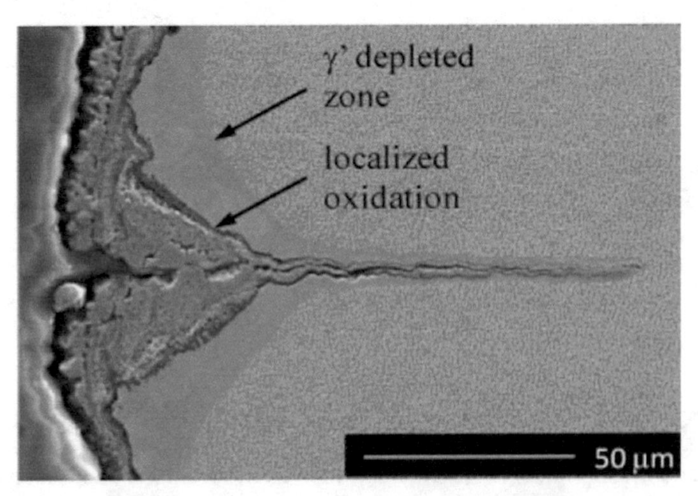

Figure 3.6. Scanning electron microscopic image showing a surface crack forming from localized region of oxide attack and formation of precipitate free zone below surface oxide layer in CMSX-2 at 1100 °C (Yandt et al. 2011).

Metal oxidation takes place when an ionic chemical reaction occurs with oxygen on a metal surface. The chemical process unfolds when electrons journey from the metal to the oxygen molecules. Then, negative oxygen ions are generated and enter the metal to form an oxide. The flux of oxygen ions and metal ions in an oxide layer are schematically shown in Figure 3.7.

Under steady-state conditions, the mass continuity in metal is given by the Fick's first law:

$$\frac{\partial c}{\partial t} = -\frac{\partial J}{\partial x} \tag{3.8}$$

where c is the concentration of the reacting element, D is the diffusion constant, x is the direction of diffusion, and J is the diffusion flux, as given by

$$J = -D\frac{\partial c}{\partial x} \tag{3.9}$$

In a homogeneous material, by matter conservation, Fick's second law holds, as:

$$\frac{\partial c}{\partial t} = D\frac{\partial^2 c}{\partial x^2} \tag{3.10}$$

When multiple elements are involved (which is often the case in complex engineering alloys and coatings), the diffusion flux of the i-th element can be expressed as

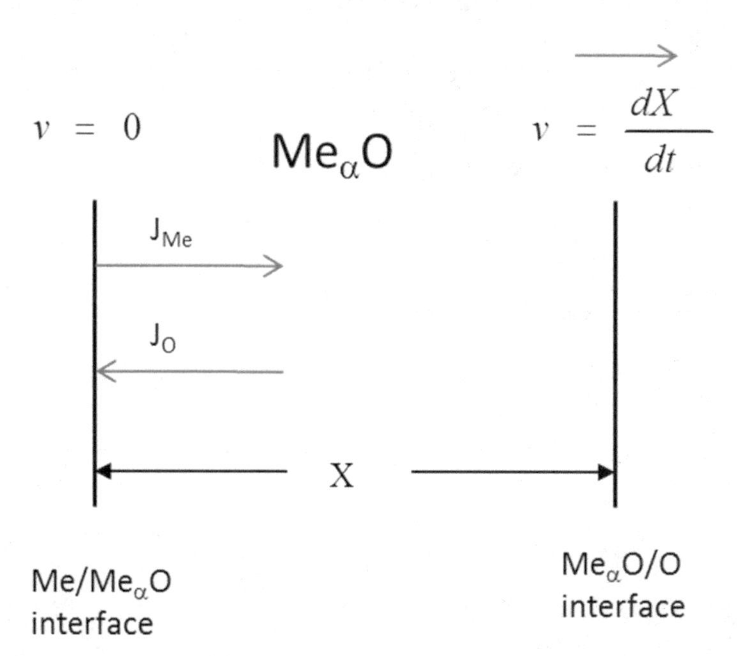

Figure 3.7. The flux of oxygen ions and metal ions through the oxide layer.

$$J_i = -\left[\sum_{j=1}^{n} D_{ij} \frac{\partial c_j}{\partial x}\right] \tag{3.11}$$

where D_{ij} are inter-diffusion constants, and hence

$$\frac{\partial c_i}{\partial t} = \sum_{j=1}^{n} D_{ij} \frac{\partial^2 c_j}{\partial x^2} \tag{3.12a}$$

Eq. (3.12a) can be written in a matrix form, as

$$\frac{\partial}{\partial t}\begin{pmatrix} c_1 \\ c_2 \\ \cdots \\ c_n \end{pmatrix} = \begin{bmatrix} D_{11} & D_{12} & \cdots & D_{1n} \\ D_{21} & D_{22} & \cdots & D_{2n} \\ \cdots & \cdots & \cdots & \cdots \\ D_{n1} & D_{n1} & \cdots & D_{nn} \end{bmatrix} \frac{\partial^2}{\partial x^2} \begin{pmatrix} c_1 \\ c_2 \\ \cdots \\ c_n \end{pmatrix} \tag{3.12b}$$

The matrix $[D]$ is positively definite, and it can be decomposed into

$$[D] = [A] \begin{bmatrix} \lambda_1 & & & \\ & \lambda_2 & & \\ & & \cdots & \\ & & & \lambda_n \end{bmatrix} [A]^{-1} \tag{3.13}$$

where $[A]$ is the transformation matrix, and λ_i (i = 1, 2, ..., n) are the eigenvalues of $[D]$.

Defining the new concentration vector $\{c'\}$ as

$$\{c'\} = [A]^{-1}\{c\} \tag{3.14}$$

Then, we have the following diagonal functions:

$$\frac{\partial c_i'}{\partial t} = \lambda_i \frac{\partial^2 c_i'}{\partial x^2} \tag{3.15}$$

The solution of Eq. (3.15) is (Smith, 2004)

$$c_i'(x,t) = c_i'(0)\,\mathrm{erfc}\left(\frac{x}{2\sqrt{\lambda_i t}}\right) \tag{3.16a}$$

where $c'_i(0)$ is the initial concentration at position x = 0 (the metal/oxide interface), and

$$\mathrm{erfc}(\xi) = \frac{2}{\sqrt{\pi}} \int_{\xi}^{\infty} e^{-t^2}\, dt \tag{3.16b}$$

Then, the element concentration $\{c\}$ can be calculated using the inverse relationship of Eq. (3.14).

At the metal/oxide interface, the metal flux is related to the rate of oxide growth as (Danielewski 1992):

$$J_{me} = c_{me} \frac{dX}{dt} = \frac{1}{2} c_{me} \sqrt{\frac{2k_{ox}}{t}} \tag{3.17}$$

where c_{me} is the metal concentration in the oxide, k_{ox} is the oxidation constant. Thus, Eq. (3.17) prescribes the boundary condition for the diffusion of reactant metal element, i.e.

$$J_{me} = \frac{1}{2} c_{me} \sqrt{\frac{2k_{ox}}{t}} = -D \frac{\partial c_a}{\partial x} = c(0) \sqrt{\frac{D}{\pi t}} \tag{3.18}$$

Then,

$$c(0) = c_{me} \sqrt{\frac{\pi k_{ox}}{2D}} \tag{3.19}$$

which defines the boundary condition for the element distribution in the metal.

The oxidation constant is defined by

$$k_{ox} = k_0 \exp\left(-\frac{Q_{ox}}{RT} \right) \tag{3.20}$$

where k_0 is the pre-exponential constant and Q_{ox} is the activation energy for oxidation.

High temperature coatings contain multiple elements to resist oxidation and melting. A typical example is MCrAlY coating, where M stands for some high-melting-point metal such as Ni and/or Co, while Al and Cr often serve as the sacrificial elements for oxidation protection. MCrAlY can be used alone as overlay coating or as the bond coat of thermal barrier coating on gas turbine components. By thermodynamics, alumina (Al_2O_3) is often the most preferred binary product to form first. When aluminum is exhausted (below certain level) in the coating, other oxides such as Cr_2O_3 and NO and spinal will form, which are detrimental to coating life. Therefore, it is important to evaluate the amount of aluminum depletion in such coatings (Che et al. 2009, Chen 2014).

CoNiCrAlY is considered as an example here. Using Eq. (3.13)-(3.16), the Al distribution profile at 1050 °C for 100 hrs, 200 hrs and 1000 hrs are evaluated and shown in Figure 3.8, in comparison with the experimental profile (Chen et al. 2012a). Note that there are significant scatters in the experimental measurement, because of the complex $\gamma + \beta$ microstructure and the high velocity oxyfuel (HVOF) thermal spray process, each point represents the average value within ±5μm. The Al concentrations as a function of time, within 10μm from the oxide/coating interface in CoNiCrAlY at 1050 °C, 1100 °C and 1150 °C are shown in Figure 3.9. The model predictions for 1050°C and 1150°C are also shown for comparison, which are in good agreement with the experimental data. Aluminum depletion can be used as a criterion of coating failure. However, as Figure 3.9 shows, after about 1000 hours at 1050 °C or even much shorter at 1150 °C, the model predicted lines

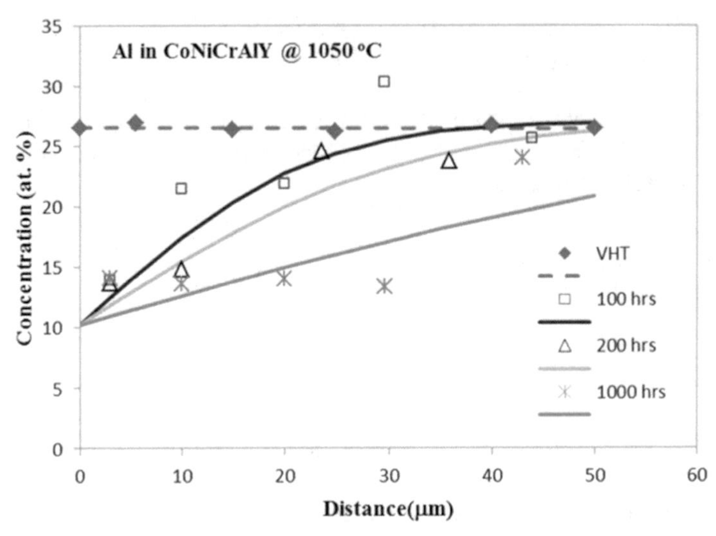

Figure 3.8. Al distribution in CoNiCrAlY after exposure at 1050 °C (Data from Chen et al. 2012).

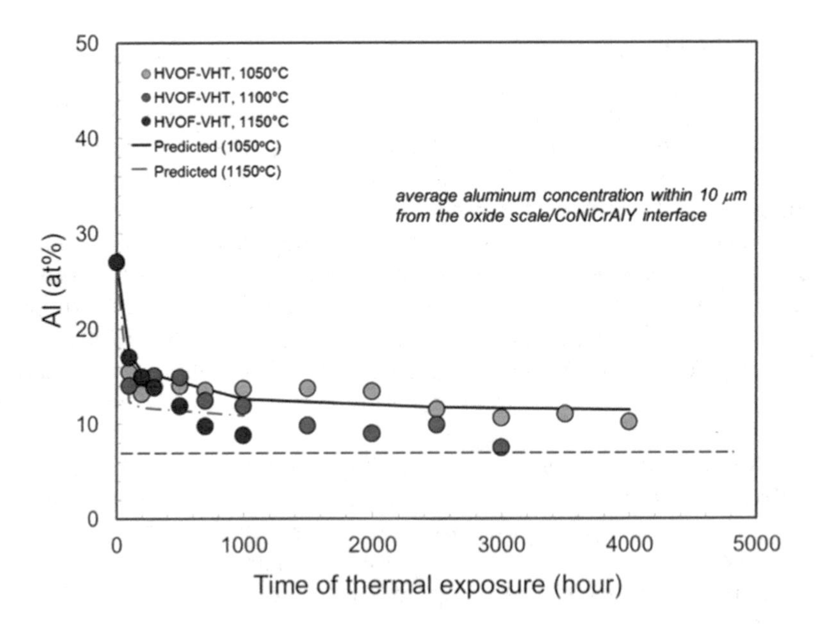

Figure 3.9. Al concentration as function of time, within 10 μm from the oxide/coating interface, in CoNiCrAlY at 1050 °C, 1100 °C and 1150 °C (Data from Chen et al. 2012).

asymptotically approach to constant levels within the experimental scatter band. Therefore, a small difference in the critical depletion level would result in a large difference in the predicted life. Considering the variability in material and processing, the Al depletion criterion alone would not be reliable in practice. Perhaps, other factors contribute to the failure process before Al is totally depleted. This will be further discussed in Chapter 12.

When an oxide scale forms continuously at the specimen surface, it will not only change the chemistry and stoichiometry of the near surface region, but also introduce a change in stress distribution. This layer of oxide will be forced to deform compatibly with the substrate until it breaks, which then leads to premature crack nucleation. The stresses in the substrate and oxide layer can be determined, by deformation compatibility:

$$\frac{\sigma_{ox}}{E_{ox}} + \alpha_{ox}(T - T_o) = \frac{\sigma_s}{E_s} + \alpha_s(T - T_o) \tag{3.21}$$

and, by force equilibrium

$$\sigma_{ox} f + (1 - f)\sigma_s = \sigma \tag{3.22}$$

where σ_{ox} is the stress in the oxide scale, E_{ox} is the elastic modulus and α_{ox} is the thermal expansion coefficient of the oxide scale (e.g. ~8μm/mK for Al_2O_3), σ_s is the stress in the substrate, E_s is the elastic modulus and α_s is the thermal expansion coefficient of the substrate material (e.g. ~14 μm/mK for Ni base superalloys), and f is the volume fraction of the oxide scale, T_0 is the reference temperature at which the oxide formation is stress free. From Eq. (3.21) and (3.22), we can deduce the stress in the oxide scale as given by:

$$\sigma_{ox} = \frac{E_{ox}}{fE_{ox} + (1 - f)E_s}[\sigma + (1 - f)E_s(\alpha_s - \alpha_{ox})(T - T_0)] \tag{3.23}$$

and the stress in the substrate as given by:

$$\sigma_s = \frac{E_s}{fE_{ox} + (1 - f)E_s}[\sigma - fE_{ox}(\alpha_s - \alpha_{ox})(T - T_0)] \tag{3.24}$$

It is interesting to see from Eq. (3.23) that the applied stress will be magnified by a factor of E_{ox}/E_s in the newly formed continuous oxide scale even at a very small volume fraction ($f \approx 0$), which explains the tendency of oxide cracking.

3.3.2 Pitting Corrosion

Pitting corrosion is a localized form of corrosion by which cavities or "holes" are produced in the metal. Pitting is considered to be more dangerous than uniform corrosion damage because it is more difficult to detect, predict and design against. Corrosion products often cover the pits. A small, narrow pit with minimal overall metal loss can lead to failure of an entire engineering system when combined with fatigue under cyclic loading. Pitting corrosion, which is almost a common denominator of all types of localized corrosion attack, may assume different shapes, as shown in Figure 3.10 (National Association of Corrosion Engineers, https://www.nace.org/ Pitting-Corrosion/).

Pitting is initiated by:

(a) localized chemical or mechanical damage to the protective oxide film; water chemistry factors which can cause breakdown of a passive film are acidity, low dissolved oxygen concentrations (which tend to render a protective oxide film less stable) and high concentrations of chloride (as in seawater)
(b) localized damage to, or poor application of, a protective coating
(c) the presence of non-uniformities in the metal structure of the component, e.g. non-metallic inclusions.

Theoretically, a local electrochemical cell that leads to the initiation of a pit can be caused by an abnormal anodic site surrounded by normal surface which acts as a cathode, or by the presence of an abnormal cathodic site surrounded by a normal surface. In the second case, post-examination should reveal the local cathode, since it will remain impervious to the corrosion attack. Most cases of pitting are believed to be caused by local cathodic sites in an otherwise normal surface.

Apart from the localized loss of thickness, corrosion pits can also be harmful by acting as stress risers. Fatigue and stress corrosion cracking may initiate at the base of corrosion pits. One pit in a large system can be enough to produce the catastrophic failure of that system. An extreme example of such catastrophic failure happened recently in Mexico, where a single pit in a gasoline line running over a sewer line was enough to create great havoc to a city, killing 252 people in Guadalajara (https://en.wikipedia.org/wiki/1992_Guadalajara_explosions).

A *pitting factor* is defined as the ratio of the depth of the deepest pit resulting from corrosion divided by the average penetration as calculated from weight loss.

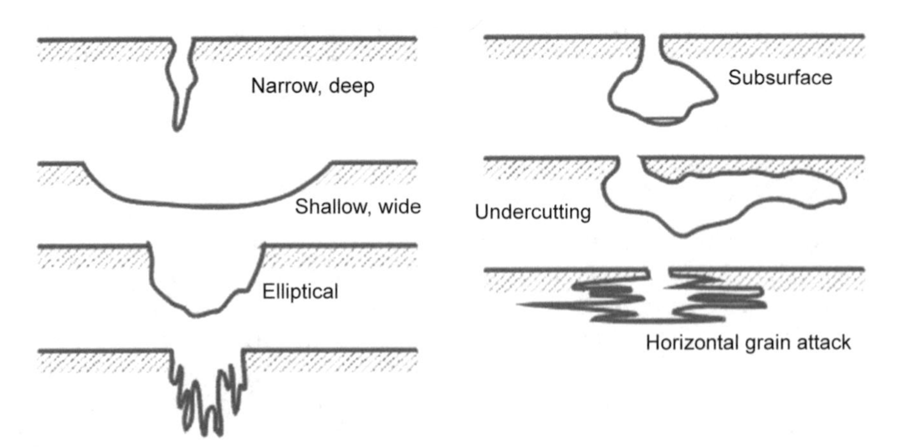

Figure 3.10. Shapes of corrosion pits (https://www.nace.org/Pitting-Corrosion/).

CHAPTER

4

The Integrated Creep-Fatigue Theory

As discussed in the previous chapters, material deformation and damage processes occur by a plethora of mechanisms. These mechanisms produce different results (behaviours) under different loading conditions, which are often categorized as creep, low cycle fatigue, high cycle fatigue and thermomechanical fatigue, and so on. Such deformation and failure processes are compounded with microstructural degradation and environmental effects such as corrosion/oxidation in service, for example, during gas turbine engine operations. Mechanical engineers mostly concern three questions, given a functional design of the component:

1. How much thermal and mechanical loads it can carry for the designed function?
2. How long will it last for the intended service?
3. What material is the best choice (performance & cost) for the above design targets?

Constitutive-damage equations are at the core of component analysis to evaluate its behaviour and life under service conditions subjected to thermal-mechanical load or magneto-electrical force variations in the working environment. Today, structural analysis can be performed using commercial finite element method (FEM), which discretizes a complex 3D-shape component into small-volume material elements of basic shapes such as tetrahedron (tet) or hexahedron (hex), etc. FEM is based on the *virtual work* or the *least energy principle*. The total energy of a mechanical component consists of two parts: i) the external work input through the boundary, and ii) the internally stored energy. The strain energy is part of the internally stored energy induced by material deformation, which requires an accurate constitutive model to evaluate. Over the past decades, many phenomenological constitutive models and empirical relationships have been proposed to describe the material behaviour and life in response to the applied load. This chapter first reviews the development of constitutive theories such as the classical theory of plasticity, the unified theory of viscoplasticity and continuum damage mechanics. For life prediction, many experimentally-established relations are widely used. Putting all these considerations together, one finds the problem that often those empirical life correlations are not linked to the constitutive parameters by mechanism(s), which cast doubts on extrapolation of the theory for untested conditions. With regards to complicated service loading conditions, some assumptions of mixing rules or interpolations and extrapolations have to be made. Experience tells us that extrapolations from empirical relations are often either dangerous or too conservative. Sometimes, even interpolations can result in erroneous results. The reason is that the controlling deformation and damage mechanisms/factors for the interpolated or extrapolated conditions are not the same as the tested condition. Either way, it has become increasingly unaffordable for industries today to rely on empirical approaches, because of the thin margin of design and costly consequences of failure. Therefore, it is important to develop a holistic constitutive-damage framework encompassing all the physical deformation and damage mechanisms as elucidated in the previous chapters. This is called the integrated creep-fatigue theory (ICFT), aiming to address the above questions all together.

4.1 Review of Constitutive Theories

In solid mechanics, a constitutive equation is a relation between stress and strain/rate as the response of the solid to external stimuli, usually as applied fields or forces. The constitutive relation for elasticity was discovered by Hooke in 1676. A brief formalism for generally anisotropic elasticity is given in Chapter 1. With regards to inelastic deformation, the problem has been addressed in two disciplines: one is material science which concerns the properties of matter in relation to its microstructure and micro-mechanisms, as discussed in Chapter 2; the other is solid mechanics which concerns the deformation responses of solids at the continuum level in 3D space. Going forward, the models from the two disciplines need to be more connected and intertwined, so to provide holistic and multi-scale solutions for design, operation and maintenance of engineering systems.

4.1.1 The Classical Theory of Plasticity

Owing to the work by Prandtl (1924), Reuss (1930), Drucker (1950), Prager (1958), Hart (1970), Hill (1998) and many others, the classical theory of rate-independent plasticity has long been developed. It requires the definition of a yield surface:

$$f(\sigma, \xi) = \sigma - \sigma_Y = 0 \tag{4.1}$$

where ξ represents a set of history dependent internal variables, σ is the von Mises stress as defined by Eq. (1.18), and σ_Y is the yield stress. An example

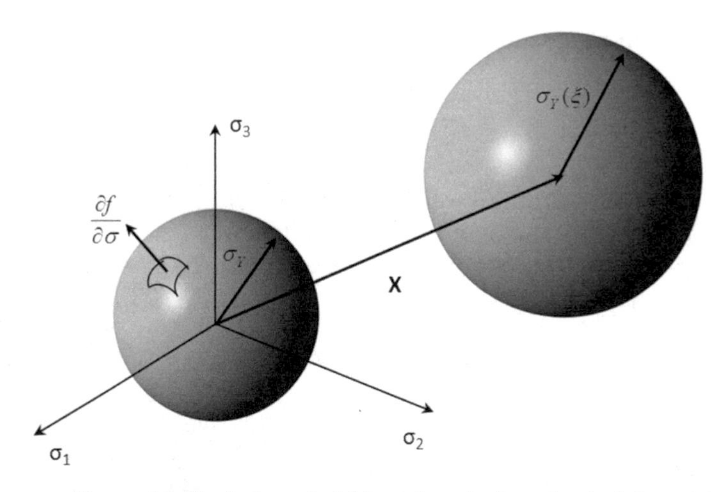

Figure 4.1. Evolution of yield surface during plastic deformation.

is schematically shown in Figure 4.1, in the principal stress space. When the stress state falls inside the yield surface, i.e. $f(\sigma,\xi)<0$, it only results in elastic deformation. Stress states outside the yield surface, i.e. $f(\sigma,\xi)>0$, do not exist. The material yields when the stress state reaches the yield surface, $\sigma = \sigma_Y$, and further loading causes plastic deformation, following the flow rule of

$$\delta\varepsilon_{ij}^{p} = \delta p \frac{\partial f(\sigma,\xi)}{\partial \sigma_{ij}} \tag{4.2}$$

where δp is the magnitude of the plastic strain increment. The derivatives $\partial f/\partial\sigma_{ij}$ indicate the normal direction of the yield surface in the stress space. As implemented in commercial FEM, the plastic strain increment is evaluated from the transient tangential of the material's stress-strain relationship. Usually, the position and shape of the yield surface evolve with plastic strain to maintain stresses either inside or on the yield surface. *Kinematic hardening* causes the centre position to shift, while *isotropic hardening* causes the yield surface to expand symmetrically. Therefore, the yield surface is continuously updated with ongoing plastic deformation. These changes would result in material loading and unloading behaviours as discussed in section 2.1.

4.1.2 Theory of Viscoplasticity

The theory of viscoplasticity basically takes the same form of Eq. (4.1) and (4.2), except that the incremental plastic strain magnitude is replaced by a scalar strain rate multiplier. The theory was first advanced for rate-dependent plasticity, and then attempted to include creep, evolving into the so-called unified constitutive theory (Miller 1987, Slavik and Sehitoglu 1986, Chaboche 1989, 2008). The general formulation is as follows.

The viscoplastic strain rate is generally expressed as:

$$\dot{\varepsilon}_{ij}^{(\text{in})} = \dot{p}\frac{\partial f\left(\sigma_{ij} - X_{ij}, R\right)}{\partial \sigma_{ij}} \quad (i, j = 1, 2, 3) \tag{4.3}$$

where $\dot{p}$ is the strain rate multiplier, X_{ij} and R denote the internal state variables, which are governed by their own evolutionary equations, following the Armstrong-Frederick hardening rule. Taking Chaboche's formulation for example (Chaboche 2008):

$$\dot{X}_{ij} = \frac{2}{3}C\dot{\varepsilon}_{ij}^{(in)} - \varepsilon^{(in)}X_{ij}\dot{p} \qquad (i, j = 1, 2, 3) \tag{4.4}$$

$$\dot{R} = b\left(Q - R\right)\dot{p} \tag{4.5}$$

where C, b, Q are temperature-dependent material constants. Specifically, the tensor X_{ij} is the back stress for kinematic hardening. The scalar variable R accounts for isotropic hardening. The plastic strain rate multiplier takes the form of either power-law (Chaboche 1989), hyperbolic sine power function (Miller 1987), or power-law/exponential function (Slavik and Sehitoglu 1986) of $(\sigma_{ij} - X_{ij})$ and R. Chaboche (2008) has shown that these equations are mathematically equivalent to describe the strain rate ranging from 10^{-8} to 10^{0} s^{-1} within an acceptable scatter band, which covers the strain rate range of practical interest under normal loading conditions, except very high strain rates under ballistic impact.

The unified constitutive theories describe the material's deformation response in lump-sum, i.e., in terms of the total viscoplastic strain ε^{in} that encompasses both plasticity and creep. The rate equation (4.3) can be written in a form associated with a certain deformation mechanism, but the supplementary back-stress equations, Eq. (4.4) and (4.5), are not derived from any deformation mechanisms aforementioned in Chapter 2, such that it is only a phenomenological theory. It is understood that deformation by dislocation glide may experience a back stress as expressed by Eq. (1.46), which is similar to the Armstrong–Frederick hardening rule, but dislocation climb and diffusion do not produce a hardening effect, while these mechanisms do produce inelastic strain during creep. Therefore, the Armstrong–Frederick hardening rule fails for these two mechanisms. Furthermore, the "unified" inelastic strain theory cannot differentiate the different physical damage generated under complex loading conditions, because it loses the clarity in delineation of plastic and creep strain. As discussed in Chapter 3, we see that cyclic plastic strain by IDG is responsible for generating PSB leading to fatigue crack nucleation and propagation causing transgranular fracture; whereas GBS is mostly responsible for intergranular fracture via grain boundary cavitation and cracking. Without mechanism delineation, the unified constitutive laws are deficient to describe evolution of physical damage for materials life prediction. Additional *damage laws* have to be supplemented, often using empirical correlations. For example, Lemaitre and Chaboche (1999) incorporated the Chaboche constitutive model with the continuum damage mechanics.

4.1.3 Continuum Damage Mechanics

The concept of continuum damage mechanics was originally proposed by Kachanov (1958) and Robotnov (1969) with a parametric damage parameter ω (value from 0 to 1) to represent the loss of internal load bearing areas due to creep cavitation, such that the true stress in the material becomes $\sigma/(1-\omega)$ during tensile creep. Later, it was expanded to include fatigue, and linear damage summation is assumed for the mixed mode of creep-fatigue interaction (Lemaitre and Chaboche 1999).

With the internal stress being modified as $\sigma/(1-\omega)$, the damage-induced flow rule may take the form (more recently, a hyperbolic-sine damage theory was developed by Dyson and McLean, 2000):

$$\dot{\varepsilon}^{(in)} = \left(\frac{\sigma}{(1-\omega)K}\right)^{n} \tag{4.6}$$

For creep,

$$\dot{\omega} = \left(\frac{\sigma}{(1-\omega)\sigma_0}\right)^{r} \tag{4.7}$$

For fatigue, in terms of the Whöler-Miner relation,

$$\frac{\delta\omega}{\delta N} = \frac{\sigma_{\max}-\sigma_e}{\sigma_u-\sigma_{\max}}\left(\frac{\sigma_a}{B}\right)^{\beta} \tag{4.8}$$

where σ_u is the material's ultimate strength; σ_e is fatigue endurance limit; σ_{max} and σ_a are the applied maximum stress and stress amplitude, respectively.

Or, in terms of Coffin-Manson relationship,

$$\frac{\delta\omega}{\delta N} = \left(\frac{\Delta\varepsilon^{(in)}}{C}\right)^{\gamma} \tag{4.9}$$

Again, in the above damage equations, the damage parameter ω itself does not differentiate fatigue and creep mechanisms that cause either transgranular or intergranular fracture. For example, if the inelastic strain were totally accumulated by the climb-mechanism during cyclic creep, would Eq. (4.9) apply, implying transgranular or intergranular fracture? Furthermore, the constitutive law in the damage-state form like Eq. (4.6) is not true in compression, because the premise of "loss of internal load-bearing area" is not valid.

4.2 Review of Life Prediction Methods

Material life prediction was not an engineering practice at all before the Industrial Age. Wöhler was the first to conduct fatigue tests in 1867, for the Prussian Railway Service, to assess the fatigue endurance limit of railway steels after the locomotive crash in Versailles. In 1910, O.H. Basquin used an power law equation to characterize the S-N curves of metals, as:

$$\frac{\Delta\sigma}{2} = \sigma'_f N_f^b \tag{4.10}$$

where, N_f is the fatigue life, $\Delta\sigma$ is the stress range, σ'_f is a fatigue strength parameter, and b is a power exponent (slope in the log-log plot).

In the early Industrial Age, the design philosophy was so conservative with a large safety factor such that structures/components could operate well below the material's fatigue endurance limit. This design philosophy is called the *safe-life approach*, i.e. materials are assumed to have no flaws from cradle (manufacturing) to grave (end of service).

After WWII, the need for high-temperature fatigue life prediction models and methods was immediately following the heels of the fast-pace development of aerospace, energy generation, and transportation equipment, such as aircraft, gas turbine engines, high-pressure vessels, and reusable rockets etc. The research was mostly conducted in industrial or government laboratories at first. L.F. Coffin of General Electric (1954) and S.S. Manson of NASA (1954) independently found a relationship between the number of cycles to fatigue failure and the plastic strain amplitude in low cycle fatigue testing, as

$$\frac{\Delta\varepsilon_p}{2} = \varepsilon_f N_c^{\,c} \tag{4.11}$$

where ε_f is the material's fatigue ductility and c is a power-index falling in the range of –0.45 to –0.65 for most metals and alloys. Eq. (4.11) is thereafter called the Coffin-Manson relationship. Instead of plastic strain, hysteresis energy was also used to correlate with fatigue life in a similar fashion (Morrow 1965).

Other fatigue models were mostly variants, extensions or combinations of Eq. (4.10) and (4.11), for examples, Smith-Watson-Topper equation (1970) which includes the stress ratio R (= $\sigma_{min}/\sigma_{max}$) effect. For high temperature applications, the Morrow equation was modified with a frequency-dependent power-exponent (Ostergren 1976). As the loading profiles of high-temperature components become more and more complicated, more sophisticated methods were proposed to accommodate the effects of inelastic strains generated under different loading profiles. For example, the strain-range partitioning (SRP) method was developed by NASA (Manson, Halford and Hirschberg 1971) to include inelastic strains generated from different types of low-cycle fatigue loading profiles including fast strain rate cycles, tensile creep rupture cycles, compressive creep ruptures, tensile-hold strain cycles, and compressive-hold strain cycles. Those strain components were obtained by analyzing the shape of the hysteresis-loops generated from those tests, rather not evaluated from mechanism-based constitutive equation. Such partitioning method is difficult to use for behaviours under complex thermomechanical loading profiles, e.g., thermo-mechanical fatigue, since the partitioning strain components are not easily delineated from a complicated hysteresis loop. Besides, the experiments to characterize the Coffin-Manson type equations for the different types of inelastic strain are very expensive.

Regarding creep, creep rupture tests are often used to evaluate the high-temperature structural durability in terms of *time to rupture*. It is still popular to use such test as an inexpensive means for screening material's resistance to high-temperature operation. For example, the generations of single crystal Ni-base superalloys are basically marked by 25 °C increase in temperature at the same rupture time under the same stress condition. Larson and Miller (1952) found that the creep rupture stress data could be collapsed with a parameter $P = T(C + \log t_r)$, called the Larson-Miller parameter (LMP), where t_r is the creep rupture life and C is a constant having a value ~20. Generally, by plotting the Larson-Miller parameter against the applied stress, one can obtain a polynomial best-fit curve. It is thus procedurally simple, but the Larson-Miller plot usually leads to a large uncertainty in creep life prediction, because the scatter on the plot would translate into exponential errors in life. Therefore, extrapolation from the short-term creep data is often either too conservative or dangerous. Other creep life relationships have also been proposed for creep life prediction, notably the Monkman-Grant relation, which correlates the creep rupture life directly with the steady-state creep rate (Monkman and Grant, 1956). These relations have been useful in analyzing short-term creep test data, however, the underlying physical mechanisms and microstructural factors are not reflected in these empirical or phenomenological relations. Therefore, the prediction of long-term creep properties from short-term experiments is still rated as one of the most important challenges in assessing structural durability, as regarded by the UK energy sector (Allen and Garwood 2007), after more than one hundred years of creep study since Andrade (1910).

With respect to complex loading profiles that may induce both fatigue and creep damage, numerous damage mixing rules were proposed, among which the simplest one is the linear damage summation rules (LDR). Time-fraction LDR is a combination of Miner-Palmgren rule and Robinson rule (Palmgren 1924, Miner 1945, Robinson 1952) which assumes that the sum of the fatigue cycle ratio and the creep time ratio with respect to failure is equal to 1, as

$$\sum \frac{N}{N_f} + \sum \frac{t}{t_r} = 1 \tag{4.12}$$

where N_f is the fatigue life and t_r is the rupture life.

Another variant of LDR is the ductility-exhaustion theory (Spindler 2007), which is a strain-fraction LDR that specifies creep damage as the ratio of current creep strain to strain at rupture, as

$$\sum \frac{N}{N_f} + \sum \frac{\varepsilon}{\varepsilon_r} = 1 \tag{4.13}$$

where ε_r is the rupture strain.

It should be pointed out that LDRs are rules of thumb, lacking a rigorous physical premise. However, the LDRs have been widely adopted in many damage formulations, e.g. the continuum damage mechanics, despite their invalidity has been reported for many real materials (Lloyd and Wareing 1981, Rees 1987, Inoue et al. 1991, Skelton 2013, Spindler 2007). Figure 4.2 shows either the time-fraction or ductility-based LDR descriptions grossly deviates from the experimental observations for type 437 weld metal at 650 °C (Spindler 2007). Halford (1991) once reviewed that there were about a few dozen models proposed up to the year of 1991, trying to account for the effect of creep-fatigue interaction during complicated loading. It is apparently out of the scope of this book to go into details of every model proposed to date. But, the fact that there are so many creep-fatigue interaction models or rules being proposed merely reflects the lack of a general theory to provide all-consistent explanations for the observed phenomena. Therefore, facing today's engineering challenges, only physics-based or mechanism-based models warrant the advantage.

4.3 The Integrated Creep-Fatigue Theory

Today, under the stringent environmental regulations, power generation or propulsion systems such as combustion engines, gas turbine engines and steam turbines, all need to burn fuels more efficiently and generate less greenhouse gases, which require higher operating temperatures pushing the materials to their operating limits. Also, for economic and safety reasons, large engineering systems or high-value assets such as aircraft and power-plants need to be managed with an integrated health management system that can make an accurate prognosis of the future state with a remaining useful life based on the current state awareness, to reduce the maintenance cost and avoid premature failure (Wu et al. 2007, Jennions 2011). This represents a shift of paradigms from the scheduled and reactive maintenance scheme to a condition-based maintenance and beyond, i.e., prognosis-based and proactive maintenance scheme. All these requirements/needs call for a robust life prediction method that is applicable to all kinds of complicated loading conditions which usually involve thermomechanical-cyclic loads with short or long dwell periods.

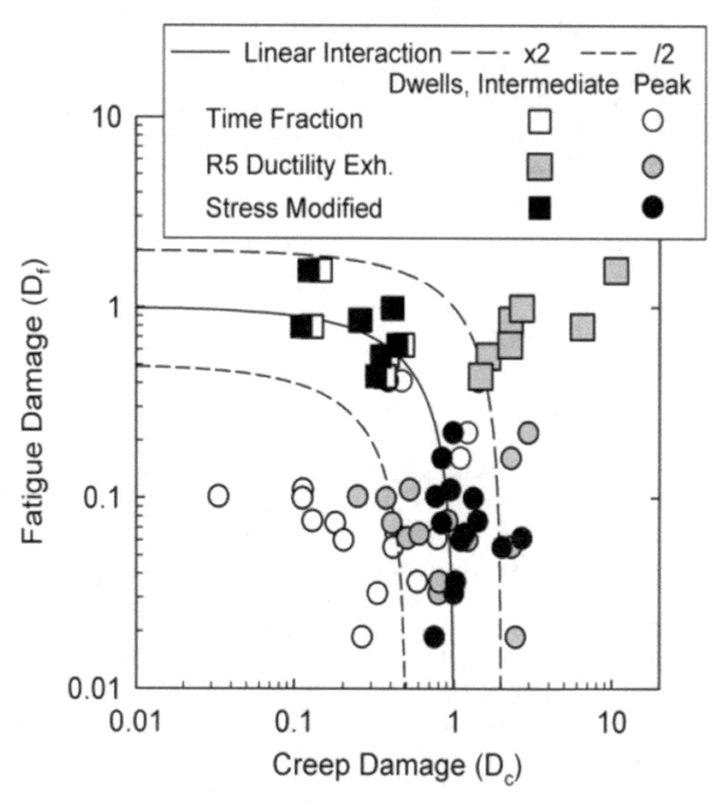

Figure 4.2. Creep–fatigue interaction diagram for Type 347 weld metal at 650 °C (Spindler 2007).

Understanding the physics of failure is a priori of material life prediction. To ensure the structural integrity, one needs to know where (failure location), how (mechanism and mode) and when (life) the component may fail under mission loads (usually a combination of vibrational, cyclic and dwell loading with simultaneous variation of temperature). In this section, the integrated creep-fatigue theory (ICFT) is introduced to tackle the material constitutive behaviour and failure problems. First, constitutive laws are formulated based on the fundamental deformation mechanisms including rate-independent plasticity, IDG, IDC and GBS, which allows the respective mechanism strain to be evaluated for characterization of the physical damage accumulation associated with the deformation process. Second, a holistic damage evolution equation is derived based on the damage accumulation process that consists of nucleation and propagation of surface/subsurface cracks in coalescence with internally distributed damage. With such understanding and implementation into FEM, the physics of failure can be rendered at the component level for engineering design-lifing analysis.

4.3.1 Physical Deformation Decomposition

It has been recognized that crystalline solids deform by a number of alternative, often competing, mechanisms that involve intragranular dislocation glide and clime, as well as grain boundary sliding, as summarized in Chapter 2. Since these deformation mechanisms operate independently, the total deformation of a crystalline material can be decomposed as:

$$\varepsilon = \underbrace{\varepsilon_e + \varepsilon_p + \varepsilon_g + \varepsilon_c}_{\text{intragranular}} + \underbrace{\varepsilon_{gbs}}_{\text{intergranular}} \tag{4.14}$$

In summary, the total deformation is formulated as the sum of intragranular deformation (ID) and grain boundary sliding (GBS) by Eq. (4.14), as schematically shown in Figure 4.3. Generally, plasticity (P) proceeds by crystallographic slip in a rate-independent manner at low temperatures. In the plasticity regime, the deformation mechanism is dislocation glide and cross-slip, shearing or looping around the obstacles along the path, giving rise to strain hardening behaviour. Thus, plastic deformation proceeds with kinematic hardening and isotropic hardening. Under static loading, the material fails at its ultimate tensile strength (UTS) by either ductile or brittle fracture; while under cyclic loading, alternating slip and slip reversal can result in fatigue failure. As temperature increases, dislocation motion acquires another degree of freedom—climb, due to vacancy flow, which assists dislocations to get around the obstacles such that time-dependent deformation, i.e. creep, is manifested. Vacancy flow at elevated temperatures is basically assisted by two diffusion processes—

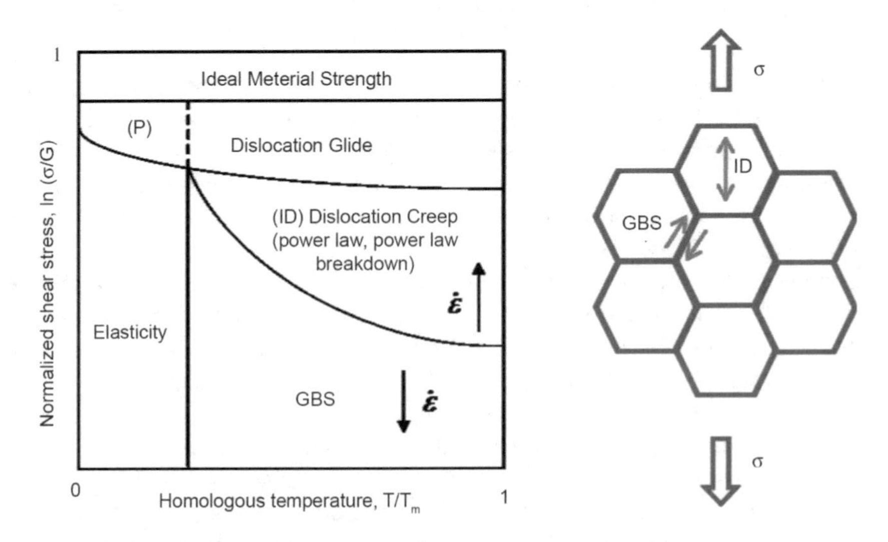

Figure 4.3. Schematic of intragranular deformation (ID) and grain boundary sliding (GBS) and their dominance regions in the deformation mechanism map.

lattice diffusion and grain boundary diffusion; the former assists dislocation climb and glide within the grain interior, resulting in intragranular deformation (ID) such as the power-law creep and power-law-breakdown phenomena, whereas the latter process assists dislocation climb and glide along grain boundaries, resulting in GBS. Therefore, using the physical deformation decomposition formulation, Eq. (4.14), the ICFT can cover all deformation behaviours of crystalline materials over the entire homologous temperature range and their lifetime, provided the microstructural changes due to precipitation coarsening and grain growth are updated by their own metallurgical laws.

In contrast to the unified constitutive theories that lump-sum plastic and creep deformation into total viscoplastic strain, the ICFT with the strain decomposition, Eq. (4.14), and deformation mechanism-based constitutive equations given in Chapter 2, can evaluate the mechanism strain produced by each individual deformation mechanism. For 3D component analysis, rate-independent plasticity can be implemented into the "plasticity" subroutine and time-dependent deformation can be implemented into the "creep" subroutine of FEM. The mechanism-constitutive equations are given in the corresponding sections of Chapter 2. This way, the mechanism strains can be evaluated in 3D tensor form. The significance of mechanism-strain delineation lies in material damage analysis and life prediction. As one understands that slip concentration via dislocation glide and dislocation pile-up is the primary cause for fatigue crack nucleation, whereas dislocation climb tend to ease dislocation pile-ups (section 1.6) but promote void growth during creep. Thus, the strains produced by different mechanisms have different impacts on damage accumulation in the material, as discussed in detail in Chapter 3.

With the options offered by the commercial FEM, the plasticity subroutine and creep subroutine can be run in parallel to evaluate the overall deformation response, where plasticity is limited with an evolving yield surface, as implied by Eq. (4.1), but creep is not, as long as the stress can activate particular deformation mechanisms at the given temperature. Indeed, given enough thermal energy at high temperature, most creep mechanisms do operate without a defined "yield surface". The mechanism decomposition formulation, Eq. (4.14), does not exclude incorporation of internal-state variables, e.g., isotropic hardening variable for cyclic hardening, or microstructural state variable for creep, but microstructural evolution laws should follow metallurgical theories.

The ICFT is consistent with the physical metallurgy of crystalline material deformation and damage. In particular, while plasticity proceeds by dislocation glide, creep deformation is assisted by dislocation climb that releases dislocation pile-ups at the obstacles in the glide plane. Because of the random nature of vacancy flow, dislocation climb would help to ease slip concentration rather than contributing to fatigue damage. On the other hand, vacancy flow promotes cavitation and void growth along grain boundaries, which forms the so-called internally distributed damage. This is why mechanical fatigue damage often results in transgranular fracture; whereas creep often causes intergranular fracture. In high-temperature deformation, all of the above mechanisms contribute to inelastic straining, but their effects on damage accumulation are different. The mechanism-based decomposition allows formulation of fatigue and creep damage with respective mechanisms (plasticity, intragranular dislocation climb-plus-glide as well as GBS), leading to either transgranular or intergranular fracture. The unified constitutive models do not provide such mechanism-based delineation.

Therefore, from the physics-of-failure point of view, it is essential to formulate material constitutive laws encompassing all possible deformation mechanisms and identify the roles of each mechanism strain in damage accumulation and interaction. This is the essence of the ICFT. It provides the framework for physics-based material life prediction, to address deformation, failure modes and lifetime holistically. This will be demonstrated for various deformation processes including creep, low cycle fatigue (LCF), high cycle fatigue and thermomechanical fatigue (TMF) as discussed in the next few chapters. It should be realized that no matter how complex the material behaviours are, they are manifestations of the fundamental deformation mechanisms operated in competition under various service loading conditions.

4.3.2 Holistic Damage Accumulation

The holistic damage accumulation process involves crack nucleation and propagation in coalescence with internally distributed damage by multiple mechanisms. For generality, consider a polycrystalline material under arbitrary thermomechanical loading, as schematically shown in Figure 4.4. In a physical sense, mechanical fatigue damage accumulates in the form of persistent slip bands (PSB) under cyclic loading, which eventually leads to crack nucleation, mostly at the surface or subsurface because of local stress concentration due to manufacturing flaws (e.g., machining scratches) or surface roughness and less material constraints. Environmental spices may also attack the material from surface, e.g., oxide scales can form at the surface upon exposure at high temperatures. These mechanisms can cause crack nucleation at surface or subsurface. On the other hand, cavitation and micro-cracks may develop inside the material, as *internally distributed damage*, either along the grain boundaries or in the grain interior, depending on the material microstructure. Internally distributed damage is mostly promoted by creep, but it can also be induced by other mechanisms pertinent to specific alloys, e.g. intergranular embrittlement by magnesium segregation in ductile cast iron. Thus, the holistic damage accumulation process consists of surface/subsurface crack nucleation and propagation in coalescence with internally distributed damage or discontinuities.

Fatigue damage can be regarded as accumulation of irreversible slip offsets on particular slip systems, which act as crack nuclei. Partially irreversible

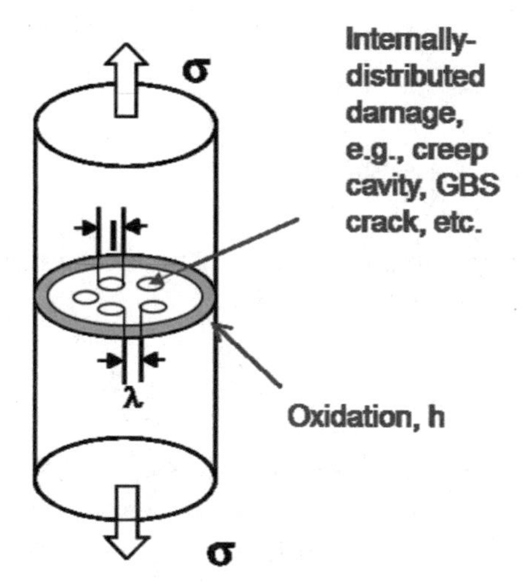

Figure 4.4. A schematic of damage development in a material cross-section.

slips are also responsible for transgranular fatigue crack propagation (Neumann 1974, Laird and Smith 1982, Wu et al. 1993). Therefore, in a holistic sense, we can use $(da/dN)_f$ to represent the rate of slip offset accumulation leading to crack nucleation as well as fatigue crack growth, even though the functional dependence of $(da/dN)_f$ on loading parameters is different for crack nucleation and crack growth. On the other hand, internally distributed damage may be promoted by creep and/or intergranular embrittlement, which develop in the forms of voids, grain boundary cavities and microcracks, depending on the material microstructure. Over the temperature range from room temperature (RT) to the melting temperature (T_m), different types of internal damage may develop. For example, Zener-Stroh-Koehler (ZSK) cracks, i.e., dislocation pile-ups, may develop at low temperature, because of limited dislocation climb activity; whereas grain boundary cavitation and cracking can develop at high temperature, because of vacancy diffusion and condensation as well as grain boundary sliding (Hull and Rimmer 1959, Raj and Ashby 1975, Harris 1965, Wu and Koul 1996). Particular treatments of such internally distributed damage should be given with due consideration of the creep damage mechanisms, as elucidated in Chapter 3.

Based on the above damage accumulation process, the holistic damage accumulation rate can be derived as follows. Consider the coalescence of internally distributed damage with a propagating fatigue crack, the overall rate is (Wu 2009):

$$\frac{da}{dN} = \left(\frac{da}{dN}\right)_f + \frac{l_c + l_z}{\Delta N} \tag{4.15}$$

where l_c is the creep cavity/crack size, l_z is the ZSK crack size, ΔN is the number of cycles during which the fatigue crack propagates between two cavities or ZSK cracks separated by an average distance of λ.

Assume that during the period of ΔN, the dominant crack only propagates by the fatigue mechanism, i.e., $(da/dN)_f \sim \lambda/\Delta N$, then we can rewrite Eq. (4.15), as

$$\frac{da}{dN} = \left(1 + \frac{l_c + l_z}{\lambda}\right)\left(\frac{da}{dN}\right)_f \tag{4.16}$$

In the presence of environmental effects, which aid the propagation of the surface crack, the total crack growth rate is

$$\frac{da}{dN} = \left(1 + \frac{l_c + l_z}{\lambda}\right)\left\{\left(\frac{da}{dN}\right)_f + \left(\frac{da}{dN}\right)_{env}\right\} \tag{4.17}$$

Suppose that the newly formed oxide scale breaks in every cycle of crack propagation, then the oxidation cracking increment per cycle is $(da/dN)_{ox} = h$ (the thickness of newly-formed oxide scale from fresh metal, $h = \sqrt{2k_{ox}\tau}$, τ is the cycle period). Finally, when the dominant crack grows into a critical crack length, a_c, fracture occurs. Here, it should be noted that under any cyclic loading profile, elasticity would be experienced first by materials. Therefore, linear elastic fracture mechanics (Chapter 10) may apply at least to the final fracture of materials containing cracks.

Under constant amplitude loading, Eq. (4.17) can be integrated into:

$$\frac{1}{N} = D\left\{\frac{1}{N_f} + \frac{h}{a_c}\right\} \tag{4.18a}$$

with

$$D = \left(1 + \sum \frac{l}{\lambda}\right) \tag{4.18b}$$

and (according to linear fracture mechanics, Chapter 10)

$$a_c = \frac{1}{\pi}\left(\frac{K_{IC}}{Y\sigma_{max}}\right)^2 \tag{4.18c}$$

where K_{IC} is the fracture toughness, Y is the crack-shape correction factor.

Eq. (4.18) reduces to pure fatigue in absence of other damaging mechanisms such as creep and oxidation. If a material fails by creep alone, i.e. only with the growth of internally distributed damage. Then, the D factor reaches a critical maximum, considering that the damage size l is increasing and the inter-void spacing λ is decreasing with creep strain at the same time. By the generality of the aforementioned mechanisms, Eq. (4.17) or Eq. (4.18) should apply to all metallic materials under general thermomechanical conditions (in any combination of fatigue and creep loading profiles).

Traditionally, material behaviours are often classified in accordance with the loading types. *Creep* is observed under constant load, undergoing the primary, secondary and tertiary stages with time until creep rupture.

Tensile deformation is observed when the material is monotonically loaded in tension until failure (the reverse is called compression). *Cyclic deformation* is termed when the material is subjected to alternating loads, which results in fatigue. *Thermomechanical fatigue* (TMF) is called when both mechanical load and temperature change simultaneously in a cyclic manner. These phenomena are used to be described separately by phenomenological equations. The empirical or phenomenological equations do not provide a holistic picture of deformation and damage evolution, because the interplays of the aforementioned fundamental deformation mechanisms are not elucidated. The ICFT, as outlined in this chapter, describes the material deformation responses as the collective outcome of interactions between fundamental deformation and damage mechanisms as recognized in Chapters 2 and 3. Thus, it will be realized that the deformation and failure mode vary with the dominance of participating mechanisms, depending on the loading and temperature profiles, but it is governed by the same rules of participation (physics). This allows the practicing engineers to apply the same rules consistently to analyze and interpret experimental observations (test data) and predict the materials behaviours under service conditions that cannot be duplicated in laboratory tests. By unveiling the interplays of deformation mechanisms, the ICFT has two significant meanings:

(i) to ensure reliable life prediction based on the controlling mechanism(s); and
(ii) to feedback information for material/component design against the undesirable failure mode, thus to improve materials' strength or durability.

4.3.3 Model Calibration and Implementation in FEM

As discussed above, for generality, the ICFT constitutive model is formulated to cover all micro-mechanism strains under arbitrary loading. It should be understood, as shown by the deformation mechanism map (Figure 4.3), that micromechanism strain could be manifested as the dominant macroscopic strain in a particular deformation regime (stress-temperature region). Thus, the corresponding mechanism parameters can be obtained from simple mechanical testing through calibration with the observed behaviour. In the simplest case, elastic modulus is determined from the linear portion of the tensile stress-strain curve. Similarly, parameters of rate-independent plasticity, Eq. (2.1), can be (easily) obtained by analysing the monotonic and/or cyclic tensile behaviour at low temperatures when creep contributions are negligible. To completely characterize creep, Eq. (4.14) has to be used together with the respective mechanism rate equations given in Chapter 2 to describe creep behaviour over a wide temperature range. Detailed analysis to determine creep mechanism parameters will be discussed in Chapter 5. By the deformation kinetics, it is always the fastest rate mechanism that causes the material to yield. Therefore, only the most significant rate mechanism, e.g.

intragranular dislocation glide and climb-plus-glide, needs to be considered for the rate-dependent low-cycle fatigue and thermomechanical fatigue deformation. If deformation occurs at intermediate stress-temperature range, perhaps GBS plays the dominant role in cyclic deformation, e.g., ratcheting.

As for incorporation with FEM, the rate-independent plasticity and creep equations can be implemented in the "plasticity" and "creep" subroutines, respectively. When chosen to run in parallel in FEM simulation, it mimics the independent operation of these micro-mechanisms. Thus, kinematic/isotropic hardening is naturally associated with plasticity, whereas creep acts as softening/stress-relaxation mechanism.

In the next few chapters, we will see how the ICFT constitutive and damage laws outlined in Chapter 1 to 4 are used to describe deformation behaviours and its life evolution under different loading conditions.

CHAPTER

5

Creep

5.1 Overview

Creep refers to the phenomena of time-dependent deformation under constant load or stress at elevated temperature. Following Andrade's first observation in 1910, creep testing has been a conventional way to characterize material's high-temperature strength. A typical uniaxial creep frame setup is shown in Figure 5.1, where the creep coupon is loaded by a dead weight through a lever arm. Such a creep test is therefore called constant-load creep test. During the test, the coupon is enclosed in a furnace set at temperature and its elongation is measured using a linear variable differential transformer

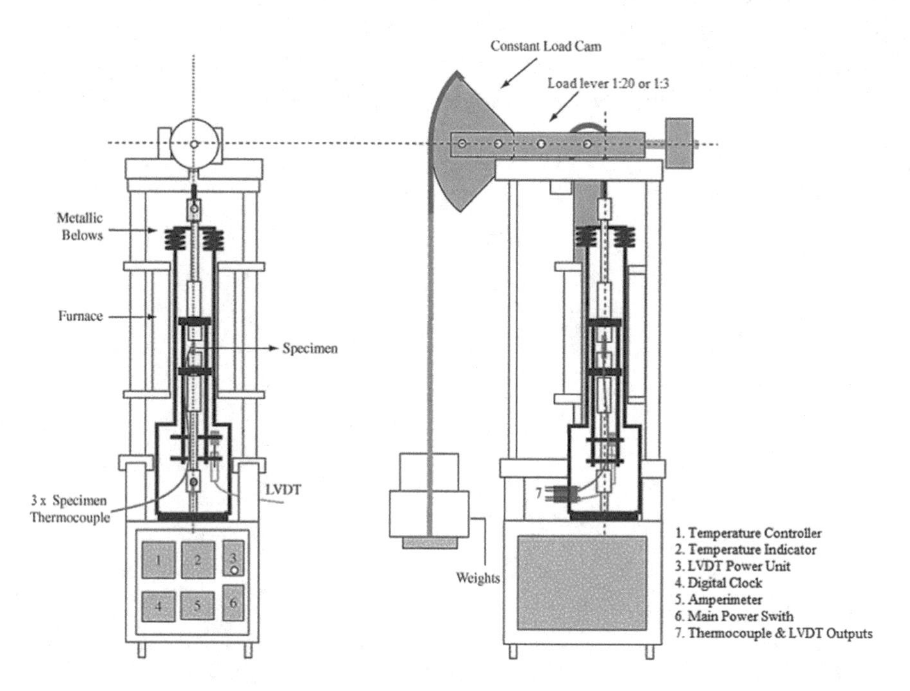

Figure 5.1. Schematic drawing of main components of creep testing machine (after Bueno, 2008).

(LVDT). The elongation-time data are recorded to the point of specimen fracture, i.e., stress rupture, and hence creep life is called the rupture life.

Creep deformation as high-temperature design limitations is widely recognized in industrial design codes, e.g., ASME Boiler and Pressure Vessel Code. The creep failure criteria can be given either in terms of creep strain to be reached, e.g. 1%, or hours of service, e.g. 100,000 hours, against which the allowable stress is determined. With regards to the above criteria for component design, a few questions and factors have to be completely understood.

1. What is the creep behaviour, including creep strain as function of time and creep rate as function of stress and temperature?
2. What are the effects of chemical composition and microstructure (via manufacturing process and heat treatment, etc.) on the material's creep resistance?
3. Will the material microstructure change during long-time service, and if changed, what would be its effect on the material's performance?
4. Does the operating environment have an effect on the material's performance?

Experimental studies have shown that creep deformation can occur under stresses well below the material's yield strength at elevated temperatures, usually above 0.3 T_m (T_m is the material's melting temperature in Kelvin degree). Figure 5.2 shows a schematic creep curve from a constant-load creep test. The creep elongation generally evolves with time through three stages: 1) primary, 2) secondary (steady-state) and 3) tertiary stages, as

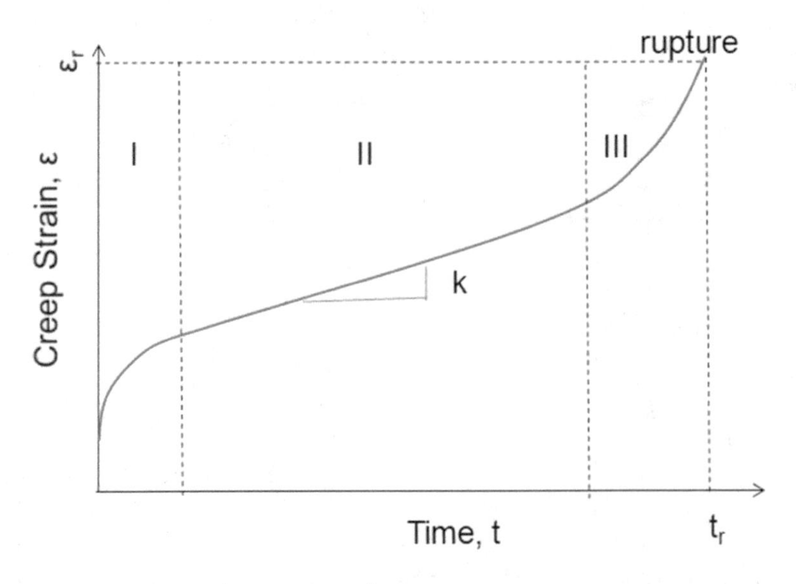

Figure 5.2. A schematic creep curve.

indicated in Figure 5.2. The primary creep exhibits a decelerating creep rate; the secondary creep stage has a nearly constant creep rate, which is usually the minimum creep rate; and the tertiary creep exhibits an accelerating creep rate to rupture. By its nature, the constant minimum creep rate is often used to characterize the material's creep property.

Many models have been proposed to describe the creep-curve. Andrade (1914) first proposed a creep strain function as time to the power of 1/3. Norton and Bailey proposed a general power-law of stress and time (Norton 1929, Bailey 1935). Graham-Walles (1955) extended those power-laws into a power-law series of creep strain equations. Evans and Wilshire (1985) proposed the θ-projection method. In addition, the continuum damage mechanics (CDM) is also developed to describe the creep strain accumulation with $\sigma/(1-\omega)$, where ω represents the loss of internal area by void growth (Kachanov 1958, Robotnov 1969, Lemaitre and Chaboche 1999, Dyson and McLean 2000, Hayhurst 2005). These models or theories can depict the creep curve, but do not indicate what mechanisms are responsible for the creep curve shape, nor do they contain failure criteria leading to intergranular or transgranular or mixed mode rupture, except parametrically, $\omega = 1$.

As far as creep rupture is concerned, many empirical approaches such as the Larson-Miller method (Larson and Miller 1952), the Monkman-Grant relation (Monkman and Grant 1956), and more recently the Wilshire equation (2003) have been proposed to formulate relationships between stress and rupture life at temperature. These relationships are mostly correlations established from short-term creep test data, which are also often attempted to be extrapolated for long-term creep life prediction. Obviously, the extrapolation is useful in design with economic considerations, because for some high-temperature equipment such as pressure vessels and steam turbines that are required to operate for over 100,000 hours, there is no time to conduct such long-term creep tests to validate the design before the product needs to be put into the market. However, many factors such as microstructural evolution and environmental effects cannot be fully taken into account in short-term creep tests, therefore empirical relationships are not totally reliable for long-term creep life prediction. Furthermore, if short-term creep tests are conducted in the mechanism regimes different from the long-term service conditions, extrapolations from such tests are not physically warranted. Prediction of long-term creep properties from short-term experiments is still rated as one of the most important challenges in assessing structural durability, as regarded by the UK energy sector (Allen and Garwood 2007). The question is how to ensure that service failures are controlled by the same mechanism(s) as in the design tests?

A significant effort has been devoted to understanding the underlying deformation mechanisms during creep (Ashby 1972, Frost and Ashby 1982, Ashby 1983, Weertman 1955, Langdon 1970, Raj and Ashby 1971, Dyson and Maclean 1983, Dyson and Gibbons 1987, Wu and Koul 1995, 1996, 1997), as summarized in Chapter 2. Through these studies, it has been understood that creep properties of complex engineering alloys depend on composition and microstructure including solute contents, the size and volume fraction of second-phase particles (precipitates) in the matrix, the grain size, grain boundary morphology, and grain boundary precipitate size and distribution. It only needs a mechanism-based creep model to connect those recognized deformation mechanisms to creep behaviour and life. This is important for advanced material design against service-induced failure, since the failure mode varies with stress, temperature, microstructure and environment. Therefore, it is called for a physics or mechanism-based creep model to capture the physical interplay of various rate-controlling deformation mechanisms with the confinement of microstructure, and to predict the creep life over a wide range of stress and temperature.

5.2 The Deformation-Mechanism based True-Stress (DMTS) Creep Model

As stated in Chapter 4.3, the total inelastic deformation is comprised of rate-independent plastic deformation and time-dependent flow contributed from intragranular dislocation glide-plus-climb and GBS, each mechanism is governed by its own constitutive equation. Here, we are discussing creep at stresses well below the yield strength, so the rate-independent plastic deformation can be ignored. Diffusion flow is also neglected because it rarely occurs in engineering alloys under conditions of engineering relevance (Wadsworth et al. 2002, Wilshire 2002). Hence, the total creep rate is given by

$$\dot{\varepsilon} = \dot{\varepsilon}_g + \dot{\varepsilon}_c + \dot{\varepsilon}_{gbs} \tag{5.1}$$

The constitutive equation of creep has to be established for the true stress-strain rate relationship to be implemented into FEM for component stress analysis, because FEM always evaluates the Cauchy stress. However, most creep tests are conducted under uniaxial constant load conditions, as schematically shown in Figure 5.1. Therefore, for constant-load creep data analysis, the mechanism-rate equations need to be reduced to engineering-stress formulations, to determine the mechanism parameters.

By the rule of incompressible flow, the uniaxial true stress-strain can be expressed as:

$$\begin{aligned} \varepsilon &= \ln(1+e) \\ \sigma &= \sigma_a \exp(\varepsilon) \end{aligned} \tag{5.2}$$

where e is the engineering strain measured from LVDT or extensometer, and σ_a is the engineering (initial or nominal) stress.

5.2.1 Model Development

For mathematical convenience, the GBS, intragranular dislocation glide (IDG) and intragranular dislocation climb (IDC) rate equations from Chapter

2 are all expressed as power-law as:

$$\dot{\varepsilon}_{gbs} = A\sigma^p = A\sigma_0^p \exp(p\varepsilon) \approx (1 + p\varepsilon)A\sigma_0^p = (1 + p\varepsilon)\dot{\varepsilon}_{s0} \tag{5.3}$$

$$\dot{\varepsilon}_g = B\sigma^n = B\sigma_0^n \exp(n\varepsilon) \approx (1 + n\varepsilon) B\sigma_0^n = (1 + n\varepsilon)\dot{\varepsilon}_{g0} \tag{5.4}$$

$$\dot{\varepsilon}_c = (1 + M\varepsilon)C\sigma^m = (1 + M\varepsilon)C\sigma_0^m \exp(m\varepsilon) \approx (1 + M\varepsilon + m\varepsilon)C\sigma_0^m$$

$$= (1 + M\varepsilon + m\varepsilon)\dot{\varepsilon}_{c0} \tag{5.5}$$

where $A = A_0 \exp\left(-\frac{Q_{gbs}}{RT}\right); B = B_0 \exp\left(-\frac{Q_g}{RT}\right); C = C_0 \exp\left(-\frac{Q_c}{RT}\right)$

Here A, B and C are Arrhenius type constants for GBS, IDG and IDC mechanisms, respectively; $\dot{\varepsilon}_{g0}, \dot{\varepsilon}_{c0}, \dot{\varepsilon}_{s0}$ are the steady-state rates at the nominal engineering stress σ_0, respectively; and $p\ (\geq 2)$, n, m are the apparent stress exponents for GBS, IDG and IDC, respectively.

Substituting Eq. (5.3)-(5.5) into Eq. (5.1), we have

$$\dot{\varepsilon} = \dot{\varepsilon}_{gbs} + \dot{\varepsilon}_g + \dot{\varepsilon}_c = (1 + p\varepsilon)\dot{\varepsilon}_{s0} + (1 + n\varepsilon)\dot{\varepsilon}_{g0} + (1 + M\varepsilon + m\varepsilon)\dot{\varepsilon}_{c0} \tag{5.6}$$

which can be integrated into:

$$\varepsilon = \frac{1}{M^*}[\exp(M^* kt) - 1] \tag{5.7a}$$

where

$$M^* = \frac{[p\dot{\varepsilon}_{s0} + n\dot{\varepsilon}_{g0} + (m + M)\dot{\varepsilon}_{c0}]}{k} \tag{5.7b}$$

$$k = \dot{\varepsilon}_{s0} + \dot{\varepsilon}_{g0} + \dot{\varepsilon}_{c0} \tag{5.7c}$$

Here, k is the total rate constant, and M^* is the tertiary shape parameter.

To construct the total strain equation, the transient creep strain of GBS from Eq. (2.29) needs to be added into Eq. (5.7a), since the steady-state GBS is already included in Eq. (5.7a), to conform with the deformation kinetics, Eq. (5.1). Thus, we obtain the total creep strain equation, as:

$$\varepsilon = \varepsilon_0 + \frac{\sigma}{\beta^2 H}\left[1 - \exp\left(-\frac{\beta^2 H \dot{\varepsilon}_{s0} t}{\sigma(\beta - 1)}\right)\right] + \frac{1}{M^*}[\exp(M^* kt) - 1] \tag{5.8}$$

where ε_0 is the initial elastic strain (in a creep experiment, this also includes the compliant strain of the specimen fixture), H is strain hardening coefficient, β is a material structure parameter, defined in section 2.4.

Then, the instantaneous total creep rate is given by:

$$\dot{\varepsilon} = \frac{\dot{\varepsilon}_{s0}}{(\beta - 1)} \exp\left(-\frac{\beta^2 H \dot{\varepsilon}_{s0} t}{\sigma(\beta - 1)}\right) + k \exp(M^* kt) \tag{5.9}$$

Here it should be noted that Eq. (5.8) is derived for polycrystalline materials where transient behaviour is dominated by GBS. In single crystal materials, primary creep is often either delayed due to dislocation multiplication (Johnston 1962), or very limited in precipitation-hardened Ni-base single crystal superalloys (Carry and Strudel 1977, 1978). Pronounced and undelayed primary creep in single crystal Ni-base superalloys can only be observed in the temperature range of 750 °C ~800 °C and at high stresses above 550 MPa, which occurs by a number of mechanisms involving formation of dislocation ribbons, cutting of γ' precipitates, relaxation of misfit stress, and in the presence of γ' rafts (Pollock and Argon 1992, Rae and Reed 2007). In single crystal Ni-base superalloys, γ'-strengthening remains to be effective up to 750 °C before significant rate-dependent deformation occurs. Therefore, stress has to be sufficiently high to form dislocation ribbons within the γ channels, and drive super-dislocations to cut γ' precipitates. At 750 °C, however, few polycrystalline materials could survive under a high stress of 550 MPa for any long time. Therefore, in polycrystalline materials, contributions from intragranular mechanisms to primary creep can be ignored; on the other hand, grain boundaries themselves are sources of abundant dislocations to initiate early deformation during creep. Thus, GBS-led primary creep can become dominant at lower temperatures and lower stresses (e.g. 500–700 °C in Waspaloy). By the weakest link postulation, the intragranular primary creep mechanisms may not necessarily be invoked, as GBS proceeds along the grain boundaries (the weakest-link paths) in a polycrystalline Ni-base superalloy. This is the very reason why grain boundaries are preferably eliminated in materials for extreme high-temperature applications such as gas turbine blades.

To use the simplest mathematics to represent the complexity of polycrystalline material deformation, it will be shown that the formulation in the context of Eq. (5.8) is sufficient. Interestingly, Eq. (5.8) takes the same mathematical form as the θ-projection method (Evans and Wilshire 1985). But here, the strain equation is derived from deformation mechanisms and its parameters are defined by the identified deformation mechanisms, whereas the θ-parameters are polynomial-fitted to the stress and temperature dependence. With mechanism-based parameters, the advantage of Eq. (5.8) is that the creep performance of a material can be tailored by controlling the dominant deformation mechanisms, and the failure mode as well as rupture life can all be predicted (see later subsections for details).

5.2.2 Creep Curve Analysis

As the DMTS creep model is derived from the involved deformation mechanisms, the next question is how to determine the mechanism parameters? In this section, it shows a step-by-step analysis of the experimentally recorded creep data, using Eq. (5.8) and (5.9) through a multiple linear regression scheme. In the following, creep of a Ni-base gas turbine disk alloy—Waspaloy is used as a case study to illustrate the analytical procedure.

Step 1—Determining the creep rate. From the creep curve data, by differentiation, creep rate can be obtained, as shown in Figure 5.3 as an example for Waspaloy at 510 MPa and 700 °C. As observed, the creep rate first decelerates, reaching a minimum point, then accelerates into the tertiary stage. Beyond the minimum creep rate point and before deformation instability near rupture, the logarithm of creep rate can be represented by a linear function of time. According to Eq. (5.9), this linear relationship can be represented by

$$\log \dot{\varepsilon} = \log k + \frac{M^{*}k}{2.3}t \tag{5.10}$$

Then, the parameters k and M^* can be determined from the intercept and slope of the linear regression, respectively. Note that the rate constant k is close to the minimum creep rate, it is the true steady-state creep rate commencing from the initial loading. Towards the end, however, deviation of the actual elongation from the model trending is due to deformation instability with severe local necking just before rupture.

Step 2—Charactering stress and temperature-dependence of the involved deformation mechanisms. Once the creep rate constants, k, under all tests conditions are obtained. Delineation of deformation mechanisms can be analyzed via piece-wise linear regression on a log k vs. log σ_a plot, Figure 5.4. Using the stress-rate plot and the Arrhenius plot, the rate equation parameters for GBS, IDG and IDC (p, n, m, A_0, B_0, C_0, Q_{gbs}, Q_g, Q_c) can all be determined, in a way as in conventional creep rate analysis. The average dislocation multiplication factor M can be estimated from the strain-rate-weighted tertiary shape parameter, M^*, via Eq. (5.7b).

An example is given here for Waspaloy, as shown in Figure 5.4. Using the three power-law equations to fit the steady-state creep rate behaviour,

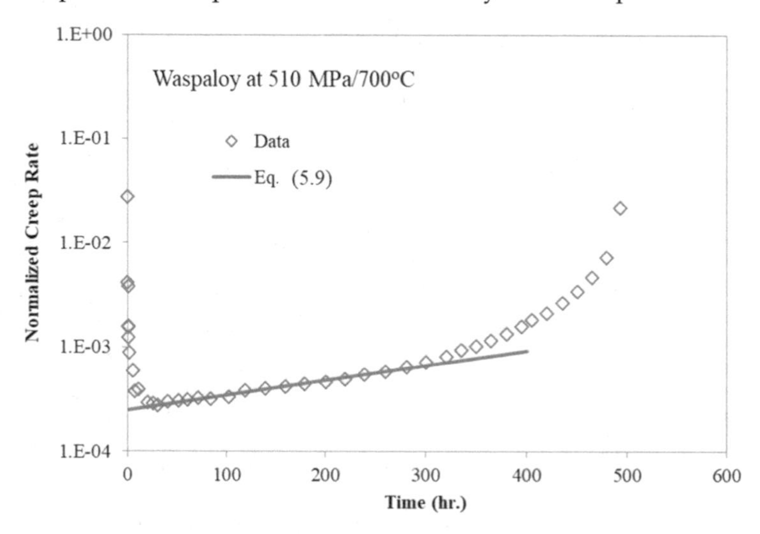

Figure 5.3. Creep rate vs. time relationship for Waspaloy at 510 MPa and 700 °C.

one can see that IDG dominates in the high stress region with a high power-law exponent ($n \approx 24$ for Waspaloy) at relatively low temperatures, 550 °C ~650 °C. At temperatures above 700 °C, the IDC mechanism becomes dominant with a much lower power-law exponent (~7 for Waspaloy). These two mechanisms have been well recognized in Ashby's deformation mechanism map. In the intermediate stress region at intermediate temperatures, i.e., between 600 °C and 700 °C, another deformation mechanism must also be operative, which turns out to be GBS, in order to make up the total creep rate. Furthermore, creep curves under these test conditions exhibit a pronounced transient behaviour, which is also indicative of the GBS-dominance (Wu and Koul 1995). Having considered all the possible deformation mechanisms, i.e., IDG, IDC and GBS, the steady-state creep rates k can be matched very well, using the DMTS model compared to the experimental data, as shown in Figure 5.4.

Step 3—Determination of the primary creep parameter. The last step is to determine the primary creep parameters associated with GBS. Choose a creep curve that exhibits a pronounced transient stage, determine the primary strain and time, from which, H and β can be determined, as discussed in section 2.4. Other examples of GBS-controlled transient creep have been shown in Chapter 2.

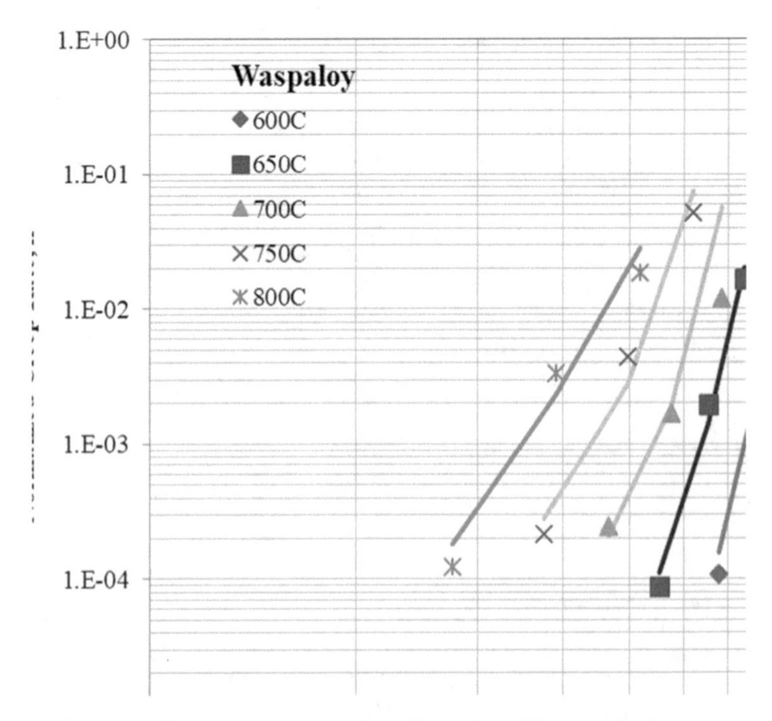

Figure 5.4. Normalized creep rates in Waspaloy. The symbols represent the experimental data and the lines represent the mechanism-based power-laws.

After going through the above three steps, all mechanism parameter values are determined, as given in Table 5.1 for Waspaloy. The apparent activation energy and power-exponent values are close to those reported by Wilshire and Scharning (2009) (an apparent activation energy value of 442~452 kJ/mol. and power exponents from 4 to 18) for the minimum creep rate in Waspaloy. Under the test conditions presently concerned, diffusion flow (with a power exponent ~1) was not observed.

As a demonstration of the DMTS creep model with parameter values given in Table 5.1, creep curves are simulated for Waspaloy under various conditions (temperature: 600–800 °C; stress: 240–990 MPa), exhibiting typically the three-stage behaviour. The predicted creep curves are in good agreement with the experimental data as shown in Figures 5.5–5.9, in a normalized scale. It should be noted that near the very end of the tertiary stage with large deformation, the true stress increases with strain exponentially rather than linearly as simplified in Eq. (5.3)–(5.5), and also sever local necking occurred in the specimens, which led to the discrepancies between the model and the observation. At the final stage of creep, internal voids or cracks may grow extensively in the creeping material such that material deformation becomes unstable. The overall excellent agreement with the experimental observations (at least up to 5% strain on the true strain scale) in a wide stress and temperature range promises the general applicability of the DMTS model for engineering applications, because the underlying deformation mechanisms are valid for all crystalline materials.

Table 5.1. DMTS Mechanism Parameters for Waspaloy

Deformation mechanism	**Parameters**				
GBS	A_0 (hr^{-1})	Q_{gbs} (kJ/mol.)	p	β	H (MPa)
	8.53 E − 9	437	10.1	1.03	13800
IDG	B_0 (hr^{-1})	Q_b (kJ/mol.)	n		
	6.58 E − 46	470	23.85		
IDC	C_0 (hr–1)	Q_c (kJ/mol.)	m	M	
	4.75 E+3	498	7.07	100	

The DMTS model in essence is a true-stress model deduced to describe creep curves under constant load conditions. Once it is calibrated with creep test behaviour, it can be used as the constitutive law in the finite element method (FEM) for component analysis. On the contrary, in a rigorous sense, engineering-stress creep models cannot be used as the constitutive law, since in FEM the stress is always the Cauchy stress. The significance of implementing a true-stress creep law is to evaluate stress re-distribution in complex 3D components, a phenomenon called "stress shake-down", which has significant implications for the component life.

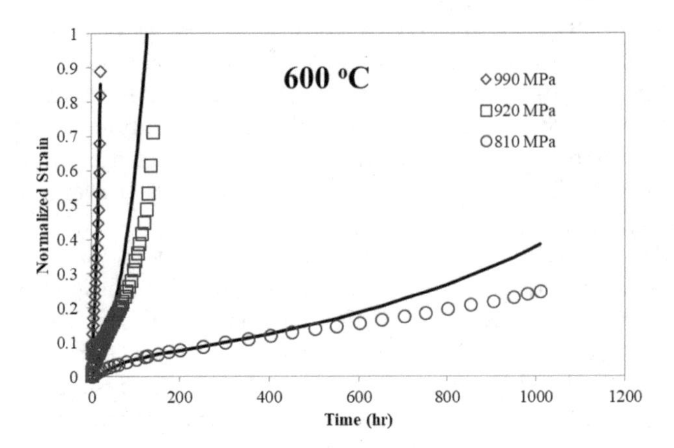

Figure 5.5. Experimental (symbols) and predicted (line) creep curves for Waspaloy at 600 °C.

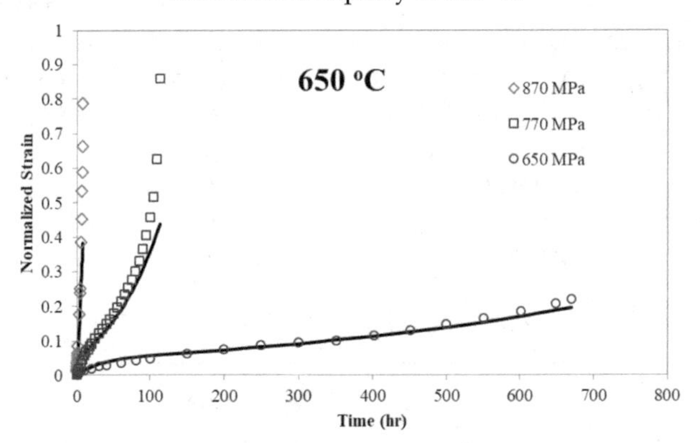

Figure 5.6. Experimental (symbols) and predicted (line) creep curves for Waspaloy at 650 °C.

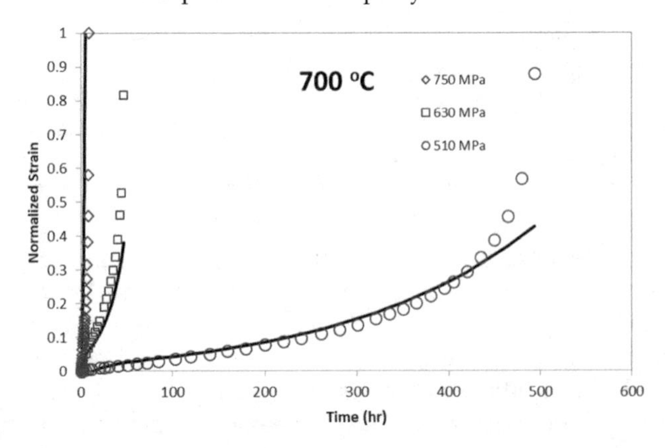

Figure 5.7. Experimental (symbols) and predicted (line) creep curves for Waspaloy at 700 °C.

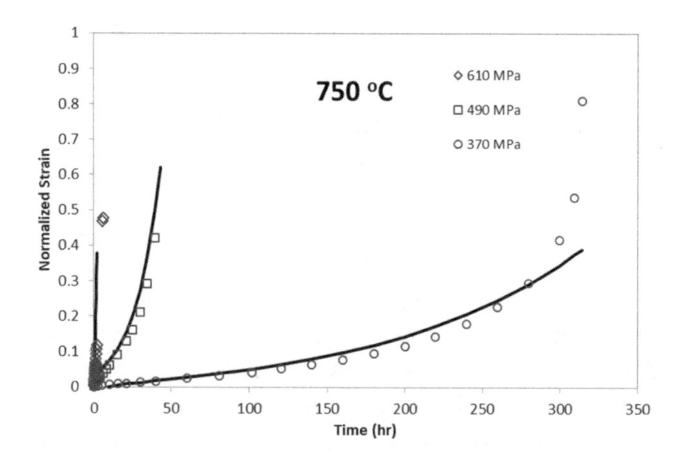

Figure 5.8. Experimental (symbols) and predicted (line) creep curves for Waspaloy at 750 °C.

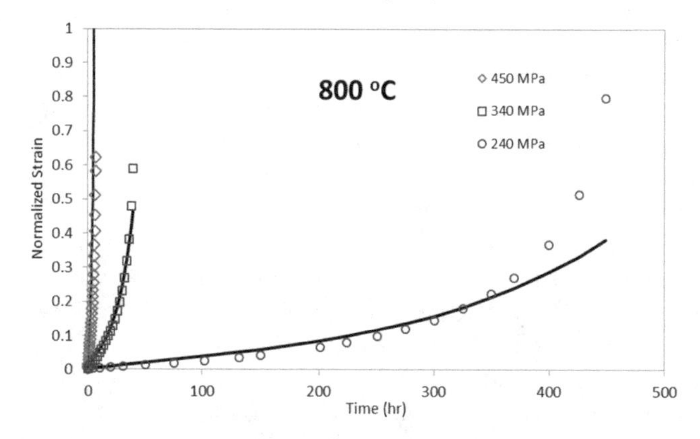

Figure 5.9. Experimental (symbols) and predicted (line) creep curves for Waspaloy at 800 °C.

5.2.3 Creep Failure Criteria and Life Prediction

It has been reviewed in Chapter 3 that creep can cause intragranular but mostly intergranular damage in the forms of cavities and microcracks along grain boundaries or at triple junctions in polycrystalline materials. Because creep damage occurs at microstructural discontinuity sites, it is called *internally distributed damage*. Apparently, the physical damage form depends on the material microstructure and the stress-temperature conditions that induce the damaging mechanism(s). Therefore, it is important to relate the specific damage form to the responsible mechanism-strain.

With regards to creep failure, we consider that creep damage proceeds in association with both intragranular deformation (ID) and GBS, leading to either transgranular, intergranular or a mix-mode fracture, for example, as observed in Waspaloy (Yao et al. 2013). In light of the above discussions, we consider that GBS and ID mechanisms compete to result in creep rupture, which occurs as an event when either GBS reaching a critical value or ID attains the intragranular ductility, whichever comes first in time. Therefore, mathematically, the creep rupture can be expressed as:

$$t_r = \min\begin{cases} t(\varepsilon_{gbs} = \varepsilon_{gbs,cr}) \\ t(\varepsilon_{g/c} = \varepsilon_{g/c,cr}) \end{cases} \tag{5.11}$$

where $\varepsilon_{gbs,cr}$ is the critical GBS strain, and $\varepsilon_{g/c,cr}$ is the intragranular ductility. $\varepsilon_{gbs,cr}$ is a microstructure dependent quantity, as GBS is dependent on the grain size, the grain boundary precipitate size and morphology (Wu and Koul 1995, 1997), whereas $\varepsilon_{g/c,cr}$ is a property of the matrix alloy.

Using Eq. (5.11), one can predicate creep rupture life as well as the failure mode. As observed from the creep curves, Figures 5.5-5.9, Waspaloy fails at a strain level close to the full normalized scale (normalized by its tensile ductility) in high-stress/low-temperature and low-stress/high-temperature regions. In the intermediate stress-temperature region, however, creep ductility is dramatically reduced by 80% (0.2 in the normalized scale). In this region transient creep is very pronounced, and therefore GBS must play a dominant role in reducing the creep ductility, leading to "brittle" creep rupture. Figure 5.10 shows the creep curve and rupture mode of Waspaloy under different creep conditions: A) at 650 °C/870 MPa, the material failed transgranularly by slip band fracture and ductile fracture, which indicates the IDG mechanism; B) at 750 °C and 350 MPa, the material failed transgranularly by ductile tearing due to dislocation climb controlled activities; C) at 700 °C/510 MPa, material failed by a mix mode with ductile tearing near the grain boundaries, which is a result of combination of GBS and IDC mechanism; and D) at 650 °C and 580 MPa, the material ruptured totally in an intergranular mode, which revealed GBS as the dominant deformation and failure mechanism. These observed failure modes corroborate the DMTS model prediction by the failure criteria Eq. (5.11).

At high temperatures, oxidation is an environmental degradation factor participating in damage accumulation in combination with ID and GBS mechanisms. Oxidation proceeds by the parabolic law:

$$x = \sqrt{2k_{ox}t} \tag{5.12}$$

Since oxides are brittle in nature, when an oxide scale forms around a cylindrical specimen, it reduces the cross-sectional area, thus effectively reduces the ductility of the material by an area fraction (obeying the incompressible flow rule):

$$\phi_{ox} = \frac{\pi D x}{\pi\left(\frac{D}{2}\right)^2} = \frac{4x}{D} \tag{5.13}$$

where D is the diameter of the specimen.

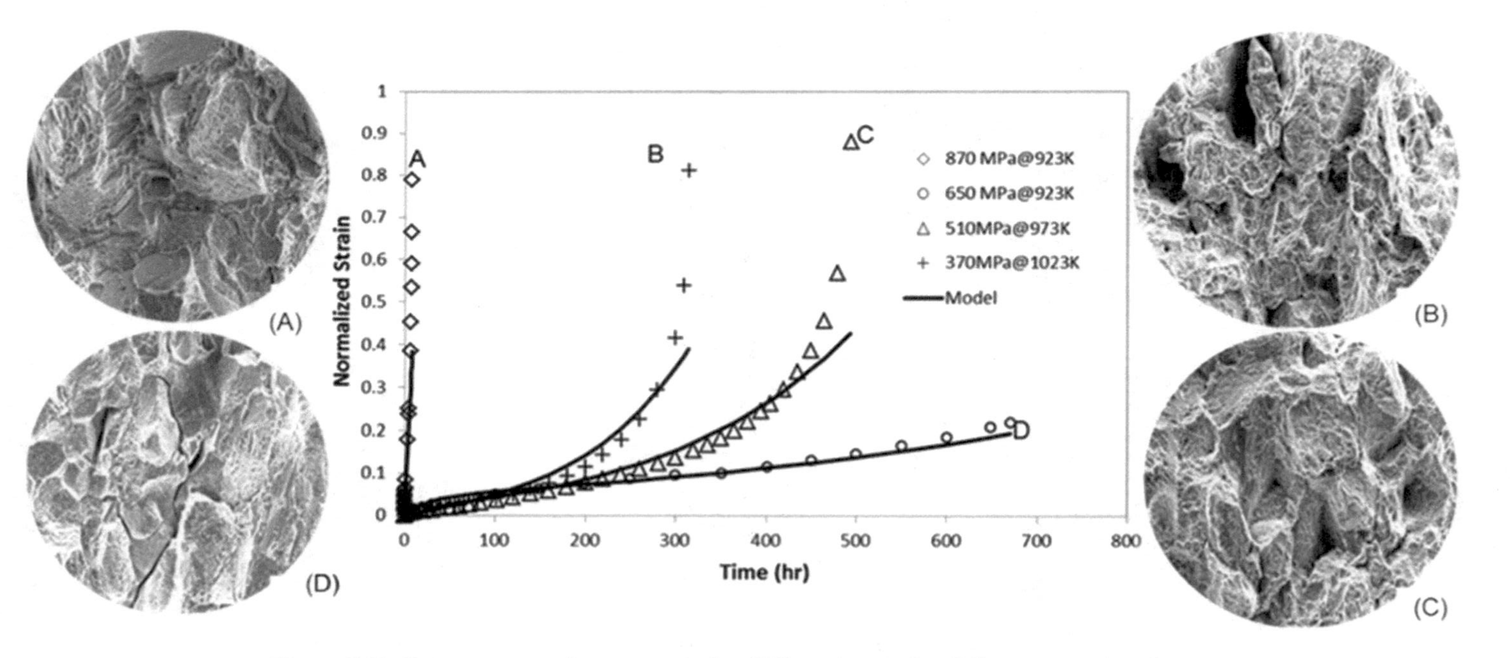

Figure 5.10. Creep curve and rupture mode of Waspaloy under different creep conditions.

With the effect of oxidation in consideration, Eq. (5.11) can be modified into:

$$t_r = \min \begin{cases} t(\varepsilon_{gbs} = \varepsilon_{gbs,cr} - \phi_{ox}) \\ t(\varepsilon_{g/c} = \varepsilon_{g/c,cr} - \phi_{ox}) \end{cases} \tag{5.14}$$

As another example, creep behaviours of Mar-M 509 under various stress and temperature conditions are simulated using Eq. (5.1)-(5.8), as shown in Figure 5.11 (Wu 2015). The model not only describes well the deformation behaviour in agreement with the experiments for this cobalt base alloy, but also explains the failure phenomena in terms of creep-oxidation interaction. Particularly for the creep cases at 871 °C/172.4 MPa and 1093 °C/48.3 MPa, it was observed that the creep rate in the former case was faster than the latter, but the creep life was much longer (983 hrs. vs. 250 hrs.), because the high temperature (1093 °C) significantly reduced the critical GBS level due to oxidation. If the Monkman-Grant relationship were used to predict the creep life of the 1093 °C/48.3 MPa case, one would obtain a longer life than the 871 °C/172.4 MPa case, because of the missing consideration of oxidation in the relationship. This is an example that an empirical relationship may infer dangerous conclusions in life prediction.

In general, with the environmental effect such as oxidation in consideration, Eq. (5.14) actually expresses a mix-mode failure criterion for creep-life prediction, which explains the variation of creep ductility with temperature and stress. With creep failure defined by mechanisms, creep life can be accurately predicted in relation to the responsible mechanism-strain.

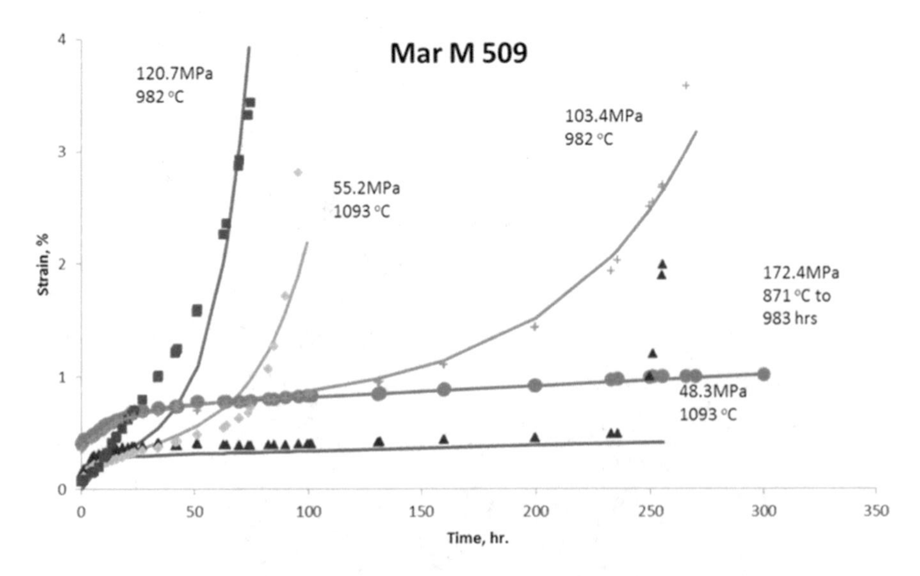

Figure 5.11. Creep curves (model) in comparison with the test data for Mar-M 509.

Figure 5.12 shows the DMTS model prediction (solid lines) of creep rupture for Waspaloy in comparison with the experiments. Figure 5.13 shows the Larson-Miller parameter (LMP) plot of both the experimental creep rupture data and the model prediction with $P = T\ (22 + \log t_r)$. Interestingly, both DMTS creep model and the LMP collapse the creep rupture data from Figure 5.12. Also, assuming constant microstructure, the creep rupture strengths of Waspaloy for 100,000 hrs at 650 °C and 750 °C are also predicted. The DMTS model and the LMP method agree well on the short-term creep

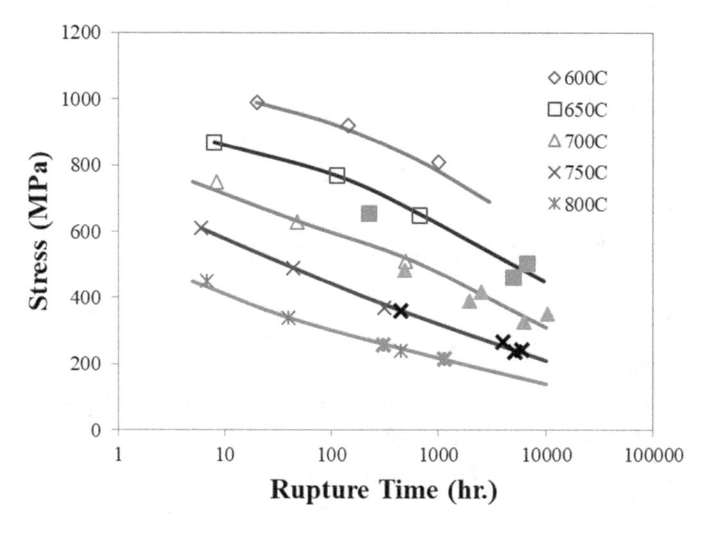

Figure 5.12. Creep rupture behaviour of Waspaloy. The open symbols are taken from Wu et al. (2012) and the solid/thick symbols are taken from Bergers et al. (2001).

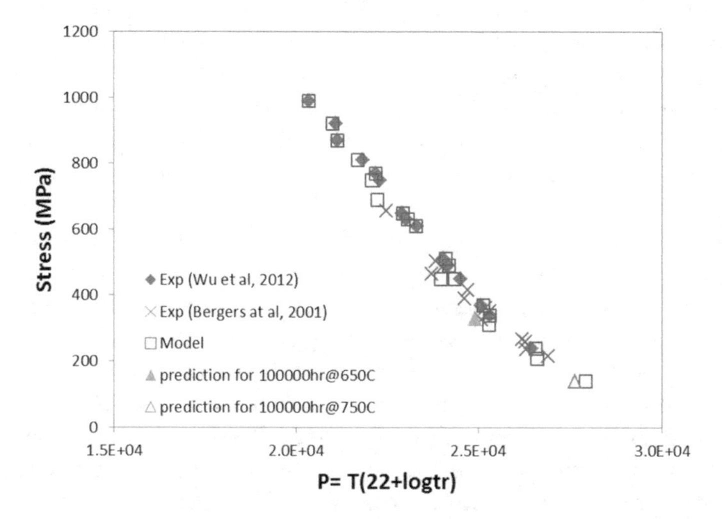

Figure 5.13. Comparison of the prediction of the mechanism-based creep model with the LMP method for Waspaloy.

rupture behaviour. In addition, the DMTS model predicted that the failure of 100,000 hrs at 650 °C would be intergranular by GBS, but that at 750 °C would be transgranular as controlled by IDC, as shown in Figure 2.15. The Larson-Miller Parameter does not provide any mechanism information.

In conclusion, the DMTS creep model in the context of Eq. (5.8) combined with the mix mode failure criteria Eq. (5.14) provides a holistic description of creep behaviour as well as lifetime and failure mode. The model predictions are in excellent agreement with the experimental observation. It also quantitatively characterizes the failure mode in term of the ratio of GBS to ID, indicating the propensity of intergranular and intragranular fracture. For long-term creep life prediction, effects of microstructural evolution and possibly oxidation should be included. By the general validity of the physical deformation mechanisms and its true-stress formulation, the model can serve as a constitutive law for component analysis.

5.3 Creep of Modified Grade 91 Steels

The 9Cr-1Mo-V-Nb (modified Grade 91) steel is a heat-resistant ferritic steel. It can be produced in various product forms such as extrusion, forging, pipes, plates and tubes, which are widely used for power-plant components such as main steam pipes and headers of ultra-supercritical steam turbines (Klueh 2005). The creep behaviour of modified Grade 91 steel is known to vary with the processing history, albeit with similar composition, as shown in the Creep Data Sheet No. 43A, National Institute for Materials Science, Japan (NIMS 2014). To understand the effects of composition, processing and microstructure, material creep behaviour needs to be analyzed using the DMTS creep model to identify the contributions of fundamental deformation mechanisms such as IDG, IDC and GBS quantitatively. This section proceeds with the mechanism quantification analysis, and it also serves as model validation for different product forms such as plate, pipe, tube and forging of Grade 91 steels. The validated model will be used to predict the long-term creep strength along with microstructural evolution and oxidation, as discussed in later sections.

The modified Grade 91 is considered as the bench mark in the 9Cr steel class, owing to its typical martensitic microstructure and precipitation hardening. Tempered martensitic 9Cr steels are generally strengthened by three mechanisms to resist creep: (1) solid-solution hardening, (2) dislocation hardening and (3) precipitation hardening. The high Cr ferric steel is normalized in an austenite regime (usually at 1040 to 1100 °C) and then cooled to room temperature. The 9-12 wt% Cr concentration enables martensitic transformation during air cooling, leading to a martensitic lath structure with a high dislocation density. The steel is then subjected to tempering at 650 to 780 °C, allowing the martensitic lath structure to change into a subgrain structure (Eggeler et al. 1989). Several types of particles such as $M_{23}C_6$, MX (M represents metal and X represents either C or N, or a combination of the two) and Laves phase $(FeCr)_2(Mo,W)$ may precipitate mostly along the lath/grain boundaries during the heat treatment. Mo and W are mostly dissolved in the solid solution, but they may also participate in the Cr-rich $M_{23}C_6$ and MX precipitation. However, after long-time exposure at elevated temperatures, a large portion of Mo or W content would be found in $M_{23}C_6$ and newly-formed Laves phase. V and Nb in 9-12% Cr steels are expected to form MX, resulting in carbides (MC), nitrides (MN), or carbonitrides [M(C,N)], which pin on the lath/grain boundaries. Metallurgical examinations have demonstrated that Laves-phase can form in 9-12% Cr steels during thermal exposure at 600- 650 °C. As suggested above, 9Cr-1Mo-V-Nb may undergo substantial microstructual changes during long-time thermal exposure, which may affect its long-term creep resistance. The growth of present precipitates and formation of new precipitates either along grain/lath boundaries or within the grain interior will draw solute atoms from the matrix, thus reducing the solution hardening effect. The research by Maruyama et al. (2001) showed the solution hardening effect of W in iron, where the creep rate of α-iron was lowered by three orders of magnitude in Fe-2.3%W alloy; both materials have no precipitates, and hence the creep resistance of the latter alloy must be attributed to solid solutioning. On the other hand, changing the grain boundary precipitates morphology would affect GBS. Therefore, it is important to understand quantitatively the contributions of intragranular deformation (ID) mechanisms as well as GBS to the overall creep resistance of the material, in order to predict its long-term creep strength along with the microstructural evolution.

NIMS has generated extensive creep data on modified Grade 91 steels (NIMS 2014). In this section, eight (8) NIMS-modified Grade 91 products and one (1) ASME SA182-01 Grade F91 forging steel are analyzed. These materials have very similar chemical composition, but undergone through different manufacturing processes and heat treatments histories as given in Table 5.2. All the NIMS materials were tempered martensitic steels. Plate MgA and MgB had the same chemical composition and manufacturing process, except the different stress relieving heat treatments: MgA was aged at 730 °C for 8.4 hours while MgB at 740 °C for 60 minutes. Tube MGD and MGF were extruded steels; F91 was forged, the microstructure of the as-received F91 is shown in Figure 5.14; all other modified grade 91 steels were hot rolled. Thermal histories and manufacturing processes endow different mechanical properties, including creep behaviours. The ultimate tensile strength (UTS) of these materials at various temperature are given in Table 5.3, which are used as the stress normalizing parameter (σ_T) in the present creep study. Creep data of the NIMS-modified 9Cr-1Mo steels were published in the Creep Data Sheet No. 43A (NIMS, 2014). The creep data of F91 were obtained from Zhang et al. (2016).

Figure 5.14. The microstructure of as-received F91 at 1000 × magnification.

Table 5.2. Heat treatment histories of NIMS samples

Type	NIMS Code	Processing and thermal history	Standard
Tube	MGD	Hot extruded and cold drawn 1050 °C/10 min AC 780 °C/40 min AC	ASME SA-213/SA-213M Grade T91
	MGF	Hot extruded 1045 °C/60 min AC 780 °C/60 min AC	
	MGG	Hot rolled 1050 °C/15 min AC 790 °C/60 min AC	
Plate	MgA	Hot rolled 1050 °C/10 min AC 770 °C/60 min AC 740 °C/8.4 h FC	ASME SA-387/SA-387M Grade 91
	MgB	Hot rolled 1050 °C/10 min AC 770 °C/60 min AC 740 °C/60 min FC	
	MgC	Hot rolled 1060 °C/90 min AC 760 °C/60 min AC 730 °C/8.4 h FC	
	MgD	Hot rolled 1050 °C/30 min AC 780 °C/30 min AC	

Pipe	MGQ	Hot rolled 1060 °C/60 min AC 780 °C/60 min AC	ASME SA-335/SA-335M Grade P91
F91		Forged 1080 °C/4 hr AC 800 °C/5 hr AC	ASME SA182-01 Grade F91

Note: AC = air cooling, FC = furnace cooling

Table 5.3. Ultimate tensile strengths (σ_T) of NIMS samples at various temperatures

NIMS reference code		**Ultimate tensile strength (MPa)**							
		450 °C	**500 °C**	**550 °C**	**575 °C**	**600 °C**	**625 °C**	**650 °C**	**700 °C**
Tube	MGD	N/A	478	412	375*	344	305*	267	199
	MGF	N/A	487	418	381*	347	309*	270	201
	MGG	N/A	489	419	379*	340	306*	266	200
Plate	MgA	508	444	399	N/A	320	N/A	266	N/A
	MgB	506	477	417	N/A	330	N/A	277	N/A
	MgC	523	480	427	385*	355	326*	290	N/A
	MgD	N/A	478	413	373*	343	308*	268	N/A
Pipe	MGQ	N/A	464	402	366*	333	300*	266	N/A
F91		N/A	475#	413#	N/A	339#	N/A	271#	N/A

Note: *values are interpolated; #mean values of the columns.

For the modified Grade 91 steels, it is found that normalization of the applied engineering stress by UTS at test temperature is convenient to categorize the stress regions from low to high: ≤0.35 is designed as the low stress region; 0.35-0.5 as the intermediate stress region; ≥0.6 as the high stress region, corresponding to the dominance of certain deformation mechanisms as shown in Figure 2.4. Modified Grade 91 steels are designed to serve under 750 °C, therefore, 600-750 °C is considered as the high temperature range, 500-600 °C is considered as the intermediate temperature range, and 450–500 °C is considered as the low temperature range.

Generally, the IDG mechanism dominates in the high-stress region at low temperatures; IDC mechanism dominates in the low-stress region at high temperatures; whereas GBS mechanism usually takes place in between. Within their dominant regions, IDC, IDG and GBS assume a constant power exponent, m, n, p, respectively. In analyzing the creep rate data, the model parameters, A_0, B_0, C_0 and Q_{gbs}, Q_g, Q_c of Eqs. (5.3) to (5.5), are obtained using the Arrhenius plots, for GBS, IDG and IDC, respectively, and are given in Table 5.4. The total steady-state creep rates are then calculated, according to Eq. (5.7c). The predicted steady-state creep rate versus stress curves are compared with the experimental data, as shown in Figure 5.15 to Figure 5.22, for various modified 9Cr-1Mo steel products. The creep rate vs. stress curves are in good agreement with the experimental data—the average coefficient of determination (R^2) of all fitting lines is 0.99. This demonstrates that the DMTS model is applicable for the NIMS-modified Grade 91 steels of various product forms.

Table 5.4. DMTS mechanisms parameters for 9Cr-1Mo-V-Nb steels (σ_T = UTS)

Type	NIMS Code	IDC			GBS			IDG		
		C_0	Q_c		A_0	Q_{gbs}	p	B_0	Q_g	n
Tube	MGD	6.48E+19	378360	6.6	6.65E+20	384916	10.1	5.02E+26	416042	18.8
	MGF	1.18E+20	413953	4	1.42E+23	392762	11.9	3.32E+27	430377	18.3
	MGG	2.65E+17	346811	5.5	2.88E+23	384808	14.5			
Plate	MgA	3.00E+23	439995	6.4	2.64E+27	428738	17.3	9.48E+32	487460	25.7
	MgB	3.55E+25	475105	6	9.65E+26	423139	16.3	1.20E+34	493867	28.8
	MgC	1.02E+17	337118	6.2	3.53E+25	413529	14.7	2.14E+29	439903	24
	MgD	6.36E+17	348076	6	1.33E+22	350738	16.5			
Pipe	MGQ	3.09E+19	380931	6	1.34E+26	413596	16.5			
F91		1.48E+23	455345	4.5	1.97E+29	465346	15.6	7.79E+33	488184	28.8

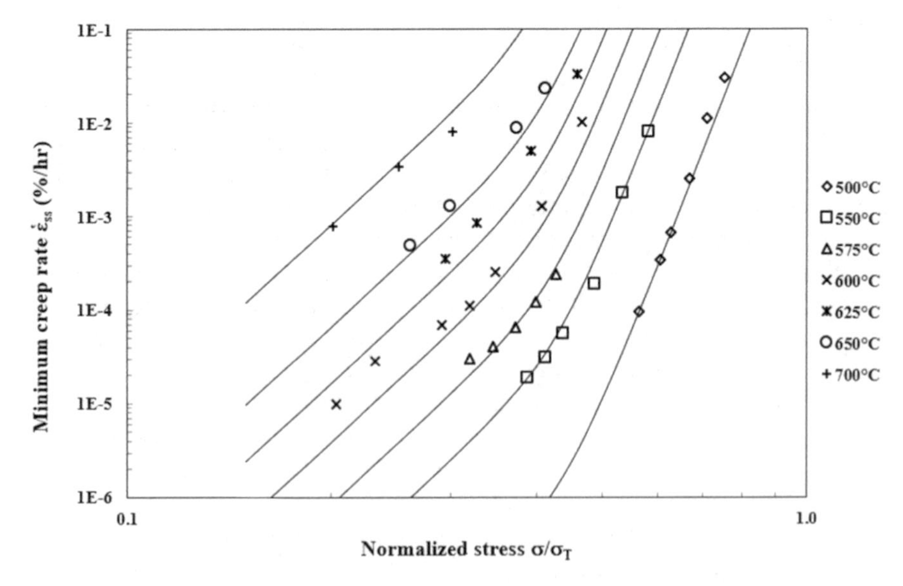

Figure 5.15. Model fitting for NIMS Tube MGD (symbols: experimental data; lines: model).

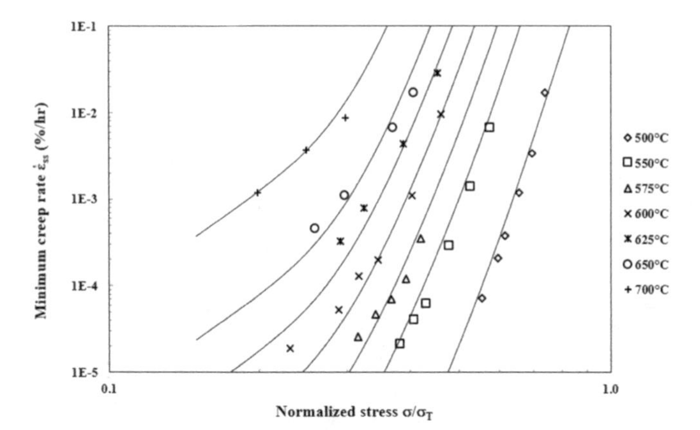

Figure 5.16. Model fitting for NIMS Tube MGF (symbols: experimental data; lines: model).

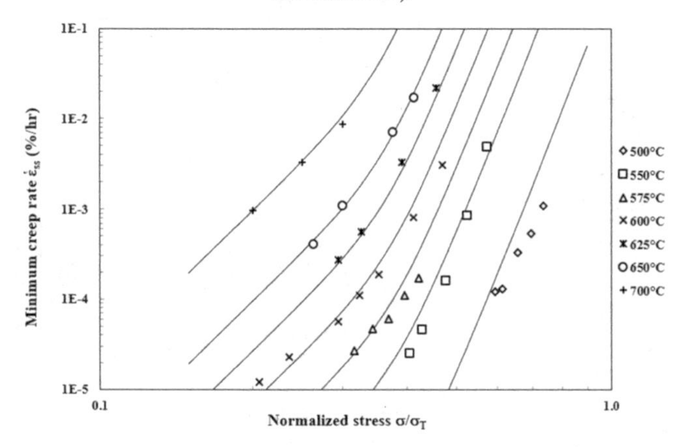

Figure 5.17. Model fitting for NIMS Tube MGG (symbols: experimental data; lines: model).

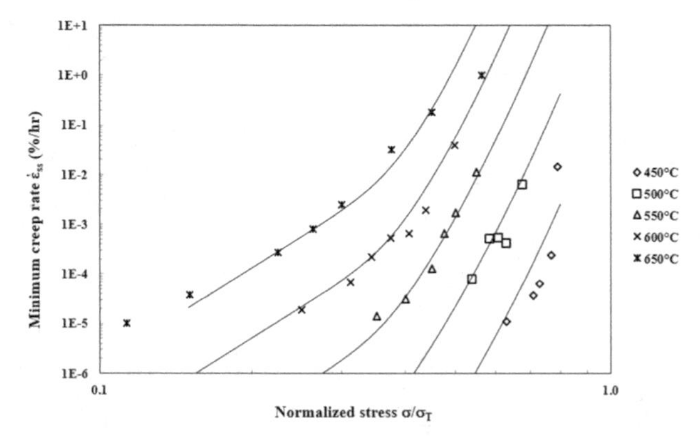

Figure 5.18. Model fitting for NIMS Plate MgA (symbols: experimental data; lines: model).

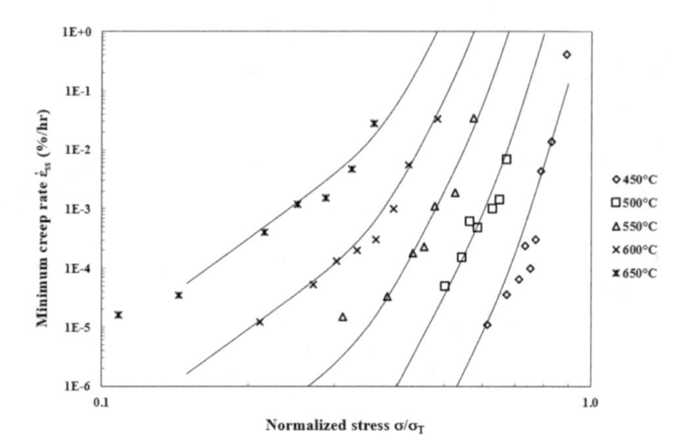

Figure 5.19. Model fitting for NIMS Plate MgB (symbols: experimental data; lines: model).

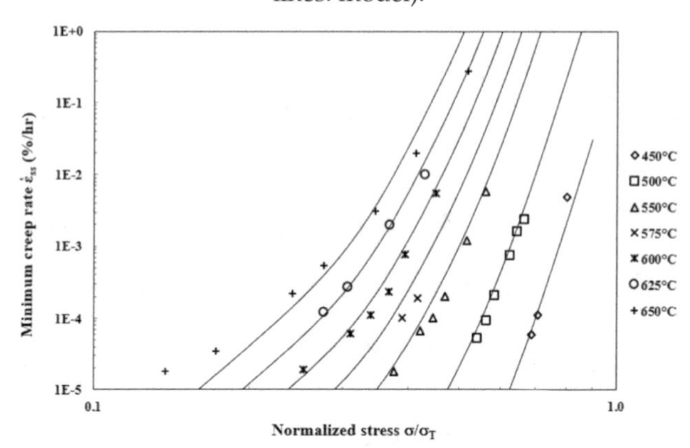

Figure 5.20. Model fitting for NIMS Plate MgC (symbols: experimental data; lines: model).

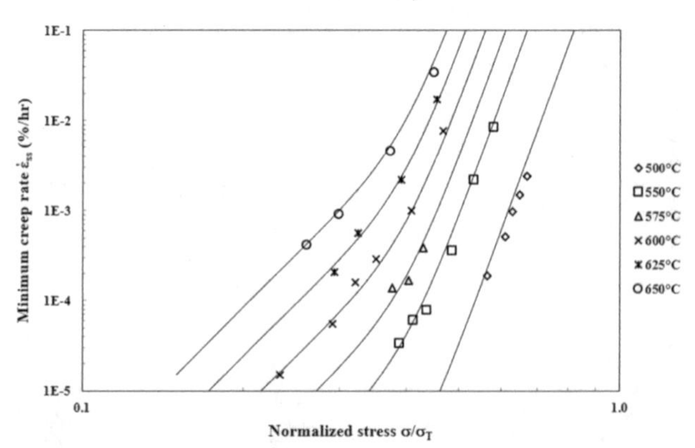

Figure 5.21. Model fitting for NIMS Plate MgD (symbols: experimental data; lines: model).

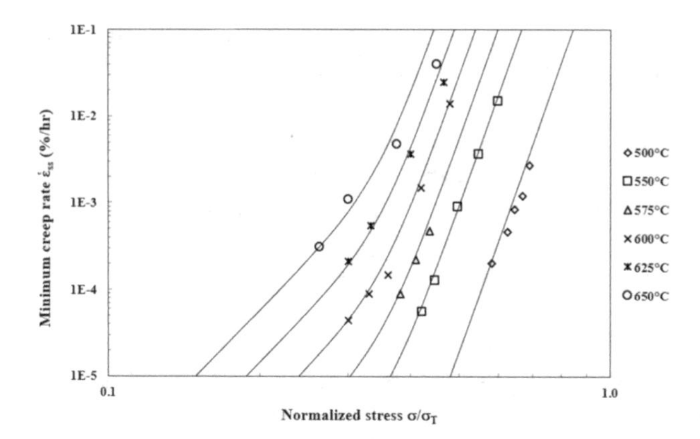

Figure 5.22. Model fitting for NIMS Pipe MGQ (symbols: experimental data; lines: model).

Among the NIMS-modified Grade 91 steels, three product forms were manufactured according to different standard codes. Tubes for boilers and heat exchangers followed ASME SA-213/SA-213M Grade T91; plates for boilers and pressure vessels followed ASME SA-387/SA-387M Grade 91, and the seamless pipes for high-temperature service were manufactured according to ASME SA-335/SA-335M Grade P91. The materials manufactured following the same standard exhibit the similar creep behaviour. By comparison, the power exponent p of GBS is about 12 for tubes, 16 for plates and the pipes; while the power exponent n of IDG is about 18 for tubes and 26 for plates. However, the power exponent m of IDC is approximately 6 for all modified Grade 91 steels. The activation energies of GBS and IDG follow the same trend, i.e., that for plates are higher than tubes. The pronounced differences in the power exponent and activation energy indicate that manufacturing process affects the creep behaviour of the same grade steels. The hot-rolled and forged microstructures are heavier "worked" containing a higher dislocation density, creating larger barriers for glide either within the grain interior (affecting IDG) or along grain or lath boundaries (affecting GBS). On the other hand, IDC is a diffusion-controlled mechanism, its stress dependence is less affected by thermomechanical processing. Furthermore, discrepancies in the parameter values are shown for the same type products with different forming processes. For instance, tube MGG is hot rolled while MGD and MGF are hot extruded, MGG does not show the high-power exponent in IDG. Pipe MGQ shows the same phenomena with only IDC and GBS to control the minimum creep rates.

All the NIMS-modified Grade 91 steels are tempered martensitic steels. The normalizing temperature range was from 1045 °C to 1060 °C, and the tempering temperature varied between 760 °C and 780 °C. Also, plate MgA, MgB, and MgC had a history of stress relieving heat treatment at 730 °C or

740 °C, from 1 hour to 8.4 hours, with furnace cooling. It appears that hot rolled products tempered at 780 °C and 790 °C do not show an IDG region. This is perhaps due to the reason that tempering at higher temperatures relieves the martensitic laths to allow more inter-lath or grain boundary sliding. According to the NIMS record, plate MgA and MgB are identical, except that MgA experienced a much longer stress relieving heat treatment than MgB at the same temperature. The creep behaviour of both is very similar, which means that the duration of the stress relief does not have significant effect on creep.

To fully characterize the creep behaviour of F91, the available experimental data on F91 are collated with the data on MgB for two reasons: (1) the two materials have similar heat treatment history; (2) the creep rates and rupture lives of the two materials under the same test conditions are the closest among various forms of Grade 91 steels. Then, the data for both materials are combined to validate the model parameters for F91. The calibrated model for F91 is illustrated in Figure 5.23. The fitting lines are in good agreement with the experimental data, with an average coefficient of determination $R^2 = 0.88$.

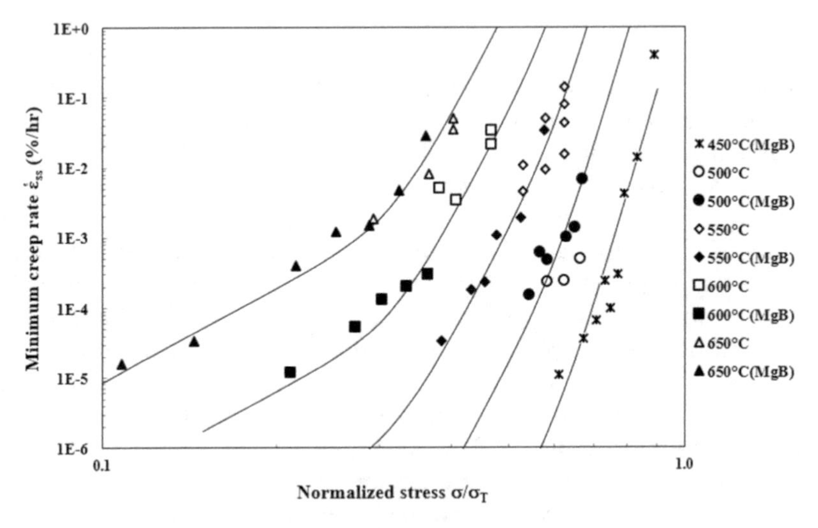

Figure 5.23. Model fitting for F91 and NIMS Plate MgB (symbols: experimental data; lines: model).

5.4 Creep of MCrAlY-Coated F91

A series of creep tests have been conducted on MCrAlY-coated F91 coupons at 550 °C and 600 °C (Zhang et al. 2019). The MCrAlY coating was deposited on F91 coupon with a coating thickness of 280 μm, using the high velocity oxyfuel (HVOF) thermal spray technique. The coating was dense and adhered to the F91 substrate (no spallation) during the entire creep testing. There was no oxidation observed in all ruptured coupons. Hence, the MCrAlY coating and F91 together can be considered as a composite system to carry the total creep load. To evaluate the stress distribution in such MCrAlY-F91 composite, the following assumptions are made:

- The coating layer deforms compatibly with F91 substrate;
- Stress in the coating layer distributes uniformly;
- Stress in the F91 substrate distributes uniformly.

In this case, upon initial loading when deformation is purely elastic, the classical composite rule applies:

$$F_{app} = F_s + F_c \tag{5.15a}$$

and

$$\frac{\sigma_s}{E_s} = \frac{\sigma_c}{E_c} \tag{5.15b}$$

where $F_s = \sigma_s \cdot A_s$, $F_c = \sigma_c \cdot A_c$; E_s and E_c are Young's modulus of F91 steel and coating, respectively; and F_{app} is the applied load.

From Eq. (5.15), the stresses in steel and coating are obtained, respectively:

$$\sigma_s = \frac{E_s}{E_s \cdot A_s + E_c \cdot A_c} \cdot F_{app} \tag{5.16}$$

$$\sigma_c = \frac{E_c}{E_s \cdot A_s + E_c \cdot A_c} \cdot F_{app} \tag{5.17}$$

Young's modulus of Grade 91 steels and MCrAlY coating at elevated temperature can be found or extracted from the literature (ASME International, 2016, Saeidi et al. 2011). For 9Cr steels, E_s = 174 GPa @ 550 °C and 168 GPa @ 600 °C, respectively. The Young's moduli of HVOF-MCrAlY coating are E_c = 66 GPa @ 550 °C and 51 GPa @ 600 °C. The calculated σ_s and σ_c by Eq. (5.16) and Eq. (5.17) along with the minimum creep rate ($\dot{\varepsilon}_m$) and time to rupture (TTR) of each test are listed on Table 5.5.

With coating protection, there is no oxidation of the substrate, so the coated coupon creep curve reflects the intrinsic creep behaviour of the substrate material under the given stress σ_s condition. The DMTS creep model is then applied to analyze the creep data of the coated F91 coupons, to determine the oxidation-free mechanism parameter. Through best-fitting to the experimental creep-curve, the work hardening coefficient H and the dislocation multiplication factor (M) are determined, which are given in Table 5.6. The β = 1.008 is determined to be a constant for all the coupons with the same microstructure (actually, all the above modified Grade 91 steels have similar microstructures).

Table 5.5. Stress distribution in coated-coupons for given creep conditions

Temp	Nominal stress on uncoated coupon	F_{app}	σ_s	σ_c	$\dot{\varepsilon}_m$	TTR
°C	MPa	N	MPa	MPa	%/hr	hr
	260	3202.14	233	88	4.70E-05	1256.4
550	240	2955.82	215	82	3.79E-05	1886.3
	220	2709.51	197	75	2.99E-05	4178.3
	160	1970.55	147	44	6.56E-04	780.1
600	140	1724.23	128	39	3.56E-04	3601.5
	130	1601.07	119	36	2.13E-04	6112.7

In the F91-coating composite, the mechanism strain rates are calculated from Eq. (5.3)-(5.5) based on the true stress in the substrate σ_s and the mechanism parameters for F91, as given in Table 5.4. Then, the creep curves are calculated using Eq. (5.8) based on the parameters given in Table 5.6. The calculated creep curves of the coated F91 coupons are shown in Figures 5.24 and 5.25 for the two temperature conditions, 550 °C and 600 °C, respectively, which are in good agreement with the experimental observations. The DMTS model also infers that GBS dominates in the stress/UTS ratio range of 0.35-0.6. Figure 5.26 shows the microcturctures of crept F91 in the unnecking region at the time of creep rupture. It is seen that at high stress/low temperature ~0.63UTS/550 °C, the microstructure contains elongated laths and large intragranular voids, which is indicative of intragranular deformation by IDG; at the intermediate stress/intermediate temperature ~0.41-0.47UTS/600 °C, the microstructure retained the original shape of the martensitic laths, as compared with the as-received microstructure shown in Figure 5.14, but small voids can be observed at the grain boundaries, which is indicative of GBS; and at low stress/high temperature ~0.4UTS/650 °C, the lath-structure becomes elongated too, indicative of intragranular deformation by IDC. The DMTS model not only matches the creep curves but also indicate the failure mode, as corroborated with the metallurgical evidence. Then, it is proven that the DMTS model is sutable to describe the creep behavior of F91 with the identified GBS, IDC, and IDG deformation mechanisms.

Table 5.6. Creep curve parameter values

Temp	Dislocation multiplication factor M	Work hardening coefficient H
°C		GPa
500	300	47.16
550	200	41.15
575	120	30.56
600	110	25.41
625	10	17.41
650	5	8.58

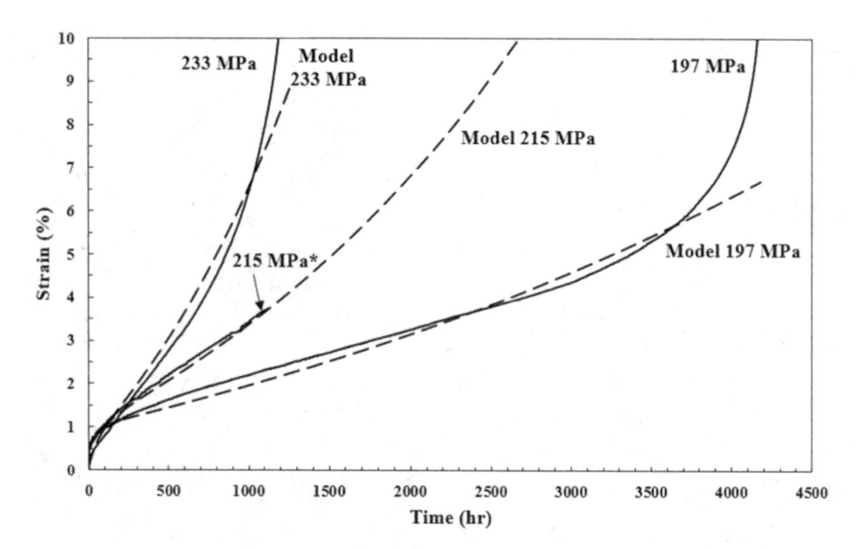

Figure 5.24. Creep behaviour of coated F91 coupons at 550 °C and the DMTS model description.

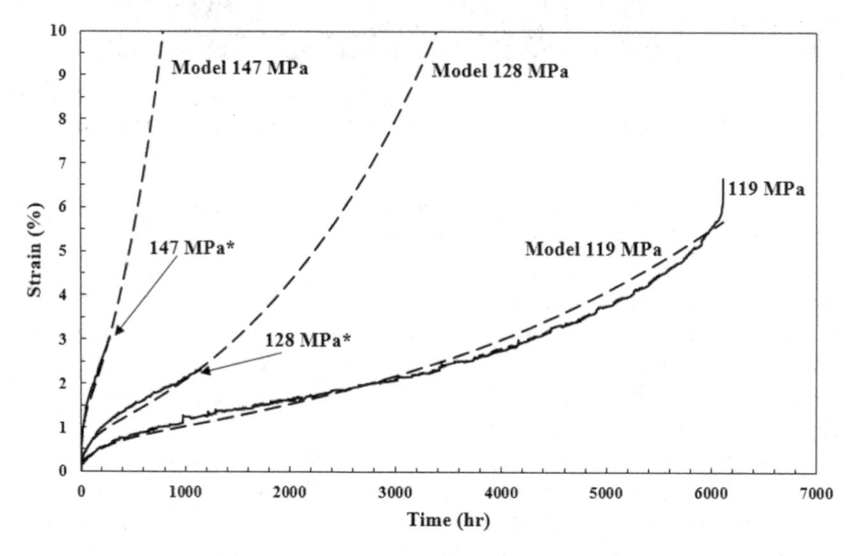

Figure 5.25. Creep behaviour of coated F91 coupons at 600 °C and the DMTS model description.

5.5 Oxidation-Modified DMTS Model and Validation

Under creep test conditions in stagnant air, metals can be oxidized, which would obviously have an impact on creep strain and life. Ashby and Dyson (1984)

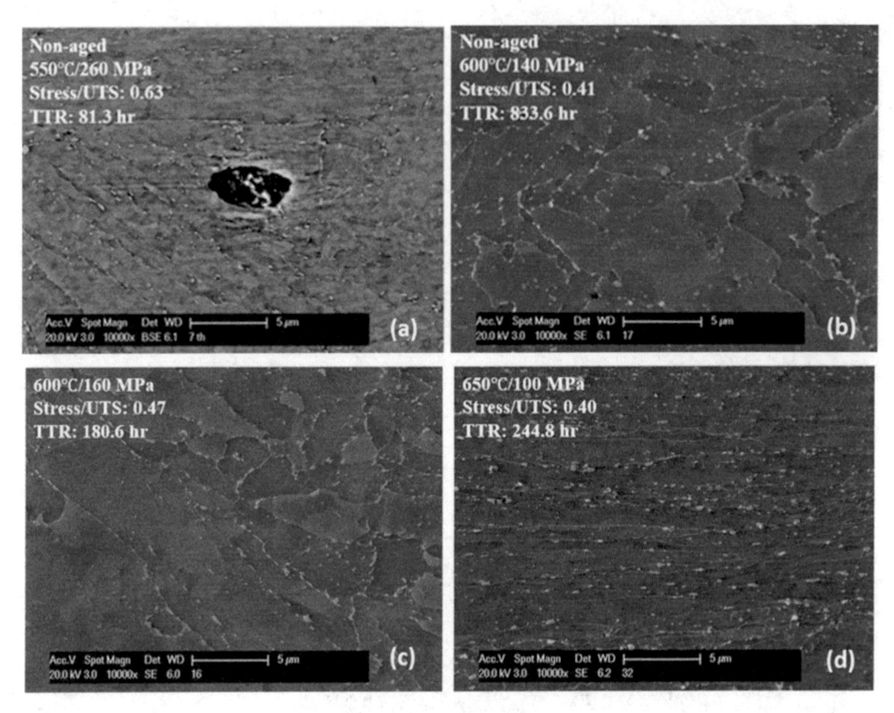

Figure 5.26. SEM microstructure of longitudinal cross-section creep-ruptured at a) 550 °C/260 MPa, b) 600 °C/140 MPa, c) 600 °C/160 MPa, d) 650 °C/100 MPa.

and Dyson and Osgerby (1995) considered creep-environment interaction in continuum damage-mechanics. A full-extent model in connection to deformation mechanisms has yet to be developed. Ideally, oxidation-free creep data are needed to know to delineate the oxidation effect. Some short-term creep tests were conducted on 2¼Cr-1Mo steel in vacuum for comparison with air (Bueno, 2008). Such oxidation-free creep data are rarely available for most materials, because long-term vacuum creep tests are cost-prohibitive. Conventional creep analyses are mostly done to creep data with coupon-borne influence of oxidation. Therefore, empirical extrapolation of short-term creep data for long-term creep life prediction without considering oxidation is questionable, because oxidation is a time-dependent process. Therefore, it is necessary to separate the oxidation effect from materials' intrinsic creep mechanisms. For example, massive creep rupture life data generated by NIMS indicated that there is a "life breakdown" from short-term creep test extrapolation for Grade 91 steels, especially at high temperatures (Kimura et al. 2011, Kimura and Tabuchi 2012, Masatsugu et al. 2016). The life breakdown phenomenon indicates the existence of other life-limiting factors for long-term creep—oxidation is apparently one. In this section, we will modify the DMTS model with oxidation-reduced cross-section to address its effect on creep rate, creep strain-time curve as well as rupture life.

5.5.1 Model Development

Metal oxidation depends on its chemical composition and temperature. Some metals such as aluminum may form an oxide (Al_2O_3) scale that is dense and protective. Some metals such as steels form iron-oxides (a mixture of Fe_3O_4, FeO and Fe_2O_3) that is loose and often spall off the specimen surface during creep testing (Yurechko et al. 2011).

The growth of oxide scale, δ, follows the parabolic law as expressed by:

$$\delta = \sqrt{2k_{ox}t} \tag{5.18}$$

where k_{ox} is the oxidation rate coefficient, which is an Arrhenius-type constant; t is the exposure time.

For generality, we assume that oxide scale formation reduces the specimen load-bearing area, as given by

$$A_x = A_0 - 2\pi r\delta \tag{5.19}$$

then Eq. (5.2) can be re-written as:

$$\sigma = \frac{P}{A_x}\exp(\varepsilon) \tag{5.20}$$

where, P is load, A_x is the cross-section area of specimen affected by oxidation, A_0 is the original cross-section, and r is the original radius of the specimen.

Substitute Eq. (5.19) to Eq. (5.20):

$$\sigma = \frac{P}{A_0 - 2\pi r\delta}\exp(\varepsilon) = \frac{P}{A_0(1-\omega_{ox})}\exp(\varepsilon) = \frac{\sigma_a}{1-\omega_{ox}}\exp(\varepsilon) \tag{5.21}$$

where ω_{ox} is the ratio of area loss due to oxidation defined as

$$\omega_{ox} = \frac{2\pi r\delta}{\pi r^2} = \frac{2\delta}{r} \tag{5.22}$$

Substituting Eq. (5.21) to Eqs. (5.3)-(5.5), we have:

$$\dot{\varepsilon}_{gbs} = A\sigma^p = A\left(\frac{\sigma_0}{1-\omega_{ox}}\right)^p \exp(p\varepsilon) \approx \frac{1+p\varepsilon}{(1-\omega_{ox})^p}A\sigma_0^p = \frac{1+p\varepsilon}{(1-\omega_{ox})^p}\dot{\varepsilon}_{s0} \tag{5.23}$$

$$\dot{\varepsilon}_g = B\sigma^n = B\left(\frac{\sigma_0}{1-\omega_{ox}}\right)^n \exp(n\varepsilon) \approx \frac{1+n\varepsilon}{(1-\omega_{ox})^n}B\sigma_0^n = \frac{1+n\varepsilon}{(1-\omega_{ox})^n}\dot{\varepsilon}_{g0} \tag{5.24}$$

$$\dot{\varepsilon}_c = (1+M\varepsilon)C\sigma^m = (1+M\varepsilon)C\left(\frac{\sigma_0}{1-\omega_{ox}}\right)^m \exp(m\varepsilon) \approx \frac{1+M\varepsilon+m\varepsilon}{(1-\omega_{ox})^m}C\sigma_0^m$$

$$= \frac{1+M\varepsilon+m\varepsilon}{(1-\omega_{ox})^m}\dot{\varepsilon}_{c0} \tag{5.25}$$

Substituting Eq. (5.23)-(5.25) into Eq. (5.1), we have

$$\dot{\varepsilon} = \dot{\varepsilon}_{gbs} + \dot{\varepsilon}_g + \dot{\varepsilon}_c = \frac{1+p\varepsilon}{(1-\omega_{ox})^p}\dot{\varepsilon}_{s0} + \frac{1+n\varepsilon}{(1-\omega_{ox})^n}\dot{\varepsilon}_{g0} + \frac{1+M\varepsilon+m\varepsilon}{(1-\omega_{ox})^m}\dot{\varepsilon}_{c0} \tag{5.26}$$

Redefine the oxidation-affected creep parameters, as:

$$M^* = \left[\frac{\frac{p}{(1-\omega_{ox})^p}\dot{\varepsilon}_{s0} + \frac{n}{(1-\omega_{ox})^n}\dot{\varepsilon}_{g0} + \frac{m+M}{(1-\omega_{ox})^m}\dot{\varepsilon}_{c0}}{k}\right] \tag{5.27a}$$

$$k = \frac{1}{(1-\omega_{ox})^p}\dot{\varepsilon}_{s0} + \frac{1}{(1-\omega_{ox})^n}\dot{\varepsilon}_{g0} + \frac{1}{(1-\omega_{ox})^m}\dot{\varepsilon}_{c0} \tag{5.27b}$$

Then Eq. (5.8) still holds for the total creep strain, as expressed by:

$$\varepsilon = \varepsilon_0 + \frac{\sigma}{\beta^2 H}\left[1-\exp\left(-\frac{\beta^2 H\dot{\varepsilon}_{s0}t}{\sigma(\beta-1)}\right)\right] + \frac{1}{M^*}[\exp(M^*kt)-1] \tag{5.28}$$

The above equations reduce to the basic DMTS model when oxidation is absent. Note here that the oxidation effect on the primary creep is ignored, because the time is too short. Thus, this section attempts to evaluate the impact of oxidation on creep life breakdown of modified Grade 91 steels, by incorporating the oxide growth into the DMTS model, henceforth called the oxidation-modified DMTS (O-DMTS) model.

5.5.2 Oxide Scale Measurement and Evaluation

When a metal is exposed to high temperature under a mechanical load, oxide scales can form but may not always adhere to the metal surface. Direct measurement of the oxide scale thickness on sectioned coupons is not always accurate, because portions of the oxide scale can spall off during creep and specimen handling. Here, an area deduction method is proposed, as schematically shown in Figure 5.27, by which the oxidized area ($2\pi r\delta$, r is the radius of the remaining cross-section, δ is the oxide layer thickness) is obtained as the original deformed specimen cross-section area minus the remaining un-oxidized area, as given by

$$2\pi r\delta = \frac{A_0}{1+\varepsilon_{cr}} - A_r \tag{5.29}$$

where ε_{cr} is defined as the critical creep strain attained at 90% of the time-to-rupture (TTR), just before necking unstability occurred.

For creep ruptured coupons, the cross-section should be chosen away from the necking region and ε_{cr} should be the strain level on the creep curve before necking takes place. For consistency, we set this to be the strain at 90% TTR when deformation of the entire specimen still remains stable. Figure 5.28 shows the cross-sectional view of selected coupons after creep tests at temperatures from 500 °C to 650 °C. The cross-sections were cut off from the gauge sections away from the necking region, such that they represent scenarios of F91 oxidation during stable creep deformation. During the creep tests, oxidation occurred all over the specimen surface in the gauge section, and it became increasingly severe as the temperature was increased. However, because the iron oxides were loose and brittle, most of the scales spalled off during either testing or sectioning. For examples, as seen in Figure 5.28, a thin oxide layer was observed at the coupon surface of 500 °C and 550 °C-exposed coupons; a thicker oxide scale was observed in the 600 °C-exposed coupon; in the 650 °C-exposed coupon, oxidation products could spall off repeatedly, which resulted in an irregular surface layer, as opposed to the perfect circular cross-section.

Once the oxide scale thickness is estimated from Eq. (5.29), the oxide rate coefficient (k_{ox}) at temperature can be determined using Eq. (5.18) for each ruptured specimen (t = TTR). The average k_{ox} at temperature is assumed to obey the Arrhenius relationship, as shown in Figure 5.29, from which the activation energy and proportional constant of k_{ox} are determined, in unit of $\mu m^2/h$, as

$$k_{ox} = 2.04 \times 10^9 \, e^{-19698/T} \tag{5.30}$$

5.5.3 Creep Strain-Time Behaviors of F91

For comparison, both the basic DMTS model, Eq. (5.8) and the oxidation-modified DMTS (O-DMTS) model, Eq. (5.28), are used to describe the uncoated F91 creep curves, as shown in Figures 5.30-5.33. By its physical meaning, the O-DMTS model represents the air-creep behaviour, and the DMTS model represents the material-intrinsic creep behaviour without influence of oxidation. The O-DMTS model is shown to describe the strain-

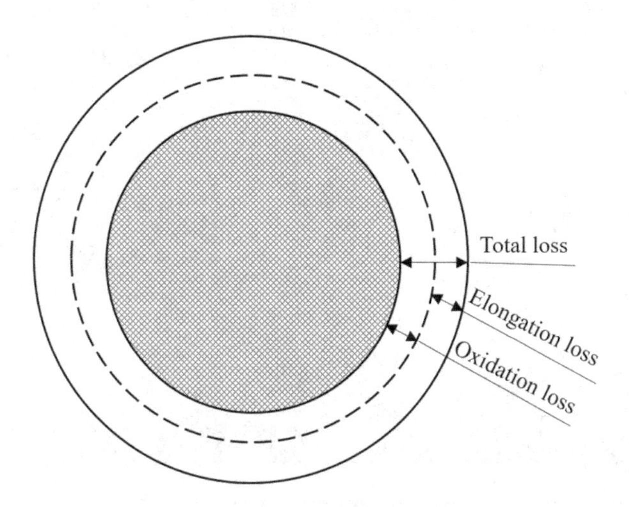

Figure 5.27. Schematic drawing of the ruptured coupon cross-section.

time behaviour of F91 during creep at all temperatures from 500-650 °C better than the original DMTS model. First of all, the creep rate is increased because of the oxidation effect, indicating that in addition to the material-intrinsic deformation mechanisms, GBS, IDC, and IDG, oxidation has a significant effect, especially for long time creep, where the oxide growth becomes more and more influential. Similar trend was observed on $\dot{\varepsilon}_v$ creep rates of 2¼Cr-1Mo steel as compared in air and vacuum (Bueno, 2008). For short times up to onset of minimum creep rate, the two models are not much different, but for long-time creep behaviour, the O-DMTS is especially better than the basic (oxidation-free) DMTS model. This means that the minimum creep rate during short-term creep tests can be regarded closely as material intrinsic behaviour, but the effect of oxidation cannot be ignored for prediction of long-term creep behaviour in air. Otherwise, the prediction can result in gross error. This point has not been emphasized enough in the literature of creep.

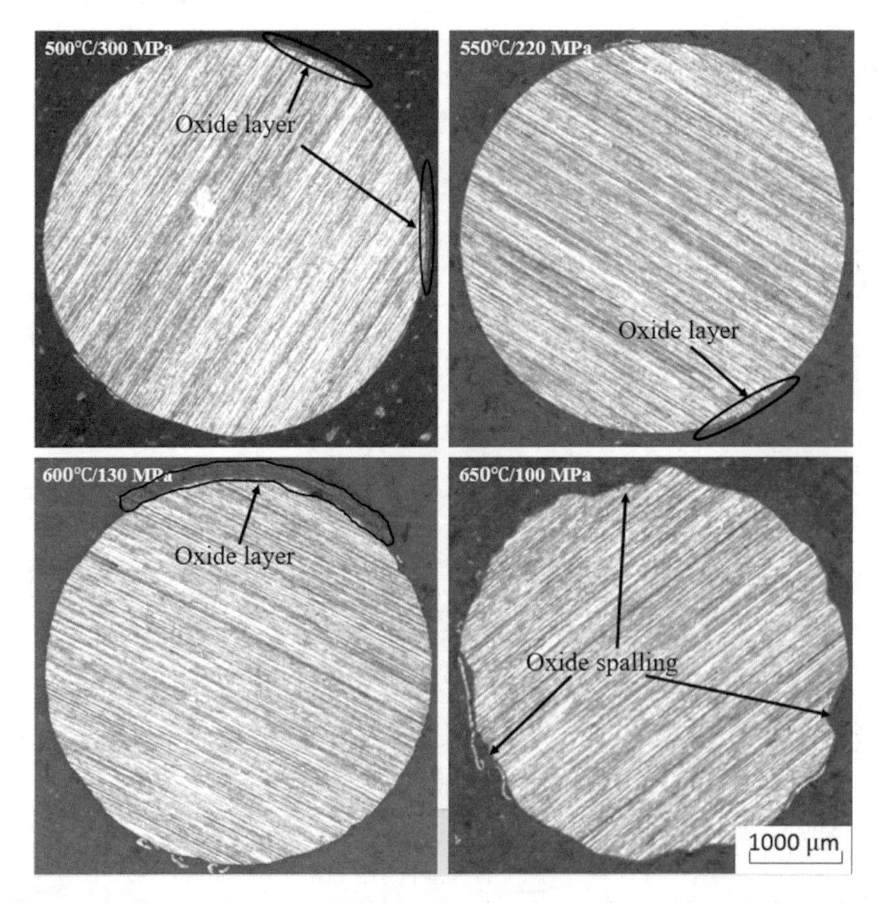

Figure 5.28. Ruptured uncoated F91 coupons showing surface oxide scale formation, observed using an optical microscope.

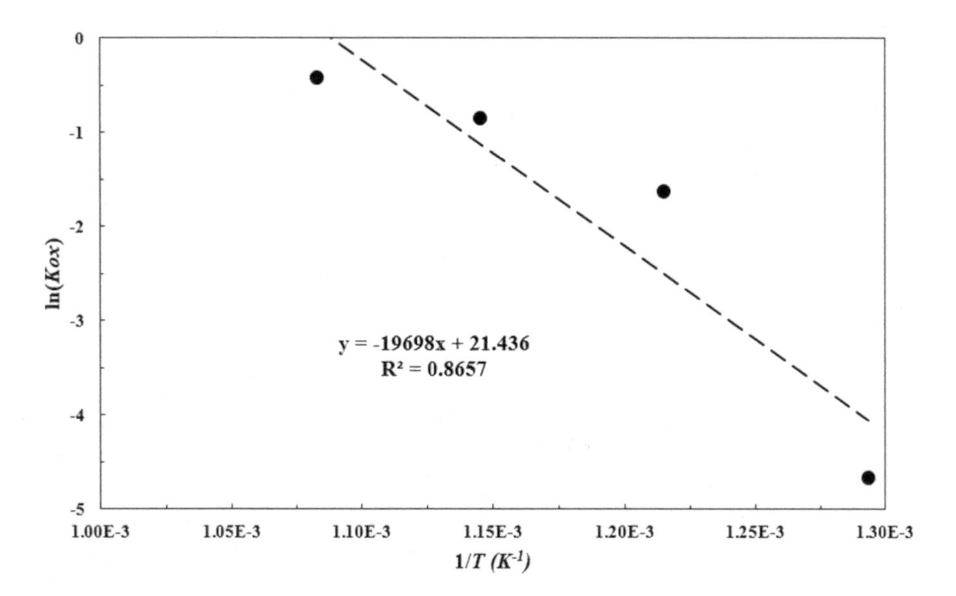

Figure 5.29. Arrhenius-plot of K_{ox} for determination of activation energy and proportinal constant.

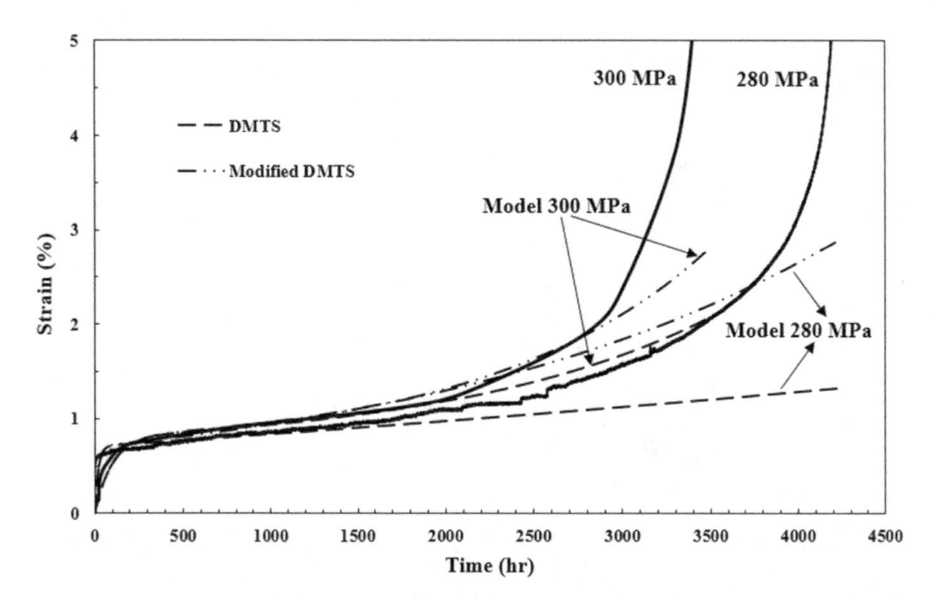

Figure 5.30. Comparison of the basic and modified DMTS models for creep strain-time curve of uncoated F91coupons at 500 °C. The solid line represents experimental behaviour.

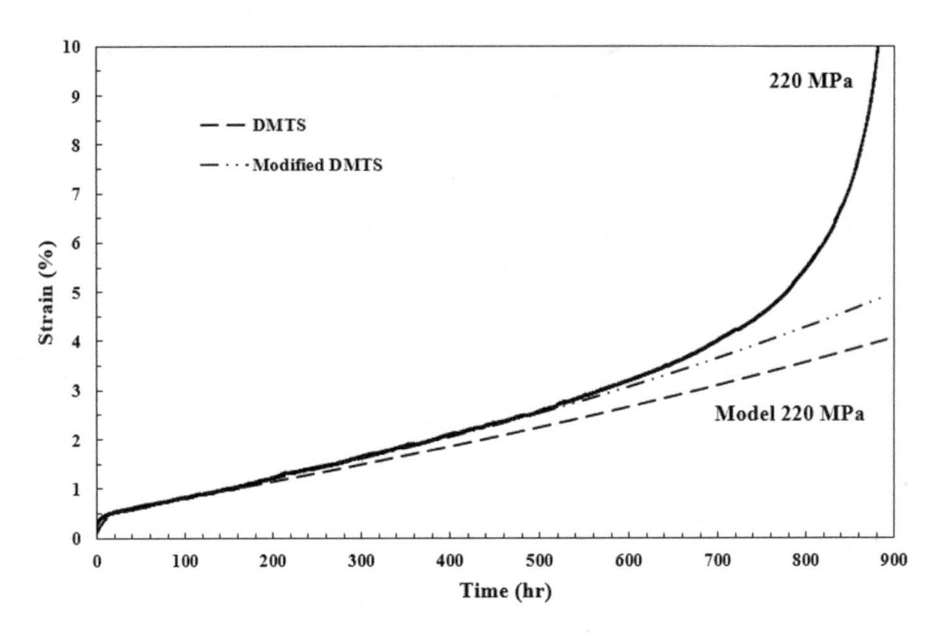

Figure 5.31. Comparison of the basic and modified DMTS models for creep strain-time curve of uncoated F91coupons at 550 °C. The solid line represents experimental behaviour.

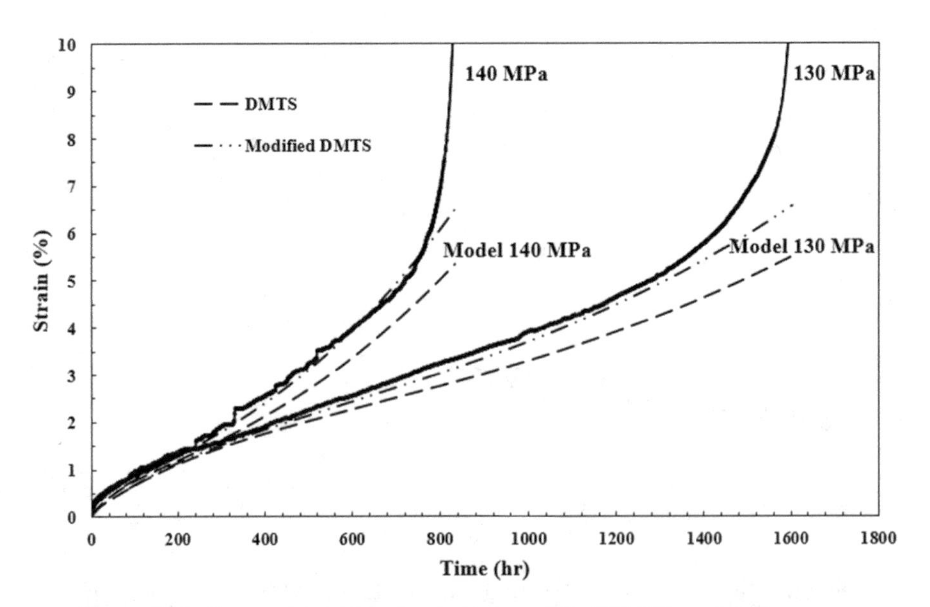

Figure 5.32. Comparison of the basic and modified DMTS models for creep strain-time curve of uncoated F91coupons at 600 °C. The solid line represents experimental behaviour.

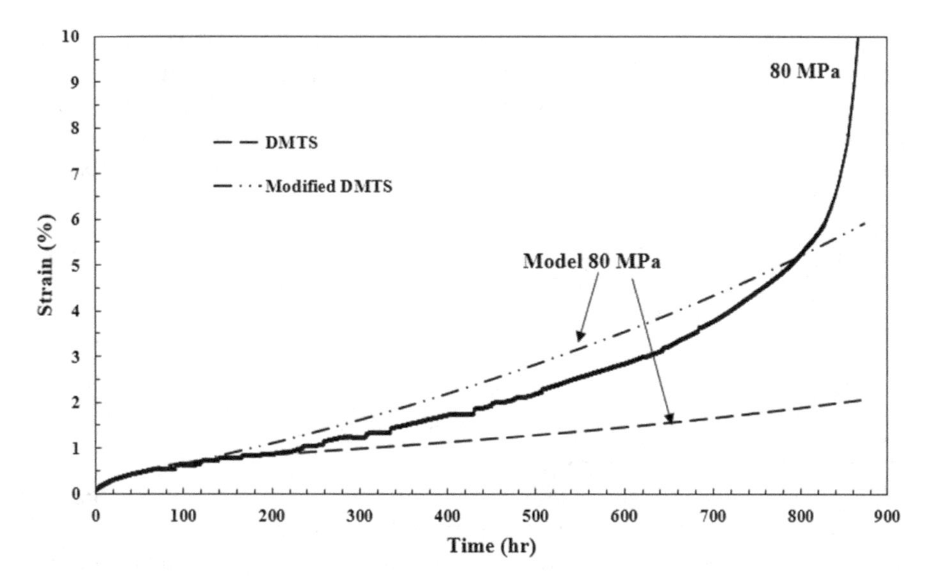

Figure 5.33. Comparison of the basic and modified DMTS models for creep strain-time curve of uncoated F91coupons at 650 °C. The solid line represents experimental behaviour.

5.6 Prediction of Long-Term Creep Life by O-DMTS Model

In the competitive world today, engineering products need to be turn-around fast and cost-effective, so to stay ahead of the competitors. Therefore, designers are often asked to predict the materials' long-term performance based on limited short-term test data. While short-term empirical methods are always questionable for long-term extrapolation, a mechanism-based model offers the promise, because it covers the physics of failure, irrespective of the observation time-length. In this section, attempts are made to use the mechanism parameters obtained from analysis of short-term creep data (<5,000 hours) to predict long-term creep lives (>10^4 hours) of Grade 91 steels. The NIMS-Grade 91 steels long-term rupture life data are used to validate the model prediction. This way, the allowable stresses of Grade 91 steels at elevated temperatures can be calculated using the mechanism-based creep model.

Since Grade 91 steels have similar microstructures in the as-received condition, it can be assumed that the structural parameters: H, β and M, are the same for all Grade 91 steels. These parameters have been determined for F91, as summarized in Table 5.6. The mechanism parameters for various Grade 91 steels are given in Table 5.4. In order to predict time-to-rupture (TTR), using Eq. (5.8) or Eq. (5.28), a failure strain needs to be defined. Since that at TTR creep rupture is often associated with deformation instability (necking), 90% TTR is found to be more appropriate to define the failure strain ε_{cr} level for modified grade 91 steels when deformation remains stable and it can be calculated using the creep-strain equation, Eq. (5.28). The failure strain ε_{cr} levels are determined from short-term F91 creep tests, as function of temperature. For long-term TTR prediction, two terms are negligible in Eq. (5.8) and (5.28): the initial elastic strain, ε_0, and the exponential term in the transient part: $\exp\left(-\dfrac{\beta^2 H\dot{\varepsilon}_{s0}t}{\sigma(\beta-1)}\right)$, because TTR >> t_{tr}^{p} (the primary time). Then, Eq. (5.8) or Eq. (5.28) can be re-written as:

$$TTR = \frac{1}{M^{*}k}\ln\left[1 + M^{*}\left(\varepsilon_{cr} - \frac{\sigma}{H\beta^2}\right)\right] \tag{5.31}$$

where ε_{cr} is the critical failure strain.

In a creep test, the critical failure criterion needs to be carefully evaluated by examining ruptured coupons and strain-time curves. As often observed, creep coupons would experience necking which led to specimen rupture in no time, i.e., the final stage of tertiary creep could occur with extremely large elongation almost instantaneously. For example, Figure 5.34 shows the creep curves at 600 °C/130 MPa and 500 °C/280 MPa, with final elongation all above 20%, but the strain levels at which necking starts are different. Creep necking at 600 °C/130 MPa occurred at 6.15% strain after 90% TTR, but at 500 °C/280 MPa it occurred at 2.25%. This indicates that F91 has a brittle-to-ductile transition with increasing temperature. Therefore, the critical failure strain, ε_{cr}, is defined to be at 90% TTR. Table 5.7 summarizes ε_{cr} at 90% TTR for F91 at various temperatures. For the present purpose, the same trend of

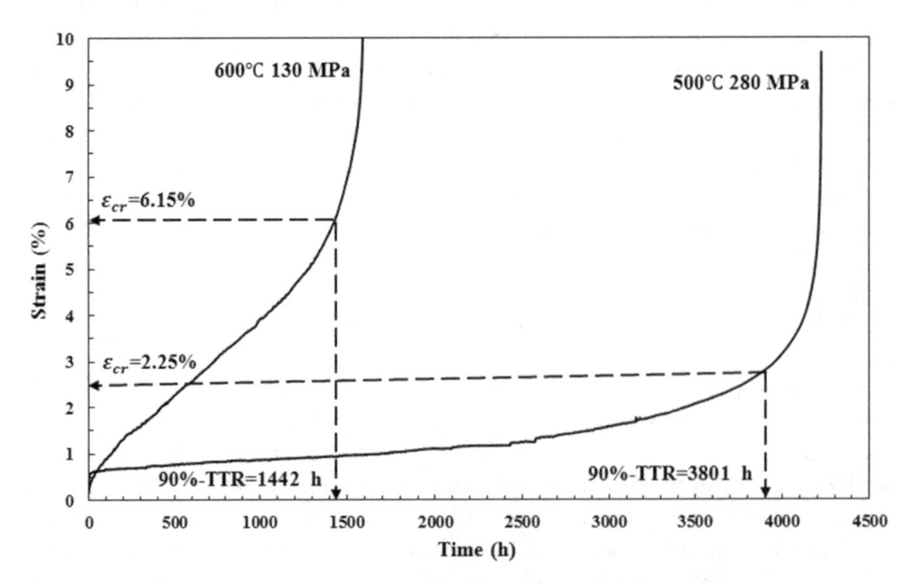

Figure 5.34. Creep failure strain at 90%-TTR for F91.

ε_{cr} for F91 is assumed to be true for all Grade 91 steels. Therefore, the creep failure criteria for Grade 91 can be assumed as

- ε_{cr} = 2-3% over the temperature range of 450-500 °C.
- Above 500 °C, ε_{cr} is in the range of 6-7%.

The life prediction procedures are summarized as follows:

1. Use β, *H* and *M* values of F91 in Table 5.6 for all Grade 91 steels.
2. Use the mechanism parameters in Table 5.4 and the oxidation equation to evaluate M^* and *k* for particular NIMS-Grade 91 steel, according to Eq. (5.27).
3. Calculate TTR using Eq. (5.31) with the ε_{cr} value for the temperature condition.

For comparison, both the basic DMTS model and the O-DMTS model with *k* and M^* are substituted in Eq. (5.31) to predict the long-term creep life of NIMS Grade 91 steels, as shown in Figures 5.35-5.38. It shows that two models have similar performance at TTR<10^4 region. The difference starts to show up above ~10^4 hours: the O-DMTS model describes the "life breakdown" phenomenon in agreement with the experimental observations, whereas the basic DMTS model represents the behaviour in "vacuum". Thus, the allowable creep stress is brought down by the oxidation effect at any given life target.

Table 5.7. F91critical strain at 90%-TTR

Temp	Stress	TTR	90%-TTR	Strain at 90%-TTR	Average ε_{cr}
°C	MPa	hr	hr	%	%
	320	2142.5	1928.25	2.05	
500	300	3503.7	3153.33	2.00	2.10
	280	4223.8	3801.42	2.25	
	260	81.3	73.17	7.15	
550	240	305.0	274.5	5.46	6.00
	220	891.4	802.26	5.38	
	160	180.6	162.54	8.40	
600	140	833.6	750.24	5.40	6.65
	130	1602.0	1441.8	6.15	
	110	78.8	70.92	7.83	
650	100	244.8	220.32	6.45	6.85
	80	873.4	786.06	6.26	

For the comparison purpose, the predicted creep lives of NIMS Plate MgB by the O-DMTS model are also plotted in the LMP diagram in comparison with the experimental 90%-TTR data. The Larson-Miller constant, C_{LM}, is 33 (Shrestha et al. 2013). Figure 5.39 shows that the predicted lives agree very well with the experimental data. While the LMP method do collapse the creep rupture data, the plot has to include long-term creep data in order establish the relationship between LMP and stress, usually via polynomial fitting. This method should be used with caution, because long-term creep at a lower temperature and short-term creep at higher temperature under the same stress may be controlled by different deformation mechanisms (see deformation mechanism maps in Chapter 2), even though both tests make up the same LMP value. On the other hand, the O-DMTS creep model can predict the creep life with the contributing fractions of GBS/IDG/IDC, which indicates the controlling creep deformation mechanism(s) and rupture failure mode. Figure 5.40 shows two microstructures at creep rupture under

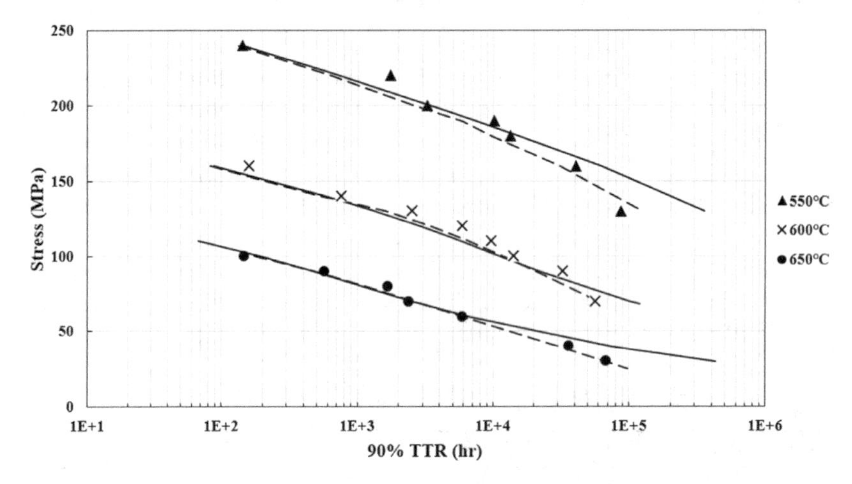

Figure 5.35. Comparison of the basic and modified DMTS models life prediction for MgB (symbols: experimental data; solid lines: DMTS; dash lines: O-DMTS).

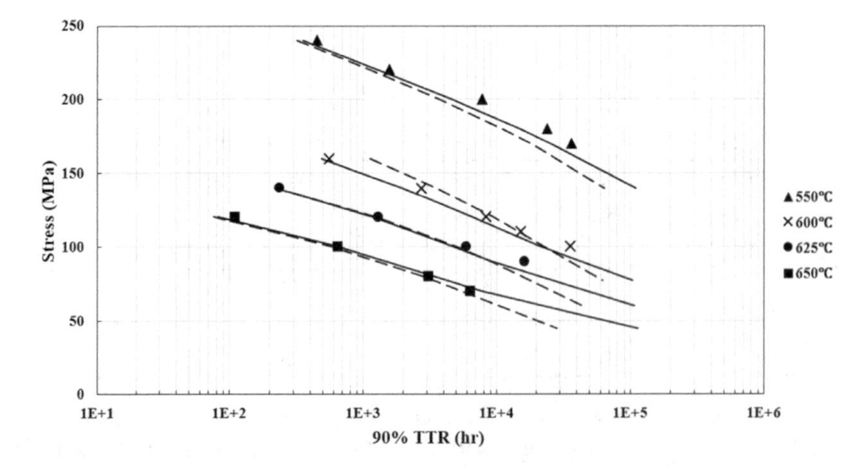

Figure 5.36. Comparison of the basic and modified DMTS models life prediction for MgD (symbols: experimental data; solid lines: DMTS; dash lines: O-DMTS).

600 °C/160 MPa and 650 °C/100 MPa, along with the mechanism pie-chart made up by the contributions from the three mechanisms to the total strain rate. In the 600 °C/160 MPa case, GBS was predominant (~94%) and the microstructure retained its original lath structure; whereas in the 650 °C/100 MPa case, as the model predicts that a significant portion (~38%) of creep deformation occurs by IDC, the lath structure indeed became more elongated. The model predication is thus supported by the metallurgical evidence. Using the DMTS model, such pie-charts can be drawn for every creep condition. This way, engineers are able to understand the physics of failure and identify the failure modes in quantitative details, whereas the LMP plot does not provide such information, from life prediction point of view.

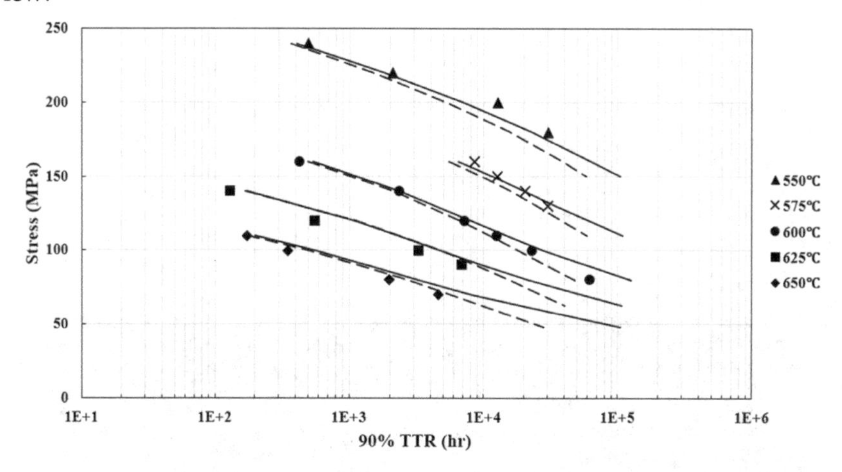

Figure 5.37. Comparison of the basic and modified DMTS models life prediction for MGD (symbols: experimental data; solid lines: DMTS; dash lines: O-DMTS).

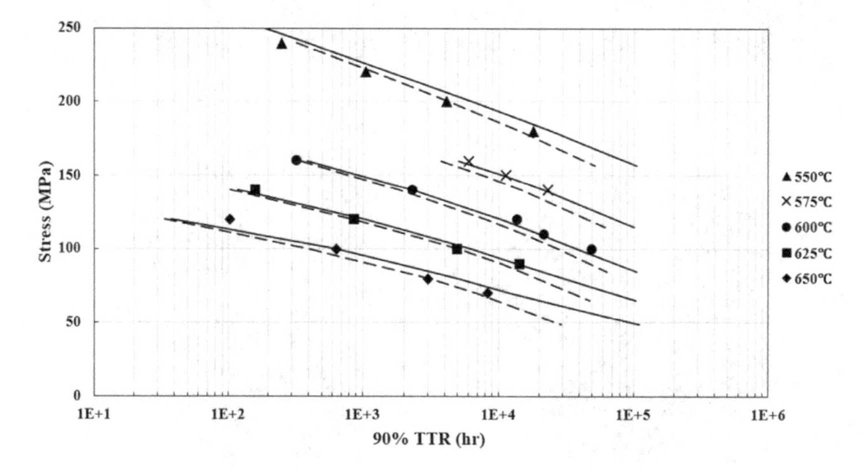

Figure 5.38. Comparison of the basic and modified DMTS models life prediction for MGQ (symbols: experimental data; solid lines: DMTS; dash lines: O-DMTS).

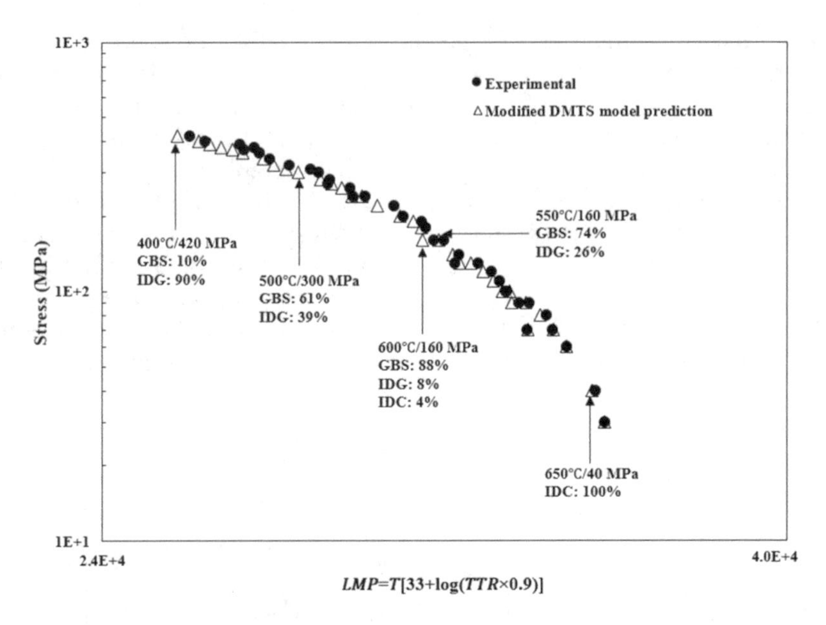

Figure 5.39. The modified DMTS model prediction of NIMS Plate MgB in LMP plot with experimental data.

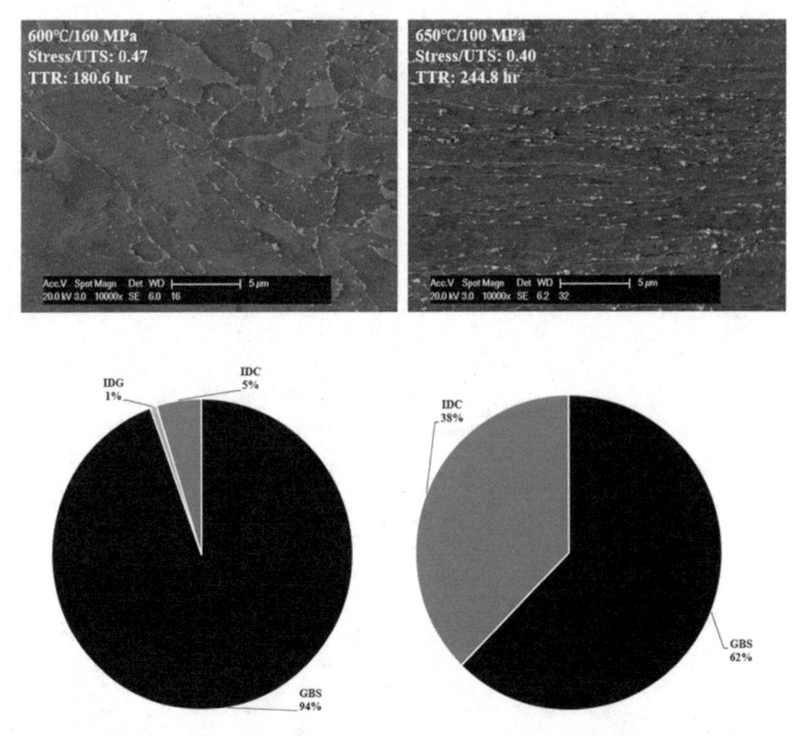

Figure 5.40. Microstructure analysis with pie-charts to indicate the participating mechanisms.

5.7 Effect of Microstructural Evolution

The effect of microstructural evolution on creep properties is an important subject, because few engineering alloys could retain their pristine microstructure during long-term creep exposure. Yet, it is complicated because of the alloy chemistry and initial microstructure specific to each material. For examples, cubic γ' precipitates in single crystal Ni-base superalloys would evolve into a rafted structure after exposure at high temperatures (Reed 2006); a discrete distribution of grain boundary precipitates in IN738 could evolve into an almost continuous network after long-time service exposure (Castillo et al. 1988); and new phase could form in addition to coarsening of the pre-existing precipitates, such as Laves phase formation in Grade 91 steels (Zhang et al. 2017b). It is almost impossible to generalize all the phenomena in the limited space of this book. Only until microstructural evolution as function of stress, temperature and time be established based on metallurgical principles, and its effects on deformation-mechanism operations be quantified, can a complete understanding be achieved for a specific material. The DMTS model provides a framework to accommodate those effects. In this section, F91 is used as an example to illustrate how microstructural evolution affects deformation mechanisms and its effect on creep properties of the material.

Grade 91 steels are first normalized (or austenitized) at 1080 °C (1975 °F) for 4 hours, followed by tempering at 650-780 °C, to allow the brittle martensitic lath structure to transform into a fine elongated micrograin structure (Eggeler et al. 1989, Maruyama et al. 2001). After tempering, the microstructure transforms form martensite to ferrite. The material is thus called "tempered martensite", merely to reflect this heat treatment history. Tempered ferritic steels are aimed to obtain secondary precipitations, such as $M_{23}C_6$ and MX, which are desirable because it enhances creep strength via precipitation hardening. However, additional thermal exposure at 550-600 °C can cause a considerable amount of Mo-rich Laves phase to form. Microstructural evolution in F91 was studied by Zhang et al. (2017b) through aging heat treatment at 600 °C for up to 5,000 hours. The Laves phase $Fe_2(Mo,W)$ was mostly found around Cr-rich carbide $M_{23}C_6$, which pinned on the prior-austenite grain boundaries and martensitic lath boundaries. This secondary precipitation hardening by Laves phase could be rapidly offset by its coarsening into large clusters during long-term thermal exposure under creep conditions, leading to earlier creep rupture than the unaged material.

The mechanism of Laves phase formation was found to consist of the following processes: first, Mo and Si solutes from the matrix segregated at micrograin boundaries to nucleate Laves phase, which coarsened rapidly to form bulky clusters during the extended thermal exposure (Isik et al. 2014). Limited empirical data and Johnson-Mehl-Avrami-Kolmogorov (JMAK) equation suggested that the volume fraction of Laves phase would change with time as a sigmoidal curve during prolonged aging at 600-650 °C (Strong and Gooch 1997). For F91, the Laves phase could be found at 600 °C but almost none found at 650 °C (Zhang et al. 2017b). As the thermal exposure continued, Laves phase could progressively cause embrittlement instead of strengthening (Klueh 2005). Laves phase formation in F91 under both aged (at 600 °C) and creep conditions (at 550 °C and 600 °C) have been examined by Zhang et al. (2017b) in detail. Figure 5.41 depicts the Mo concentration in the particle clusters in the aged and creep-ruptured aged coupons at 550 °C and 600 °C, respectively. First, during aging at 600 °C, the Mo content increased almost linearly with the aging time under no stress. Subsequently under the creep conditions, the Mo content apparently increased at higher rates by the effect of stress. The average growth rate can be estimated by the final amount of Mo subtracting the initial amount (after aging but before creep), then divided by the TTR. Figure 5.42 shows the Laves phase (Mo concentration) growth rates as a function of stress at 550 °C and 600 °C respectively, and both have a stress exponent of approximately 10.

The creep strain-time curves of the aged and non-aged coupons at 550 °C and 600 °C are compared in Figures 5.43 and 5.44, respectively. At 550 °C, the aged coupons exhibited higher minimum creep rate and earlier tertiary creep than the unaged coupon. At 600 °C, although the minimum creep rates of the two types of coupons were close in the transient stage, the aged coupons exhibited an earlier transition to tertiary creep. Given the evidence of Mo segregation and associated void growth, it is reasonable to believe that the formation of Laves phase caused the increase of creep rate and reduction of creep rupture life. One exception is at 600 °C/130 MPa,

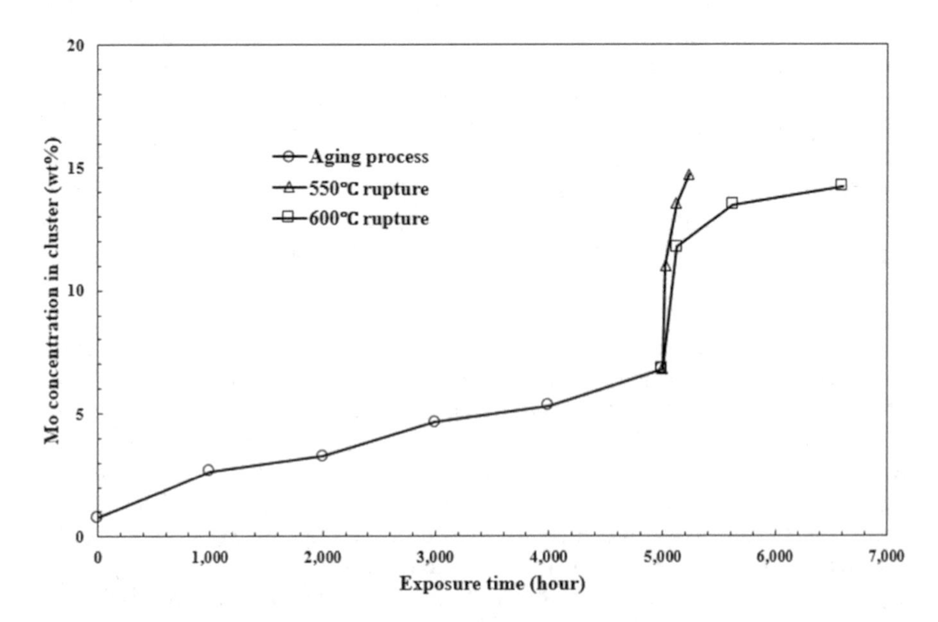

Figure 5.41. Variations of Mo concentration during aging and in creep ruptured aged-coupons.

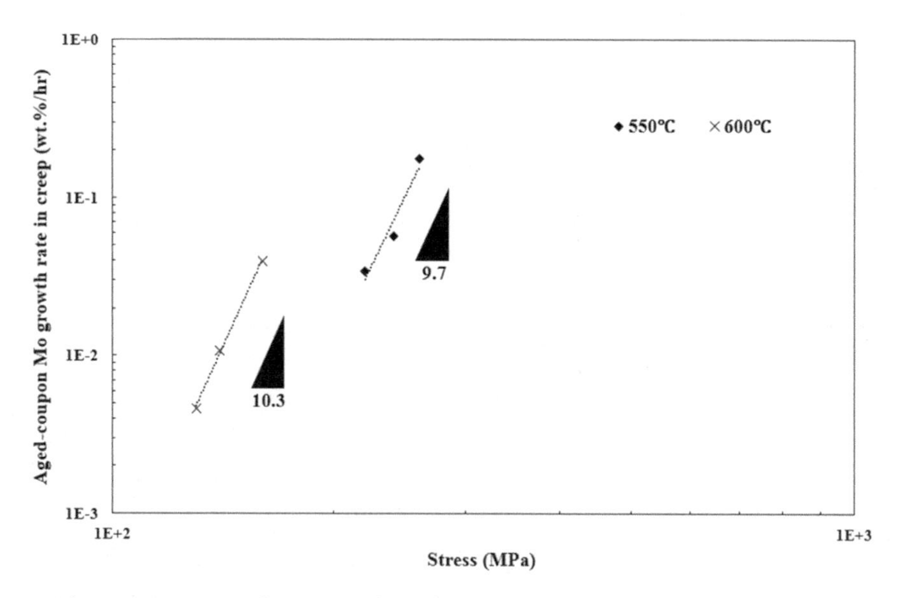

Figure 5.42. Laves phase growth rates as a function of stress in aged-coupons during creep tests.

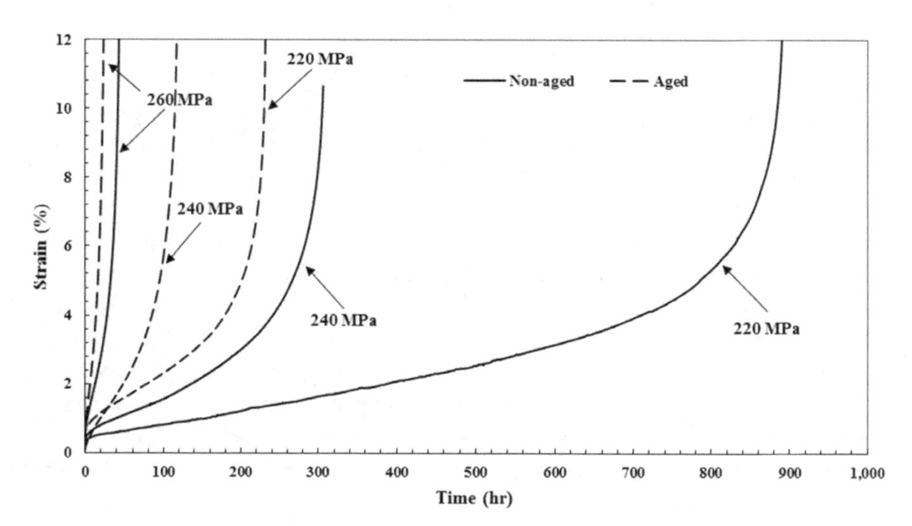

Figure 5.43. Creep strain-time curves of non-aged and aged coupons tested at 550 °C.

the two types of coupons have the same lifetime around 1,600 hours. In this case, a large amount of Laves phase particles and a few Z-phase particles also formed in the non-aged coupon after 1,600 hours of creep. Z-phase (size ~50 nm) is a type of Cr-rich nitride compound (sometimes, with Nb and/or V as substitutes of Cr).

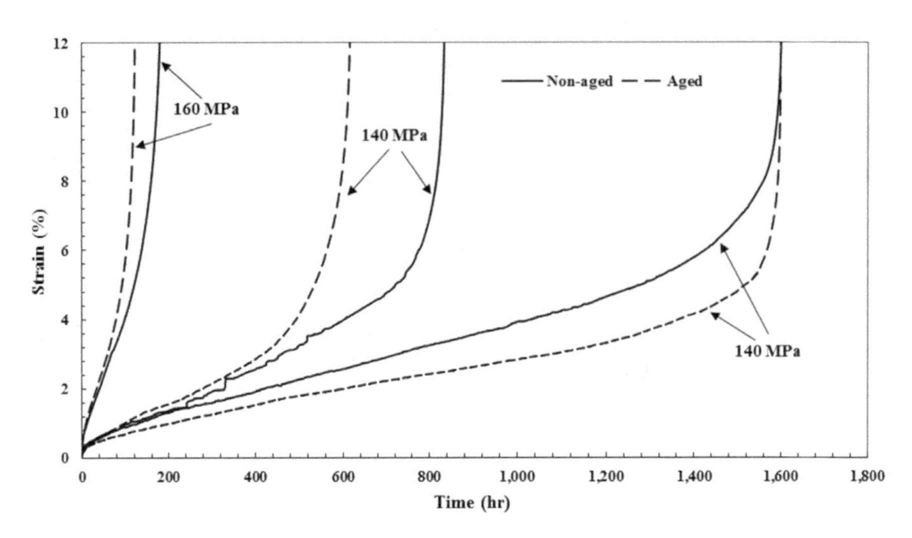

Figure 5.44. Creep strain-time curves of non-aged and aged coupons tested at 600 °C.

It has been discussed that in modified Grade 91 steels, IDC, IDG, and GBS assume constant power exponents for each of the mechanism: IDC has an exponent of 6, GBS has an exponent in the range of 10 to 17, and IDG has the highest exponent, ranging from 18 to 29. The power exponents of non-aged and aged F91 are shown on a minimum creep rate versus normalized stress in log-log scale, as shown in Figure 5.45. At 550 °C, the creep rate power-exponent of unaged F91 is 19.1, but that of aged F91 reduces to 16.2. At 600 °C, power-law exponents of the two materials are very close, ~11.7 to 12.9. Considering that the Laves phase mostly formed at lath and grain boundaries, it is reasonable to assume that Laves phase would primarily affect GBS, not IDG. Thus, subtracting the same IDG component from the total minimum creep rate, the GBS+IDC rates are shown in Figure 5.46. It is noticed that the aged F91 creep rate exhibits a power-law exponent almost identical to GBS in the non-aged F91, but with a larger proportional constant. It is also noticed, interestingly, this power-exponent is very close to that of the Laves phase formation. Therefore, it can be concluded that Laves phase formation has proportionally increased GBS, since it weakens the pinning effect of $M_{23}C_6$/MX on lath/grain boundaries.

Figure 5.47 compares the microstructures from the two types of coupons crept under 550 °C/260 MPa and 600 °C/140MPa. It can be seen that under both creep conditions more and larger voids have grown in association with formation of Laves phase clusters (the Laves phase appear to be bright in SEM, because it has higher concentration of Mo) along the grain/lath boundaries, which attest the predication of the DMTS model. This demonstrates how the effect of microstructural evolution is connected to the deformation mechanisms. More experiments should be conducted to examine the effect of Laves phase formation on IDC, to complete the study.

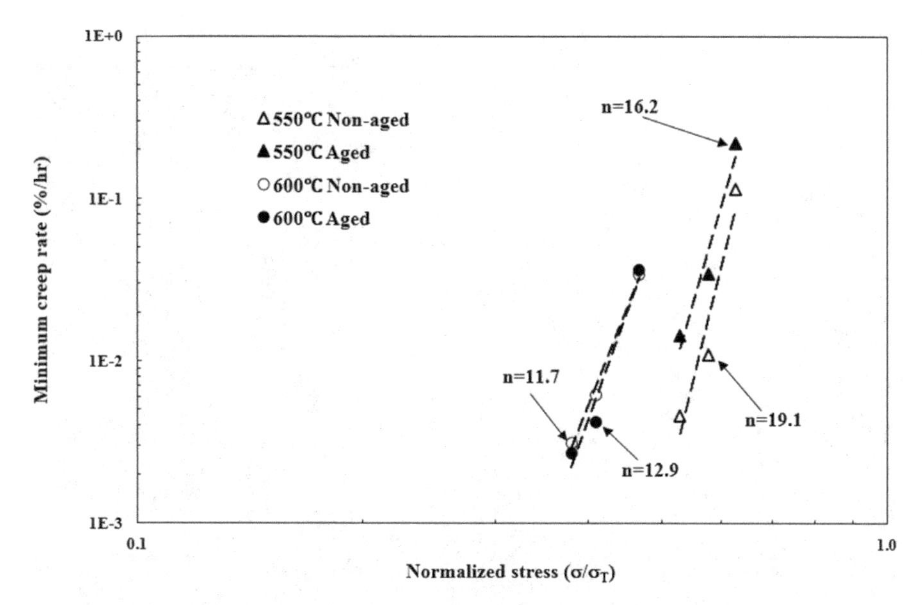

Figure 5.45. Minimum creep rate vs. normalized stress.

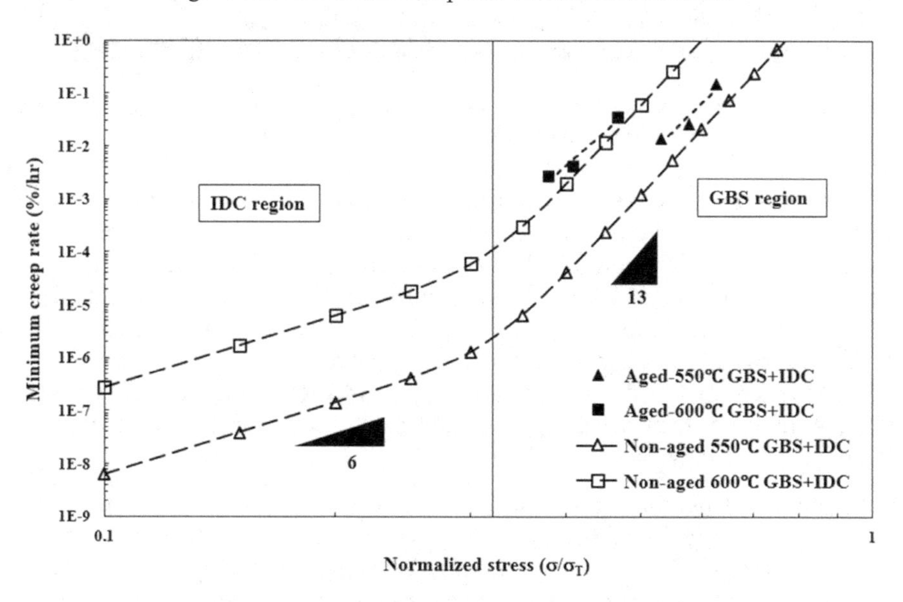

Figure 5.46. Comparison of minimum creep rate between two types of coupon without IDG.

5.8 Summary

With the development of the deformation-mechanism based true-stress (DMTS) creep model and the oxidation modified DMTS (O-DMTS) creep

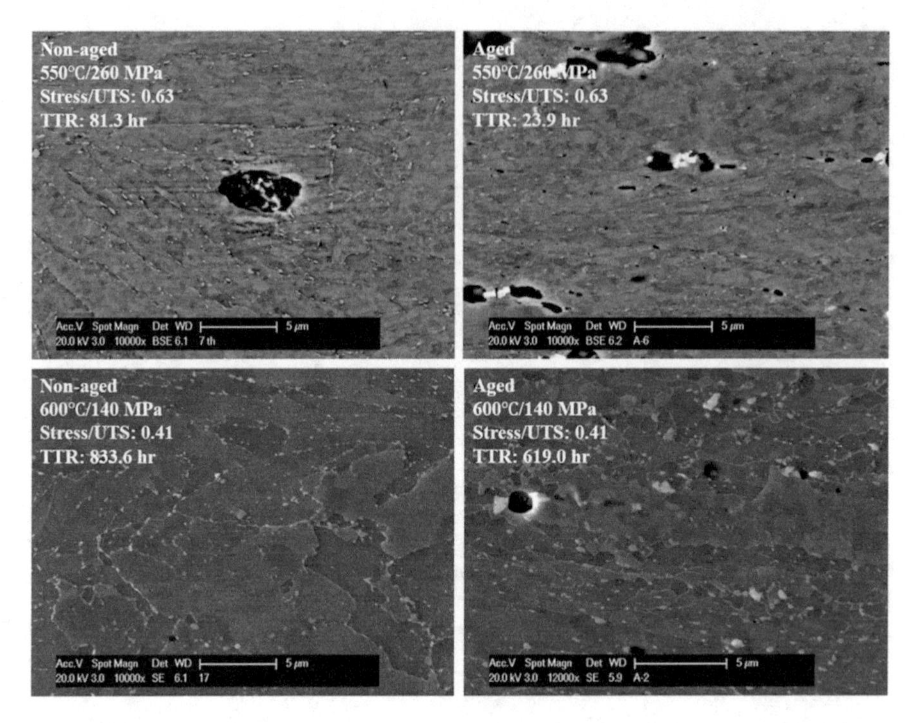

Figure 5.47. Comparison of microstructures between non-aged and aged coupons after creep rupture at 550 °C and 600 °C.

model, this chapter formulates a systematic approach to modeling creep behaviour and predicting the long-term creep life and failure mode based on the identified deformation mechanisms within the ICFT framework. Over the past one hundred years, these phenomena were treated with various empirical equations, but empirical equations do not connect to each other, such that the understanding cannot be complete. Especially, the empirical long-term creep life prediction was not warranted with mechanism and failure mode information. Some significant points of this Chapter are summarized below.

- The fundamental deformation mechanisms such as GBS, IDC and IDG are built into the DMTS creep model, which has been validated for Grade 91 steels, Ni-base superalloy Waspaloy and Co-base superalloy Mar-M 509. By the general validity of the physical mechanisms involved, the DMTS model can be applicable to all crystalline materials. It can be extended to include the effects of processing (as discussed for Grade 91 steels in section 5.3), microstructure (as discussed in Chapter 1 and 2) and its evolution (as discussed for F91 in section 5.7).
- The DMTS creep model is further modified to include the effect of oxide scale formation on the true stress during creep process. This model gives a quantitative description of the environmental effect separated from

the material-intrinsic deformation mechanisms: IDG, IDC and GBS. It then predicts the long-term creep performance with true oxidation contribution, as oppose to other empirical methods where short-term data are extrapolated without this consideration. As a demonstration, the oxidation-modified DMTS (O-DMTS) model is validated with the NIMS creep data for modified 9Cr-1Mo steels and F91 steels. In short, this chapter demonstrates a mechanism-delineated approach for prediction of long-term creep behaviour and failure mode from short-term creep data with the identified deformation mechanisms, which is the first time over the last one hundred years of creep study. In principle, it can be applied to all metallic materials which experience IDG, IDC and GBS as common deformation mechanisms and suffer oxidation at high temperatures.

- Using the DMTS or O-DMTS model (the latter reduces to the former in absence of environmental effect), a complete characterization of creep behavior can be achieved in terms of mechanism (IDG, IDC, and GBS) parameters. With such models, the effects of manufacturing process and heat treatment history on elevated-temperature creep behaviour such as creep rate, creep rupture mode and creep life can be quantified by comparison. The results can not only be used to predict the structural integrity for safe operation of the high temperature component, but also to feedback to material design tailoring the material creep properties for optimized performance.
- This chapter, through studying the effect of MCrAlY coating on creep performance of the modified Grade 91 steel (F91), also introduces a mechanistic approach to deal with creep of coated components, using the composite rules, which is useful for design of oxidation-resistant components for high-temperature applications.
- In this chapter, an area-deduction method is introduced, for the first time in creep study, to quantitatively assess the oxide scale growth during creep. This method can be helpful to separate the oxidation effect from conventional creep testing in air. Since vacuum creep test is expensive, and therefore oxidation-free creep data are not readily available, this method can be effectively used to estimate the oxidation growth rate coefficient, k_{ox}, for metals that suffer oxidation during creep. With this method, the DMTS model is extended to O-DMST model for long-term creep life prediction with environmental effects.
- Using the O-DMTS creep model, the Larson-Miller parameter method has been analyzed and furbished with mechanism partition information, providing insights into the controlling deformation mechanisms and potential failure mode. This is a significant improvement to the state-of-the art creep design methodology. In addition, it saves a significant testing effort to generate massive experimental data as the Larson-Miller parameter method requires.

- The effect of Laves phase formation on creep of F91 has been studied, which sets an example linking the effect of microstructural evolution to the controlling deformation mechanisms. Microstructural evolution would especially influence long-term creep performance of materials. Similar studies should be carried out for every high-temperature structural material over a wider stress and temperature range, incorporating quantitative description of phase transformation and growth. Linking microstructural evolution with deformation mechanisms should be the focus of future creep studies. Only this way can a complete description of material creep behaviour be achieved.
- This chapter presents the DMTS or O-DMTS creep model that describe the three-stage (the primary, the secondary and the tertiary stage) creep behaviour, predicts the time to rupture and failure mode with explicit stress and temperature-dependence. This is extremely important for high temperature component stress analysis and life prediction, because creep deformation can induce stress redistribution and thus change the dominant failure mechanism at critical locations over time. Therefore, it is important to delineate each mechanism contribution through creep data analysis. The following analytical procedures need to be taken:
    1. Determine the IDG, IDC and GBS mechanism parameters of the DMTS model from the minimum creep rates of short-term creep strain vs. time data.
    2. Determine the β, H and M values from creep curve analysis.
    3. Determine the critical total rupture strain ε_{cr}, if necking occurs at the final stage of creep rupture; or calibrate the critical GBS strain $\varepsilon_{gbs,cr}$ and the intragranular ductility $\varepsilon_{g/c,cr}$ to intergranular and transgranular mode of creep rupture, respectively.
    4. Determine the oxidation rate, if any, using the area-deduction method.
    5. Apply the oxidation corrected DMTS model, i.e., the O-DMTS model to the air-creep data.
    6. Perform creep life and failure mode prediction using O-DMTS model (which reduces to the baseline DMTS model in absence of oxidation).
    7. Incorporate microstructural evolution laws applicable to the material.
    8. Use the true-stress form of O-DMTS model for FEM analysis of component performance under creep conditions.

CHAPTER

6

Low Cycle Fatigue

Fatigue generally refers to material failure under cyclic loading. It accounts for 90 percent of in-service component failures. In engineering, fatigue failure is often divided into i) low-cycle fatigue (LCF) with an appreciable amount of cyclic plasticity in less than 10^4 cycles, and ii) high-cycle fatigue (HCF) under cyclic stress below the yield point with failure cycles greater than 10^4. Even though HCF deformation is macroscopically in the elastic regime, it is believed that strain concentration at microstructural discontinuities such as grain boundaries and inclusions is responsible for nucleating cracks. Therefore, HCF can be regarded as microstructural LCF crack nucleation plus propagation. Therefore, LCF is the most important subject in fatigue study. A schematic LCF test setup is shown in Figure 6.1, which is usually conducted in strain-control to mimic the deformation condition on an engineering component where localized plastic deformation is restrained by the surrounding elastic material in stress concentration regions such as notches and holes, or where steep thermal gradient exists. In this sense, LCF failure problems are most critical to component durability.

Experimentally, Coffin-Manson relation, Eq. (4.11), has been found to exist between the LCF life and the plastic strain range. A room-temperature LCF master curve was proposed by Manson (1965) for a variety of materials. However, at high temperatures, both the fatigue ductility and power-exponent could change significantly with test variables such as the frequency and temperature, because time-dependent deformation and environmental effect are also involved simultaneously. This has motivated various empirical modifications to the Coffin-Manson equation, but without linking to the deformation and damage mechanisms. Therefore, a mechanism-delineation approach is needed to unveil the physical interplay of the participating deformation and damage mechanisms in LCF processes. In this chapter, LCF phenomena are discussed using ICFT with regards to mechanisms of pure mechanical fatigue and internally distributed damage and oxidation.

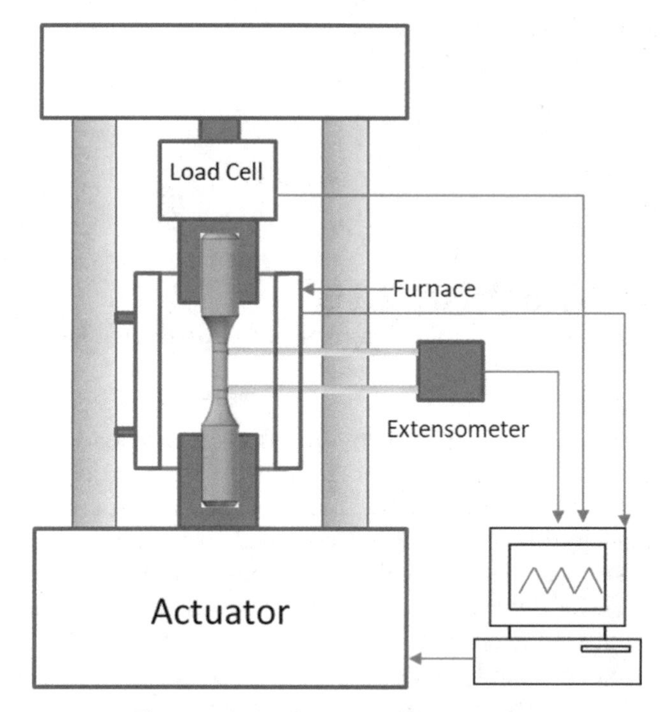

Figure 6.1. A schematic LCF test setup.

6.1 Mechanical Fatigue

Fatigue failure has been studied crossing two centuries, since Wöhler in 1867. Mechanical fatigue usually occurs under cyclic loading at low temperature, e.g., room temperature, in absence of environmental effects. Thus, it is controlled purely by cyclic plasticity via alternating slip, as discussed in section 3.1. In the 1950s, Coffin and Manson independently found that a power-law relation, Eq. (4.11), exists between the cyclic plastic strain range and fatigue life with the exponent falling in a narrow range of –0.45 to –0.65 for most metals and alloys. But, engineers have to perform fatigue tests to determine the fatigue ductility (the proportional constant) and exponent for each material ever studied. In order to understand the LCF behaviour under complex loading conditions, a thorough understanding of mechanical fatigue with regards to its controlling deformation mechanism must be obtained first.

6.1.1 Tanaka-Mura's Model Re-Visited

Alluding to the formation of persistent slip bands and surface extrusions/intrusions (section 3.1), a micromechanics model of fatigue crack nucleation was developed by Tanaka and Mura (1981), which contains inverted dislocation pile-ups along two-layer slip bands. They envisaged that the irreversibility of dislocation motion in the two adjoining layers results from different levels of back stresses in the forward direction on layer I and in the reversed direction on layer II, as shown in Figure 6.2. Suppose that the first tensile loading causes dislocation pile-up on layer I. The positive back stress (which opposes the stress causing dislocation motion) due to positive dislocations on layer I facilitates the pile-up of negative dislocations on layer II during reversed loading, creating a series of vacancy dipoles. The back stress due to dislocations on layer II helps further pile-up of dislocations on layer I during the following forward cycle. This process leads to dislocation pile-up accumulation with increasing number of fatigue cycles, eventually leading to fatigue crack nucleation.

Then, Tanaka and Mura obtained a theoretical fatigue crack nucleation life as given by (Eq. 35 and Eq. 36 in Tanaka and Mura, 1981)

$$N_c = \frac{4\pi(1-\nu)w_s a^3}{\mu}\Delta\gamma^{-2} \tag{6.1}$$

or, in terms of stress, as

$$N_c = \frac{4\mu w_s}{\pi(1-\nu)a}(\Delta\tau - 2k)^{-2} \tag{6.2}$$

where w_s is the surface energy, μ is shear modulus, ν is Poisson's ratio, k is friction stress, and a is half grain size.

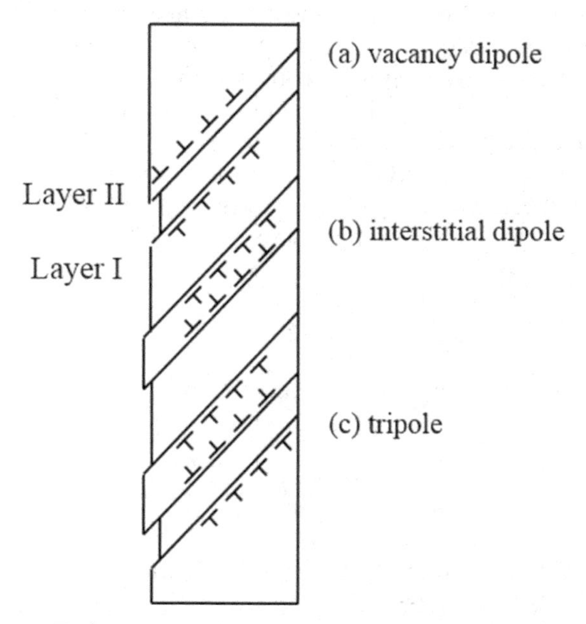

Figure 6.2. Dislocations in (a) vacancy dipoles (forming an intrusion), (b) interstitial dipoles (forming an extrusion) and (c) tripoles (forming an intrusion-extrusion pair) at the surface.

This model receives its popularity because it captures the essence of crack nucleation via dislocation slip and it predicates the dependence of fatigue crack nucleation life N_c on the cyclic plastic strain range $\Delta\gamma$ with a power-exponent – 2, which agrees with the Coffin-Manson relationship. However, the strain-based version, Eq. (6.1), is rarely used in practice, because in the original Tanaka-Mura model $\Delta\gamma$ bears a physical dimension of $[m]^2$ (see the discussion later), which cannot be experimentally determined from strain measurements. If the true strain were used in Eq. (6.1), one would find an additional physical dimension of $[m]^4$ on the right hand side of the equation, which would not be correct. Actually, the stress-based version, Eq. (6.2), is most often used in engineering analyses for real materials. But, in those analyses, the surface energy w_s is termed as the "specific fracture energy" that is often given arbitrary values (other than independently assessed) to fit the S-N fatigue curve, e.g., $w_s = 440$ kJ/m^2 for stainless steel (Tryon and Cruse 1998), and $w_s = 2$ kJ/m^2 for a martensitic steel (Kramberger et al. 2010, Bruckner-Foit and Huang 2008, Jezernik et al. 2010). These values are orders-of-magnitude higher than the surface energies of metals, reported by Tyson and Miller (1977). Besides, the lattice friction stress k is also difficult to estimate from stress-strain measurements on material coupons. Therefore, in a physically rigorous sense, Eq. (6.1) and/or Eq. (6.2) have never been experimentally validated.

Recently, the Tanaka-Mura model has been re-visited to correct its dimensional problem (Wu 2018). In this section, we will follow Tanaka-Mura's mathematical derivation to the point where they made the mistake, and correct the dimensional problem of Eq. (6.1). Then, we will proceed to validate the corrected model by comparison with experimental fatigue properties of pure metals and alloys. In the evaluation, nothing but material physical properties, w_s, μ, v, and b, are used.

As it shows in Tanaka-Mura's model, under the first loading of stress τ_1 greater than the frictional stress k, the dislocation distribution is produced on layer I with $D_1(x)$, satisfying the force-balance equation:

$$\int_{-a}^{a} \frac{\mu b D_1(\xi) d\xi}{2\pi(1-\nu)(x-\xi)} + \tau_1 - k = 0 \tag{6.3}$$

Using the Muskhelishvili (1953)'s inversion formula, Eq. (6.3) is solved to yield

$$D_1(x_1) = \frac{2(1-\nu)(\tau_1 - k)}{\mu b} \frac{x}{\sqrt{a^2 - x^2}} \tag{6.4}$$

Note that the dislocation distribution is asymmetrical about $x = 0$. This is typical of Bilby-Cottrell-Swinden distribution with equal number but opposite sign of dislocations distributed on the two sides, which would lead to formation of a centre crack or a surface edge crack (with half of the configuration for real).

The total number of dislocations between $x = 0$ and a is (Eq. 5, Tanaka and Mura 1981)

$$N_1 = \int_0^a D_1(x')dx' = \frac{2(1-\nu)(\tau_1 - k)a}{\mu b} \tag{6.5a}$$

The plastic displacement caused by the motion of dislocations is given by the integral (Eq. 5a, Tanaka and Mura 1981)

$$\phi(x) = \int_x^a bD_1(x')dx' \tag{6.5b}$$

In Tanaka-Mura's model, the "total plastic displacement" (later called "plastic strain") is calculated by (Eq. 6, Tanaka and Mura, 1981)

$$\gamma_1 = \int_{-a}^{a} \phi(x)dx = \int_{-a}^{a} bD_1(x')x\,dx' = \frac{\pi(1-\nu)(\tau_1 - k)a^2}{\mu} \tag{6.6}$$

Note that Eq. (6.6) is actually integration of displacement, which results in a dimension of $[m]^2$. Therefore, the physical meaning and dimension of γ_1 as either "the total plastic displacement" or "strain" is incorrect!

By definition, strain is displacement over the distance it is measured. In this case, the slip distance is evaluated as the number of dislocations (given by Eq. 6.5a) times the Burgers vector, i.e., $\Delta l = N_1 b$, which occurs over the distance a. Thus, the plastic strain due to the dislocation pile-up should be equal to $N_1 b/a$, or given by the integral:

$$\gamma_1 = \frac{1}{a}\int_0^a bD_1(x)dx = \frac{2(1-\nu)(\tau_1 - k)}{\mu} \tag{6.7}$$

Then, we follow Tanaka and Mura's subsequent derivation procedure as follows.

First, the stored energy associated with the dislocation pile-up on layer I is calculated to be

$$U^{(1)} = \frac{1}{2}(\tau_1 - k)\gamma_1 \tag{6.8}$$

On loading reversal, another pileup occurs in layer II, satisfying

$$\int_{-a}^{a} \frac{\mu b D_2(\xi)d\xi}{2\pi(1-\nu)(x-\xi)} + \int_{-a}^{a} \frac{\mu b D_1(\xi)d\xi}{2\pi(1-\nu)(x-\xi)} + \tau_2 + k = 0 \tag{6.9}$$

The distribution function for the pileup in layer II is thus given by

$$D_2(x) = -\frac{2(1-\nu)(\Delta\tau - 2k)}{\mu b}\frac{x_1}{\sqrt{a^2 - x^2}} \tag{6.10}$$

where $\Delta\tau = \tau_1 - \tau_2$ is the stress range.

The plastic strain associated with the pile-up in layer II is given by

$$\gamma_2 = \frac{1}{a}\int_0^a bD_2(x)dx = -\frac{2(1-\nu)(\Delta\tau - 2k)}{\mu} \tag{6.11}$$

And hence, the stored energy associated with the dislocation pile-up in layer II is given by

$$U^{(2)} = \frac{1}{2}(\Delta\tau - 2k)\gamma_2 \tag{6.12}$$

On the k-th reversal, the dislocation distribution, $D_k(x)$, the strain γ_k, and the stored energy $U^{(k)}$ are obtained in a similar manner:

$$D_k(x) = (-1)^{k+1}\frac{2(1-\nu)(\Delta\tau - 2k)x}{\mu\sqrt{a^2 - x^2}}, \gamma_k = (-1)^{k+1}\Delta\gamma, U^{(k)} = \Delta U \tag{6.13}$$

where

$$\Delta\gamma = \frac{2(1-\nu)(\Delta\tau - 2k)}{\mu} \tag{6.14}$$

$$\Delta U = \frac{1}{2}(\Delta\tau - 2k)\Delta\gamma \tag{6.15}$$

The index k takes $2N$ at the minimum and $2N + 1$ at the maximum stress after N cycles.

By the Griffith energy criterion (Griffith 1921), the entire pileup bursts into a crack once the stored energy in the material volume (ba) becomes equal to the energy to form new crack surfaces ($2a$) (Eq. (34) in Tanaka and Mura 1981):

$$N\Delta Uba = 2aw_s \tag{6.16}$$

where w_s is the surface energy, J/m^2.

Then, the cycle to crack nucleation can be obtained, by substituting Eq. (6.15) into Eq. (6.16), as

$$N_c = \frac{8(1-\nu)w_s}{\mu b}\frac{1}{\Delta\gamma^2} \tag{6.17a}$$

or, in terms of stress,

$$N_c = \frac{2\mu w_s}{(1-\nu)(\Delta\tau - 2k)^2 b} \tag{6.17b}$$

The most striking difference between Eq. (6.17a) and Eq. (6.1) is that Eq. (6.1) obtained by Tanaka and Mura contains extra terms of ba^3. Unless strain could be measured with a dimension of [m]^2, Eq. (6.1) cannot be directly used for fatigue life analysis. Comparing Eq. (6.17b) to Eq. (6.2), Eq. (6.17b) asserts that the fatigue life is proportional to w_s/b instead of w_s/a as given by Eq. (6.2), despite there may exist a Hall-Petch type relationship in the lattice resistance k. Now, we have shown that the extra grain-size dependence in Tanaka-Mura's original model was introduced by evaluation of strain from the displacement integration. It is also shown that Eq. (6.17) stands no matter what the length scale of a really is, which means a can be the length of any region where persistent slip band spreads, be it within a fine grain in a polycrystalline material or in a large single-crystal turbine blade.

6.1.2 Model Validation

To physically validate the model, we shall proceed with the strain-based version, Eq. (6.17a), since both the fatigue life N_c and plastic strain range $\Delta\gamma$ are observable quantities, and the w_s, μ and b are known material properties and parameters, at least for pure metals. Now that the dimensional problem of the Tanaka-Mura model has been corrected, fatigue life should be evaluated analytically, using Eq. (6.17a) with material physical properties, w_s, μ, v, and b.

Tyson and Miller (1977) have obtained the surface energies for pure metals. The surface energy $\gamma_{SV}(= w_s)$ at temperature T is given by the following equation:

$$\gamma_{SV} - \gamma_{SV}(T_m) = \int_T^{T_m} \frac{S_{SV}}{A} dT = \phi(T)\frac{RT_m}{A} \tag{6.18}$$

where $\gamma_{SV}(T_m)$ is the surface energy at the melting temperature, A is the surface area per mole of surface atoms, S_{SV} is the entropy. The parameter ϕ is obtained from the integration as function of the homologous temperature, T/T_m, as shown in Figure 6.3. The values of $\gamma_{SV}(T_m)$ and RT/A for pure elements are given in Table 6.1 (Tyson and Miller 1977).

In engineering practice, fatigue life is assessed through testing of coupons of certain surface finish by machining. We apply a surface roughness factor R_s, and use the Taylor factor relationship $\gamma = \sqrt{3}\varepsilon$ to convert Eq. (6.17a) into

$$N_c = \frac{8(1-\nu)R_s w_s}{3\mu b}\frac{1}{\Delta\varepsilon_p{}^2} \tag{6.19}$$

In the following, we will discuss the influence of various factors, including R_s on fatigue life.

Low cycle fatigue (LCF) lives of type 316 stainless steel with different surface finish are shown in Figure 6.4. The theoretical prediction of Eq. (6.17a) exactly matches the experimental data for the electropolished surface obtained from Wareing and Vaughan (1979), which represents an "ideal" case. By comparison, the machined surface roughness has an effect of $R_s \sim 1/3$.

Uniaxial LCF life vs. plastic strain relations for several metals and alloys at room temperature are evaluated using Eq. (6.19) as shown in Figure 6.5. For these cases, the homologous temperature is ~ 0.25, $\phi = 0.85$, and $R_s = 1/3$ (assuming the same machined condition). The material property parameters and the calculated values of the fatigue life coefficient are given in Table 6.2,

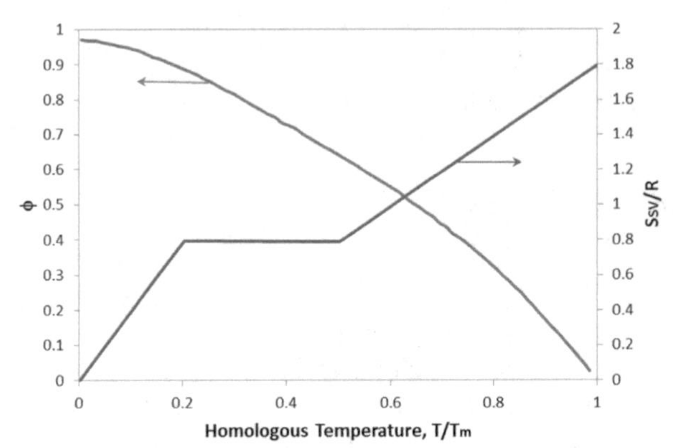

Figure 6.3. Variation of surface free energy and entropy with homologous temperature, after Tyson and Miller (1977).

Table 6.1. Values of surface energy and RT_m/A

Element	γ_{sv} (J/m²)	RT_m/A (J/m²)	Element	γ_{sv} (J/m²)	RT_m/A (J/m²)
Ag	1.086	0.160	Na	0.234	0.027
Al	1.020	0.123	Nb	2.314	0.342
Au	1.333	0.173	Nd	0.812	0.090
B	1.060	~0.55	Ni	2.080	0.300
Ba	0.326	0.054	Os	2.950	0.489
Be	1.298	0.330	Pb	0.540	0.053
Bi	0.446	0.043	Pd	1.743	0.260
Ca	0.425	0.077	Pt	2.203	0.286
Cd	0.696	0.066	Rb	0.104	0.013
Co	2.218	0.304	Re	3.133	0.493
Cr	2.006	0.348	Ru	2.655	0.388
Cs	0.084	0.011	Rh	2.325	0.334
Cu	1.566	0.224	Sb	0.461	0.136
Fe	2.123	0.294	Si	0.940	0.195
Ga	0.845	0.036	Sn	0.661	0.048
Ge	0.748	0.129	Sr	0.358	0.061
Hf	1.923	0.270	Ta	2.493	0.409
Hg	0.580	0.025	Ti	1.749	0.240
In	0.658	0.042	Tl	0.550	0.052
Ir	2.658	0.393	U	1.780	0.159
K	0.129	0.016	V	2.301	0.321
Li	0.472	0.050	W	2.765	0.500
Mg	0.688	0.097	Zn	0.896	0.097
Mn	1.298	0.245	Zr	1.687	0.222
Mo	2.510	0.397			

all materials are assumed to have a Poisson's ratio of 0.3. The theoretical predictions of Eq. (6.19) are in very good agreement with experimental data for Type 316 stainless steel (Wareing and Vaughan, 1979), copper (Mughrabi

Table 6.2. Calculation of fatigue coefficient for a number of metals/alloys

Material and Reference	E (GPa)	b (10^{-10} m)	$\frac{8(1-\nu)R_s w_s}{3\mu b}$
Cu	112	2.56	0.099
Ti	54.5	3.21	0.181
W	286	2.74	0.066
Fe (Type 316 stainless steel)	199	2.48	0.117
Ni (Waspaloy)	211	2.48	0.072
Co (Mar-M 509)	211	2.48	0.077

and Höppel 2010), titanium (Zhang et al. 1998), tungsten (Schmunk and Korth 1981), Waspaloy (Lerch and Jayaraman 1984), and Mar-M 509 (Reuchet and Remy 1979 1983). As shown in Figure 6.5, the fatigue life lines for different metals are not far apart. For Waspaloy, data from coarse grained (CG) microstructure with the grain size of 125 μm and fine grained (FG) microstructure with the grain size of 16 μm do not show much difference either in the LCF life vs. plastic strain plot. In terms of stress, however, the Hall-Petch effect of grain size as well as precipitates may affect the lattice friction resistance k such that the relationship, Eq. (6.17b), may indeed be affected by the microstructure. Also, in alloys, solute atoms may have an effect on the surface energy, as compared to the pure metal. Further studies

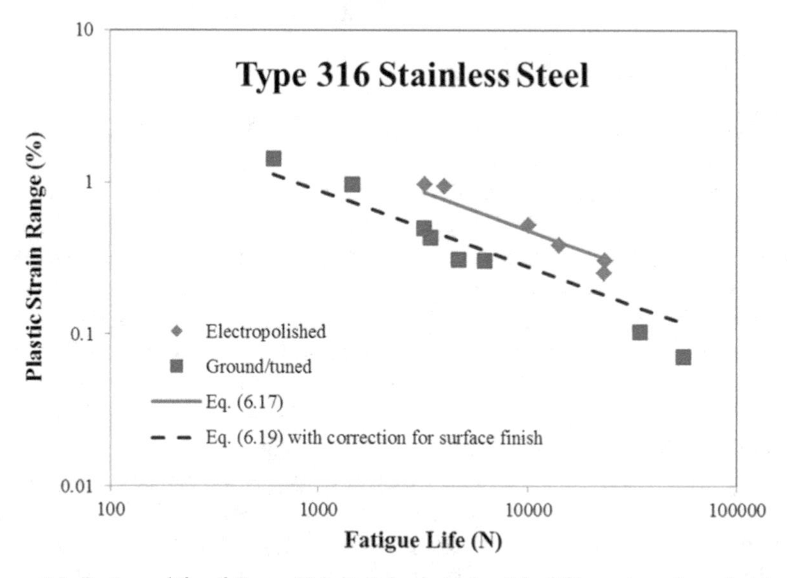

Figure 6.4. Fatigue life of Type 316 stainless steel with different surface finish. The symbols represent the referenced experimental data. The lines represent theoretical predictions of Eq. (6.17) and Eq. (6.19).

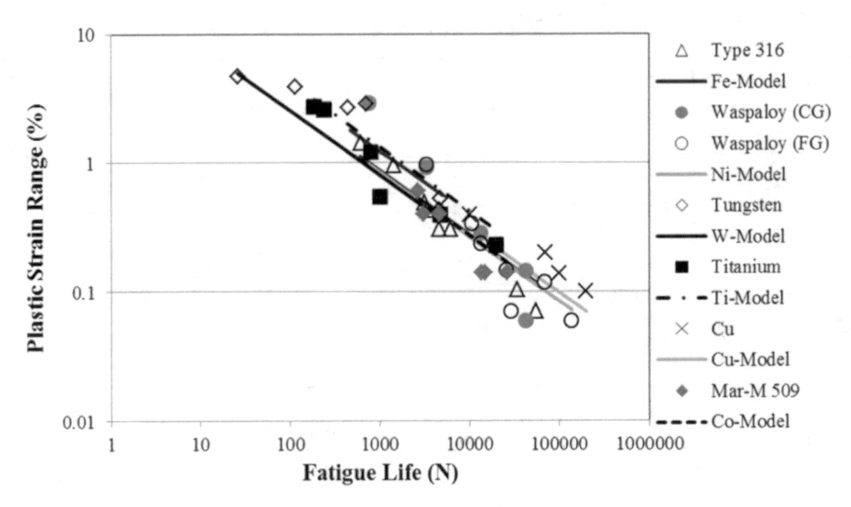

Figure 6.5. Predicted fatigue curves in comparison with experimental data.

are needed to evaluate the effect of solute concentration (at. %) on the surface energy.

In conclusion, Eq. (6.19) is very promising for fatigue design for the first approximation without resorting to experiment, which has a huge advantage over empirical approaches. Other factors such as high-temperature damage mechanisms and environmental effect will be delineated in the studies discussed in next few sections.

6.2 Cyclic Deformation

Materials cyclic deformation is complicated by involvement of multiple deformation mechanisms and microstructure evolution. The trace of stress-strain under LCF condition results in a hysteresis loop. The hysteresis loops may evolve within the controlled strain range as the cycle continues. When the hysteresis loops evolve with increasing stress, the material is *cyclically hardened,* which means the material becomes more resistant to the imposed straining. If the loops evolve with decreasing stress, the material is *cyclically softened*, which means the material becomes less resistant to straining. Most often, the hysteresis loop will become cyclically stabilized after a short period of cycling. For this case, a cyclic stress-strain relationship can be established by connecting the tips of all the stabilized loops at different strain levels, the relationship is called the *cyclic stress-strain curve,* which represents the material response during the entire cyclic deformation process, just as the stress-strain relationship under monotonic loading. For cyclic hardening/softening cases, however, no unique hysteresis behaviour can represent the material response for the cyclic deformation process; usually, the mid-life cyclic stress-strain response is taken as the characteristic one.

Cyclic deformation at high temperature involves both plasticity and creep that usually lead to a rate-dependent cyclic stress-strain or hysteresis behaviour. The phenomena on different materials have been described by many constitutive theories such as the unified constitutive theory (e.g., Chaboche, 1989) and the *two-layer viscoplasticity theory* (Seifert et al. 2014). It has been pointed out in Chapter 4 that the unified theories of viscoplasticity do not specifically delineate the contributions of plasticity and creep mechanisms, even though they can describe the total stress-strain response. The two-layer theory assumes plasticity and creep deformation are exhibited in different volume fractions of materials. However, from the microstructure and deformation mechanism point of view as discussed in Chapter 1&2, such volumes cannot be physically defined in terms of microstructural constituents (grain, precipitates, and grain boundaries). Therefore, from the physics of failure point of view, such theories are not able to describe physical damage evolution in association with the responsible deformation mechanism at specific microstructural locations. The metallurgical understanding achieved to date has clarified that cyclic plasticity via irreversible slip causes PSB leading to fatigue damage, whereas creep deformation induces microcracks, cavity/void growth, mostly along grain boundaries. Particularly, creep deformation arises mainly due to dislocation climb that releases pile-up dislocations at the obstacles in glide, and because of the random nature of vacancy flow, dislocation climb would help to ease slip concentration, thus not directly inflicting fatigue damage. On the other hand, vacancy flow promotes cavitation and void growth along grain boundaries, which forms the so-called internally distributed damage and may adversely affect the creep rate. This is why mechanical fatigue damage often results in transgranular fracture; whereas creep often causes intergranular fracture. In high-temperature deformation, all of the above mechanisms may contribute to inelastic straining, but their respective roles on damage accumulation are different, leading to a mix-mode failure. Therefore, it is important to use a mechanism-delineated approach to characterize cyclic deformation phenomena of engineering materials. In the following sections and later chapters, it will be demonstrated that various hysteresis behaviours are actually manifestation of different deformation mechanisms.

6.2.1 Cyclic Deformation by Plasticity

Similar to monotonic plastic deformation, the cyclic stress-strain curve by rate-independent plasticity can be expressed as:

$$\varepsilon = \frac{\sigma}{E} + \left(\frac{\sigma - \sigma_0}{K'}\right)^{n'} \tag{6.20}$$

where K' and n' are cyclic plastic strength and cyclic stress-exponent, which can be different from the monotonic parameters for the same material.

Figure 6.6 shows the cyclic stress-strain curves of ductile cast iron (DCI) at room temperature (RT) and 400 °C (Wu et al. 2014). The experimental

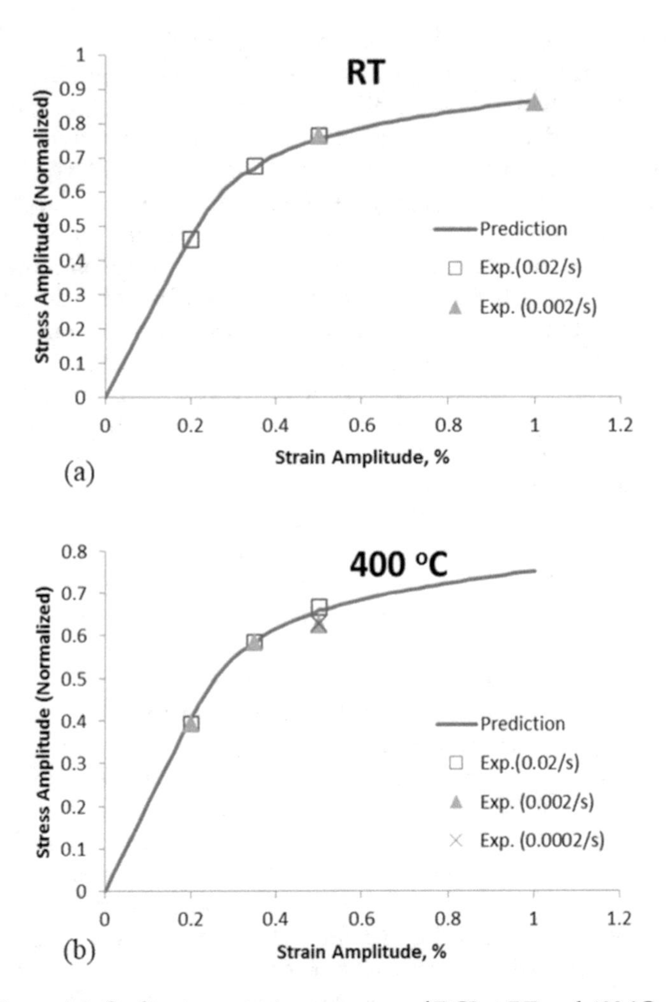

Figure 6.6. Cyclic stress-strain responses of DCI at RT and 400 °C.

stress-strain data at different strain rates all fall on one curve for each temperature condition, as described by Eq. (6.20), which proves that the deformation is controlled by rate-independent plasticity.

When a material's hysteresis loops are displaced with the compressive tips at the zero point of the stress range vs. strain range plot, as shown in Figure 6.7 for an austenitic cast steel at room temperature (RT), if the upper branches of all hysteresis loops coincide together, the cyclic stress-strain behaviour is called the Masing behaviour (Masing 1927) (If not, non-Masing). For the Masing behaviour, the upper branches of the hysteresis loops happen to fall on the cyclic stress-strain curve:

$$\Delta\varepsilon = \frac{\Delta\sigma}{E} + \left(\frac{\Delta\sigma - 2\sigma_0}{2\sqrt[n']{2}K'}\right)^{n'} \tag{6.21}$$

Figure 6.7 shows the hysteresis loops of an austenitic cast steel cycled at a strain rate of 0.02 s^{-1}. The hysteresis loops of 1% cyclic strain range at strain rates of 0.002 s^{-1} and 0.0002 s^{-1} are also shown in Figure 6.7, which demonstrates that the behaviour of this austenitic cast steel is rate-independent and Masing at room temperature (Wu et al. 2017). The upper branches of the hysteresis loops are well described by Eq. (6.21), from which the cyclic plasticity mechanism parameters σ_0, K' and n' are determined. For the Masing materials, the cyclic plasticity parameters are equal to the monotonic.

Generally, a material can be either cyclic hardened or softened, depending on its initial dislocation structure. A material in an annealed conditions often cyclic hardens, but a material in the cold-worked condition can cyclic soften, even for materials of the same composition. The cyclic stress response of the austenitic cast steel at RT is shown in Fig. 6.8. It undergoes slight hardening and then softening within the first 100 cycles, because the cast steel has a low initial dislocation density. After 100 cycles, the material's cyclic behaviour becomes fairly stabilized. Plumtree and Raouf (2001) studied the relation between the cyclic stress-strain behaviour and substructure for several alloys. They concluded that metals with fine dispersion particles and single phase low stacking-faulty-energy (SFE) exhibited Masing behaviour, but the high SFE materials exhibited non-Masing behaviour with dislocation cell formation. In high SFE materials the formation of a cellular structure effectively stores dislocations. On loading, a higher stress must be imposed for dislocation sources to operate within the cell walls. Once these sources are operational, however, there will be less resistance for dislocations to

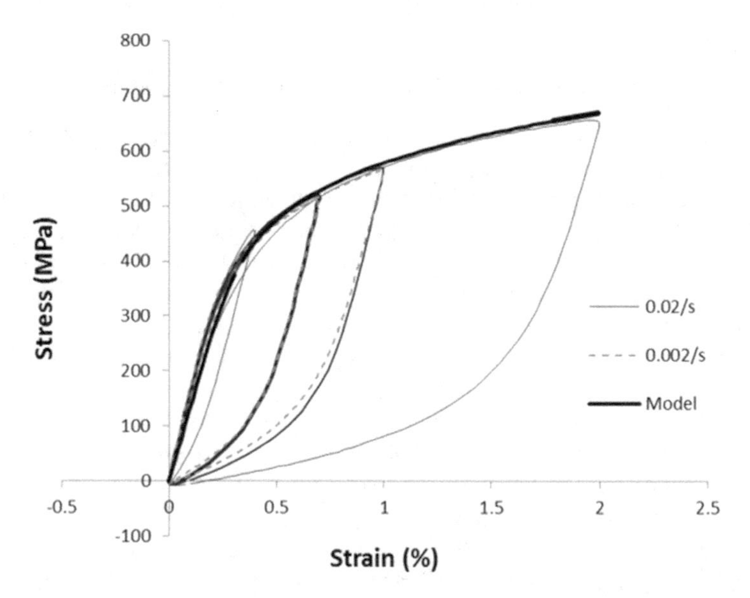

Figure 6.7. Stabilized hysteresis loops of austenitic cast steel at room temperature.

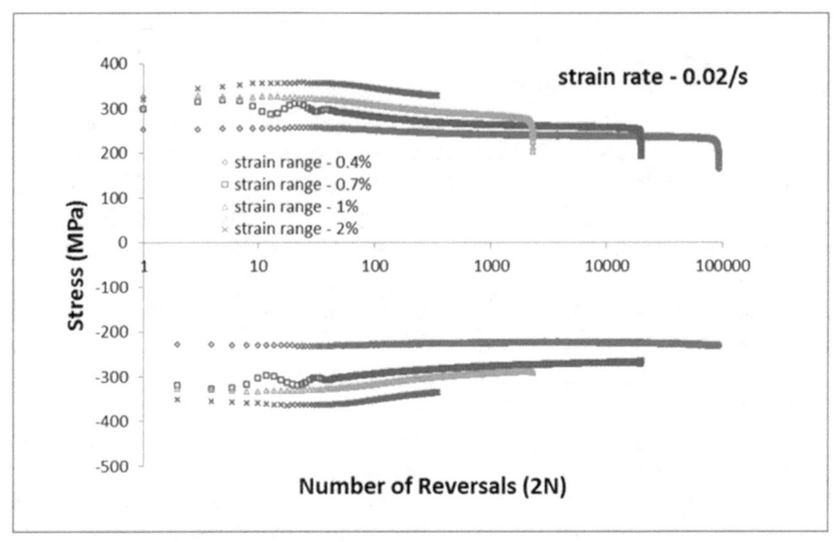

Figure 6.8. Cyclic peak-valley stresses of austenitic cast steel at RT.

move in the relatively dislocation-free cell until they encounter the adjacent walls where interactions and annihilations occur. Therefore, the flow stress of the hysteresis branches lies above the cyclic stress-strain curve.

Rate-independent cyclic plasticity can still predominate up to an elevated temperature before creep mechanisms take effect. For examples, DCI exhibits rate-independent plasticity up to 400 °C (Wu et al. 2014); Cobalt-base superalloy Mar-M 509 up to 600 °C (Reuchet and Remy, 1983); and single crystal Ni-base superalloys can be up to 750 °C (Reed 2006).

6.2.2 Cyclic Plasticity with Dynamic Strain Aging

Some materials such as austenitic stainless steels can exhibit a rate-dependent cyclic hardening behaviour, because of dislocation motion-induced hardening mechanisms affecting plasticity. For example, austenitic stainless steel Type 316L and austenitic cast steel 1.4848 were observed to experience dynamic strain aging (DSA) in the mid-temperature range of 250–600 °C(Alain et al. 1997, Hong et al. 2007, Wu et al. 2017). DSA is a result of interaction between diffusing solute atoms and mobile dislocations during plastic deformation. It is manifested by pinning of slow moving dislocations, restricting the cross-slip of screw dislocations, and hence enhancing slip inhomogeneity, thus resulting in an increased stress (hardening) to reach the prescribed strain level. In a cyclic process, this means continued evolution of the hysteresis loops with the number of cycles. When dynamic strain aging (DSA) intervenes, the cyclic behaviour becomes non-Masing.

Figure 6.9 shows that, at 400 °C, the austenitic cast steel 1.4848 exhibited significant cyclic-hardening until fracture, especially at a slow strain rate, $\dot{\varepsilon} = 0.002\ s^{-1}$. The hardening rate appears to be increasing with the cyclic strain

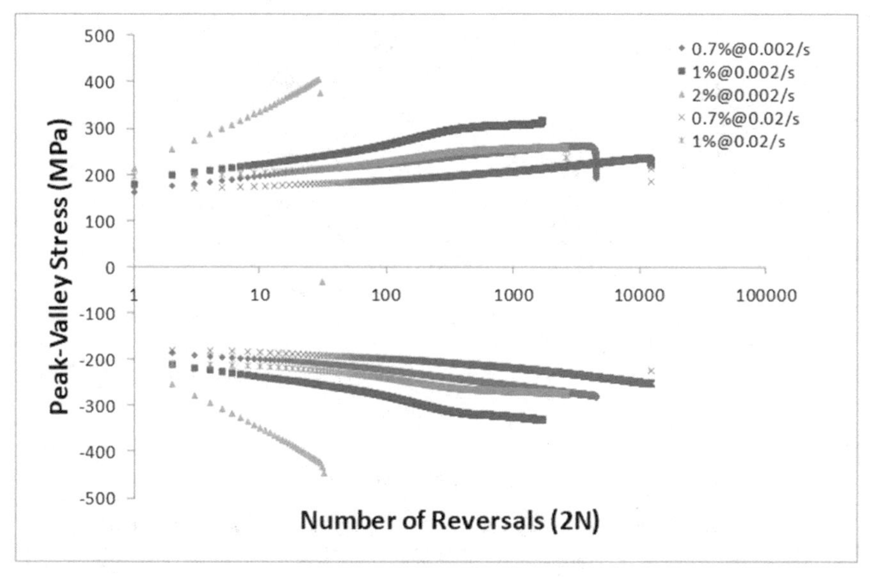

Figure 6.9. Cyclic peak-valley stresses of austenitic cast steel at 400 °C.

range. When cycling within the same strain range, lower strain rate resulted in stronger cyclic hardening. The phenomenon is called *negative strain rate sensitivity*, typical of DSA. The negative strain rate sensitivity of cyclic hardening has also been observed in Type 316L austenitic stainless steel (Hong et al. 2007). When the mid-life hysteresis loops are displaced with the compressive tips at the zero point of stress range-strain range plot, as shown in Figure 6.10, it is found that the 1.4848 austenitic cast steel exhibits non-Masing behaviour at 400 °C. The first cyclic stress-strain curve, the mid-life cyclic stress-strain curve, and the last cyclic stress-strain curve (the curve that connects the peaks of the last hysteresis loops at different strain ranges, before load drop due to crack initiation) are also shown, calculated using Eq. (6.21), to see the evolution of the cyclic stress-strain behaviour.

The mechanism of dislocation motion dragging solute atoms is called the Cottrell solute atmosphere (Cottrell 1953). The drag force depends on the ratio of dislocation velocity, v, to the diffusivity of the solute atom, D_i, in a function as $F \propto (v / D_i)\ln(D_i / v)$. In keeping with the plasticity formulation, we postulate that DSA induces additional drag stress as expressed by

$$K_{DSA} = \propto \frac{\dot{\varepsilon}^m}{D_i} \ln\left(\frac{D_i}{\dot{\varepsilon}^m}\right) \tag{6.22a}$$

which should be added to the baseline lattice drag stress, resulting in the total drag stress, as:

$$K = K_0 + k_1 T + k_2 \left\langle \frac{\dot{\varepsilon}^m}{D_i} \ln\left(\frac{D_i}{\dot{\varepsilon}^m}\right) \right\rangle \tag{6.22b}$$

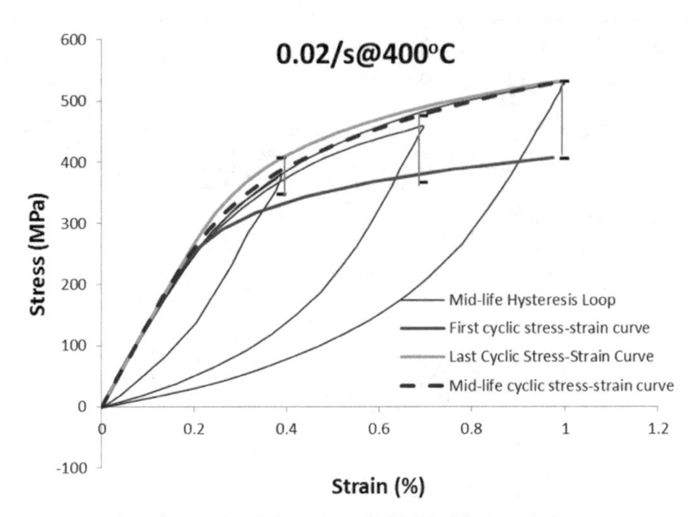

Figure 6.10. Mid-life hysteresis behaviours of austenitic cast steel at 400 °C, at strain rate of 0.002 s^{-1}.

where K_0, k_1, k_2 are material constants; m is a power-law exponent; $\langle\cdot\rangle$ signifies that $\langle x\rangle = x$, if $x > 0$, and $\langle x\rangle = 0$, if $x < 0$. Here, the baseline drag stress is formulated as a linear function of temperature, for simplicity. Eq. (6.22b) is compared with the experimental best-fit values at mid-life, as shown in Figure 6.11.

If one wants to describe cyclic hardening continuously, an internal state variable may be introduced. Here, we modify Chaboche's internal state variable R, Eq. (4.5), as

$$\dot{R} = \left[b\left(Q - R + cp\right) + c\right]\dot{p} \tag{6.23}$$

where b, Q, c are constants; and p is the accumulated plastic strain. A new term cp is introduced to describe the steady increase of isotropic hardening. When $c = 0$, it reduces to Chaboche's original formulation, which describes isotropic hardening approaching a constant level. The integration of Eq. (6.23) lead to

$$R = Q(1 - e^{-bp}) + cp \tag{6.24}$$

And it adds to the drag stress as $K = K_0 + R$, with K_0 being the drag stress in absence of cyclic (isotropic) hardening.

As cyclic plastic strain accumulates under DSA conditions, dislocation structure evolves, which also leads to change in the power-law exponent, n. By the same token, we formulate n as

$$n = n_0 - Q_n[1 - \exp(-b_n p)] + c_n p \tag{6.25}$$

where b_n, Q_n, c_n and n_0 are also best-fit constants.

With the above two equations to describe the evolution of plasticity constants K and n under DSA condition as shown in Figure 6.12 for the

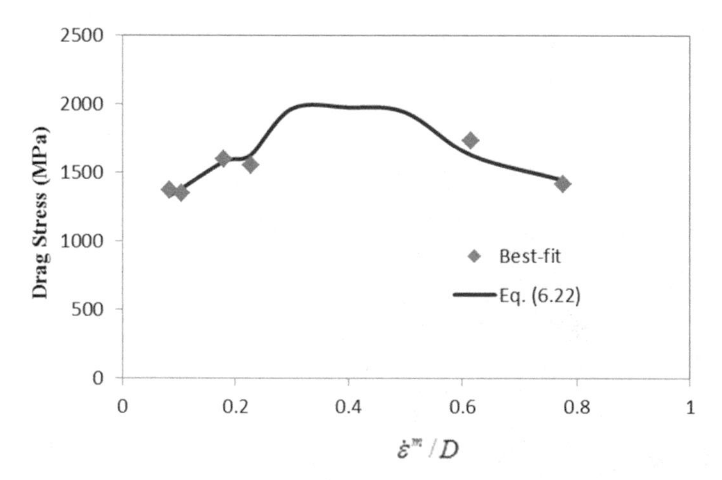

Figure 6.11. Drag stress for plasticity considering the effect of DSA.

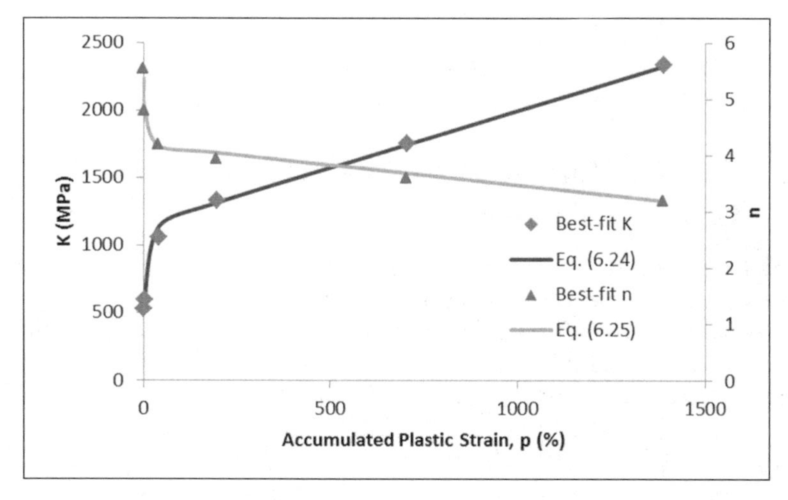

Figure 6.12. Plasticity parameters as evolved with accumulated plastic strain for the conditions between ±0.34% strain at strain rate of 0.002 s^{-1}, 673K (400 °C).

best-fit to the cyclic behaviour with strain range of $\varepsilon = \pm 0.34\%$ and strain rate of $\dot{\varepsilon} = 0.002\ s^{-1}$. Assuming constant kinematic hardening for each reversal, the hysteresis loops can be simulated, as shown in Figure 6.13 (lines). In principle, Eq. (6.24) and (6.25) could be further characterized in a similar fashion of Eq. (6.22). But, for component analysis, it would be computationally cumbersome to update the accumulated plastic strain p in every material element cycle by cycle until final failure for thousands or even millions of cycles. Some engineering simplification have to be made whether to use the "mid-life" or "cycle-by-cycle" characterization approaches, depending on the engineering problem in hand.

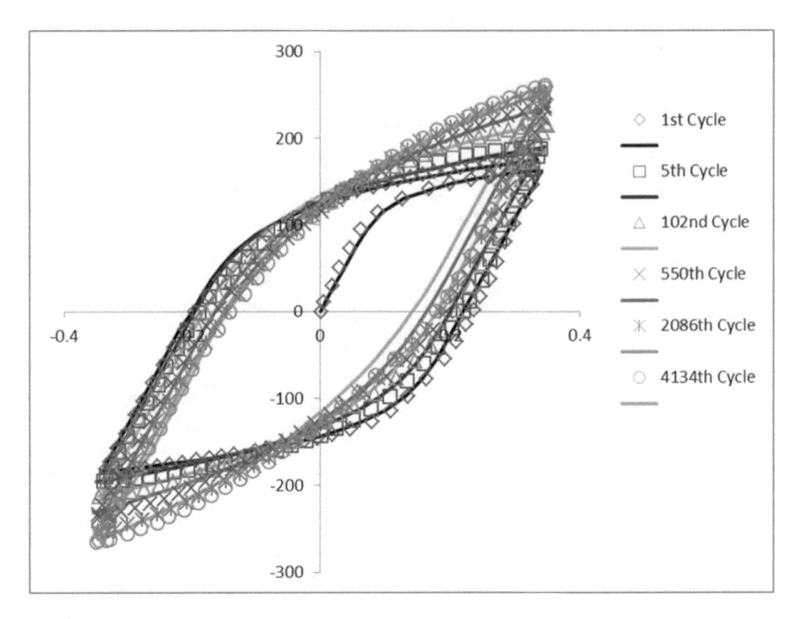

Figure 6.13. Experimental and simulated hysteresis loops of cast austenitic steel cycled between ±0.34% strains at strain rate of 0.002 s^{-1}, 673K (400 °C).

6.2.3 Cyclic Deformation by Plasticity Coupled with Dislocation Climb

At high temperatures, creep mechanisms inevitably participate in the deformation process. High silicon-molybdenum (SiMo) DCI exhibits power-law creep, as shown in Figure 2.3. The material is widely used for automotive exhaust manifolds and turbocharger housings. Its microstructure contains uniformly graphite nodules and Mo-rich eutectic phases distributed along grain boundaries, which suppresses GBS (Wu et al. 2014, 2015). Therefore, deformation of DCI is comprised of plasticity and power creep as:

$$\dot{\varepsilon} = \dot{\varepsilon}_e + \dot{\varepsilon}_p + \dot{\varepsilon}_v = \left[\frac{1}{E} + \frac{n}{K} \left(\frac{\sigma - \sigma_0}{K} \right)^{n-1} \right] \dot{\sigma} + C\sigma^m \tag{6.26}$$

Strain-controlled low-cycle fatigue (LCF) tests were conducted on DCI at strain rates of 0.02 s^{-1}, 0.002 s^{-1} and 0.0002 s^{-1} in the temperature range from RT to 800 °C (Wu et al. 2014). The material exhibited cyclic-stable behaviour, even though with slight cyclic softening at high temperatures. The rate-independent cyclic stress-strain curves are manifested by pure plasticity at RT and 400 °C, as shown in Figure 6.6, because under these conditions creep is negligible. As temperature is increased to 600 °C, rate-dependent behaviour is manifested due to power-law creep. Eq. (6.26), is used to predict the cyclic stress-strain curves for DCI at 600 °C and 800 °C, as shown in Figures 6.14 and 6.15, respectively.

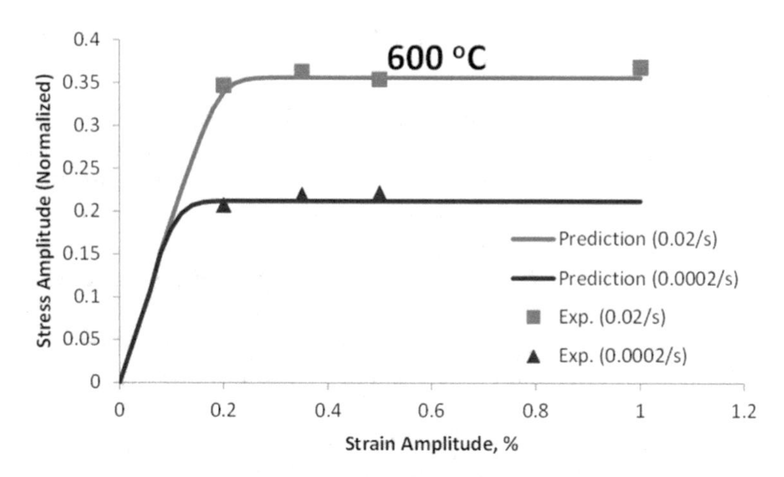

Figure 6.14. Cyclic stress-strain curve for DCI at 600 °C.

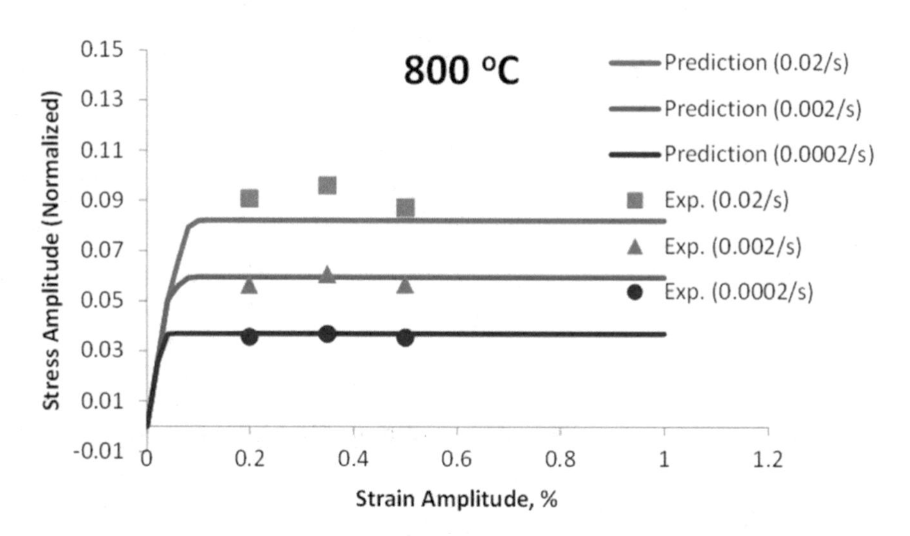

Figure 6.15. Cyclic stress-strain curve for DCI at 800 °C.

The ICFT model, Eq. (6.26), agrees well with the observed cyclic stress-strain behaviour of DCI. It naturally describes the transition from rate-independent stress-strain behaviour at low temperature to time-dependent behaviour at high temperature. Particularly, the absence of strain hardening in the high-temperature behaviour attests the manifestation of dislocation climb creep, which represents the material's ability to deform with time under constant stress. As the model evaluated, at high temperatures >600 °C, the cyclic deformation of DCI proceeds mostly by power-law creep, rather than plasticity.

6.2.4 Cyclic Deformation by Plasticity Coupled with Dislocation Glide

During monotonic or cyclic loading, while elastic-plastic deformation occurs instantaneously, rate-dependent behaviour is manifested by creep mechanisms. According to the deformation kinetics, it is always the highest rate mechanism in effect to cause material yielding. Under strain-controlled loading conditions, deformation is forced to keep up with the strain loading rate. In many materials under high strain-rate loading, dislocation glide dominates over dislocation climb.

According to Eq. (4.14), under constant strain-rate loading conditions, the material obeys the deformation constraint condition:

$$\dot{\varepsilon} = \dot{\varepsilon}_e + \dot{\varepsilon}_p + \dot{\varepsilon}_v = \left[\frac{1}{E} + \frac{n}{K}\left(\frac{\sigma - \sigma_0}{K}\right)^{n-1}\right]\dot{\sigma} + \dot{\varepsilon}_v = \frac{\dot{\sigma}}{E_t} + \dot{\varepsilon}_v = const. \tag{6.27}$$

where E_t is the instantaneous tangent modulus defined as

$$\frac{1}{E_t} = \frac{1}{E} + \frac{n}{K}\left(\frac{\sigma - \sigma_0}{K}\right)^{n-1} \tag{6.28}$$

When dislocation glide dominates the time-dependent deformation process, as discussed in section 2.2,

$$\dot{\varepsilon}_v = 2A_g \sinh \Phi \tag{6.29a}$$

where

$$\dot{\Phi} = \frac{V(\dot{\sigma} - H\dot{\varepsilon}_v)}{kT} \tag{6.29b}$$

Similar to the monotonic case described in section 2.2, Eq. (6.27) and (6.29) can be combined into a first-order differential equation of Φ, as given by

$$\dot{\Phi} = \frac{E_t V}{kT}\left[\dot{\varepsilon} - 2A_g\left(1 + \frac{H}{E_t}\right)\sinh \Phi\right] \tag{6.30}$$

This differential equation can be solved in infinitesimal steps as follows. We assume that the evolution of the energy, Φ, undergoes a series of infinitesimal isothermal steps, for each i-th step, the energy state evolves from Φ_{i-1} to Φ_i over the time interval $\Delta t_i = t_i - t_{i-1}$. Within this interval, the transient tangent E_t is constant. Then, following the solution procedures of section 2.2, we obtain:

$$\ln\left(\frac{e^{-\Phi} - a}{e^{-\Phi} + b}\right)\Bigg|_{\Phi_{i-1}}^{\Phi_i} = -\frac{VE_t\dot{\varepsilon}\sqrt{1+\chi^2}}{kT}\Delta t_i \qquad (i = 1, 2, \ldots) \tag{6.31}$$

Summing up all these infinitesimal steps, we obtain:

$$\sum_{i=1}^{N} \ln\left(\frac{e^{-\Phi}-a}{e^{-\Phi}+b}\right)\Bigg|_{\Phi_{i-1}}^{\Phi_i} = -\sum_{i=1}^{N} \frac{VE_t\dot{\varepsilon}\sqrt{1+\chi^2}}{kT}\Delta t_i \tag{6.32a}$$

where

$$\chi = \frac{2A_g}{\dot{\varepsilon}}\left(1+\frac{H}{E_t}\right), a = \frac{\sqrt{1+\chi^2}-1}{\chi}, b = \sqrt{1+\chi^2}+1 \tag{6.32b}$$

Let N→∞, the left-hand side of Eq. (6.32a) will be equal to the logarithmic difference between the final state and the initial state, and the right-hand side is an integration over the loading period. After mathematical rearrangement, we have:

$$\left(\frac{e^{-\Phi}-a}{e^{-\Phi}+b}\right) = \left(\frac{1-a}{\chi+b}\right)\exp\left[-\int_{t_0}^{t} \frac{VE_t\dot{\varepsilon}\sqrt{1+\chi^2}}{kT}dt\right] \tag{6.33}$$

where t_0 is the time to reach the elastic limit, or in other words, for plastic flow to commence.

The elastic limit is defined by

$$\Psi_0 = \frac{(\sigma-\sigma_0)}{kT} = 0 \tag{6.34}$$

From Eq. (6.33), we can obtain the stress-strain response as follows:

$$\sigma - H\varepsilon_v - \sigma_0 = -\frac{kT}{V}\ln\left(\frac{a+\omega b}{1-\omega\chi}\right) \tag{6.35a}$$

where ω is the response function defined by

$$\omega(\varepsilon) = \left(\frac{1-a}{\chi+b}\right)\exp\left\{-\frac{V\int_{\varepsilon_0}^{\varepsilon} E_t d\varepsilon\sqrt{1+\chi^2}}{kT}\right\} \tag{6.35b}$$

where $\varepsilon_0 = \sigma_0/E$.

The mid-life hysteresis loops of 1.4848 austenitic cast steel at 900 °C are shown in Figure 6.16 (a) and (b), for strain rates, 0.002/s and 0.0002/s, respectively, with mechanism parameter values given in Table 6.3. The cyclic behaviour appears to be closely Masing, and the upper branches of the hysteresis loops are described well by dislocation-glide creep with linear work-hardening, Eq. (6.35). Strain hardening occurs when the loading rate exceeds the dislocation glide rate, forcing dislocations to pile-up at obstacles, raising the flow stress. At 900 °C, the material appears to be cyclic stable right from the beginning and the stress response is higher at higher strain rate.

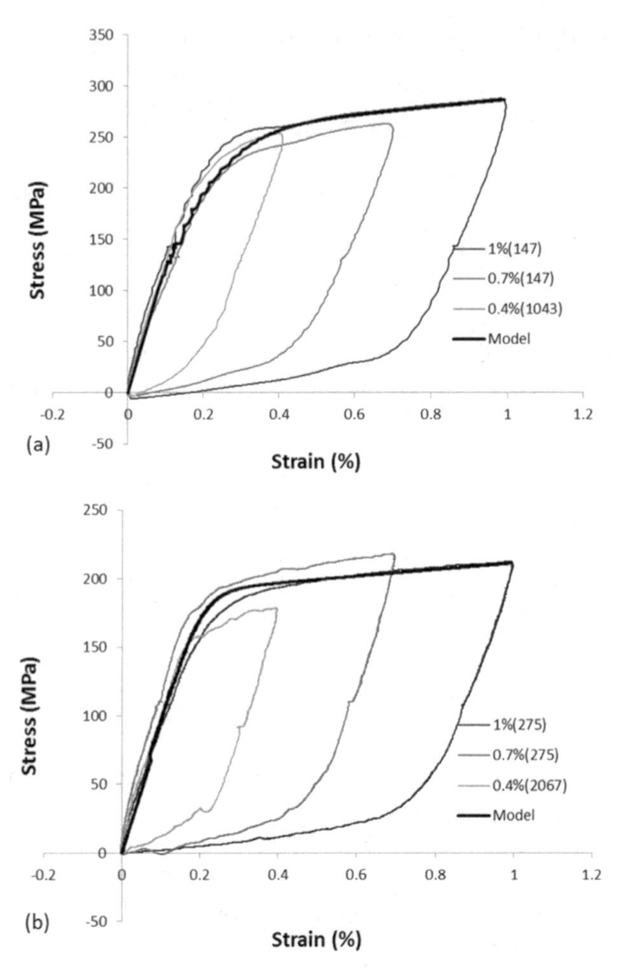

Figure 6.16. Mid-life hysteresis behaviours of austenitic cast steel at 900 °C, (a) at strain rate of 0.002/s, (b) at strain rate of 0.0002/s.

At larger strain amplitude ~ 1%, the cyclic behaviour is slightly softened, as shown in Figure 6.17. Apparently, at high temperature, the positive rate sensitivity arises from dislocation glide. When temperature is decreased and creep effect is negligible, Eq. (6.35) reduces to pure plasticity.

Another case of plasticity coupling with dislocation glide during cyclic deformation is shown for Mar-M 509, which is a carbide-strengthened cobalt base superalloy, and it exhibits cyclically stable behaviour, as observed by Reuchet and Remy (1983). The cyclic stress-strain behaviours of Mar-M 509 are shown in Figure 6.18 with the curves described by Eq. (6.35). Below 600 °C, deformation proceeds by pure plasticity. At 900 °C, the flow stress is significantly reduced and linear strain hardening is observed due to dislocation glide (the creep deformation model is calibrated to the creep behaviour shown in Figure 5.11).

Table 6.3. Deformation mechanisms and parameters

Temperature		293 K	673 K	873 K	1073 K	1173 K
Plasticity	σ_0 (MPa)	100	100	100	100	70
	n	5.54	4.75	4.75	4.75	4.75
	K (MPa)	1217	-	-		922
DSA	K (MPa)	$K = 922 + \left(\frac{1217-922}{1173-293}\right)(1173-T) + 1800\left\langle \frac{\dot{\varepsilon}^{0.1}}{185e^{-\frac{30000}{RT}}} \ln\left(\frac{185e^{-\frac{30000}{RT}}}{\dot{\varepsilon}^{0.1}}\right)\right\rangle$				
Creep	H (MPa)	-	-	-	3734	3734
	A_0 (s^{-1})			1.47×10^{34}		
	Q (kJ)			891		
	V (m^3)			5.6×10^{-28}		

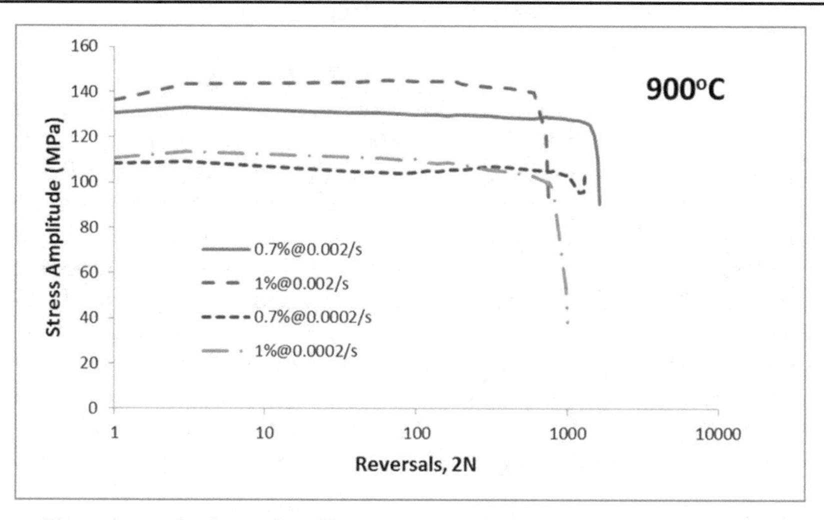

Figure 6.17. Cyclic peak-valley stresses of austenitic cast steel at 900 °C.

6.3 Low Cycle Fatigue Life Prediction

Section 6.1 has discussed LCF crack nucleation life by dislocation dipole pile-up distribution. The corrected Tanaka-Mura model, hereafter called the Tanaka-Mura-Wu model, has been shown to be in excellent agreement with experimental observations (the Coffin-Manson relationship) with a physically defined fatigue ductility (in terms of the shear modulus μ, Poisson's ratio v, Burgers vector b, and surface energy w_s of the material) and a power-index of -1/2. This is typical mechanical fatigue in absence of internally distributed damage and environmental effect. In Chapter 4, a mechanism-based holistic life evolution model, in the context of Eq. (4.18), is presented, within the

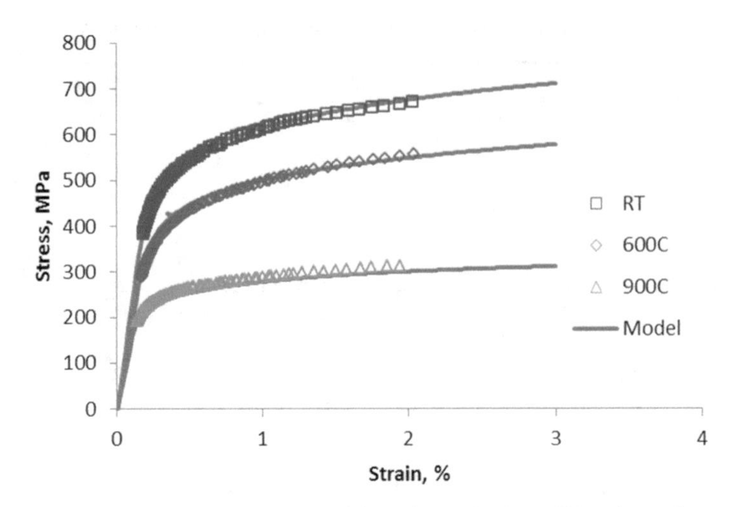

Figure 6.18. Cyclic stress-strain behaviours of Mar-M 509, data after Reuchet and Remy (1983).

framework of ICFT. In this model, the rate-independent plastic strain ε_p and time-dependent creep strain ε_v are delineated, so that mechanical fatigue and the effect of creep are evaluated separately, the latter mostly contributes to the growth of internally distributed damage (IDD). In addition, oxidation effect is also included. Hence, the life-limiting factor can be traced to particular controlling mechanism(s) and the respective strain component(s). According to Eq. (4.18), the LCF life is inversely related to the sum of fatigue and oxidation damage magnified by an internal damage factor D, where the factor *D* has to be formulated with due consideration of the physics of failure for the material concerned. Some case-study examples are given below to illustrate how various types of IDD interact with fatigue and oxidation in LCF, as described by Eq. (4.18).

6.3.1 Fatigue Life of Cast Irons with Different Graphite Morphology

Cast irons are widely used in automotive combustion engines and exhaust systems. These materials can be made with microstructures containing either flake graphite (FG), nodular graphite (NG) and/or vermicular graphite (VG). Figure 6.19 shows the fatigue crack nucleation and propagation path in a NGI at room temperature. Small cracks tend to nucleate from the graphite/iron interface, which then propagate transgranularly, cutting through the graphite nodules in the path (Seifert and Riedel 2010, Wu et al. 2014). Fatigue crack nucleation and propagation in VGI has also been studied in detail. Especially, in-situ observations on VGI fatigue samples revealed the crack nucleation and propagation mechanisms in the vermicular graphite iron, as shown in Figure 6.20 (Qiu et al. 2018). The VGI material contains ferrite

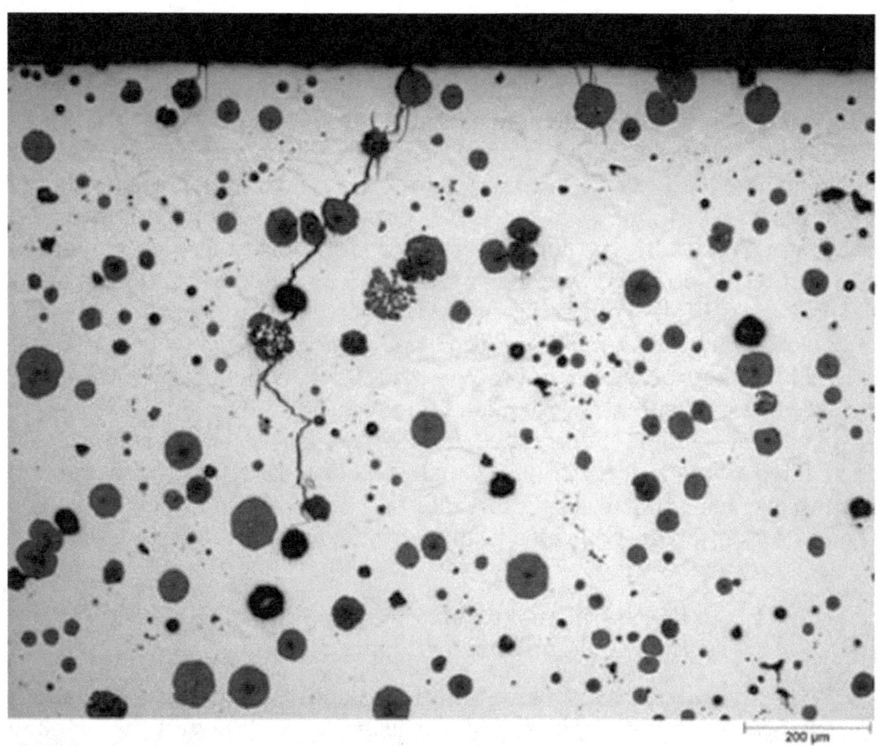

Figure 6.19. An optical micrograph of high Silicon SiMo DCI with the cell boundaries (diffused gray contrast) resolved by etching. The arrows indicate the Mo-rich eutectic carbide phase.

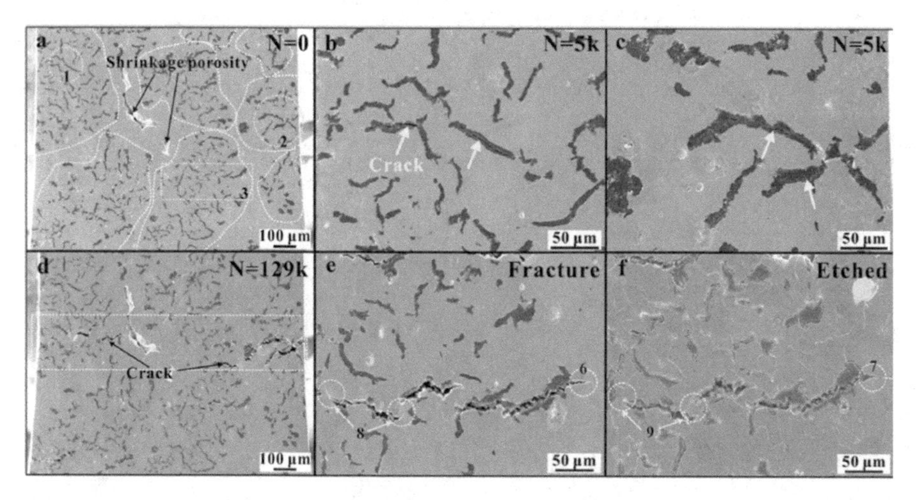

Figure 6.20. Micrographs of in-situ observation on fatigue of a VGI after different number of cycles (a-d) and after fracture (e-f).

zones (the dash line enclosed region) with VG clusters and pearlite regions (outside the dashed curve), as shown in Figure 6.20a. During fatigue, graphite particles start to de-bond, causing crack initiation (pointed by arrows) in Figure 6.20b and 6.20c, which are magnified from regions 1 and 2 in Figure 6.20a. With increasing number of cycles, more and more graphite particles de-bond, eventually leading to formation of the main fatigue crack, as shown in Figure 6.20d. By the weakest link hypothesis, crack propagation would occur along the path with longest graphite and shortest ferrite ligament in between. The ratio of the vermicular graphite length to the ferrite ligament is ~ 9 on average. The magnified zones at position 3 in Figure 6.20a before and after etching are shown in Figures 6.20e and 6.20f, respectively. It is seen that internal fatigue cracks initiated by VG particle debonding, which would be linked together resulting in the main fatigue crack.

As far as fatigue is concerned for the cast irons, the graphite can be considered as internally distributed microstructural defect, since the graphite itself may crack or the iron-graphite interface may debond under stress. The ratio of maximum graphite size to the ferrite ligament is roughly estimated by measurement:

- $l/\lambda = 1{:}1$ for the NG microstructure (Figure 6.19)
- $l/\lambda = 9{:}1$ for the VG microstructure (Figure 6.20)

Then, the mechanical fatigue life (in absence of creep and oxidation) for the two cast irons can be calculated using Eq. (4.18), with the baseline fatigue property for iron as given in Table 6.2 and the microstructural IDD factor $D = (1 + l/\lambda)$. The predicted fatigue curves are shown in Figure 6.21 in comparison with the experimental data: the agreement is excellent.

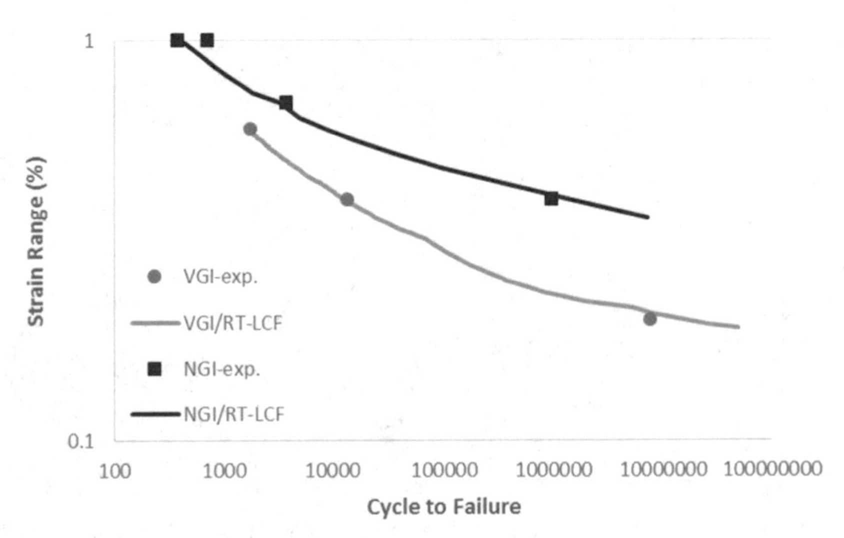

Figure 6.21. Fatigue life of NGI and VGI. The experimental data for NGI is taken from (Wu et al. 2014) and for VCI from (Zou et al. 2018). The lines represent Eq. (4.18).

6.3.2 LCF Life Coupled with Creep

Creep induced IDD has been extensively discussed in section 3.2. Creep as a failure mode alone is also discussed in Chapter 5. In this section, the effect of creep-IDD on LCF life will be discussed for Mar-M 509.

First of all, it should be recognized that, even though cyclic deformation of MAR-M 509 at temperatures below 600 °C is controlled by pure plasticity, fatigue crack nucleation occurs by two distinct mechanisms, as observed by Reuchet and Remy (1979, 1983): i) at low amplitude ($\Delta\varepsilon_p/2 < 0.5\%$), fatigue crack nucleation occurred along crystallographic planes in the matrix; and ii) at high amplitude ($\Delta\varepsilon_p/2 > 0.5\%$), it occurred by cracking of MC carbides at or near the surface. These two mechanisms are independent to each other, and hence the total mechanical fatigue damage follows the kinetics of parallel rate mechanisms:

$$\frac{1}{N_f} = \frac{1}{N_{slip}} + \frac{1}{N_{MC}} \tag{6.36}$$

Each fatigue damage follows the Coffin-Manson relationship, as

$$\frac{\Delta\varepsilon_p}{2} = \varepsilon_f N^{\,c} \tag{6.37}$$

with $c = -0.5$ and $\varepsilon_f = 0.12$ for slip-induced matrix cracking (N_{slip}), and $c = -0.085$ and $\varepsilon_f = 0.00685$ for MC cracking (N_{MC}). Note that fatigue life exponent $c = -0.5$ and the fatigue ductility $\varepsilon_f = 0.138$ are theoretically predicted by the Tanaka-Mura-Wu model, Eq. (6.19), which are very close to the experimental best-fit values. This confirms that the matrix crack nucleation mechanism indeed occurs by plasticity via slip. On the other hand, MC carbides are hard particles, the cracking of MC carbides appears to be brittle, i.e., within the elastic regime of carbide itself, and thus having a low fatigue ductility and exponent (for this part, we use the empirical Coffin-Manson relationship, because there is no information available on the surface energy of MC carbide-matrix interface). The LCF data from Reuchet and Remy (1983) are shown in Figure 6.22. The LCF life of MAR-M 509 at 600 °C also follows the same curve as the room temperature, because creep and oxidation contributions are negligible at this temperature.

As temperature is increased to 900 °C, the effect of creep deformation is manifested, Figure 6.22. Reuchet and Remy (1979) observed that creep cavities mostly formed at eutectic phase and MC carbide interfaces in Mar-M 509 at 900 °C. Therefore, in accordance with Eq. (3.7) and Eq. (4.18b), we formulate the IDD in Mar-M 509, as

$$D = (1 + \beta\varepsilon_v) \tag{6.38}$$

where ε_v is the creep strain, and β is the creep-effect factor. With the creep contribution in consideration, at 900 °C in vacuum, the fatigue life is described by Eq. (4.18) without oxidation ($h = 0$). In this case, the life in vacuum at 900 °C is only marginally lower than RT.

At 900 °C in air, however, oxidation effects become significant. The oxidation increment at crack-tip per cycle is $h = \sqrt{2k_{ox}\tau} = 2\sqrt{k_{ox}\Delta\varepsilon / \dot{\varepsilon}}$. Then, combining the above considerations into Eq. (4.18), we obtain

$$\frac{1}{N} = (1+\beta\varepsilon_v)\left\{\frac{1}{N_f} + 2\pi\left(\frac{Y\sigma_{max}}{K_{IC}}\right)^2 \sqrt{k_{ox}\frac{\Delta\varepsilon}{\dot{\varepsilon}}}\right\} \tag{6.39}$$

As it can be seen in Figure 6.22, the LCF life of MAR-M 509 at 900 °C in air is greatly reduced by oxidation, as compared to that in vacuum. The

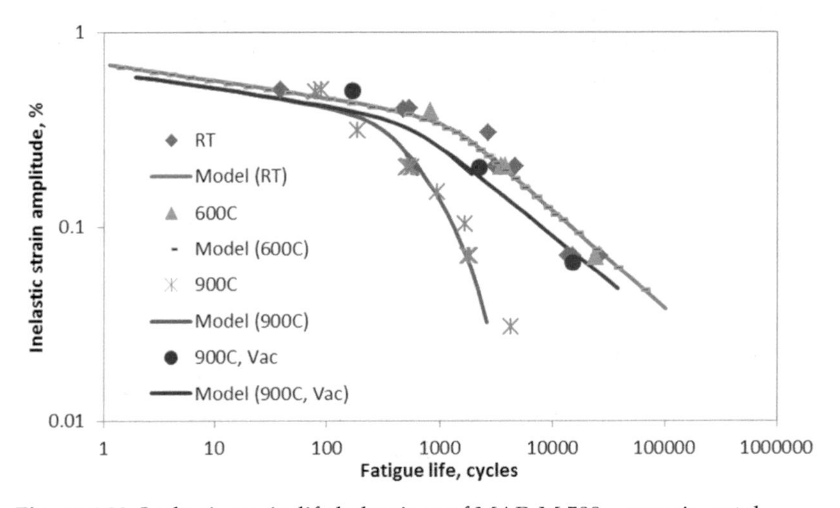

Figure 6.22. Inelastic strain-life behaviour of MAR-M 509—experimental and prediction of Eq. (6.39).

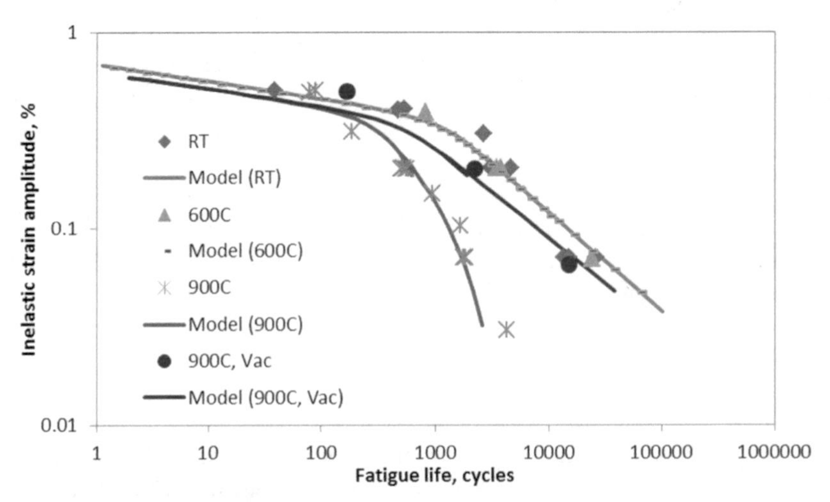

Figure 6.23. Total strain-life behaviour of MAR-M 509—experimental and prediction of Eq. (6.39).

LCF life as function of the total strain for Mar-M 509 is shown in Figure 6.23. In the total strain-life plot, because the plastic strain (including dislocation glide) approaches zero, as the total strain falls into the elastic regime ($\sigma \rightarrow \sigma_0$), the fatigue life becomes infinite. The above analysis has demonstrated that Eq. (6.39) can appropriately take into account the effect of matrix slip, MC carbide cracking, oxidation and creep mechanisms, and the description agrees very well with the experimental observations on Mar-M 509. It is appreciated that only the ICFT delineates the contributions from various physical mechanisms, whereas other empirical or phenomenological models do not.

6.3.3 LCF Life Coupled with Intergranular Embrittlement and Creep

Intergranular embrittlement (IE) is recognized as an additional damage mechanism in ductile cast iron (DCI), which is particularly responsible for causing the material's brittleness at medium temperature (BMT). At 400 °C, BMT occurs by an intergranular embrittlement mechanism associated with dissolution of the Mg/P constituent particles, which has a deleterious effect on grain boundary strength. Since BMT is related to Mg/P dissolution, which is a kinetics process, therefore BMT is rate-dependent. In this section, a special treatment is given to this mechanism, because it introduces intergranular fracture to LCF. This mechanism-based approach can be applicable to other materials suffering intergranular cracking such as by hydrogen embrittlement, etc.

With regards to BMT of DCI, Kobayashi et al. (1989) found that the minimum-elongation temperature T_ε is dependent on the strain rate, as shown in Figure 6.24, satisfying the Arrhenius relationship:

$$\dot{\varepsilon} = 1.09 \times 10^{19} \exp\left(-\frac{290000}{RT_\varepsilon}\right) \tag{6.40}$$

To describe the BMT phenomenon, we define an intergranular embrittlement (IE) damage parameter, ϕ, as:

$$\phi = \exp\left[-\left(\frac{T - T_\varepsilon}{T_0}\right)^2\right] \tag{6.41}$$

where T_ε is the minimum ductility temperature defined by Eq. (6.40), and T_0 is the scaling temperature that characterizes the range of IE influence. The elongation (proportional to $1-\phi$) at different strain rates are shown in Figure 6.24. The curves represent Eq. (6.41), which depicts a symmetrical upside down bell-shape curve, even though the test behaviour is a little bit skewed. The main point here is to capture the characteristics of BMT with a minimum ductility. Therefore, Eq. (6.41) can be used to deduce the effect of IE on LCF.

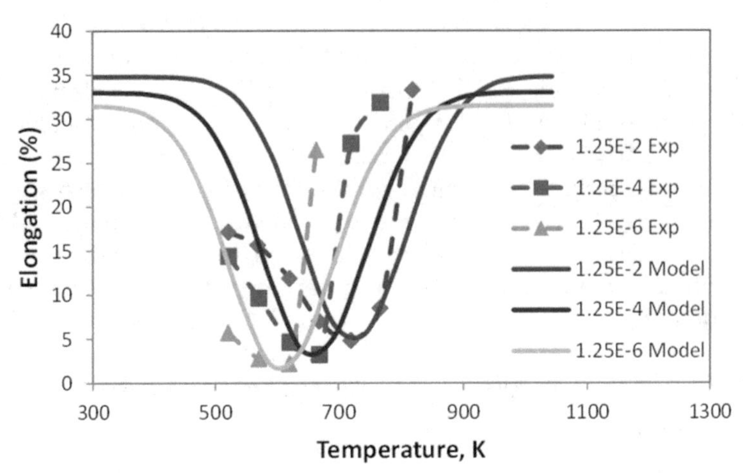

Figure 6.24. Effects of strain rate and temperature on tensile elongation of DCI.

Since the fracture toughness of a ductile material depends on its ductility, we assume that the fracture toughness of DCI also depends on ϕ, as given by

$$K_{IC} = K_0 + (1 - \phi)K_1 \tag{6.42}$$

where K_0 and K_1 are material constants. When $\phi = 1$, $K_{IC} = K_0$, which assumes IE fracture; and when $\phi = 0$, $K_{IC} = K_0 + K_1$, which assumes the full toughness of the material.

Naturally, IE is a form of internally distributed damage that enters into Eq. (4.17), where l equals to the eutectic cellular size multiplied by ϕ, and λ represents the thickness of the Mo-rich eutectic phase. The eutectic cellular size in DCI is in the order of $l = 100$ μm, while the grain boundary Mo-rich eutectic phase is about λ (2-7 μm) thick. Therefore, the IE contribution to the internal damage is $(l/\lambda)\phi$, which would have a maximum effect in the order of ~50.

In addition, as temperature is increased, creep occurs and the nodular voids in DCI become enlarged. During the LCF tests on DCI, graphite nodules were all detached from the matrix at high temperatures, and no intergranular creep cavities were observed along the grain boundaries, which meant that the Mo-rich eutectic phases was effective to suppress grain boundary sliding (Wu 2014). Hence, we assume that the creep contribution to the internally distributed damage is $\beta\varepsilon_v$, where β is the void growth factor. Therefore, the overall damage factor D for DCI can be written as:

$$D = 1 + \frac{l}{\lambda}\phi + \beta\varepsilon_v \tag{6.43}$$

With all the above in consideration, the LCF life for DCI can be expressed as:

$$\frac{1}{N} = \left(1 + \frac{d}{\lambda}\phi + \beta\varepsilon_v\right)\left\{\frac{1}{N_f} + 2\pi\left(\frac{Y\sigma_{max}}{K_{IC}}\right)^2 \sqrt{k_{ox}\frac{\Delta\varepsilon}{\dot{\varepsilon}}}\right\} \tag{6.44}$$

The normalized LCF life data for DCI are shown in Figure 6.25. The best-fit RT-LCF life has a power exponent of $c = -0.52$ and fatigue ductility of 0.213, which are very close to the theoretical exponent value of –0.5 and fatigue ductility value of 0.2418, as predicted by the Tanaka-Mura-Wu model, Eq. (6.16) with the NG microstructural feature (section 6.3.1). The fatigue fracture mode is transgranular at RT, as shown in Figure 6.26 (a).

At 400 °C, however the LCF life was significantly reduced by a factor of 55, apparently due to IE. The fracture mode at this temperature was almost entirely intergranular, but with some ligaments of transgranular ductile fracture, in contrast to that at RT, as shown in Figure 6.26 (b). This number corresponds to the ratio of the eutectic cell size to its thickness when ϕ becomes close to 1 such that the damage factor D reaches its maximum. Manson (1965) proposed a "universal" fatigue equation, where the LCF life is directly related to the ductility of the material. But, this apparently cannot explain the drastic reduction of LCF life by 55 times at 400 °C, since the reduction in the ductility of DCI is only by ten (10) times, as compared to RT. Also, in this case, there is a change in fracture mode from transgranular to predominantly intergranular. The IE effect has not been captured by other LCF failure models. This further validates the ICFT's physical premise of fatigue crack nucleation and propagation in coalescence of with internally distributed damage (IE-induced grain boundary fracture facets, in this case).

At high temperatures above 600 °C, an interesting cross-over phenomenon is observed: the LCF life is even longer than the RT life above certain strain range level. This behaviour cannot be explained by the linear damage rule (LDR), by which high temperature fatigue life would always be less than the room temperature life, because the creep and oxidation damage add up positively at high temperatures. But, this is not true for DCI. The ICFT provided a mechanism-based explanation as follows.

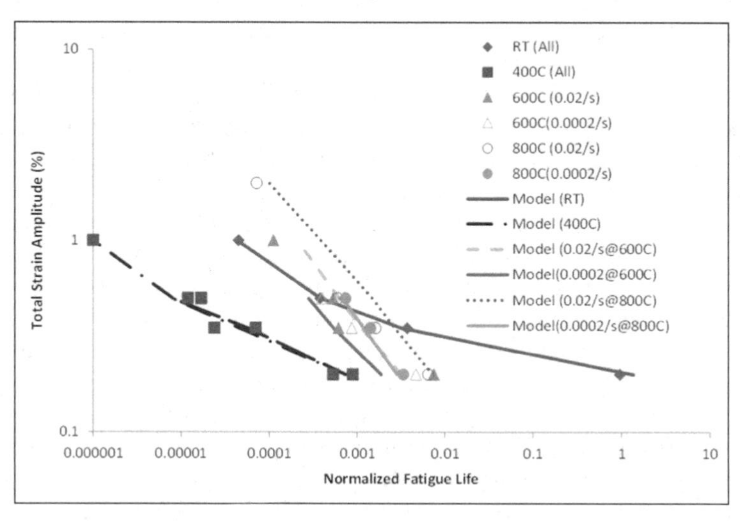

Figure 6.25. LCF life vs. the total strain amplitude for DCI.

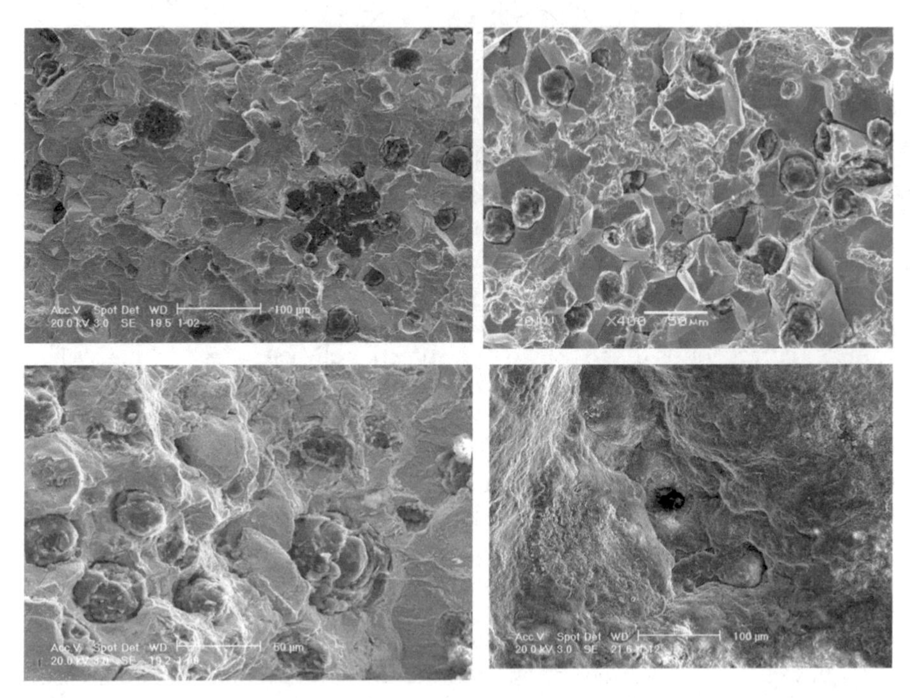

Figure 6.26. Fracture surfaces of specimens with 1% total strain range (a) at room temperature, (b) at 400 °C, (c) at 600 °C and strain rate of 0.02/s, and (d) at 800 °C and strain rate of 0.02/s.

At temperatures above 600 °C, rate-dependent stress-strain behaviour was manifested because of creep, Figures 6.14 and 6.15. Thus, at a given total strain level, the contribution of plasticity would be reduced and creep strain would become predominant, compared to the low temperature cases. This trend would proceed with decreasing strain rate and/or increasing temperature. Since mechanical fatigue occurs by irreversible slip, reduction of plastic strain ($\Delta\varepsilon_p$) would lead to a longer fatigue life, if the amount of internal damage by creep or IE was small. Looking at a constant fatigue life in Figure 6.25, additional inelastic (presumably creep) deformation does not contribute to fatigue at temperatures >600 °C. This agrees with the dislocation climb mechanism in physical essence. By the classification of fatigue and internally distributed damage, the ICFT model attributes the creep effect on damage evolution to the enlargement of voids occupied by graphite nodules in DCI. It is evident as shown in the micrographs, Figure 6.26 (c), which shows that graphite nodules all separated from the matrix at temperatures >600 °C. Intergranular embrittlement became less significant and even negligible, except at very high strain rate at 600 °C, which could be inferred from Eq. (6.40) and (6.41), but oxidation became the life-limiting factor, as the ratio h/a_c increased with temperature.

Another aspect of creep effect on LCF life lies in dynamic stress relaxation as deformation proceeds at a constant strain rate. It reduces the flow stress significantly. This seems to provide a beneficial effect on LCF life by both delaying oxide-crack nucleation and allowing the specimen to tolerate a larger crack, since the critical crack length is inversely proportional to stress to the 2nd power. Therefore, at temperatures >600 °C, the cross-over behaviour occurred, which means that even with participation of two commonly-perceived detrimental mechanisms—creep and oxidation, LCF of DCI could even be longer than at room temperature when those mechanisms were absent. The ICFT can not only explain the cross-over behaviour, but also why LCF at 800 °C were consistently higher than that at 600 °C, as shown in Figure 6.25, even it seems to be contradictory to the conventional thinking. By the same token, in LCF of DCI at 800 °C, because the stress was much reduced with the ease of extensive creep deformation, as shown in Figure 6.15, the increase of the critical crack length would be much larger than the increase of oxide penetration by the parabolic law such that the ratio h/a_c became smaller than that at 600 °C, resulting in a longer LCF life at 800 °C. At a given temperature, decreasing strain rate has an effect to reduce LCF life, as expected, since it allowed more time for oxidation per cycle. But, again, decreasing strain rate also lowered the flow stress, which in turn had a beneficial effect on LCF life. Hence, the overall effect was somewhat neutralized, as seen in Figure 6.25. At low strain ranges (amplitudes), however, the fatigue mechanism permitted long-elapsed cycles (time), and oxidation would eventually become the controlling failure mechanism of high temperature LCF process, resulting in LCF life shorter than that at RT, as shown in Figure 6.25. Indeed, the ICFT model delineates the LCF behaviour in terms of mechanism strain components and environmental factors and thus provide insightful understanding of the roles by various physical mechanisms. The empirical Coffin-Manson equation itself could not provide such mechanism-delineation.

The fractographic evidence supports the model predication on the failure mechanisms. Figure 6.26 (a) shows the fracture surface morphology of RT fatigue fracture, which consists of ridges and river patterns. This indicates that fatigue of DCI at room temperature proceeded mostly by a transgranular fracture mode. On the fracture surface, some graphite nodules appeared to be cut by the propagating crack, and some were by-passed by the crack, leaving shallow pulling-out marks, which suggest that the iron/graphite interface was weakened by cyclic deformation, but still remained an effective bonding between graphite nodules and the matrix. The fracture mode at 400 °C was predominantly intergranular, as can be seen from Figure 6.26 (b). It appeared that debonding of graphite nodules from the iron matrix initiated intergranular cracks, and these nodule-nucleated cracks coalesced along grain boundaries. Intergranular fracture could be attributed to intergranular embrittlement (IE) in DCI due to segregation of magnesium and phosphorus, which was believed to cause low ductility at medium temperatures. Similar grain boundary fracture features have also been observed in tensile tests (Kobayashi et al. 1989). As the test temperature was increased to 600 °C, oxidation damage could be seen on the fracture surface. At a high strain rate of 0.02 s^{-1}, the surface was lightly oxidized, and intergranular fracture still took place in a large portion, Figure 6.26 (c). At 800 °C, oxidation was presumably the major damage mechanism with oxide scales covering the entire fracture surface, as shown in Figure 6.26 (d) where graphite nodules cannot be seen any more. This mode of fracture was prevalent at all strain rates tested at this temperature. The overall fracture surface was more flattened and featureless due to oxidation and reverse creep deformation during compression. It could be concluded that crack growth at this temperature was totally controlled by the kinetics of oxidation of Si-Mo iron.

The intricate fatigue-creep-oxidation interaction may be better summarized in a LCF mechanism map, as shown in Figure 6.27, for DCI at strain rate 0.02 s^{-1}. The ordinate shows the life fraction of fatigue, N/N_f, and oxidation, Nh/a_c. The abscissa shows the homologous temperature (T/T_m, where T_m is the melting temperature in degree Kelvin). For illustration of the relative contribution of fatigue, IE, creep and oxidation, Eq. (4.18) is re-written as

$$\frac{1}{D}=\left\{\frac{N}{N_f}+\frac{Nh}{a_c}\right\} \tag{6.45}$$

Eq. (6.45) simply states that the sum of fatigue fraction (due to plasticity alone) and oxidation fraction for a material should always be equal to or less than 1. For an ideal material with no internally distributed damage and discontinuities, D = 1, and Eq. (6.45) resumes to the linear summation rule

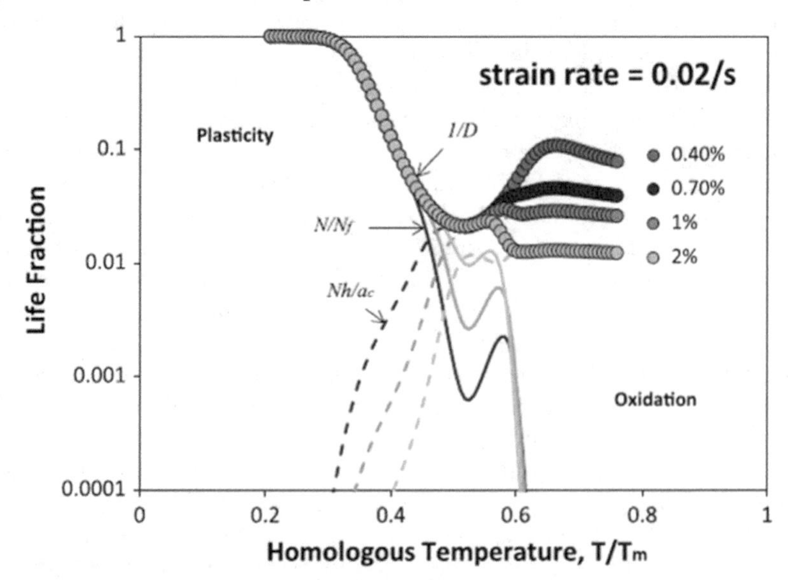

Figure 6.27. LCF mechanism map for DCI at strain rate of 0.02 s^{-1}.

for fatigue and oxidation. However, for a material that inevitably contains internally distributed damage and discontinuities such as promoted by IE or creep, D will be greater than 1, and the "damage tolerance" for fatigue and oxidation will be reduced to a number less than 1 as $1/D$.

It can be noticed that starting from 250 °C IE has a significant effect on fatigue life with a drastic increase of the internal damage factor D, mostly contributed by IE in the medium temperature range. Below 400 °C, the total life fraction is almost exclusively comprised of rate-independent plasticity-induced fatigue. Above 400 °C, with the onset of time-dependent deformation, the life fraction decreases drastically, and at the same time the oxidation life fraction, Nh/a_c, rises up to dominance. Interestingly, Nh/a_c approaches a constant level as temperature increases at a given cyclic strain range. With increasing strain range at high temperatures, the internal damage factor D further increases with creep strain accumulation, and the total fatigue portion is reduced. The LCF mechanism map may have significant implications in guiding material design for application under high temperature LCF conditions: any means to combat oxidation by alloying or coating will increase the fatigue life, as Nh/a_c approaches a constant at a given strain range. Of course, at the same time, alloy designer will also have to consider limiting the accumulation of internally distributed damage in the material. Along this line, aluminum-containing DCI seems to be promising.

6.3.4 LCF Life Coupled with DSA and Creep

As discussed in section 6.2.2, austenitic stainless steels deform by plasticity with DSA at intermediate temperatures, 400–600 °C, until creep is manifested at high temperatures, 800 °C and 900 °C. In this section, we will discuss how IDD develops in association with these processes. First, recall that the austenitic cast stainless steel 1.4848 exhibits rate-independent Masing behaviour at RT, but when DSA intervenes, the material cyclic-hardens continuously, exhibiting a non-Masing behaviour (Wu et al. 2017). Continuous cyclic hardening results in increasing peak-valley stresses and decreasing plastic strain range cycle by cycle, but the hysteresis energy remains fairly constant, as shown in Figure 6.13. At high temperatures, the material becomes cyclic stable again as creep mechanism operates.

Normally, the LCF life is correlated with the plastic strain range, if the material is cyclic stable, as discussed in section 6.1. Actually, the LCF behaviour of austenitic steel 1.4848 can be characterized with the Coffin-Manson relation with $\varepsilon'_f = 0.22$ and $c = -0.5$ at RT when the material is cyclic stable (Wu et al. 2016), which agrees very well with the theoretical value by the Tanaka-Mura-Wu model, Eq. (6.19). However, with continuous cyclic hardening, fatigue damage assessment in terms of plastic strain range cycle by cycle would be very tedious in practice. Hence, the energy-based Morrow relation can be used as an alternative to characterize fatigue life, as (Morrow 1965):

$$W_p = W_0 N_f^d \tag{6.46}$$

where W_p is the plastic strain energy of the hysteresis loop, W_0 and d are material constants.

At intermediate temperatures, 400 °C and 600 °C, DSA promotes slip inhomogeneity with dislocation pile-ups, forming concentrated slip bands or dislocation walls. For example, a planar slip structure of DSA was observed in 316L stainless steel (Hong et al. 2005). Dislocation pile-ups are embryos of cracks, the internal damage by DSA can be formulated as proportional to the excess forest dislocation density. Therefore, the D-factor for austenitic stainless steel can be written as:

$$D = 1 + \alpha\left[\left(\frac{\Delta\sigma_H}{\mu b}\right)^2 - \rho_0\right] + \beta\varepsilon_v \tag{6.47}$$

where $\Delta\sigma_H$ is the amplitude of cyclic hardening (maximum attainable peak stress minus the peak stress of the first cycle), ρ_0 is the dislocation density level below which there is no instantaneous crack nucleation, α is the proportional constant for dislocation-nucleated cracks, and β is proportional constant for creep damage.

With consideration of Eq. (6.47), Eq. (4.18) can be rewritten into

$$\frac{1}{N} = \left\{1 + \alpha\left[\left(\frac{\Delta\sigma_H}{\mu b}\right)^2 - \rho_0\right] + \beta\varepsilon_v\right\}\left\{\frac{1}{N_f} + 2\pi\left(\frac{Y\sigma_{max}}{K_{IC}}\right)^2\sqrt{k_{ox}\frac{\Delta\varepsilon}{\dot{\varepsilon}}}\right\} \tag{6.48}$$

Figure 6.28 shows the LCF life of austenitic cast steel 1.4848 in the temperature range from RT to 1173 K (900 °C). Eq. (6.46) is calibrated against the RT-LCF life for pure mechanical fatigue. At 673–873 K (400–600 °C), the LCF life is indeed significantly reduced at high strain amplitudes with significant cyclic hardening, as compared to the RT baseline. The DSA effect is more pronounced at higher strain amplitudes (higher hysteresis energy) and lower strain rates at 673–873 K (400–600 °C), and hence this results in an increased *D*-factor as given by Eq. (6.47), given that the creep strain at these intermediate temperatures is minimal. As temperature increases to 1073 K (800 °C) and 1173 K (900 °C), the DSA term in Eq. (6.47) becomes negligible, since the cyclic hardening are balanced by the creep softening, but the creep damage term remains in Eq. (6.47). In addition to mechanical fatigue and creep, at high temperatures, oxidation plays an important role to reduce the fatigue life at 1073 K (800 °C) and 1173 K (900 °C), as compared to RT.

Considering the effects of all above mechanisms, the ICFT predictions are shown (as lines) in comparison with the experimental data. The model parameters for life prediction are given in Table 6.4. As indicated by Eq. (6.48), the interaction between creep and oxidation is non-linear. Creep deformation on one hand reduces the LCF life by a factor of D, but on the other hand, it lowers the peak stress, which allows the specimen to tolerate a longer crack, thus increasing the crack propagation life. Similarly, lowering

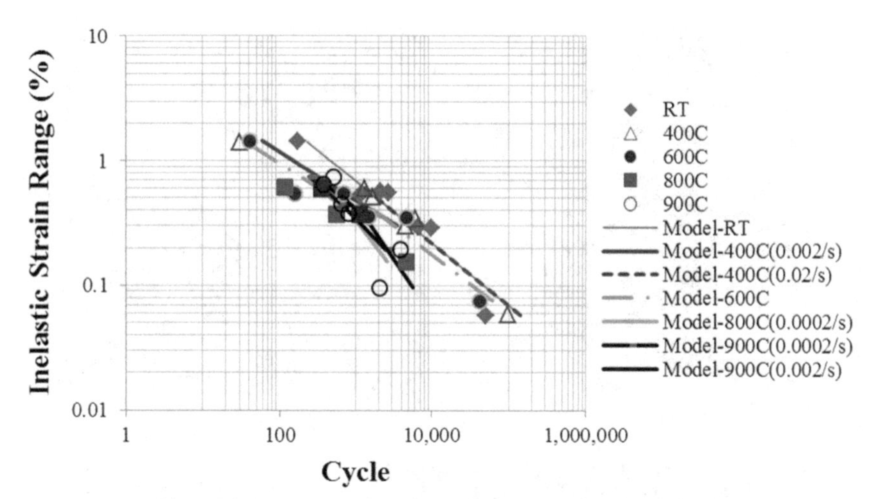

Figure 6.28. The inelastic strain energy vs. fatigue life for 1.4848 stainless steel.

the controlling strain rate allows more creep deformation to relax the peak stress, but it also allows more time of oxidation per cycle. All such conflicting effects result in a rather frequency-insensitive behaviour of LCF life of 1.4848 steel at high temperature, which is reflected in the model curves for 1073 K (800 °C) and 1173 K (900 °C) at different strain rates. Another similar austenitic cast steel 1.4949 was studied by Seifert et al. (2010b). They used a crack-tip plastic blunting model to describe the LCF and TMF behaviour of 1.4849 steel up to 600 °C. This fracture mechanics-based model does not consider oxidation explicitly.

The RT fracture surface of the austenitic cast steel contains typical striations that indicates transgranular fatigue fracture, as show in Figure 6.29 (a). Fatigue fracture at 673–873 K (400–600 °C) was also transgranular, however, DSA led to a concentrated slip band pattern, as shown in Figure 6.29 (b). At 1073 K (800 °C), the fatigue specimen failed predominantly by an intergranular fracture mode, as shown in Figure 6.29 (c). On the fracture surface of the specimen failed at 1173 K (900°C), extensive oxidation and formation of voids or cavities can be observed, as shown in Figure 6.29 (d). The above metallurgical evidence supports that creep has a significant effect on fracture at 1073–1173 K (800–900 °C). These fractographic observations corroborate the ICFT prediction.

Table 6.4. Life prediction parameter

Mechanism	**Fatigue**		**DSA**		**Creep**	**Oxidation: $k_{ox}=k_0\exp(-Q/RT)$**		
Parameter (unit)	W_{pl0} (10^4Nm^{-2})	d	α (m^2)	ρ_0 (m^{-2})	β	k_0 (m^2/s)	Q (kJ)	K_{IC}/Y (MPa$\sqrt{m}$)
Value	11336	-0.508	2.28×10^{-13}	3.71×10^{12}	1200	0.017	231	32

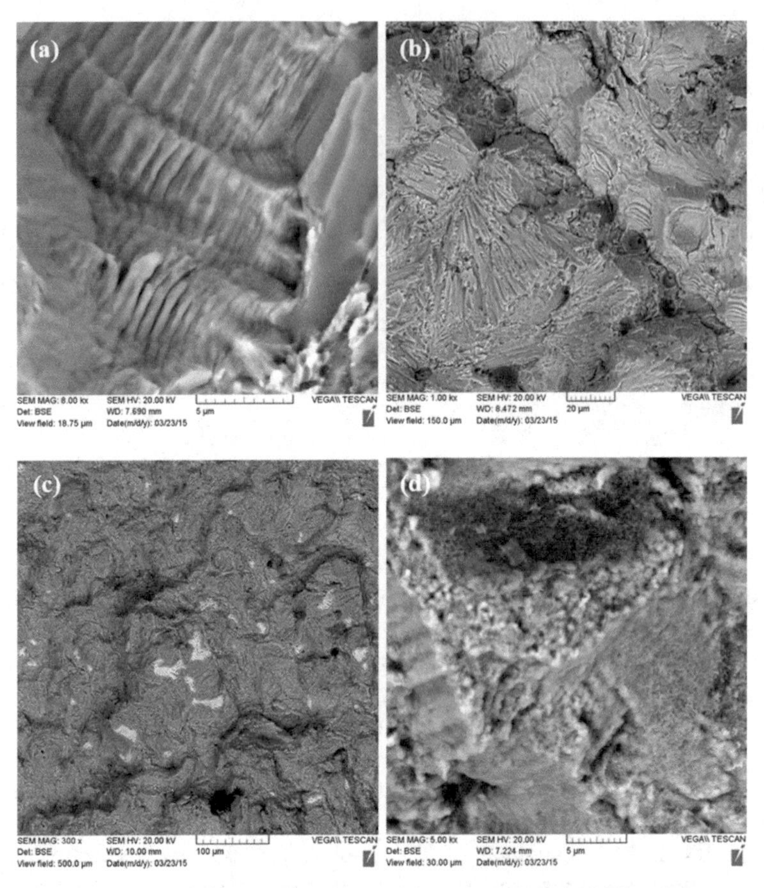

Figure 6.29. (a) Striated fatigue fracture surface at RT with 0.7% strain range; (b) fracture surface with slip-bands, fatigued with 0.7% strain range at 0.002 s^{-1}, 673 K (400 °C); (c) intergranular fracture with 0.7% strain range at 0.002 s^{-1}, 873 K (800 °C), the white spots are areas of high Ni content; and (d) cavitated and oxidized fracture surface with 0.7% strain range at 0.002 s^{-1}, 1073 K (900 °C).

6.4 LCF under Cyclic-Dwell Combined Loading

The above sections have shown cyclic deformation and LCF life under constant-amplitude fully-reversed loading such that the hysteresis behaviour is symmetrical in tension and compression. Under a complex loading profile as often seen in service, the cyclic strain range may not be imposed symmetrically about the zero strain position. Under strain control, asymmetrical accumulation of plastic strain would cause kinematic hardening (Bauschinger effect) shifting towards one direction, thus the mean stress will be gradually shaken down to a new stabilized position. The phenomenon has been discussed by Manson and Halford (2006).

Most engineering components such as gas turbine components, automotive turbo charger and exhaust components operate under complex loading profiles which can be translated into combinations of cyclic loads and dwell-hold periods at high temperatures. Inevitably, this will induce complicated material hysteresis behaviour and variable life, as compared to constant-amplitude loading. The phenomenon is generally referred to as *creep-fatigue interaction*, basically from a loading-profile point of view. By mechanisms, it is actually *creep-fatigue-environment interaction*, as delineated by the ICFT. In the following, creep-fatigue interactions under complex loading profiles are discussed with ICFT.

6.4.1 The Strain-Range Partitioning Approach

In 1970s, NASA used six types of characterization cycle to investigate creep-fatigue interaction as follows.

- *High Rate Strain Cycle (HRSC)*: It consists of ramping in tension and compression at a fast constant rate.
- *Compressive Cyclic Creep Rupture (CCCR)*: It consists of ramping to a stress level in compression plus a creep hold period, followed by reverse ramping to equal tensile strain.
- *Balanced Cyclic Creep Rupture (BCCR)*: It consists of equal creep holds in both tension and compression until specific strain reached.
- *Tensile Cyclic Creep Rupture (TCCR)*: It is opposite to CCCR with tensile creep-hold.
- *Tensile Hold Strain Cycle (THSC)*: It consists of ramping to specific strain, stress relaxation followed by reversed ramping to equal compressive strain.
- *Compressive Hold Strain Cycle (CHSC)*: It is opposite to THSC with compressive stress relaxation.

The typical material hysteresis behaviours arising from the above test conditions are shown in Figure 6.30. HRSC produces plastic strain reversed by plasticity, $\Delta\varepsilon_{pp}$; CCCR produces plastic strain reversed by creep, $\Delta\varepsilon_{pc}$; BCCR produces creep strain reversed by creep, $\Delta\varepsilon_{cc}$; TCCR produces creep strain reversed by plasticity, $\Delta\varepsilon_{cp}$; THSC produces partly $\Delta\varepsilon_{pp}$ and partly $\Delta\varepsilon_{cp}$; and CHSC produces partly $\Delta\varepsilon_{pp}$ and partly $\Delta\varepsilon_{pc}$. A strain-range partitioning (SRP) method was proposed as (Halford et al. 1977)

$$\frac{\Delta\varepsilon_{in}}{N} = \frac{\Delta\varepsilon_{pp}}{N_{pp}} + \frac{\Delta\varepsilon_{cc}}{N_{cc}} + \frac{\Delta\varepsilon_{pc}}{N_{pc}} + \frac{\Delta\varepsilon_{cp}}{N_{cp}} \tag{6.49}$$

where N_{ij} is the number of cycles to failure when the entire inelastic strain $\Delta\varepsilon_{in}$ is comprised of the named strain only. N_{ij} generally follows the Coffin-Manson relation for each type of strain component.

In the ideal case, each partitioning strain component could represent a type of mechanism strain. However, in SRP, each inelastic strain component was not evaluated using mechanism-based constitutive law, but graphically extracted from the observed hysteresis loop for the corresponding

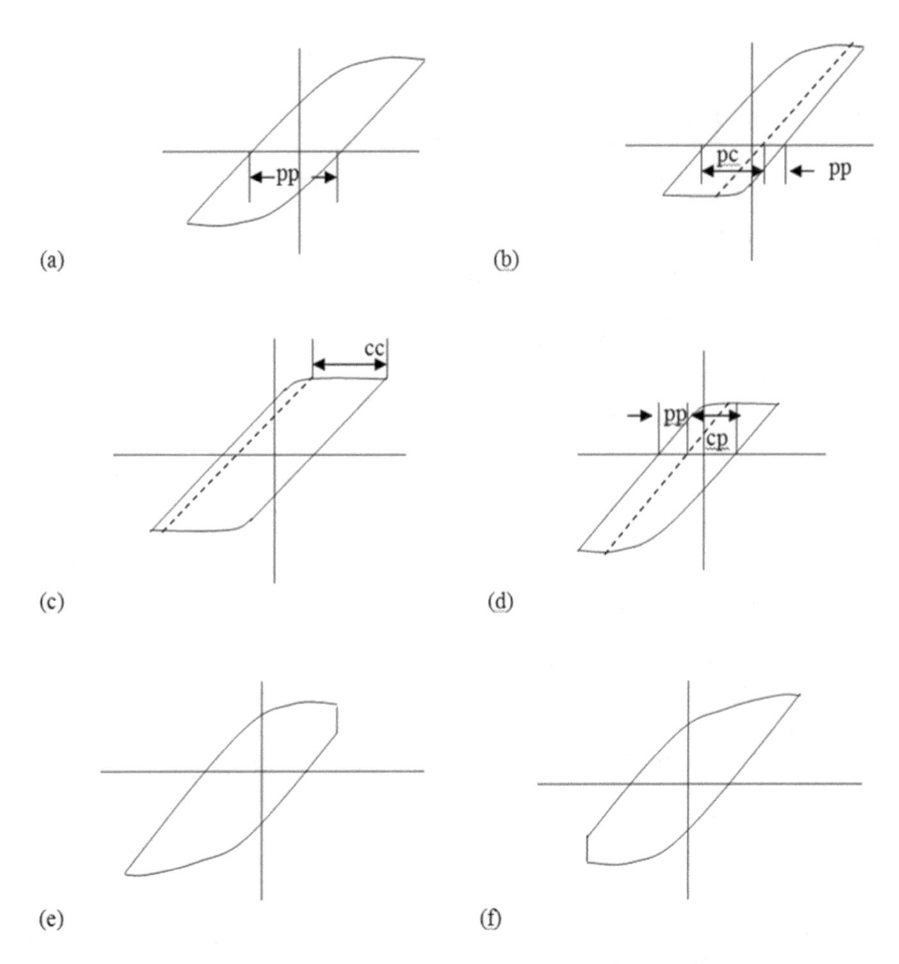

Figure 6.30. Schematic hysteresis behaviours arising from the SRP test conditions.

characterization cycle. It is almost impossible to apply the SRP method for component analysis, because the above four types of inelastic strain components cannot be clearly defined from an arbitrary hysteresis behaviour under complex loading that can be any combination of the above six cycles. Besides, establishing the four types of Coffin-Manson relations is also quite a costly experimental undertaking. Despite that, the SRP method was the most comprehensive approach dealing with creep-fatigue interaction at the time. In the following, we will use ICFT to rationalize the SRP method on NASA's test data, to gain a physics-based understanding.

Taking the Rene 80-vacuum-LCF data and the coated IN100-LCF data from the NASA contract report (Romannoski 1982) for analysis, given the test conditions, it can be considered that there is no environmental effect on the bulk failure of these two materials. Therefore, Eq. (4.18) can be reduced to:

$$N = \frac{N_f}{1 + \frac{\Delta\varepsilon_v d}{\lambda}} \tag{6.50}$$

where the damage size in the D factor is substituted by Eq. (3.7) and pure fatigue life is evaluated by the Coffin-Manson relation.

In the application of Eq. (6.50) to asymmetrical loading conditions as CCCR, TCCR, THSC and CHSC, it should be realized that, due to fully reversed plasticity, the entire inelastic strain range $\Delta\varepsilon_g$ should contribute to fatigue, but the creep strain $\Delta\varepsilon_v$ contribute to the effect of additional intergranular fracture, because GBS occurs by shearing. Table 6.5 summarizes the mechanism strain partitioning of $\Delta\varepsilon_g$ and $\Delta\varepsilon_v$ for the different creep-fatigue interaction tests.

Predictions of Eq. (6.50) for Rene 80 and IN100 are also given in Tables 6.6 and 6.7, respectively. Figure 6.31 shows the strain-life plot for Rene 80 in vacuum, and Figure 6.30 shows for the coated IN-100. Eq. (6.50) correlates well with the experimental observations for both alloys. It is demonstrated that the ICFT model, Eq. (6.50), can explain the creep-fatigue interaction phenomena that the strain range partitioning method would use four Coffin-Mansion relations to do.

Table 6.5. ICFT equivalence to strain-range partitioning

Test type	$\Delta\varepsilon_{in}$	$\Delta\varepsilon_v$
HSRC	pp	0
CCCR	pp+pc	pc
TCCR	pp+cp	cp
BCCR	pp	cc
THSC	pp+cp+cc	$\Delta\sigma/E^*$
CHSC	pp+pc+cc	$\Delta\sigma/E^*$

*Note that $\Delta\sigma$ is the range of stress drop during stress relaxation in this test.

6.4.2 Cyclic Deformation under Compressive Strain-Hold

Standard cyclic deformation has been discussed in section 6.2. Here, the cyclic deformation under CHSC condition is given some attention, because it is pertinent to the material loading profile on gas turbine components. Compressive dwell-fatigue behaviours of single crystal Ni-base superalloys with or without rhenium (Re) were investigated (Yandt et al. 2011, 2012). For discussion here, we only consider CMSX-2 as an example, which is a first-generation single crystal Ni-base superalloy used for gas turbine blades.

Table 6.6. Rene 80 at 871 °C (d/λ = 8)

Spec. ID	Test	$\Delta\varepsilon_{in}$	$\Delta\varepsilon_v$	N_f	N	Exp.
74-U-pp-13	HRSC	0.605	0	175	175	145
21U-pp-8	HRSC	0.322	0	617	617	642
41U-pp-10	HRSC	0.179	0	1997	1997	1410
22U-pp-9	HRSC	0.026	0	94675	94675	163533
42U-pp-11	HRSC	0.051	0	24606	24606	217620
92U-pc-13	CCCR	0.554	0.46	209	45	41
28U-pc-9	CCCR	0.378	0.283	448	137	149
91U-pc-12	CCCR	0.257	0.209	969	363	356
98U-pc-16	CCCR	0.258	0.183	961	390	396
29U-pc-10	CCCR	0.204	0.164	1538	665	1415
112U-cp-11	TCCR	0.385	0.308	432	125	101
86U-cp-9	TCCR	0.289	0.306	766	222	147
30U-cp-5	TCCR	0.289	0.254	766	253	193
31U-cp-6	TCCR	0.208	0.202	1479	565	530
36U-cp-7	TCCR	0.111	0.092	5194	2992	3705

Table 6.7. IN 100(Coated) at 900 °C (d/λ = 50)

Spec. ID	Test	$\Delta\varepsilon_{in}$	$\Delta\varepsilon_v$	N_f	N	Exp.
7	HRSC	0.129	0	796	796	635
6	HRSC	0.121	0	905	905	900
1	HRSC	0.138	0	696	696	1260
2	HRSC	0.086	0	1792	1792	2120
3	HRSC	0.059	0	3806	3806	3670
4	HRSC	0.05	0	5300	5300	9460
5	HRSC	0.031	0	13788	13788	12210
10	HRSC	0.026	0	19601	19601	17340
8	HRSC	0.028	0	16901	16901	27260
11	HRSC	0.014	0	67602	67602	48320
N12	CHSC	0.196	0.03375	345	128	250
N10	CHSC	0.105	0.02	1202	601	764
N9	CHSC	0.102	0.019375	1274	647	944
39	THSC	0.18	0.026875	409	174	239
N8	THSC	0.08	0.016875	2070	1123	1495
54	BCCR	0.09	0.168	1636	174	159
N5	BCCR	0.085	0.16	1834	204	200
56	BCCR	0.054	0.11	4544	699	383

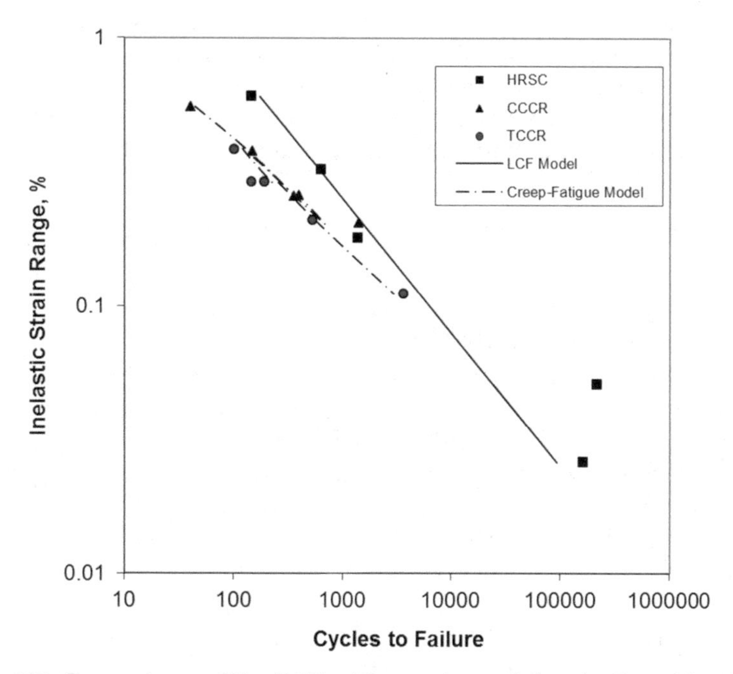

Figure 6.31. Comparisons of Eq. (6.50) with experimental data for Rene 80 at 871 °C.

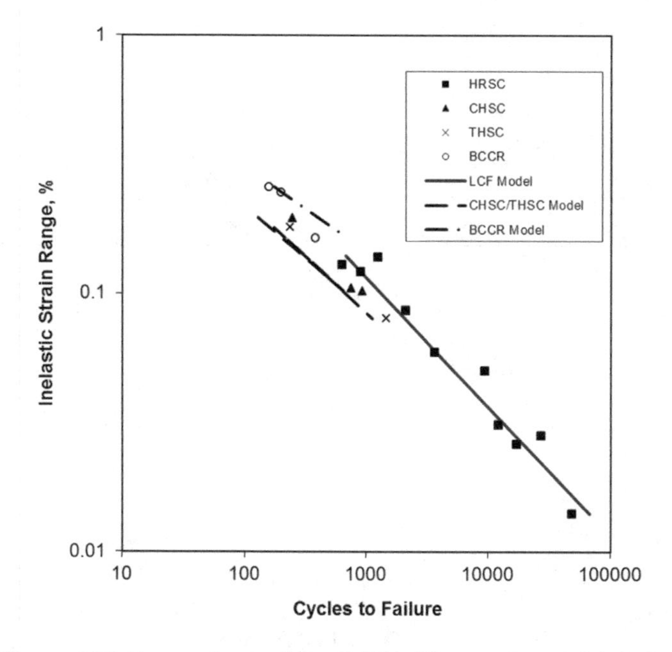

Figure 6.32. Comparisons of Eq. (6.50) with experimental data for IN 100 (coated) at 1000 °C.

Generally, during cycling with strain-hold, stress relaxation occurs. When the total strain is held constant, the material starts to creep and the creep strain incrementally replaces the elastic strain, causing stress to relax. The material was tested along the <001> direction, which is the direction of centrifugal force for the blade. The deformation equation of stress relaxation can be written as:

$$\dot{\varepsilon} = \frac{\dot{\sigma}}{E} + \dot{\varepsilon}_v \tag{6.51}$$

In single crystal Ni-base superalloys under strain-controlled LCF conditions such as the case of CMSX-2 at 1100 °C, the dominant deformation mechanism is dislocation glide. The CHSC cycle was conducted as follows: first, the material was compressed to the maximum strain at strain rate of $0.004s^{-1}$; then, a 120s-hold period was kept, followed by strain reversing back to zero. The recorded material hysteresis behaviour is shown in Figure 6.33.

According to the mechanism equation, section 2.2, the first loading part is described by Eq. (2.10). The solution for stress relaxation can be obtained as follows.

$$\dot{\Phi} = \frac{V\dot{\sigma}}{kT} = -\frac{VE\dot{\varepsilon}_v}{kT} = -\frac{2EVA_g}{kT}\sin h\,\Phi \tag{6.52}$$

Integration of Eq. (6.52) leads to

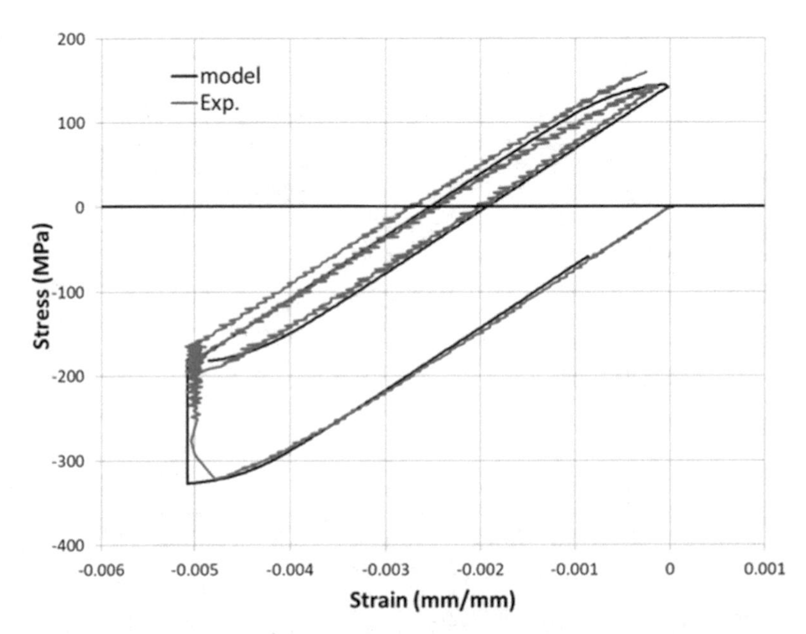

Figure 6.33. Hysteresis behaviour of CMSX-2 with a 120s-hold at 1100 °C.

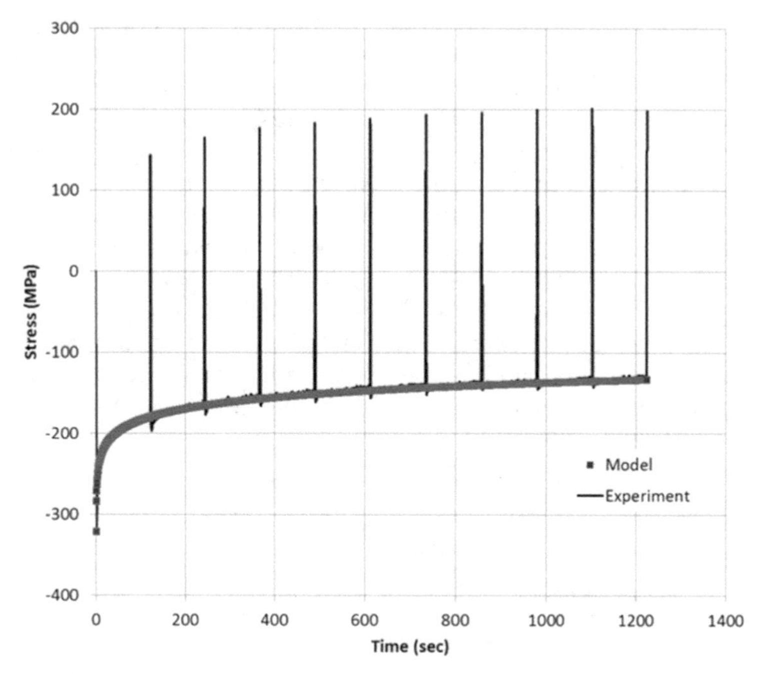

Figure 6.34. Cyclic stress evolution with time in CMSX-2 cycled between –0.005 to 0 strain with 120s-hold. Data are taken from (Yandt et al. 2011).

$$\tan h\left(\frac{\Phi}{2}\right) = \tan h\left(\frac{\Phi_p}{2}\right)\exp\left(-\frac{2EVA_g t}{kT}\right) \tag{6.53a}$$

where

$$\Phi_p = \left(\frac{V(\sigma_{max} - H\varepsilon_v - \sigma_0}{kT}\right) \tag{6.53b}$$

From Eq. (6.53), stress relaxation can be expressed, as

$$\sigma = \sigma_0 + \frac{2kT}{V}\tan h^{-1}\left[\tan h\left(\frac{\Phi_p}{2}\right)\exp\left(-\frac{2EVA_g t}{kT}\right)\right] \tag{6.54}$$

Eq. (6.54) indicates that stress relaxation starts with a potential energy Φ_p determined by the prior loading level, which controls the subsequent behaviour. The mechanism parameters are given Table 6.8. Continuous stress relaxation is shown in Figure 6.32 in comparison with the experimental observation.

Table 6.8. Mechanism parameters for CMSX-2 at 1100 °C

Materials	E (GPa)	H (MPa)	A (s^{-1})	V/kT (MPa^{-1})	σ_0 (MPa)
CMSX-2	75	1000	3.5E-10	0.05	5

It is interesting to see that even the material is strain controlled to cycle in a compressive range, positive mean stress can be developed due to stress relaxation from the maximum compression. Since a single crystal material does not have grain boundaries, the creep strain developed during the compressive hold is not expected to cause internal damage (in the form of cracks), despite that precipitate coarsening and rafting did occur (Yandt et al. 2011, 2012). Therefore, it is expected that the LCF life is mostly limited by the fully reversed inelastic strain range. On the total strain range level, the life with strain-hold had a significant debit as compared to the no-hold condition, as shown in Figure 6.35 (a). This was because that without strain-hold there was no additional creep strain to be reversed by plasticity. When the inelastic strain range is plotted against the life, the correlations for the two test conditions merge into one, as shown in Figure 6.35 (b).

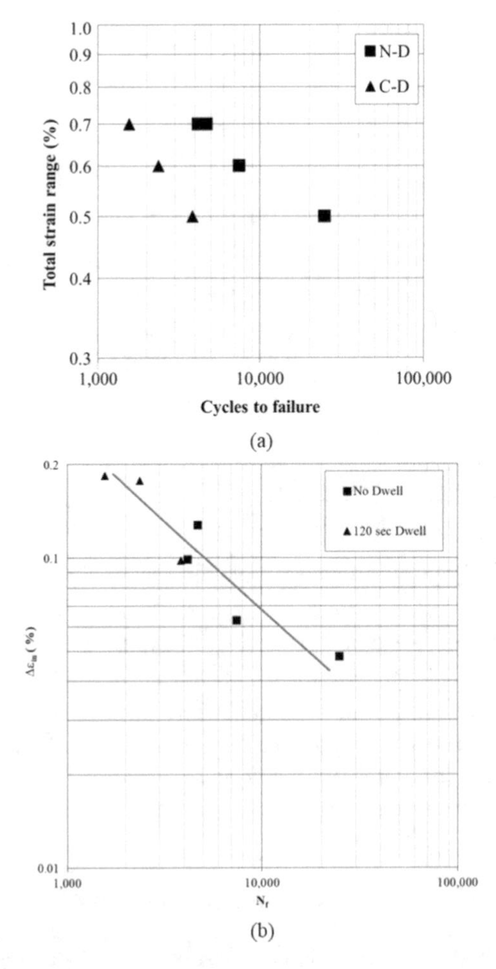

Figure 6.35. LCF life of CMSX-2 at 1100 °C vs. (a) total strain range; and (b) inelastic strain range. Data are taken from (Yandt et al. 2011).

6.5 Notched LCF

At the component level, LCF mostly occurs by localized plastic deformation at notch root, which is surrounded by the elastic region, as schematically shown in Figure 6.36. Since notches are important features of engineering components, it is important to understand the notched LCF behaviour. Neuber (1961) first proposed a simplified method to analyze notch fatigue, where the elastic stress concentration factor K_t is related to the plastic stress and strain concentration factors, K_σ and K_ε, as:

$$K_t = \sqrt{K_s K_e} \tag{6.55}$$

where $K_\sigma = \sigma/S$, and $K_\varepsilon = \varepsilon/e$, S and e are the remote elastic stress and strain.

The Neuber's rule, Eq. (6.55), is actually based on the equivalence of strain energies represented by the two triangles in Figure 6.37:

$$\frac{(K_t S)^2}{2E} = \frac{K_\sigma K_\varepsilon S^2}{2E} \tag{6.56}$$

Molski and Glinka (1981) modified the Neuber's rule, arguing that the strain energy due to elastic stress concentration, $K_t S$, should be totally transformed to elastic-plastic strain energy as represented by:

$$\frac{(K_t S)^2}{2E} = \int \sigma \, d\varepsilon \tag{6.57}$$

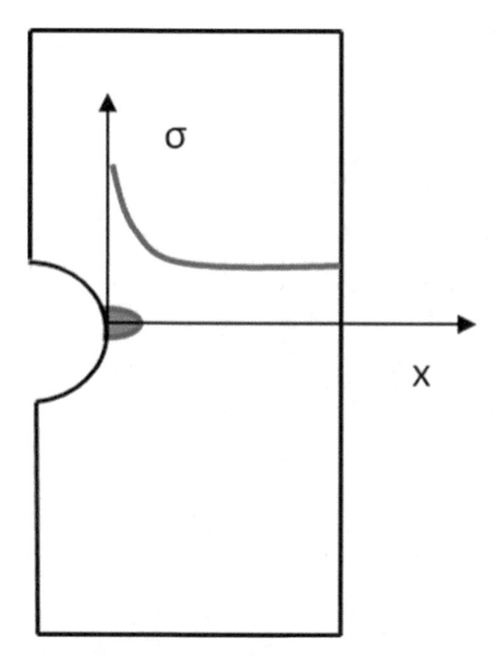

Figure 6.36. Stress concentration at notch root.

The difference between the Neuber's rule and Molski and Glinka's energy rule is shown in Figure 6.37. By strain energy equivalence, the original elastic energy (the solid line triangle) should not be transformed to the Neuber's energy (the dashed line triangle), which is really the elastic strain energy after yield. Instead, it should be transformed to the total elastic-plastic strain energy (the shaded area under the stress-strain curve). Neuber's rule leads to a larger plastic strain than the strain defined by the integration in Eq. (6.57), which would result in a shorter life. Hence, Neuber's rule is unnecessarily conservative.

Molski and Glinka (1981) only discussed the transformation of strain energy with rate-independent plasticity. As discussed in section 6.2.3 and 6.2.4, when creep mechanisms intervene such that time-dependence is manifested, the stress-strain behaviour is governed by plasticity coupled with creep mechanisms, e.g. Eq. (6.35).

As an example, the elastic stress concentration at the serration root of a gas turbine disc is shown in Figure 6.38. The elastic modulus of the material is 197 GPa at the operating temperature of 427 °C. By pure elasticity, the maximum stress at notch root is 1200 MPa, which exceeds the ultimate tensile strength of the material. The Molski-Glinka's method is applied using the cyclic stress-strain curve as shown in Figure 6.37. Thus, the equivalent stress-strain state at the serration root is found to be 919 MPa and 0.66% strain. At the operating temperature of 427 °C, both creep and oxidation are negligible for the disc material. Using the theoretical fatigue equation, Eq. (6.19) for Ni, the LCF life is calculated to be 19,300 cycles. The disc life was spin-rig validated (Beres et al. 2008).

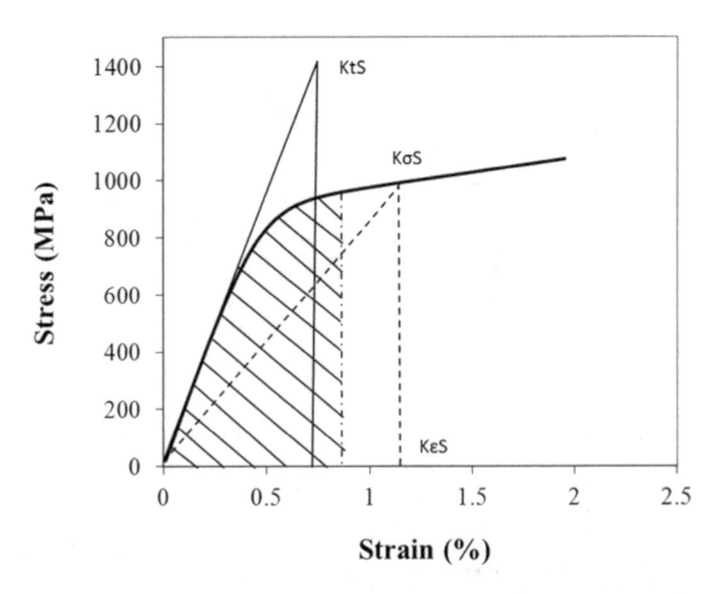

Figure 6.37. Energy interpretation of notch effect on stress concentration.

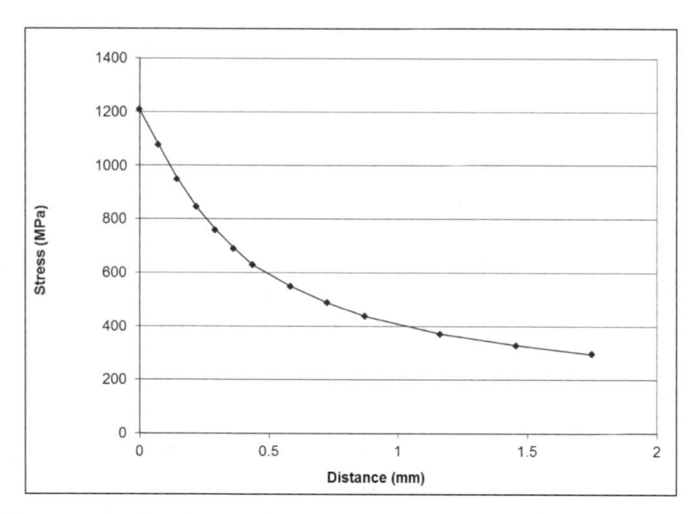

Figure 6.38. Elastic stress concentration at the notch feature of a gas turbine disc, evaluated using FEM.

6.6 Summary

In summary, the integrated creep-fatigue theory treats low cycle fatigue as a process occurring by multiple deformation and damage mechanisms. It is the interaction of these mechanisms that results in complicated LCF behaviour in a particular material.

Generally, fatigue at room temperature occurs by cyclic plasticity via alternating slip and cross-slip. Depending on the initial dislocation structure (with a low dislocation density as in casting or with a high dislocation density as in a cold-worked state), a material may initially exhibit cyclic hardening or softening for a short period of cycles but will be stabilized as the dislocation structure saturates. It is shown that RT-LCF life can be well-estimated by the Tanaka-Mura-Wu dislocation dipole pileup model, just based on the material's Burgers vector, Poisson's ratio, elastic modulus and surface energy, without resorting to experiments. Predicting low cycle fatigue life analytically is the first in the 150-year history of fatigue study. As fatigue depends on the microstructure where there is always local micro-level plastic strain concentration under nominal loading, computational microstructure fatigue analysis may shed light on fatigue design that can save tremendous testing effort for material development and application.

As temperature increases, some kinetic processes start to intervene the deformation and fracture process, depending on the alloy chemistry, causing different intermediate and high-temperature phenomena. For example, in DCI, dissolution of Mg/P constituent particles causes intergranular embrittlement, which results in very low fatigue life of DCI at ~400 °C. Also, at intermediate temperatures, 673–873 K (400–600 °C), DSA occurs in austenitic stainless steel, which causes cyclic hardening continuously until failure. The low stacking energy of the f.c.c. iron-base alloy seems to encourage segregation of alloying elements to stacking faults and interact with moving dislocations, leading to DSA. Cyclic deformation with DSA is highly non-Masing and has a negative rate sensitivity. Whereas in high stacking-fault-energy materials such as Co or Ni-base superalloys, such phenomena are not observed, and the material's LCF life up to 600 °C is comparable to RT. At high temperatures, i.e. above homologous temperature 0.6, both creep and oxidation are significant enough to affect LCF deformation and life. Creep mechanisms are responsible to induce the positive rate-dependence of the cyclic stress-strain behaviour and also cause internally distributed damage to grow, while oxidation assists surface crack nucleation and propagation. The ICFT considers all forms of internally distributed damage in the multiplying damage factor D, as summarized in Eq. (4.17) or Eq. (6.38), (6.43) and (6.47).

The ICFT also considers the environmental effect such as oxidation on overall fatigue damage, as characterized by the ratio of oxidation penetration to the critical crack length, h/a_c. The derived formulation, Eq. (4.18), reflects creep-fatigue interaction in two-folds: (i) creep physically creates internal damage that would knockdown the fatigue life by a factor of D; and ii) creep reduces the flow stress and hence reducing the h/a_c ratio, leading to increasing LCF life. The effect of cycle time or frequency is manifested as the interplay between creep and oxidation effects. By the ICFT, each damage form is formulated based on its underlying physical mechanism. The overall effect of these operating mechanisms on the LCF life occurs as a result of non-linear interactions between the nucleation and propagation of a surface crack with internally distributed damages (microcracks and voids), in the context of Eq. (4.18). It thus describes the many-faceted roles of plasticity and creep in both stress-strain behaviour and damage accumulation processes, including the crossover-behaviour in LCF life. With regards to this, it must be pointed out that the linear damage summation rule fails to rationalize these complicated phenomena. The LDR is not a physical rule, since it does not cover all observations.

In summary, the ICFT with delineation of the underlying deformation and damage mechanisms can faithfully quantify the contributions from those participating mechanisms. Such understanding can be used to guide material/component design for optimized performance. For example, adding a certain amount of Al to DCI without affecting the manufacturability is a feasible way to increase the high-temperature fatigue performance of the cast exhaust manifolds (Wescast, private communication).

CHAPTER

7

Thermomechanical Fatigue

Most engineering components for high temperature applications actually operate under severely alternating temperature/mechanical loads, such as gas turbine vanes and blades, automotive exhaust system components and even shoulder joints in circuit boards, when experiencing start-up and shutdown operations. The materials then suffer thermomechanical fatigue (TMF). Therefore, modeling TMF is particularly relevant to component design for optimized performance with the desired durability (life).

7.1. TMF General

During TMF, the total strain (ε_{tot}) of a solid material is equal to the sum of thermal and mechanical strain:

$$\varepsilon_{tot} = \varepsilon_{th} + \varepsilon_{mech} = \alpha_T(T - T_0) + \varepsilon_{mech} \tag{7.1}$$

where ε_{th} is the thermal strain, T_0 is the reference temperature, T is the current temperature, and α_T is the coefficient of thermal expansion (CTE).

Now, standard TMF test is specially designed to characterize material's response to thermomechanical loading (ASTM E2368, 2010). In such a test, the mechanical load is independently superimposed on thermal load with a definite phase angle φ. Then, the TMF cycle is characterized as: in-phase (IP) for $\varphi = 0°$, out-of-phase (OP) for $\varphi = 180°$, or diamond phase (DP) for $0° < \varphi < 180°$. A typical TMF test setup and typical TMF cycles are schematically shown in Figure 7.1. The OP cycle represents the condition where the maximum mechanical strain is reached at the "cold" end of the temperature cycle. The IP cycle represents the condition where the maximum mechanical strain is reached at the "hot" end of the temperature cycle. The DP cycle represents the condition where the maximum mechanical strain is reached at mid-temperature of the cycle.

The TMF specimens are usually made to be hollow cylinders for effective heating and cooling without thermal gradient across the cross-section. The specimen is heated with an induction coil and cooled with forced air going through inside as well as from the outside valves. The specimen temperature

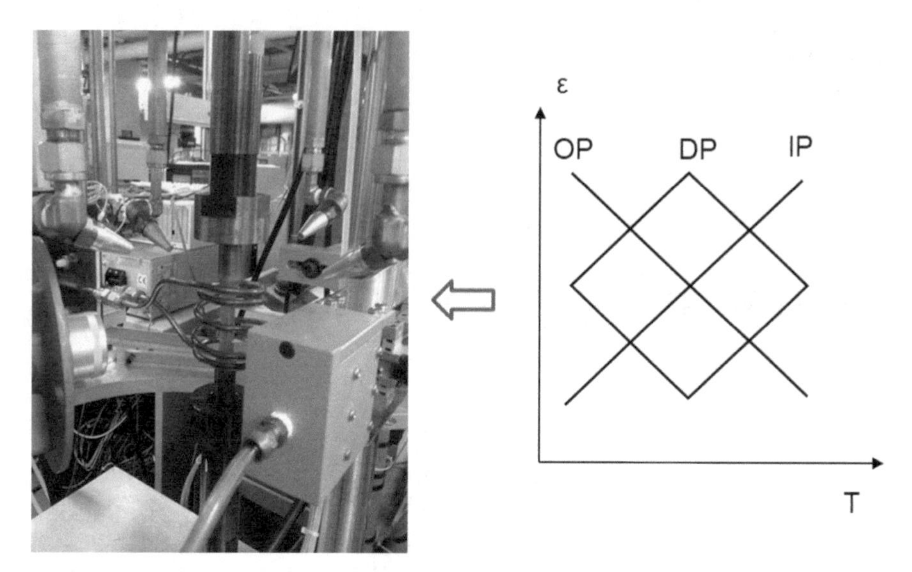

Figure 7.1. A TMF test setup and strain-temperature cycles.

is measured by a calibrated infrared pyrometer, and specimen elongation is measured using an extensometer. Both signals are feedback to a digital controller to regulate the power and force applied on the specimen using a closed loop control algorithm.

While a structural component is experiencing temperature variations, part of the mechanical strain arises from constraint of thermal expansion due to the existence of local thermal gradients or mismatch of thermal expansion coefficients of interfacing materials. The constraint causes stress in the component, and with cyclic variations of temperature, it leads to thermal fatigue (TF). Therefore, in TF, the mechanical strain is not independent of thermal loads, but rather depends on the local constraint condition of the structure or assembly. For the convenience of characterizing TF, a constraint ratio is defined, as (Wu et al. 2015)

$$\eta = \frac{\varepsilon_{th} - \varepsilon_{tot}}{\varepsilon_{th}} \tag{7.2}$$

Apparently, η is a thermal-structural parameter that characterizes the local constraint condition of a thermally loaded structure/component such as nozzle guide vanes and exhaust manifold. To one extreme, when the material is fully constrained, i.e., $\varepsilon_{tot} = 0$, then $\eta = 1$. In general, in a thermally loaded component, η takes values between 0 and 1 at different locations, which can be determined by thermal stress analysis of the component. In TF, the mechanical strain always arises in the opposite direction of thermal expansion strain but with a magnitude depending on the thermal strain, and therefore TF is a special case of OP-TMF. TMF testing under constraint condition (at a fixed constraint ratio) is called the *constrained* TMF. The constrained TMF represents the behaviour of a structural point under purely thermal loading.

An engineering component may experience both types of TMF under service conditions. For example, a gas turbine blade is subjected to centrifugal (mechanical) force in operation, but it may also experience local constrained TMF at cooling features, such that the actual TMF cycle is the superposition of the two types of TMF. Therefore, understanding both standard and constrained TMF behaviours is necessary, in order to predict the component life under service conditions.

7.2 TMF Cyclic Deformation

Without losing generality, we consider TMF occurs with plasticity coupled with hyperbolic-sine creep (note that power-law creep is also no issue). According to Eq. (4.14), under constant strain-rate loading conditions, the material obeys the deformation constraint condition, similar to Eq. (6.27):

$$\dot{\varepsilon} = \dot{\varepsilon}_e + \dot{\varepsilon}_p + \dot{\varepsilon}_v = \frac{\dot{\sigma}}{E_t} + 2A_g \sin h\,\Phi = const. \tag{7.3}$$

where the plasticity-accommodated tangent modulus E_t is defined by Eq. (6.28) and the energy evolution equation is given by Eq. (6.29).

Similar to the LCF case described in section 6.2.3, TMF is also governed by the first-order differential equation of Φ, as

$$\dot{\Phi} = \frac{E_t V}{kT}\left[\dot{\varepsilon} - 2A_g\left(1 + \frac{H}{E_t}\right)\sin h\,\Phi\right] \tag{7.4}$$

This differential equation can be solved in infinitesimal steps as follows. We assume that the evolution of the energy, Φ, undergoes a series of infinitesimal isothermal steps, for each i-th step, the energy state evolves from Φ_{i-1} to Φ_i over the time interval $\Delta t_i = t_i - t_{i-1}$. Within this interval, the transient tangent E_t is constant. Then, following the solution procedures of section 2.2, we obtain:

$$\ln\left(\frac{e^{-\Phi} - a}{e^{-\Phi} + b}\right)\Bigg|_{\Phi_{i-1}}^{\Phi_i} = -\frac{VE_t\dot{\varepsilon}\sqrt{1+\chi^2}}{kT}\Delta t_i \qquad (i = 1,2,\ldots) \tag{7.5}$$

Summing up all these infinitesimal steps, we obtain:

$$\sum_{i=1}^{N}\ln\left(\frac{e^{-\Phi} - a}{e^{-\Phi} + b}\right)\Bigg|_{\Phi_{i-1}}^{\Phi_i} = -\sum_{i=1}^{N}\frac{VE_t\dot{\varepsilon}\sqrt{1+\chi^2}}{kT}\Delta t_i \tag{7.6a}$$

where

$$\chi = \frac{2A_g}{\dot{\varepsilon}}\left(1 + \frac{H}{E_t}\right), a = \frac{\sqrt{1+\chi^2} - 1}{\chi}, b = \sqrt{1+\chi^2} + 1 \tag{7.6b}$$

Let N→∞, the left-hand side of Eq. (7.6a) will be equal to the logarithmic difference between the final state and the initial state, and the right-hand side is an integration over the loading period. After mathematical rearrangement, we have:

$$\left(\frac{e^{-\Phi}-a}{e^{-\Phi}+b}\right)=\left(\frac{1-a}{\chi+b}\right)\exp\left[-\int_{t_0}^{t}\frac{VE_t\dot{\varepsilon}\sqrt{1+\chi^2}}{kT}dt\right] \tag{7.7}$$

where t_0 is the time to reach the elastic limit, or in other words, for plastic flow to commence.

Further considering that $\sqrt{1+\chi^2}\approx 1$ since χ is usually very small, and the variation of the tangent modulus is moderate within the TMF cycle such that E_t can be replaced with an average $\bar{E}$, we integrate the right-hand-side of Eq. (7.7) under triangular loading wave form, as,

$$\omega=\left(\frac{1-a}{\chi+b}\right)\exp\left(\mp\frac{V\bar{E}\Delta\varepsilon}{k\Delta T}\ln\frac{T}{T_0}\right) \tag{7.8}$$

where $\Delta T = T_{max} - T_{min}$ is the temperature range, $\Delta\varepsilon$ is the total strain range, the −/+ sign applies to the temperature rising or declining portion of the cycle, respectively, and T_0 denotes the temperature point at which the stress reaches the elastic limit, i.e., $\sigma = \sigma_0$, and it is determined through the thermomechanical cycle T vs. ε relationship.

Then, similar to Eq. (2.10), we obtain the stress-strain relation during TMF, as

$$\sigma - H\varepsilon_v - \sigma_0 = -\frac{kT}{V}\ln\left(\frac{a+\omega b}{1-\omega\chi}\right) \tag{7.9}$$

The stress-strain response of the coarse-grained IN738LC under an out-of-phase thermomechanical fatigue condition is predicted using Eq. (7.9), as shown in Figure 7.2, with the parameter values given in Table 2.1, but with a reduced σ_0 = 40 MPa for the coarse-grained (d ~ 5 mm) material. The predicted hysteresis loop is in good agreement with the experimentally measured, which demonstrates that the mechanism parameters determined for isothermal LCF are also applicable for depicting the TMF behaviour.

The above analytical model gives users an explicit expression of the stress-strain relation as function of strain rate and temperature. Actually, it is more robust to implement the ICFT constitutive model in FEM for component analysis. Commercial FEM software such as Abaqus and MSC. Marc provides the option of coupled plasticity and creep. Users may just modify the existing subroutines of plasticity and creep to implement the ICFT model with the appropriate mechanism(s). The following sections give examples of FEM-implemented ICFT analysis. It should be realized that the mechanism-based constitutive law is independent of the loading profile, but the material behaviour is a result of the underlying mechanism under the given loading profile.

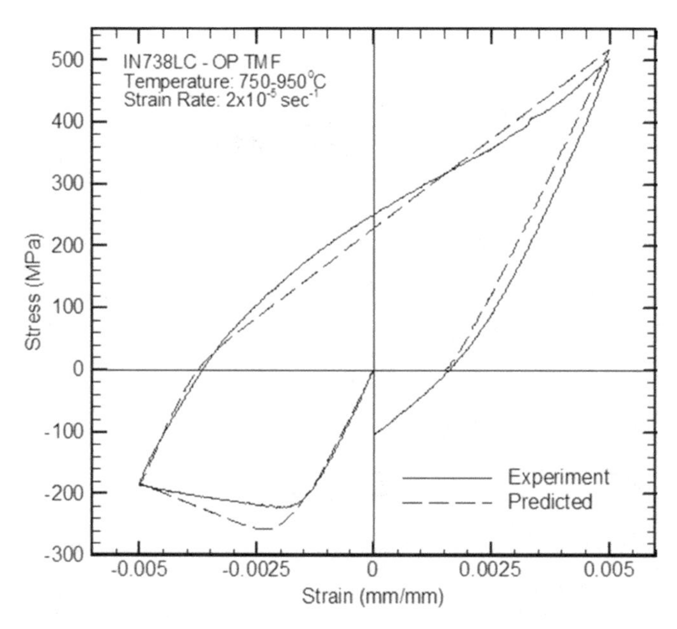

Figure 7.2. Stress-strain response of IN738LC (coarse-grain) during out-of-phase thermomechanical fatigue (Wu et al. 2002).

7.2.1 Hysteresis Behaviour of IP-TMF

The IP-TMF refers to the TMF cycle where the mechanical strain and temperature are applied in-phase. For ductile cast iron (DCI), the constitutive law is given in the context of Eq. (6.26), which has been incorporated into MSC. Marc for simulation of various TMF hysteresis behaviours (Wu et al. 2015). This section gives an example of DCI IP-TMF in the temperature range of 300-800 °C.

Figure 7.3 (a) shows the FEM simulation in comparison with the experimental 1st cycle and the mid-life hysteresis loops of DCI during 300–800 °C IP-TMF under sinusoidal wave loading. The simulated hysteresis loop is in good agreement with the experimental observation. As the mechanical loading goes in-phase with temperature, creep deformation is predominant during the loading half-cycle, which results in the loop shifting towards a positive strain from the initial origin. On reverse loading, going toward low temperature, deformation is mainly reversed by plasticity. The accumulation of both creep and plastic strains during the IP-TMF cycle are shown in Figure 7.3 (b). Indeed, the FEM simulation shows that, from the "hot" end, creep deformation is first reversed but increases to a greater level when the temperature cycle returns back to the maximum, whereas the plastic strain mainly arises during the "cold" portion of the cycle in compression. The two mechanism strains add up to be the total inelastic strain, which is in good agreement with the experimental measurement.

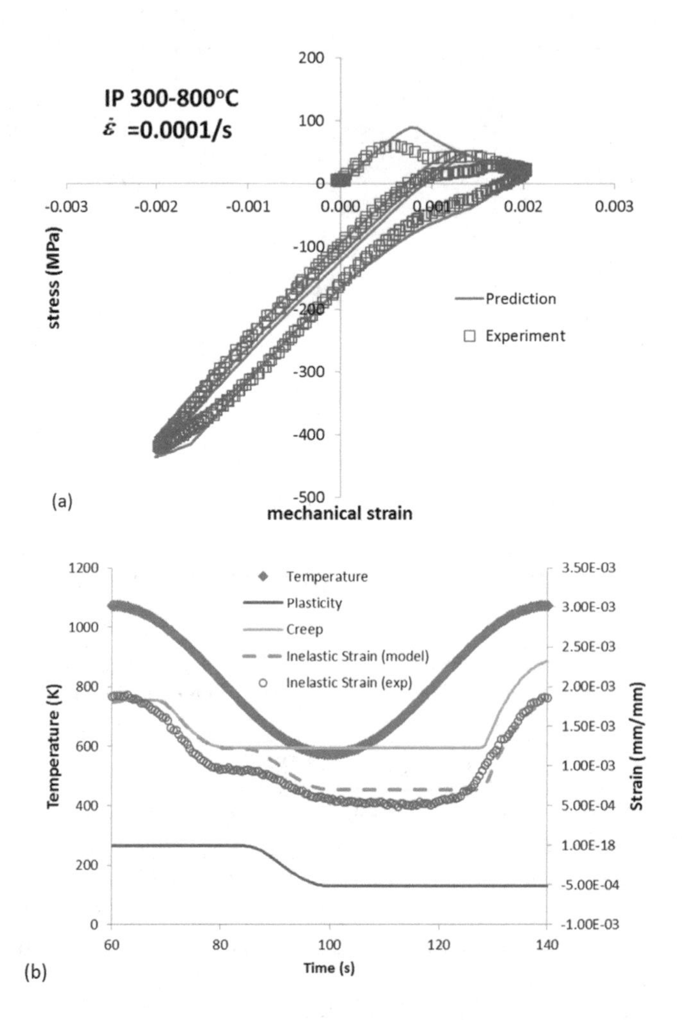

Figure 7.3. (a) The hysteresis behaviour, and (b) strain accumulation during IP-TMF in the range of 300–800 °C.

7.2.2 Hysteresis Behaviour of OP-TMF

Figure 7.4 (a) shows the FEM simulation in comparison with the experimental 1st cycle and mid-life hysteresis behaviours of DCI during 300–800 °C OP-TMF under sinusoidal wave loading. The simulated hysteresis loop is in good agreement with the experimental observation, which is the reverse of that of IP-TMF. The accumulation of both plastic and creep strains during the OP-TMF cycle are shown in Figure 7.4 (b). Apparently, as mechanical loading goes out-of-phase with temperature, creep deformation is accumulated in compression during the "hot" portion of the cycle, which shifts the hysteresis loop towards a negative strain from the initial origin. On load reversal, plastic strain arises during the "cold" portion of the cycle,

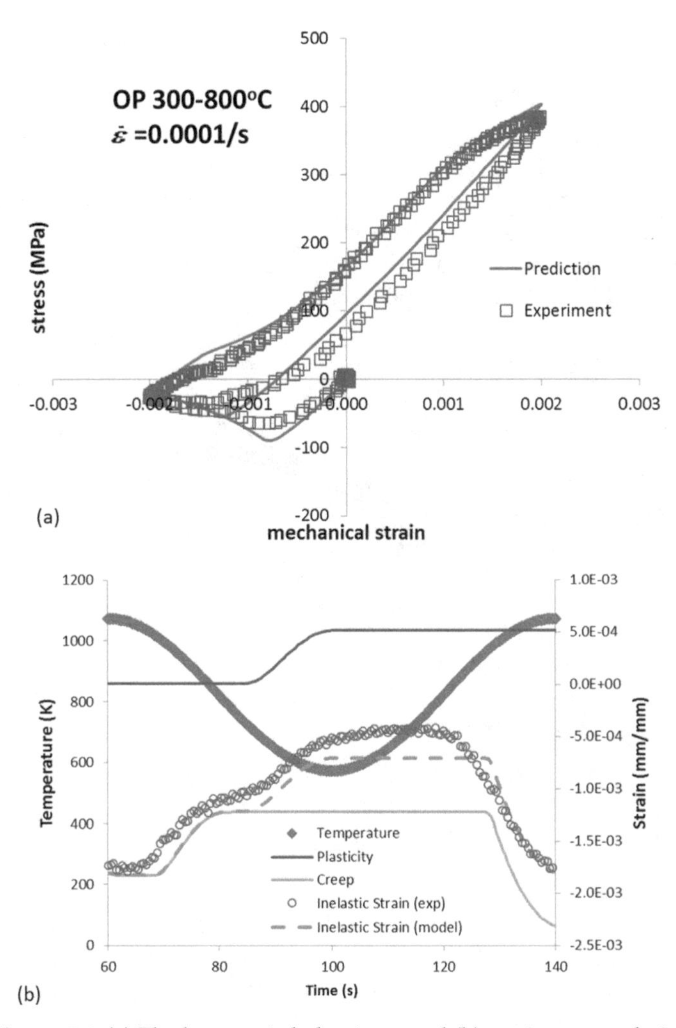

Figure 7.4. (a) The hysteresis behaviour, and (b) strain accumulation during OP-TMF in the range of 300–800 °C.

reversing the compressive creep deformation. The resultant of the two mechanism strains is equal to the net inelastic strain, which results in an asymmetrical hysteresis behaviour, as observed. Once again, the ICFT model not only provides a good description of the total strain-stress response, but also further delineates plasticity and creep strain quantitatively in the TMF cycle.

7.2.3 Hysteresis Behaviour of Constrained TMF

For the constrained TMF, Figure 7.5 shows the stress-strain response and accumulation of inelastic strain during TMF with a constraint ratio of 100% in the temperature range of 160–600 °C with 1 min. hold. FEM simulation for the hysteresis behaviour of constrained TMF matches with the experimental observation for the first and the mid-life cycle, Figure 7.5 (a). Constrained TMF is out-of-phase in nature; however, its strain range is controlled by the constraint condition, rather than independently set as in the standard TMF test. As seen from Figure 7.5 (b), the ICFT model predicts that the negative inelastic strain occurs predominantly due to creep during the temperature hold period, which causes compressive stress relaxation. This creep strain is subsequently reversed by plastic stain as the temperature decreases to the minimum, and thus the cycle repeats until failure.

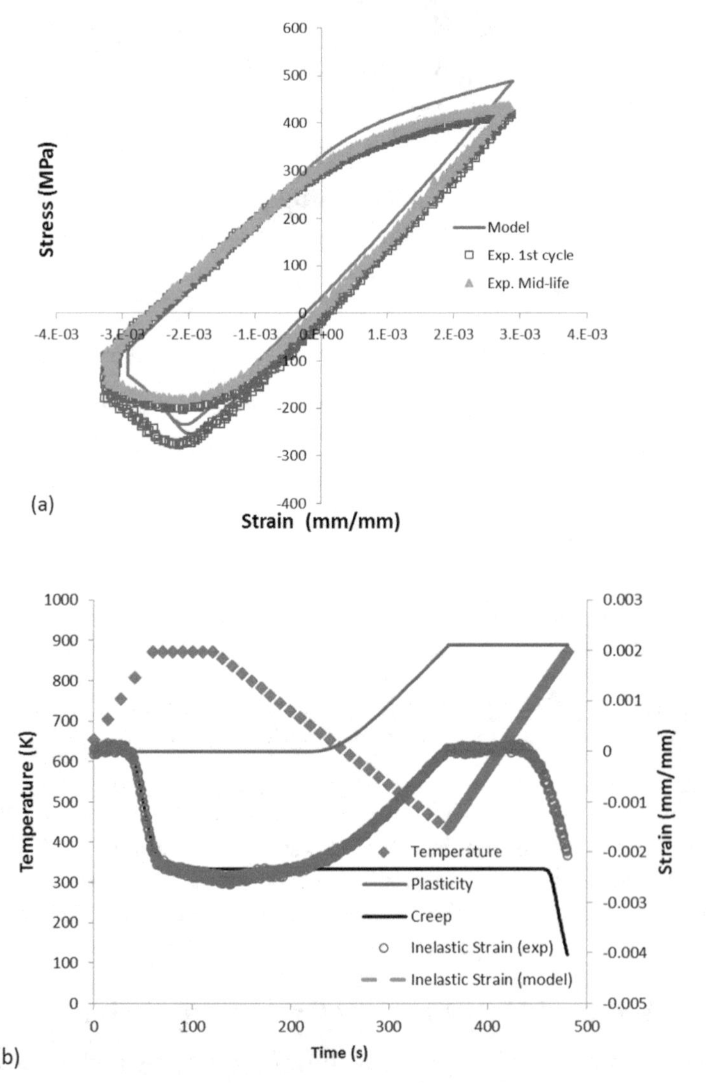

Figure 7.5. The hysteresis behaviour, (a) and (b) strain accumulation during constrained TMF with the constraint ratio of 100% in the range of 160–600 °C.

7.3. TMF Life Prediction

As discussed in Chapter 6, unified constitutive theories do not delineate creep and plasticity. Neu and Sehitoglu (1989) proposed a linear damage summation rule (LDR) to account for fatigue, creep and oxidation contributions. It has been pointed in Chapter 4 that LDR can give gross error for creep-fatigue interaction. Besides, as pointed out in Chapter 6, LDR failed to explain the cross-over phenomenon of creep-fatigue-oxidation interaction during isothermal LCF for DCI, Figure 6.25. The current state-of-the-art TMF life prediction using the "unified" approach employs more than 20 constitutive and damage parameters to fit the test results, but the interplay of the physical factors is not clearly elucidated with these empirical parameters. In this section, the ICFT is used to describe life evolution during TMF in a quantitative mechanism-delineated way.

For TMF life prediction, we adopt the same holistic damage accumulation equation, Eq. (4.18), as for LCF, except that the internal damage factor D and oxidation scale are evaluated per TMF cycle, as:

$$\frac{1}{N} = \bar{D}\left\{\frac{1}{N_f} + \frac{h}{a_c}\right\} \tag{7.10}$$

where $\bar{D}$ is the damage factor including IE, DSA, as well as creep as defined by

$$D = 1 + \frac{l}{\lambda}\bar{\phi} + \alpha\left[\left(\frac{\Delta\sigma_H}{\mu b}\right)^2 - \rho_0\right] + \beta\varepsilon_v \tag{7.11}$$

where $\bar{\phi}$ is the average intergranular embrittlement factor,

$$\bar{\phi} = \frac{1}{T_{\max} - T_{\min}} \int_{T_{\min}}^{T_{\max}} \exp\left[-\left(\frac{T - T_\varepsilon}{T_0}\right)^2\right] dT \tag{7.12}$$

During a TMF cycle, an oxide film of thickness h breaks up repeatedly at the crack tip, resulting in additional crack advance. The newly formed oxide scale can be evaluated as

$$h = \int_0^\tau \sqrt{\frac{k}{2t}}\,dt \tag{7.13}$$

and the critical crack length a_c is given by

$$a_c = \frac{1}{\pi}\left(\frac{K_{IC}}{Y\sigma_{s,\max}}\right)^2 \tag{7.14}$$

The present model assumes an ideal condition that a dense layer of oxide film forms on the smooth material surface and it breaks at the maximum stress of a TMF cycle. When such an oxide film forms on the specimen surface, the entire specimen is effectively a two-material composite. The coupon (substrate) stress is thus modified by the composite rule of strain compatibility, as given by Eq. (3.24), as:

$$\sigma_s = \frac{E_s}{fE_{ox} + (1-f)E_s}[\sigma - fE_{ox}(\alpha_s - \alpha_{ox})(T - T_r)] \tag{7.15}$$

where σ is the specimen nominal stress, T_r is the reference temperature at which the oxide forms without a mismatch stress, E_s and E_{ox}, and α_s and α_{ox} are the elastic modulus and the coefficient of thermal expansion (CTE) of the substrate and the oxide, respectively. Evidently, under isothermal conditions, the mismatch is always zero, because $T = T_r$. However, under TMF conditions, if the oxide film-breaking temperature is different from its formation temperature, the coupon stress is modified according to Eq. (7.15). For OP-TMF, this effect is the maximum, since the film forms mostly close to and at the maximum temperature, but breaks at the minimum temperature; and for IP-TMF, it is nearly null, since the film-breaking temperature (when the stress is positively maximum) is the same as the formation temperature.

7.3.1 Cyclically-Stable TMF

Taking DCI for example, again, when DCI is exposed at high temperature, a mixed oxide layer containing Fe and Si formed on a very rough material surface with graphite nodule voids. Figure 7.6 shows the oxide film formed on the surface of a DCI specimen during TMF. For computational simplicity, we assume that the oxide scale formed on DCI is purely Fe_2SiO_4, which has an elastic modulus of 126GPa and CTE of 26×10^{-6}/K at room temperature, whereas the CTE of DCI is 13×10^{-6}/K at RT. The modified stress in Eq. (7.15) is calculated with the RT properties of the Fe_2SiO_4. But, to account for the effect of mixed oxide and local stress raisers such as nodule voids, we consider f is an adjustable parameter. In fitting the OP-TMF life data, $f = 0.32$, which introduced a maximum scale stress ~276 MPa in the temperature of 573–1073 K, and ~190 MPa in the temperature range of 723–1073 K. This effective oxide volume fraction is high compared to the real oxide volume fraction (by a magnifying factor), but it is necessary to modify the aforementioned effect of oxidation such that the mixed oxide scale could break at a stress much lower than the peak stress during an OP-TMF cycle, given that plasticity and creep are all evaluated under the nominal stress.

Because the IE factor ϕ becomes maximum at T_ε, corresponding to a minimum ductility, and it diminishes rapidly at temperatures away from T_ε, the average integration of Eq. (7.13) is fairly constant for both IP and OP cycles within the same temperature range. With an average $\bar{\phi}$ greater than zero for TMF, it is expected that the TMF life, whether IP or OP, is less than the LCF life at RT. Also, it is noted that for cyclic stable materials such as DCI, the term for dislocation cracks vanishes because $\Delta\sigma_H = 0$.

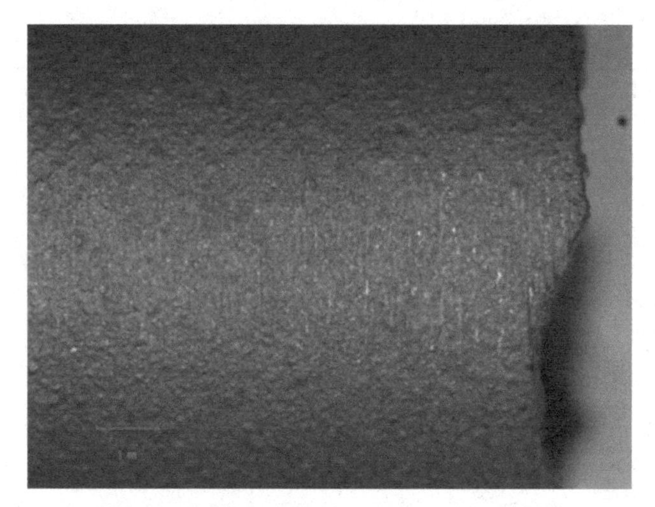

Figure 7.6. Oxide film on DCI during TMF.

Combining Eq. (7.10)–(7.15), we obtain:

$$\frac{1}{N} = \bar{D}\left\{\frac{1}{N_f} + \pi\left(\frac{Y\sigma_s}{K_{IC}}\right)^2 \int_0^{\tau} \sqrt{\frac{k}{2t}}dt\right\} \tag{7.16}$$

When the loading cycle is isothermal, Eq. (7.16) returns to that for LCF, which has been validated in the case studies as shown in Chapter 6. Now, the ICFT is used to predict the TMF life with the model parameters previously established for DCI under LCF loading. The average IE factor and the oxide scale under TMF condition are evaluated through the integrals in Eq. (7.12) and (7.13). The creep and plastic strains are evaluated using the ICFT-incorporated FEM, as shown in Figure 7.3–7.5 for DCI under various TMF conditions (IP, OP and 100%-constrained TMF with hold, respectively). By metallurgical arguments, alternating plasticity contributes to fatigue, but only tensile creep strain (here "tensile" means relative to the previous compression state) facilitates nodular void growth. The TMF life fractions per mechanism can also be calculated, according to Eq. (7.16), and the results are given in Table 7.1.

The experimental TMF life data are plotted against the mechanical strain in Figure 7.7, the solid lines just indicate the trend for the argument sake. The room temperature (RT) LCF data are also shown as the baseline for comparison. Comparing the TMF life on the same mechanical strain level, the OP-TMF lives all fell lower than the IP-TMF life; and both IP and OP lives were significantly lower than the baseline (RT) at a given strain level. In TMF, many factors such as the maximum and minimum temperature, the temperature range, the strain rate, the loading waveform, and hold time, all could influence the TMF behaviour. The ICFT model provide a physics-based explanation of the observed TMF phenomena with respect to the

Table 7.1. Fatigue, oxidation and internal damage fractions for DCI during TMF

Specimen ID	Test condition	N/N_f	Nh/a_{cr}	1/D
A03	723-1073K-OP $\Delta\varepsilon_m = 0.3\%$	3.99E-05	6.68E-02	5.34E-02
A05	573-1073K-OP $\Delta\varepsilon_m = 0.4\%$	8.53E-03	7.35E-02	3.69E-02
A07	573-1073K-OP $\Delta\varepsilon_m = 0.5\%$	9.46E-03	3.84E-02	3.18E-02
A08	723-1073K-OP $\Delta\varepsilon_m = 0.5\%$	8.87E-04	2.93E-02	3.19E-02
A09	573-1073K-OP $\Delta\varepsilon_m = 0.6\%$	1.36E-02	4.61E-02	2.79E-02
A10	573-1073K-IP $\Delta\varepsilon_m = 0.4\%$	2.39E-02	9.69E-03	3.26E-02
A11	573-1073K-IP $\Delta\varepsilon_m = 0.5\%$	4.57E-02	3.06E-03	3.30E-02
A12	573-1073K-IP $\Delta\varepsilon_m = 0.6\%$	4.55E-02	1.69E-03	2.04E-02
B01	433-873K 100% constraint	1.20E-02	1.69E-02	4.20E-02
B03	433-873K 70% constraint	8.98E-03	2.89E-02	4.20E-02
B04	433-873K 60% constraint	1.01E-02	6.55E-02	4.19E-02

underlying mechanisms. Figure 7.8 shows the prediction of the ICFT for TMF experiments within ±2 scatter band with the correlation factor of R^2 ~0.97.

Comparing the TMF lives within the same temperature range, e.g., 573–1073 K (300–800 °C), the intergranular embrittlement should have a similar effect on both IP and OP, the internal damaging factor *D*-effect for IP was slightly larger than OP because more creep strain accumulated to promote void growth, but the striking difference between OP and IP lives was because that OP exhibited large peak tensile stresses that significantly

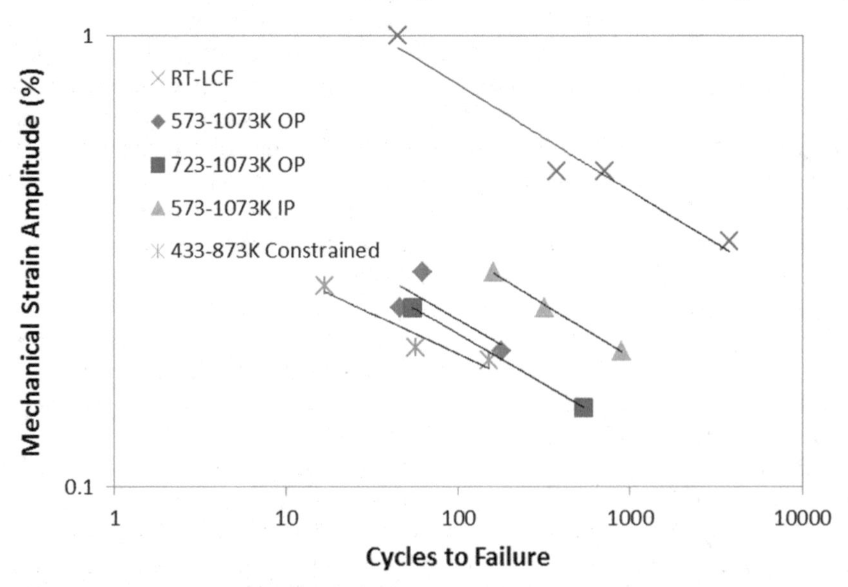

Figure 7.7. Experimental strain vs. life relationships (symbols) for various TMF conditions.

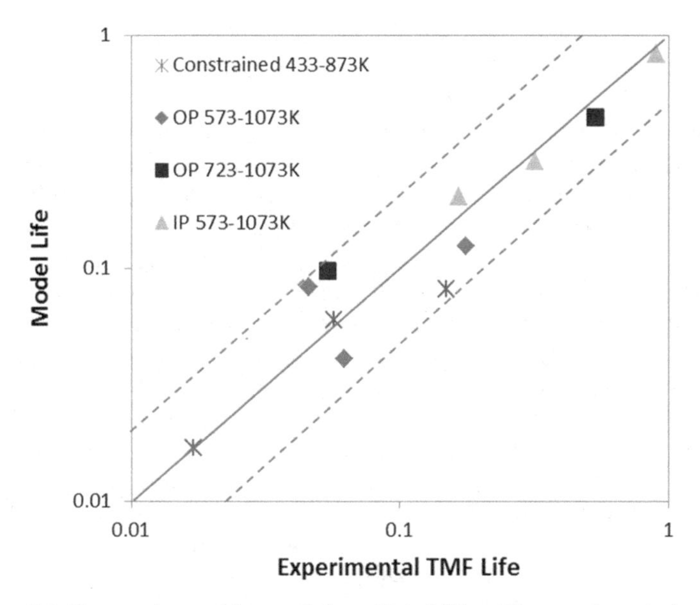

Figure 7.8. Comparison of the model predicted life with experiments for DCI.

increased the h/a_{cr} ratio. For the OP cycles of the same period, raising the minimum temperature resulted in more oxidation and more creep strain accumulation; hence OP-TMF in the range of 723–1073 K (450–800 °C) was slightly lower than that in the range of 573-1073 K (300-800 °C) due to the creep-oxidation interaction. Comparing at the same mechanical strain range level, the constrained TMF with a 1-min. hold period in the temperature range of 433–873 K (160–600 °C) resulted in the worst TMF life, albeit the maximum temperature was much lower than the other TMF cases. What was remarkable in this case was the large amount of creep strain accumulated in compression during the holding period, which was subsequently reversed by plasticity, translated to fatigue damage. Also in this case, even though the oxide scale thickness was thin compared to that formed in the range of 573–1073 K (300–800 °C), the breaking stress was higher because of the lower minimum temperature, this increased the oxide cracking ratio at fracture, which was also contributing to the shortening of TMF life.

The ICFT model offers a quantitative delineation of the roles of various factors in TMF. It not only relates the material's behaviour and life to the underlying deformation and damage mechanism but also infers the failure mode by the dominant failure mechanism(s). The life fractions (Table 7.1) show that OP-TMF life is limited by competition of plasticity-induced fatigue and oxidation, because the failure occurs at a high stress; and IP-TMF is more damaged by creep-plasticity interaction, due to the growth of internally distributed damage via creep. IE plays a role in both IP and OP-TMF. The present model can be further modified by introducing a critical stress for surface oxide scale breaking. Then, for IP-TMF, when the peak stress is below this level, the damage process will consists only creep-fatigue interaction, growing internally without oxide cracking; and the life will last longer, as the IP-TMF tests seemed to imply.

Figure 7.9 shows the fracture surface of the specimen A03 failed by OP-TMF with ±0.15% mechanical strain in the temperature range of 450–800 °C. The model predicted that both "tensile" creep and plasticity were so low in this case that fatigue would last for a large number of cycles to failure, but meanwhile, it allowed time for oxidation to contribute to the failure process significantly, in combination with the IE effect. The fractographic evidence shows that the specimen surface was covered with a continuous oxide scale, and the fracture surface was also heavily oxidized, which had erased the features of IE and mechanical fatigue.

Figure 7.10 shows the fracture surface of specimen A08 failed by OP-TMF with ±0.25% mechanical strain in the temperature range of 450–800 °C. In this case, because of the large mechanical straining, multiple cracks have formed on the specimen surface. On the fracture surface, abundant grain boundary fracture facets were observed, which are evidence of IE mechanism, since the minimum temperature was very close to BMT ~ 400 °C. With increased mechanical strain, as compared to the A03 case, oxidation fraction was lower, but the internal damage effect was larger. Because of this combination of oxidation and internal damage, the fracture surface contains a mixture of clean grain boundary facets and isolated oxides.

Figure 7.11 shows the fracture surface of specimen A12 failed by IP-TMF with ±0.3% mechanical strain in the temperature range of 300–800 °C. The fracture surface contained mostly intergranular fracture facets and enlarged and distorted nodular voids. The observation indicates large creep

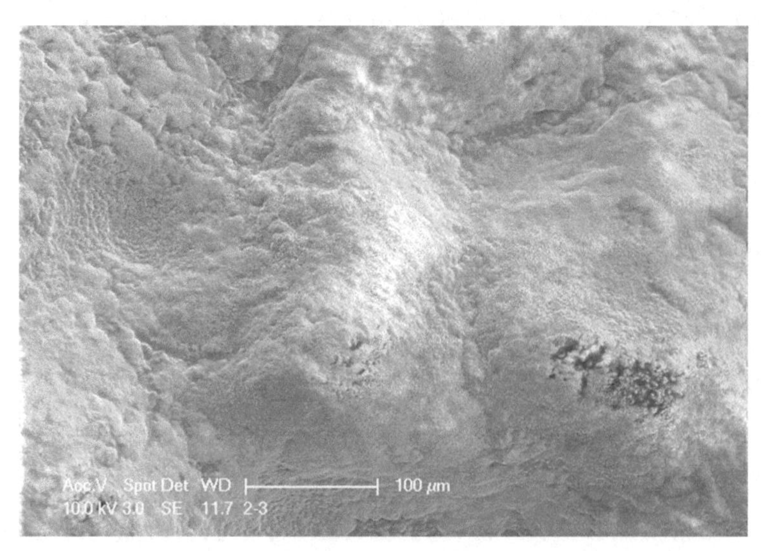

Figure 7.9. Specimen A03 failed by OP-TMF with ±0.15% mechanical strain in the temperature range of 723–1073 K (450–800 °C).

Figure 7.10. Specimen A08 failed by OP-TMF with ±0.25% mechanical strain in the temperature range of 723–1073 K (450–800 °C).

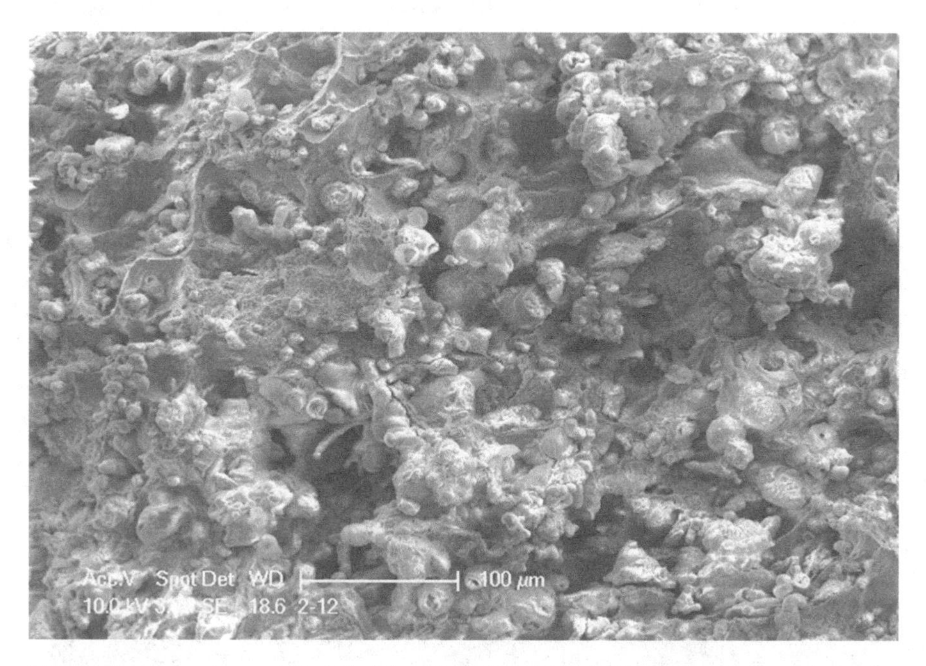

Figure 7.11. Specimen A12 failed by IP-TMF with ±0.3% mechanical strain in the temperature range of 573–1073 K (300–800 °C).

deformation, as the model evaluated. As a result, this case has the largest D factor.

Figure 7.12 shows the fracture surface of specimen B01 failed by TMF with a constraint ratio of 100% in the temperature range of 433–873 K (160–600 °C). As the temperature cycle covered the BMT, it is expected to see a lot of grain boundary fracture facets on the fracture surface; and otherwise, the failure was mainly caused by plasticity with light oxidation, since the constrained TMF was OP in nature.

The ICFT constitutive and damage model are shown to work for both LCF and TMF, because the underlying physical mechanisms are the same for the same material (in the present case, DCI). It naturally incorporated the four damage mechanisms: i) plasticity-induced fatigue, ii) intergranular embrittlement, iii) creep and iv) oxidation. The model delineates the deformation and damage mechanisms and their contributions in the asymmetrical TMF hysteresis behaviour, and thus provides good descriptions of the stress-strain responses and life under various TMF conditions (OP and IP, and constrained TMF with hold). The overall agreement of the ICFT model with the experimental life is within factor of 2.

7.3.2 Cyclically Unstable TMF

We have known that (in section 6.2) austenitic cast steel 1.4848 TMF exhibits cyclic hardening behaviour at intermediate temperatures. TMF

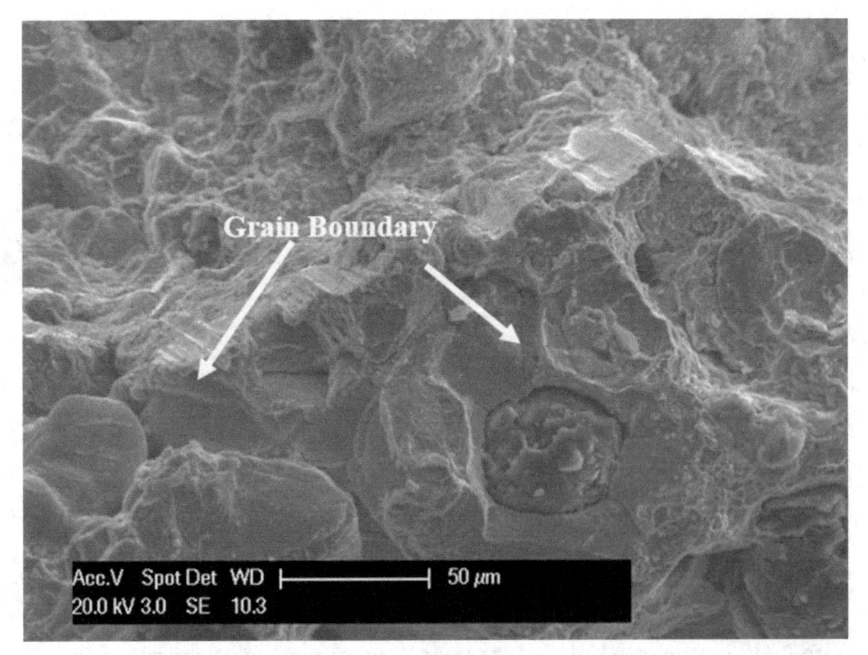

Figure 7.12. Fracture surface of specimen B01 failed by TMF with a constraint ratio of 100% in the temperature range of 433–873 K (160–600 °C).

tests conducted on this material also showed cyclic hardened hysteresis behaviour, depending on the temperature range, all the cycle profiles had a 2 min. ramping from minimum to maximum temperature, 1 min. hold at the maximum, and 4 min. ramping down to minimum. Figure 7.13 shows the cyclic peak-valley stresses of the material under 473K-873 K (200–600 °C) TMF with constraint ratios from 50% to 100%. As the TMF cycle falls in the DSA range, the material exhibited significant cyclic hardening continuously to failure (the TMF with 50% constraint ratio was a run-out). Figure 7.14 shows a) the hysteresis loops of the 1st and the 450th cycles with constraint ratio of 100%, to illustrate the cyclic evolution processes. The 1 min. hold at the maximum temperature of 873 K (600 °C) did not have any effect on the hysteresis behaviour, i.e., causing no stress relaxation, which indicates that creep deformation was minimal at this temperature. Therefore, plasticity with DSA effect is the dominant mechanism in this temperature range.

The cyclic peak-valley stresses under 473–973 K (200–700 °C) TMF with constraint ratios from 50% to 100% are shown in Figure 7.15. In this temperature range, as expected, the austenitic stainless steel also exhibited significant cyclic hardening until failure. The 100% constrained coupon failed at 128 cycles, still exhibiting cyclic hardening. The hysteresis behaviours of the 1st cycle and 102nd cycle at constraint ratio of 100% are shown in Figure 7.16. Appreciable stress relaxation could be observed during the 1 min. hold at the 102nd cycle, which indicates creep deformation at 973 K (700 °C). With smaller mechanical strain, i.e., at constraint ratios of 70% and 50%, cyclic behaviour tends to be stabilized, perhaps due to creep balancing the DSA effect.

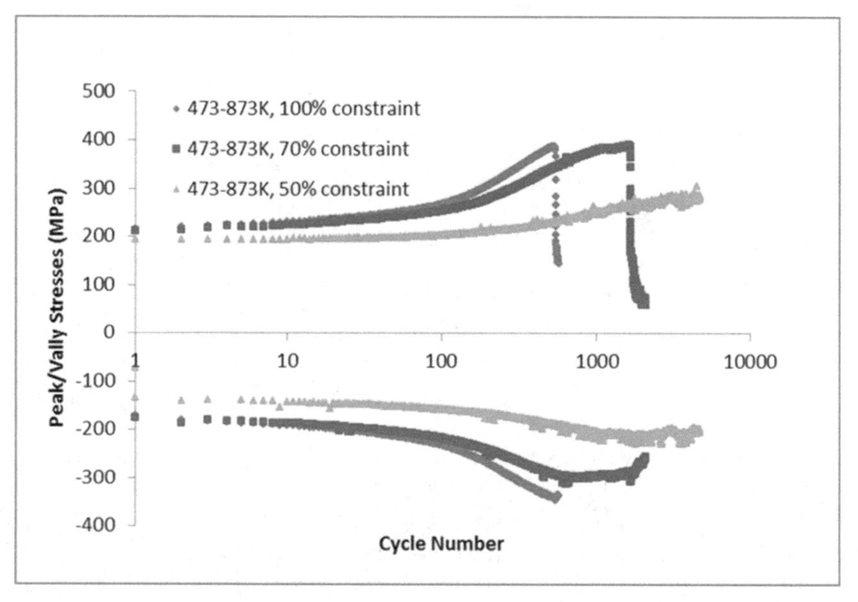

Figure 7.13. Cyclic peak-valley stresses under 473–873 K (200–600 °C) TMF.

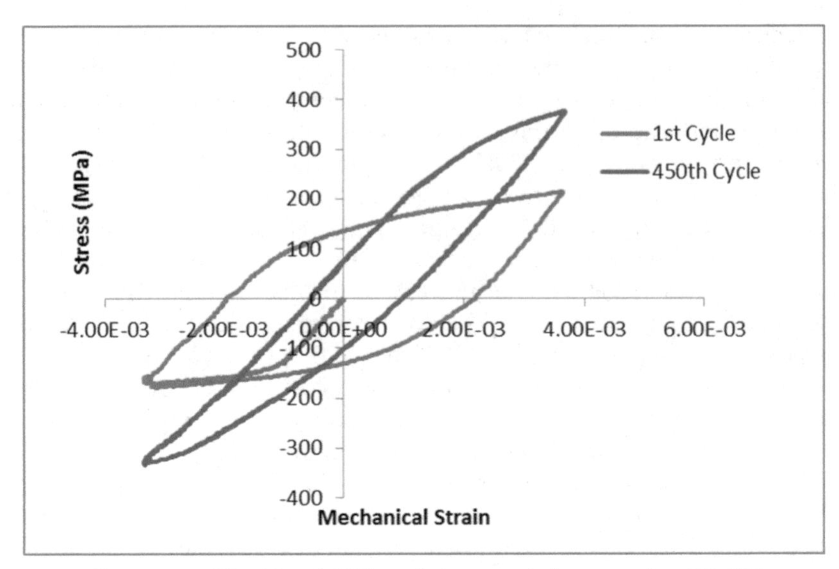

Figure 7.14. The 1st and 450th cycle hysteresis loops under 473–873 K TMF with a constraint ratio of 100%.

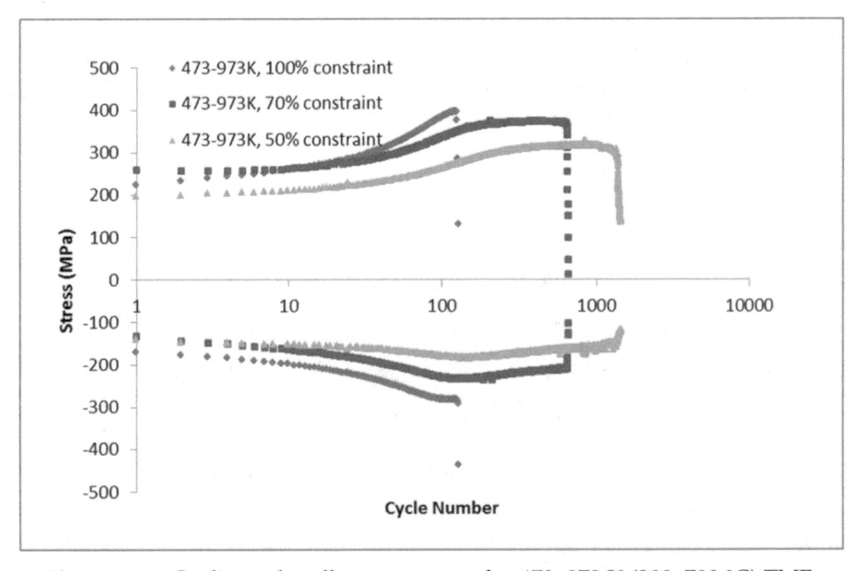

Figure 7.15. Cyclic peak-valley stresses under 473–973 K (200–700 °C) TMF.

The cyclic peak-valley stresses under 473-1073 K (200–800 °C) TMF with constraint ratios of 70% and 50% are shown in Figure 7.17. The material became cyclically stable after an initial short period of hardening. The hysteresis loops of the 1st cycle and 101st cycles with constraint ratio of 70% are shown in Figure 7.18. Apparently, compared to the TMF hysteresis behaviours with lower maximum temperatures, significant stress relaxation

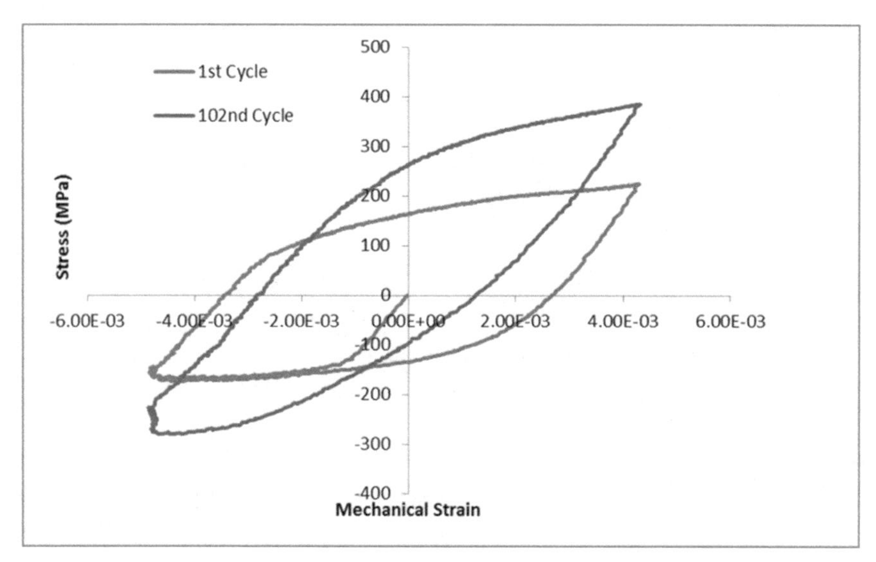

Figure 7.16. The 1st and 102nd cycle hysteresis behaviours under 473–973 K TMF with a constraint ratio of 100%.

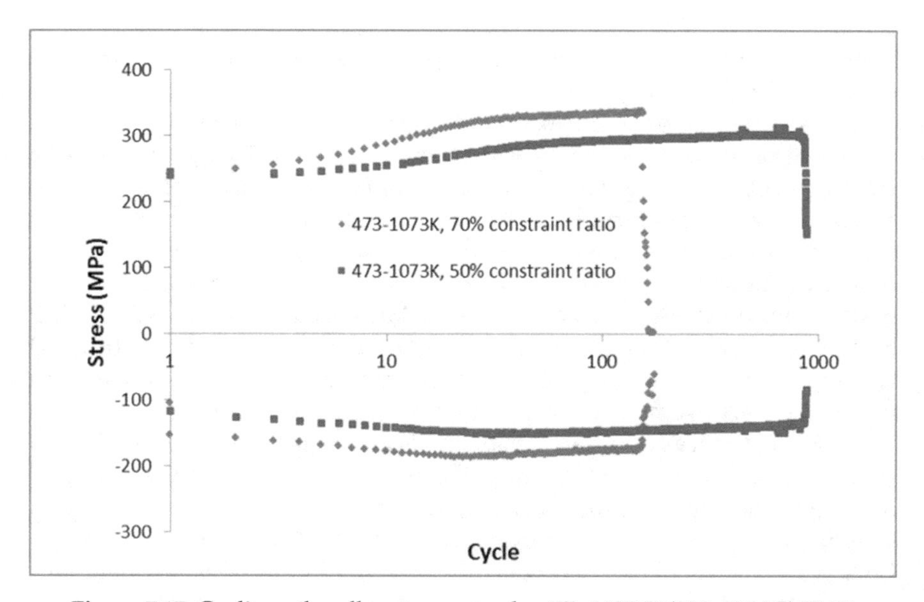

Figure 7.17. Cyclic peak-valley stresses under 473–1073 K (200–800 °C) TMF.

occurred during the 1 min hold period, starting from the 1st cycle. Because the stress-free condition was set at the mid-temperature, i.e., 773 K (500 °C) for the 473–1073 K (200–800 °C) cycle, elastic deformation mainly occurred in the 673–873 K (400–600 °C) cycle such that DSA was not manifested. It is also noticed that the hysteresis loops under 473–1073 K TMF were more "off the balance" towards tension, i.e., having a positive stress ratio, presumably

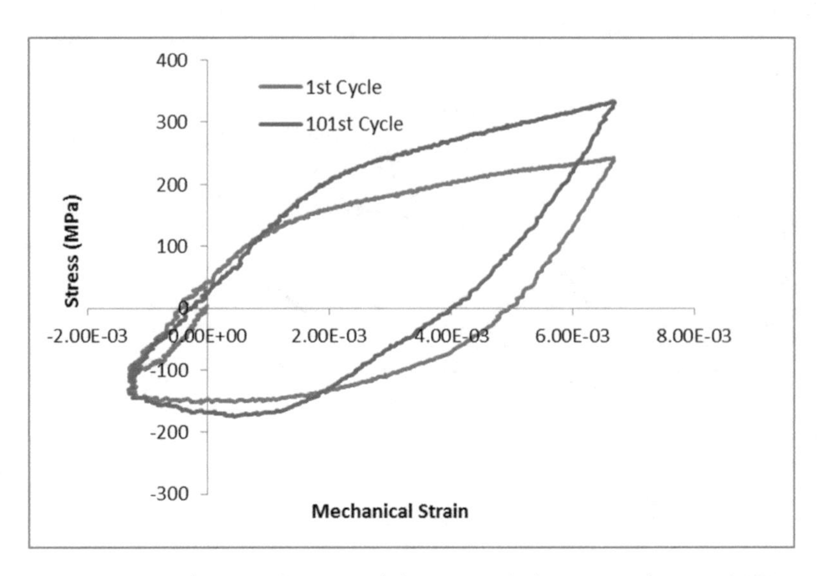

Figure 7.18. The 1st and 101st cycle hysteresis behaviours of 473–1073 K TMF with a constraint ratio of 70%.

because more softening by creep occurred in compression at temperatures ~1073 K (800 °C). In this case, the hysteresis energy remained roughly constant with the cycle.

The cyclic peak-valley stresses under 473–1173 K (200–900 °C) TMF with constraint ratios of 50% and 70% are shown in Figure 7.19. As shown, the material became cyclically stable after an initial short period of hardening. The hysteresis behaviours of the 1st cycle and 202nd cycle with constraint ratio of 70% are shown in Figure 7.20. Again, significant stress relaxation occurred during the 1 min hold period at 1173 K (900 °C), starting from the 1st cycle. It is also noticed that the degree of initial cyclic hardening in 473–1173 K TMF became lesser than that in 473–1073 K TMF. This effect could be attributed to creep-induced softening at higher temperatures. Again, in this case, the hysteresis energy remained roughly unchanged with the cycle.

TMF of cyclically-unstable material is much more complicated from simulation point of view, because the hysteresis loop evolves cycle by cycle. For engineering purposes, its TMF life can still be formulated in a simplified manner. Taking 1.4848 austenitic cast steel for example, in section 6.3.4, ICFT analysis has established the LCF life in terms of the material's cyclic deformation characteristics: the plastic hysteresis energy, the cyclic hardening range, maximum stress, and plastic and creep strain at mid-life. This correlation is an engineering simplification, because these characteristic quantities can be conveniently extracted from FEM, using the first, mid-life and last cyclic stress-strain relations of plasticity, rather than continuous simulation of thousands of cycles. Hence, TMF life prediction for 1.4848 is based on experimentally-measured cyclic data, Table 7.2, using Eqs. (7.10-7.14) with the same parameters as given in Table 6.4 for LCF (For austenitic

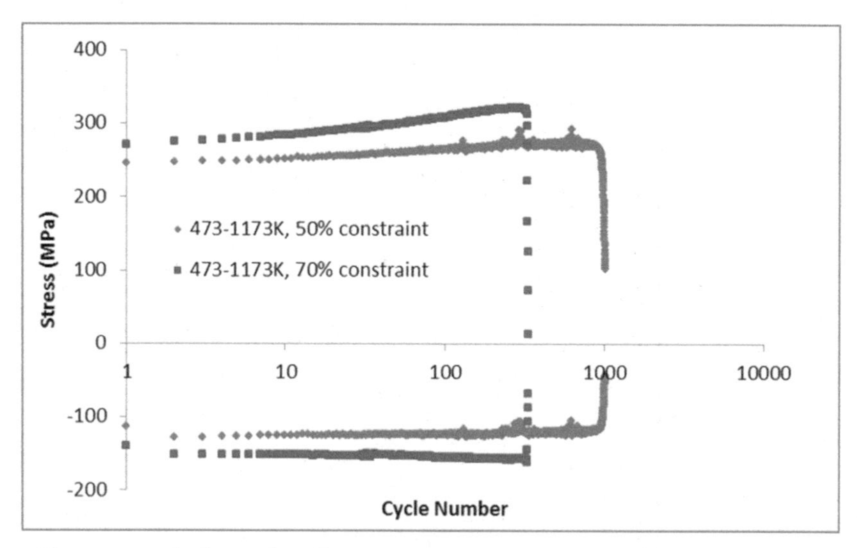

Figure 7.19. Cyclic peak-valley stresses under 473–1173 K (200–900 °C) TMF.

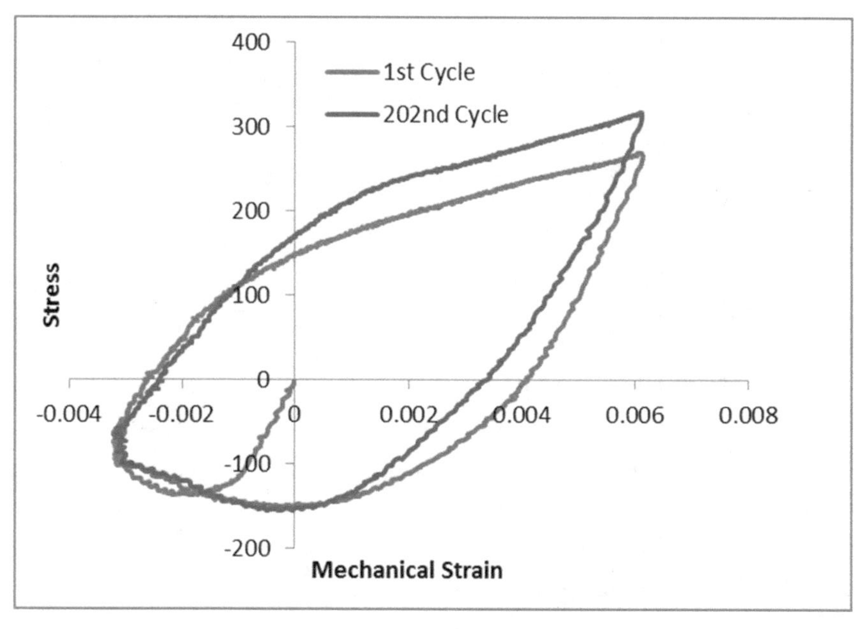

Figure 7.20. The 1st and 202nd cycle hysteresis behaviour of 473–1173 K TMF with a constraint ratio of 70%.

steel, IE is absent, $\phi \equiv 0$). The predictions are shown as lines in Figure 7.21. The correlation coefficient of the prediction/experiment is 0.96, which proves that the ICFT is indeed able to predict complicated TMF with dwell and constraint from LCF-calibrated mechanism information.

Table 7.2. TMF lives of austenitic stainless steel 1.4848

Specimen I.D	Max Temp C	Max Temp C	Restraint %	$\Delta\varepsilon t$ (%)	Plastic hysteresis	Creep strain %	mid- life σmax MPa	hardening ΔσH MPa	oxidation (m)	Failure Cycles N	Prediction N
D1-26100	200	600	100	0.7	167	0	341	171	1.60E-07	546	642
D2-26070	200	600	70	0.51	92.42	0	370	178	1.60E-07	1658	1692
D3-26050	200	600	50	0.35	39.4	0	273	94	1.60E-07	5000*	23768
D4-27100	200	700	100	0.88	242.6	0.54	335	173	1.07E-06	128	134
D5-27070	200	700	70	0.6	170	0.4	368	111	1.08E-06	650	387
D6-27050	200	700	50	0.47	56	0.24	317	120	1.05E-06	1407	1788
D7-28070	200	800	70	0.63	230	0.49	335	92	4.09E-06	155	241
D8-28050	200	800	50	0.48	121	0.33	294	51	4.09E-06	870	1080
D10-29070	200	900	70	0.63	210.4	0.6133	320	50	1.24E-05	333	203
D9-29050	200	900	50	0.47	118	0.416	273	28	1.23E-05	1000	623

*run-out

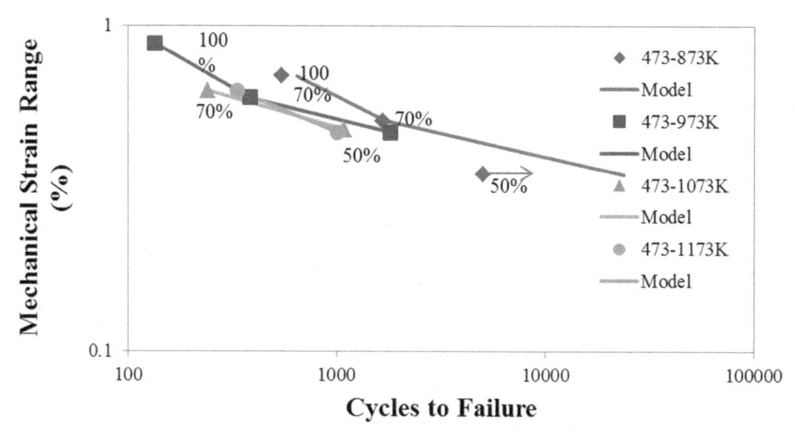

Figure 7.21. Comparison of the predicted TMF life with experimental data. The arrow indicates the run-out.

7.4 Summary

In this chapter, deformation and damage mechanisms in TMF are discussed, which are essentially the same as in isothermal LCF, but just the manifestation under thermomechanically cyclic conditions. The same constitutive law and the same damage accumulation equations apply to TMF as to LCF, which has been validated on DCI tested in the temperature ranges of 573-1073 K (300-800 °C) under OP, IP, and constrained strain-control conditions; and also on austenitic stainless steel in the temperature range of 473-1173 K (200-900 °C) under constrained TMF conditions with 1 min. dwell.

For DCI, the ICFT constitutive and damage model naturally incorporated four damage mechanisms: (i) plasticity-induced fatigue, (ii) intergranular embrittlement, iii) creep and (iv) oxidation. For austenitic stainless steel, the ICFT constitutive and damage model considered (i) plasticity-induced fatigue, (ii) dynamic strain aging, (iii) creep and (iv) oxidation. Basically, this model allows incorporation with additional damage mechanism(s) verifiable through metallurgical examination. Furthermore, the ICFT model delineates the deformation and damage mechanisms and their contributions in the asymmetrical TMF hysteresis behaviour and life, in terms of mechanism strains accumulated under complex TMF conditions: IP, OP, constrained with dwell. The overall agreement of the ICFT model with the experimental life is within factor of 2.

This has proven the assertion that material mechanical behaviours are merely manifestation of the underlying physical deformation and damage mechanisms under the given loading profiles (or boundary conditions). With this assured, it is convenient to conduct simple standard mechanical tests to validate the mechanism-model and use it to predict the behaviour under more complicated loading conditions, saving experimental efforts. This mechanism-based approach can certainly speed up component durability design.

CHAPTER

8

High Cycle Fatigue

High cycle fatigue (HCF) is a failure mode occurring under low-amplitude cyclic stresses at high frequencies, which is often characterized in the cycle range >10^4. Vibration is apparently a major source of HCF, especially at resonance with a frequency of thousands of Hertz, a structure may fail in less than an hour. Therefore, in fatigue design, HCF needs to be mitigated with allowable vibrational stresses below the fatigue endurance limit. The fatigue limit of a material is usually assessed through stress-controlled fatigue testing of smooth specimens to reach 10^7 cycles. However, in gas turbine engines during service, many factors such as mistuning, foreign object damage, prior creep exposure, and interactions with LCF may significantly reduce HCF life, posing serious threats to structural integrity. Therefore, HCF must also be treated holistically with consideration of the above factors. In this chapter, it will be shown that ICFT provides such a framework.

8.1 Pure HCF

Fatigue as a failure mode was first studied by Wöhler in 1867. Since in the HCF range little plasticity is appreciable in the materials' cyclic deformation, HCF life is therefore often correlated with the applied stress amplitude, called the S-N curve as represented by the Basquin equation, Eq. (4.10) (Basquin 1910). Under generalized loading but without macroscopic plasticity, the material is seen to be cycled between maximum and minimum stresses in the elastic regime. It is believed that at the microscopic level, due to microstructural inhomogeneity and discontinuity, localized stress concentration can cause localized plasticity to trigger fatigue damage process, albeit in very small scale initially. As discussed in sections 3.1 and 6.1, mechanical fatigue is related to cyclic plasticity, in the metallurgical sense. LCF occurs when plasticity spreads in the material body at a macroscopic level, whereas HCF starts with localized plasticity at the microstructure level while the macroscopic response of the material remains to be predominantly elastic. A deeper understanding of HCF needs simulation of cyclic plasticity at the microstructure level (Wu 2018a, 2018b). For now, we just proceed with macroscopic characterization of HCF with stress-based formula.

For the unification of terminology, a typical stress-loading profile is shown in Figure 8.1, where σ_{max} is the maximum applied stress, σ_{min} is the minimum stress, $\Delta\sigma = \sigma_{max} - \sigma_{min}$ is the applied stress range, σ_m is the mean stress, and σ_a is the stress amplitude. Under this type of load profile, the stress ratio, $R = \sigma_{min}/\sigma_{max}$, has an influence on fatigue life. There have been many S-N curve models proposed to correlate the fatigue data such as the Walker model (Walker 1970) and the Smith-Watson-Topper model (Smith, Watson and Topper 1970), etc., among which the SWT model is most popular and it takes the form:

$$\sigma_{max}\varepsilon_a = \frac{(\sigma'_f)^2}{E} N_f^{2b} + \sigma'_f \varepsilon'_f N_f^{b+c} \tag{8.1}$$

There is no particular advantages of one empirical formula over the other with regards to understanding the fatigue mechanism, but just its convenience in fitting for a particular set of experimental data.

Another way to represent HCF data is the Goodman diagram (Goodman 1919). It is constructed with the following relation for a constant life:

$$\sigma_a = \sigma_{fl}\left(1 - \frac{\sigma_m}{\sigma_u}\right) \tag{8.2}$$

where σ_{fl} is the fatigue strength at zero mean stress, and σ_u is the ultimate tensile strength of the material. If the fatigue condition given by the mean stress and the alternating stress lies under the curve given by Eq. (8.2), the material will survive. If the condition is above the curve, the material will fail. The general trend given by the Goodman relation is one of decreasing fatigue life with increasing mean stress for a given level of alternating stress. A Goodman diagram is shown in Figure 8.2 for 409 steel at RT, 400 °C, and 700 °C.

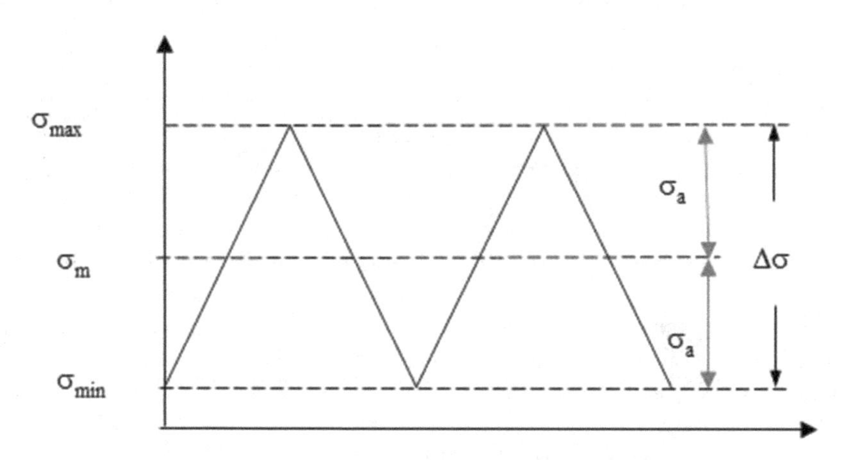

Figure 8.1. Cyclic stress loading profile.

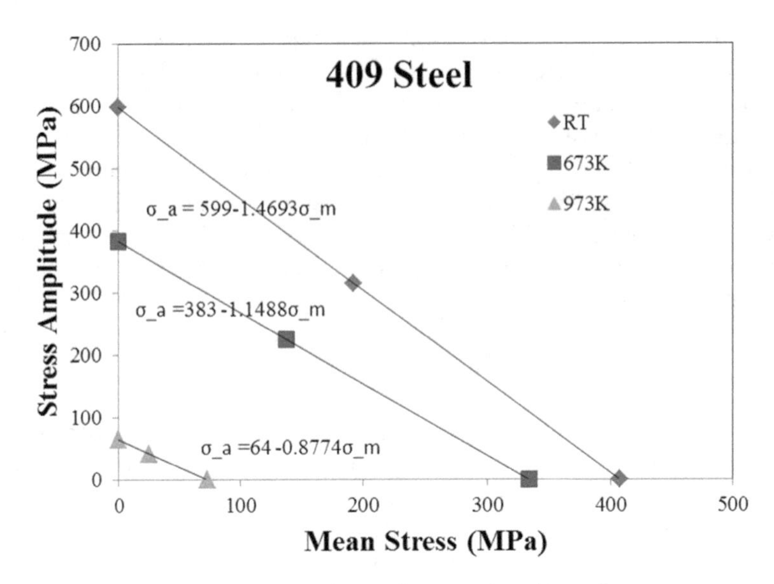

Figure 8.2. Goodman diagram for 409 steel.

8.2 HCF with Foreign Object Damage

Fan and compressor blades in gas turbine engines are often impacted by ingested hard objects during engine operation, which creates damage commonly referred as foreign object damage (FOD). FOD is a major concern for the safety and maintenance of military and commercial gas turbine engines.

FOD is mostly characterized by the size and shape of indentation produced by the impact of hard objects, which can alter the local stress distribution and the microstructure, leading early crack nucleation (Hutchings 1983, Peters et al. 2000). FOD is difficult for designers to deal with, because of the extent of FOD depends on many factors such as the colliding object size, mass, hardness, and impact velocity, which are all unpredictable in the service. Experiments have to be carried out to understand the key features of FOD and their effect on fatigue strength of the FODed material. In such a test, FOD is usually simulated via ballistic impact on material coupons at a given speed. For example, Figure 8.3 shows an impact crater created using a pneumatic gun and sabot spherical steel pellets on Ti-6Al-4V (Merati et al. 2011). At high magnification, adiabatic shear bands (ABS) are seen to form beneath the dent. On high-speed impact, deformation proceeds at extremely high rate without sufficient time for heat to dissipate, therefore the deformation process can be considered adiabatic (Yang et al. 1996). In addition to local shape and microstructural change, residual stresses may be built up. The residual stress measured from the crater rim is shown in Figure 8.4, which was generated by impact of a steel pellet (3.2 mm in diameter and a mass of

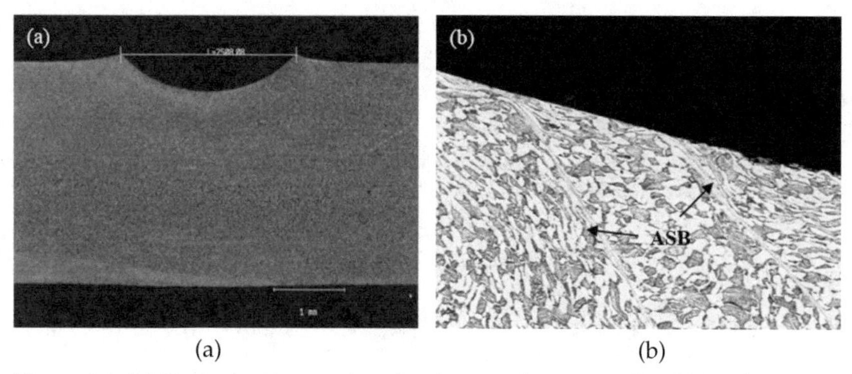

Figure 8.3. (a) Optical micrographs of an impact dent created at 300 m/s impact, 90° angle, and (b) adiabatic shear bands (ASBs), as indicated by arrows.

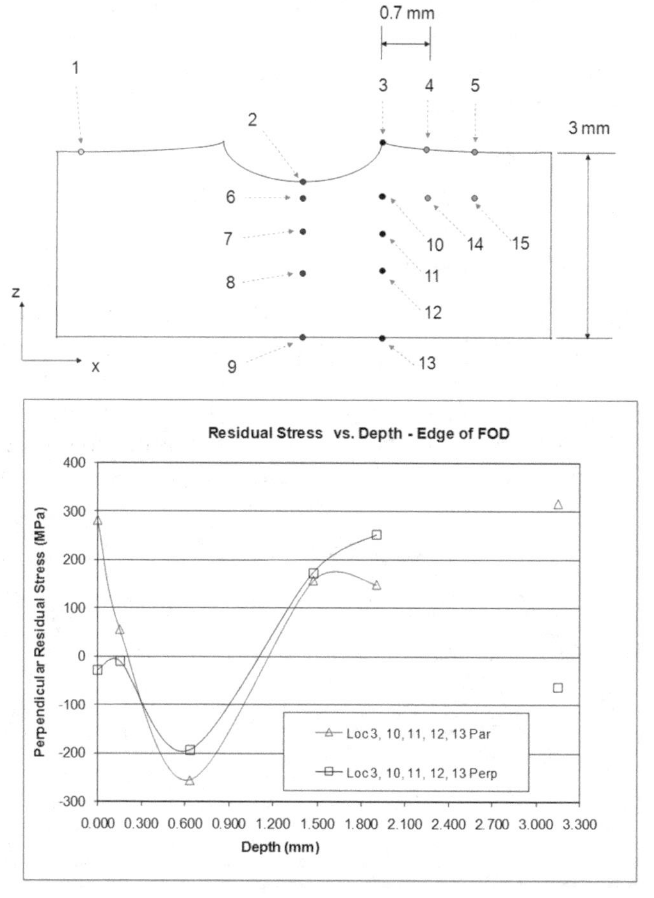

Figure 8.4. Residual stresses measurements in depth from the FOD crater edge (triangular symbols represents the stress in the loading direction, and square symbols represent stress perpendicular to the loading direction).

0.195 g) at a nominal speed of 300 m/s. As the figure shows, FOD results in a positive residual stress within 300 μm, which will be adding to the applied load. Therefore, it is not surprising that fatigue cracks are mostly found to nucleate at the FOD crater rim location.

We assume that $\sigma = \sqrt{3}\,\tau$, the stress-based fatigue equation, Eq. (6.17b), can be re-written into the following form:

$$\frac{1}{N_f} = \frac{(1-\nu)b}{6\mu w_s}(\sigma_{\max} - \sigma_{th})^2 \tag{8.3a}$$

where σ_{th} is the threshold stress which takes the form of the Walker stress, defined as:

$$\sigma_{th} = \sigma_0(1-R)^{-\gamma} \tag{8.3b}$$

For Ti6Al4V, E = 120 GPa, σ_0 = 280 MPa, γ = 0.9 (the Walker parameters are determined through best-fitting to data).

For the impacted material, the actual stress ratio R is modified by the presence of residual stress, as

$$R' = \frac{(\sigma_r + \sigma_{\min})}{(\sigma_r + \sigma_{\max})} \tag{8.4}$$

and

$$\sigma'_{\max} = (\sigma_r + \sigma_{\max}) \tag{8.5}$$

Then, the S-N behaviours of Ti-6Al-4V with and without FOD are shown in Figure 8.5. First of all, the undamaged material is described using Eq. (8.3). Then, the FODed material is predicted with consideration of the residual stress, as in Eq. (8.4) and (8.5). Using Eq. (8.3), the average fatigue life is obtained by integration over the positive residual stress field near the surface, in the depth of 0.3 mm. It can be seen that the residual stress has a

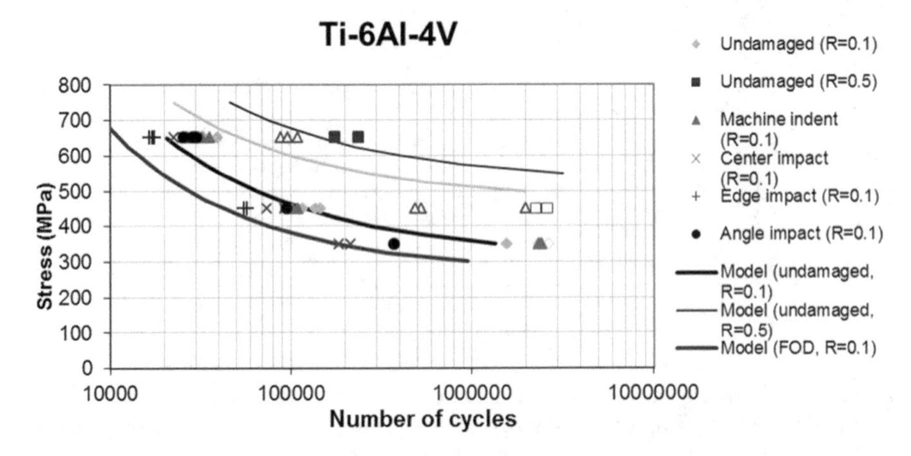

Figure 8.5. Fatigue test data and the S-N curves predicted using Eq. (8.3)-(8.5).

significant effect on fatigue life. On the other hand, the machined dent had little influence (no residual stress). The theoretical fatigue equation gives a good description for the HCF phenomena.

8.3 Cold-Dwell Fatigue

A component in service is often subjected to cyclic-dwell loadings. Some material may even suffer a life debit from dwell at ambient temperature. The detrimental effect of room-temperature dwell was first learned in late 1972 from the uncontained failure of two titanium alloy fan discs in Rolls-Royce RB211 engines which powered Lockheed Tristar aircraft at the time (Pugh 2001). Subsequently, numerous studies have been conducted on high strength titanium alloys such as IMI 685, IMI 829 and IMI 834, to investigate the dwell sensitivity (Song and Hoeppner 1988, Bache et al. 1997, Bache 2003, Wu and Au 2007). These titanium alloys typically have a bimodal $\alpha+\beta$ microstructure, where α grain has an h.c.p. structure, and transformed β has a b.c.c. structure. Experiments have revealed that these titanium alloys tend to fracture with "quasi-cleavage facets" formed on the basal planes of α grains during dwell fatigue at near ambient temperatures. Therefore, the phenomenon is also known as *cold dwell fatigue*. At first, hydrogen embrittlement was thought to be responsible for the phenomenon (Hack and Leverant 1982). Later, researchers denied this mechanism, because it was found that IMI 685 heat-treated to extract hydrogen below 10 ppm still exhibited the dwell effect (Evans 1987). Wu and Au (2007) investigated the problem and characterized the cold dwell sensitivity as function of stress, temperature and time, using a dislocation pile-up model conforming to the ICFT, which will be elucidated below.

As mentioned above, dwell-fatigue fracture of titanium alloys is often accompanied with cleavage fracture along the basal plane of the α grain or lath in the alloys. The faceted fracture is believed to occur during the dwell period, due to accumulation of dislocation pile-ups. Bache (2003) invoked Stroh's dislocation pile-up model to explain the condition of fracture facet formation under high stresses at the leading position of the dislocation pile-up. However, the Stroh's model only considered static dislocation pile-up, the kinetics (time-dependence) of dwell damage process was not considered (Stroh 1954).

A kinetics treatment of dislocation pile-up during dwell fatigue has been given by Wu and Au (2007). The build-up of dynamic dislocation pile-up is similar to the situation discussed in section 1.6.2. At constant strain rate, the number of dislocations in a pile-up is derived as a function of time, as Eq. (1.41):

$$n = \frac{\dot{\gamma}_p}{\kappa}\left(1 - e^{-\kappa\tau}\right) \tag{8.6}$$

Dislocation pile-up forming a Zener-Stroh-Koehler (ZSK) crack is treated in Chapter 9 for generalized anisotropic materials. The crack nucleation condition is defined by the Griffith energy criterion:

$$\frac{\bar{F}_{22} b_T^2}{8\pi a} = 2w_s \tag{8.7}$$

where $\bar{F}_{22}$ is an elastic constant, a is half-crack size, w_s is the surface energy, and b_T is the total Burgers vector in the pile-up.

In Chapter 4, we have derived that under constant-amplitude cyclic loading the fatigue life (without oxidation, $h = 0$) can be expressed as Eq. (4.18):

$$\frac{1}{N} = \left(1 + \frac{l_z}{\lambda}\right) \frac{1}{N_f} \tag{8.8}$$

Considering that $l_z = 2a$ and $b_T = nb$, substituting Eq. (8.6) and (8.7) into Eq. (8.8), we obtain:

$$N_d = \frac{N_0}{1 + \frac{\bar{F}_{22} b^2}{8\pi\lambda w_s}\left(\frac{\dot{\gamma}_p}{\kappa}\right)^2 \left(1 - e^{-\kappa\tau}\right)^2} \tag{8.9}$$

where N_d is the dwell-fatigue life, and N_0 is the fatigue life with no dwell. Here, λ represents the average spacing between α grains in the α + β microstructure. The microstructure of an IMI 834 disk material is shown in Figure 8.6.

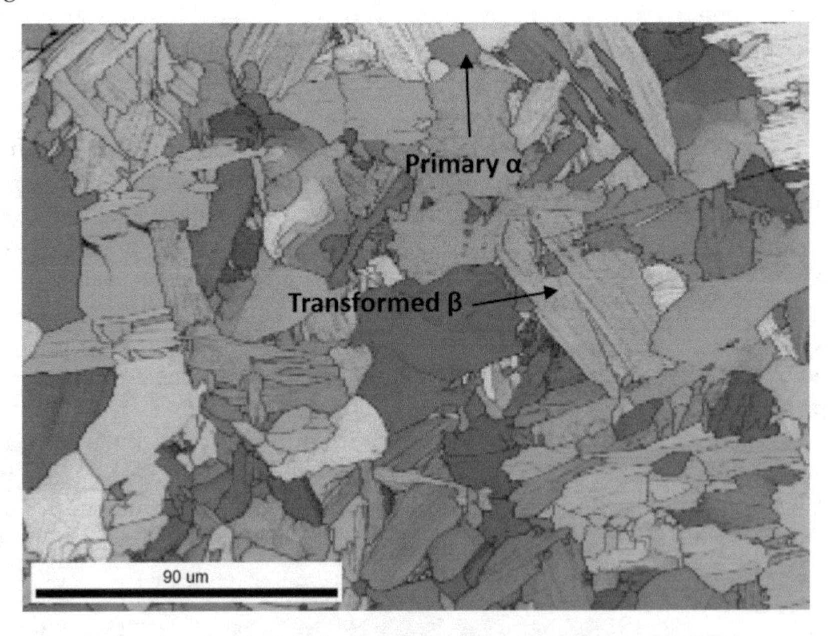

Figure 8.6. An electron back-scatter diffraction image of IMI 834.

Color version at the end of the book

Equation (8.9) shows that the fatigue life is knocked down by a factor greater than one, when a dwell period is imposed on the cyclic loading. This "knock down" factor depends on the material properties such as elastic constants, surface energy, microstructure, and most importantly it is controlled by the ratio of dislocation glide to the climb rate. This means that damage in the form of ZSK cracks is more prone to dwell at low temperatures with limited dislocation climb activities, particularly in a material with fewer active slip systems. As temperature increases, dislocation climb activities will overwhelm glide such that dislocation pile-up can hardly form, and hence the dwell damage will become minimal, but creep cavities may start to form, instead. Basically, this is the physical essence of "cold dwell" vs. "hot creep".

For engineering analysis, Eq. (8.9) may be conveniently re-written into the following form:

$$N_d = \frac{N_0}{\left(1+\left(\frac{v}{\kappa}\right)^2[1-\exp(-\kappa\tau)]^2\right)} \tag{8.10a}$$

which is used to describe the experimental observations, Figure 8.7, with

$$v = v_0 \exp\left(-\frac{Q_g}{RT}\right)\sigma^n = 4.4\times10^{-16}\exp\left(-\frac{7450}{RT}\right)\sigma^{5.4} \quad \text{for IMI 834} \tag{8.10b}$$

and

$$\kappa = \kappa_0 \exp\left(-\frac{Q_v}{RT}\right) = 1.48\exp\left(-\frac{8680}{RT}\right) \quad \text{for IMI 834} \tag{8.10c}$$

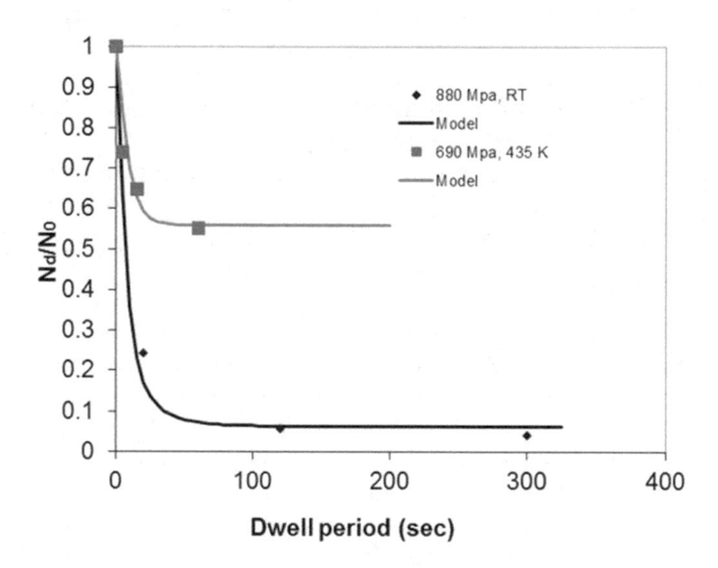

Figure 8.7. Comparison of Eq. (8.10) with the experimental data. (The RT data are from Bache et al. 1997)

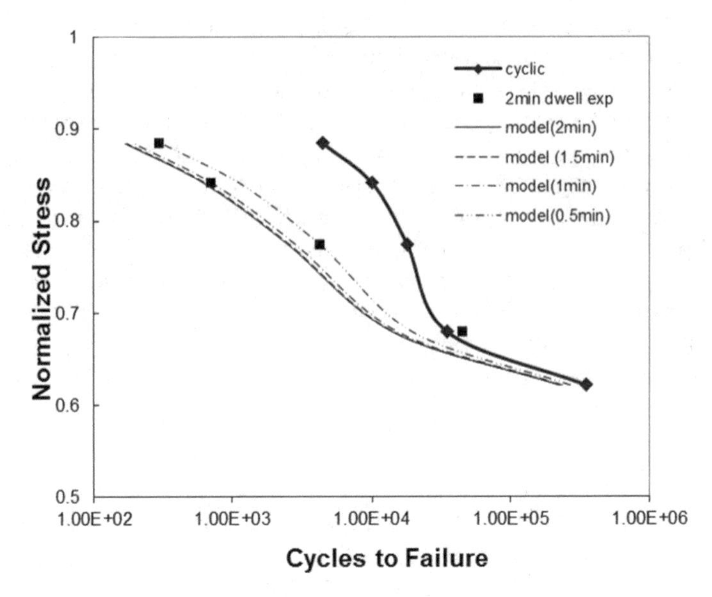

Figure 8.8. The S-N curves with/without 2 min hold at RT. The experimental data are taken from Bache et al. (1997).

Figure 8.7 shows the dwell sensitivity (ratio of no-dwell fatigue life to dwell fatigue life) increases exponentially with the dwell time, but approaches a constant level when the dwell period is long enough for IMI 834. The dwell sensitivity at 880 MPa/RT is apparently larger than that at 690 MPa/435 K, because of increasing dislocation climb activity at elevated temperature. The theoretical model, Eq. (8.10), agrees very well with the experimental observations. This behaviour suggests that the LDR does not work for dwell fatigue either. The dwell sensitivity decreases with increasing temperature. The S-N curves of dwell fatigue with different hold times are predicted by Eq. (8.10) and compared with the experimental data from Bache et al. (1997), as shown in Figure 8.8. The cold-dwell sensitivity decreases with stress. With dwell time greater than 30 sec., the dwell sensitivity became nearly constant.

Figure 8.9 shows the fracture surface of a dwell-fatigue failed IMI 834 specimen. It contains cleavage-like facets aligning perpendicular to the loading axis. At a higher magnification, fatigue striations can be seen in the regions between those fracture facets. This metallurgical evidence supports the fracture mechanism envisaged by Eqs. (8.8-8.10). The ICFT infers that titanium alloys having small α grains and large transformed β grains (large λ) will be less dwell-sensitive, whereas materials having large α grains or strongly textured will be more dwell-sensitive. This was indeed observed to be the main reason causing the dwell fatigue scatter of titanium alloys. Usually, the bar stock material has finer α grains and more equiaxed transformed β grains. Therefore, IMI 834 bars are less dwell sensitive

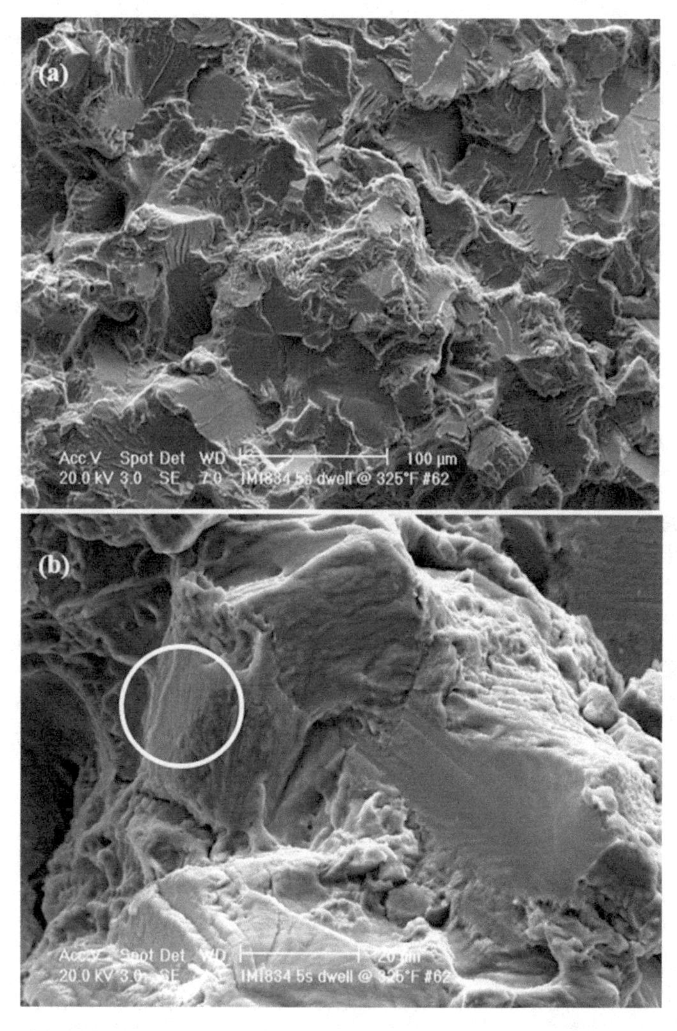

Figure 8.9. (a) A SEM micrograph of the dwell-fatigue fracture surface, and (b) at a higher magnification, showing fatigue striations (in the circled region) between brittle fracture facets.

than disc forgings (Bache et al. 1997). Also, the single mode (α colony in retained β) Ti 6246 does not exhibit dwell debit as compared to the bimodal ($\alpha + \beta$) Ti 6242 (Bache 2003). Recently, Zheng et al. (2016) performed a discrete dislocation crystal plasticity analysis for the two materials and showed that the bimodal material containing some rogue grain combination could have a very high stress in the α grain, thus increasing the propensity of fracture along the basal plane.

As Eq. (8.10) implies, healing of dwell damage may occur when dislocation climb activity is abundant at high temperatures. Thermomechanical dwell fatigue tests have been carried out under the loading profile as shown in

Figure 8.10, which involves dual loading stages: Stage I at σ_1 = 828 MPa, T_1 = 65 °C; and Stage II at σ_2 = 573 MPa, T_2 = 370 °C; both stages have the same hold period: $\tau_1 = \tau_2$ = 30 sec.

In this case, the load cycle involves one cold dwell period at a high stress σ_1 and hot dwell at a lower stress σ_2. According to Eq. (8.6), the dwell damage can be defined as the equivalent dislocation number, as given by

$$n^* = \frac{v}{\kappa}[1-\exp(-\kappa\tau)] \tag{8.11}$$

Assume that the material has nil damage initially, $n^*_0 = 0$, after experiencing the first dwell, its internal damage has reached to:

$$n_1^* = \frac{v_1}{\kappa_1}[1-\exp(-\kappa_1\tau)]\Big|_{\sigma=828MPa,T=338K,\tau=30s} = 2.28$$

During the second dwell period at a high temperature, because of dislocation climb, the dwell damage accumulated during the prior stage is mitigated to:

$$n_2^*(1) = n_1^* \exp(-\kappa_2\tau)\Big|_{T=644K,\tau=30s} = 0.00035$$

At the same time, new damage is also accumulated, as:

$$n_2^*(2) = \frac{v_2^*}{\kappa_2}[1-\exp(-\kappa_2\tau)]\Big|_{\sigma=573MPa,T=644K,\tau=30s} = 0.292$$

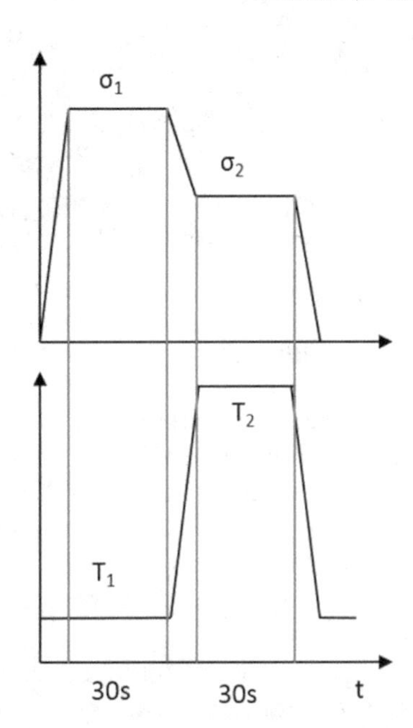

Figure 8.10. A thermomechanical dwell-fatigue loading profile.

Therefore, after the two stages of thermomechanical dwell, the total damage is given by

$$n^*_{TMDF} = [n^*_2(1) + n^*_2(2)] = 0.292$$

This results in a reduction of dwell sensitivity from 6.21 to 1.09, which has significantly reduced the dwell damage from previously accumulated at the low temperature.

8.4 HCF with Prior-Creep Strain

Creep-fatigue interaction may occur in another way as HCF with prior creep deformation. This could happen in gas turbine components such as turbine discs and blades operating at high temperatures, where creep strain will unavoidably accumulate during long-time engine operation. How the prior creep deformation affects the fatigue life is a question concerned by both gas turbine designers and operators, but only limited studies on this aspect are reported in the open literature.

A test program was conducted to investigate the effect of prior-creep strain on fatigue life of Waspaloy (Wu et al. 2010). First, creep tests were performed under stress from 600 to 680 MPa at 650 °C to a pre-determined strain level, followed by fatigue testing under a stress amplitude of 478.8 MPa ($R = -1$) at ambient temperature. Figure 8.11 shows the creep curves under various stress levels at 650 °C. Even though the creep behaviours were still in the transient stage, metallurgical examination revealed that creep damage was present in the form of grain boundary cracks/voids as shown in Figure 8.12. Those cracks and voids appeared to nucleate at grain boundary

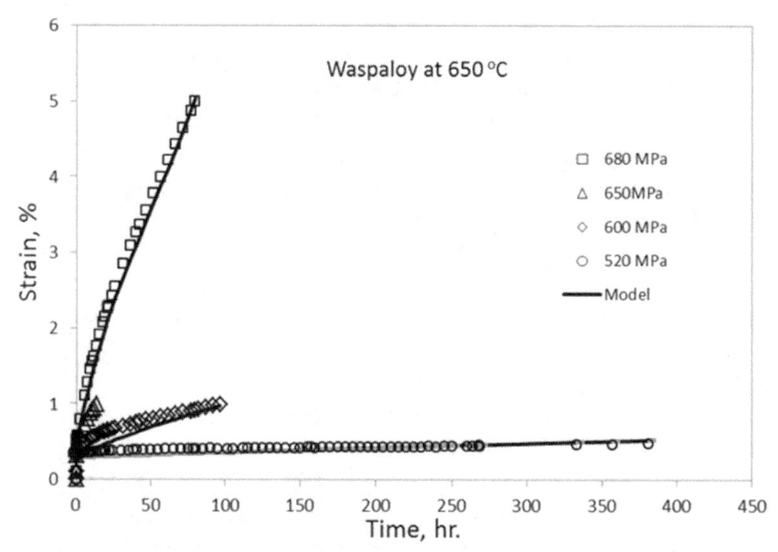

Figure 8.11. Creep curves of Waspaloy under various stresses at 650 °C.

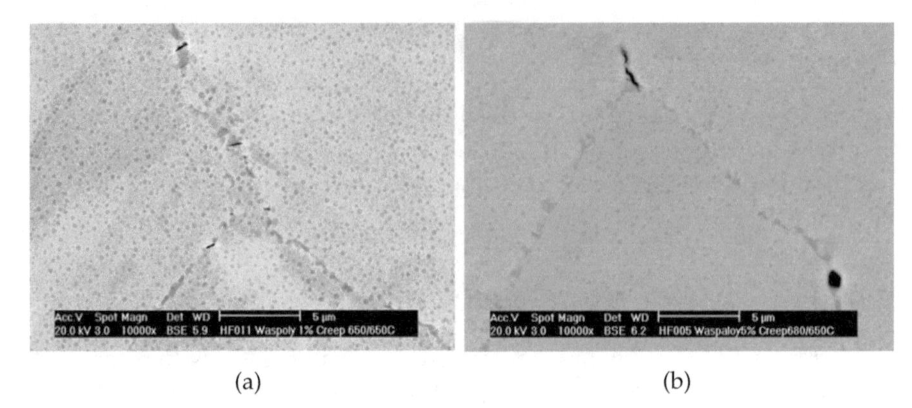

Figure 8.12. Microcracks in specimen prior-crept to (a) 1% strain at 650 MPa/650 °C, and (b) 5% strain at 680 MPa/650 °C.

precipitates and or triple junctions, and their size appeared to be proportional to the creep strain, which is consistent with the GBS mechanism typically dominant at this temperature, as discussed in section 5.2. Therefore, it can be assumed that the size of internally distributed damage is proportional to creep strain, $l = d\varepsilon_v$. Then, according to Eq. (4.18),

$$N = \frac{N_f}{1 + \frac{\varepsilon_v d}{\lambda}} \tag{8.12}$$

where N_f is the fatigue life without prior-creep, d is the grain size, λ is the spacing between grain boundary precipitate. The grain size of Waspaloy is ~100 μm, and the GB precipitate spacing is ~0.7 μm.

The fatigue life ratio with prior-crept strain at 650 °C is shown in Figure 8.13 (a). Eq. (8.12) is used to draw the trend line, which agrees with the experimental observation with a reasonable scatter in the HCF regime. When the elapsed creep time is normalized with the creep rupture life obtained for the alloy in section 5.3, the fatigue life ratio is plotted against the creep life fraction, as shown in Figure 8.13 (b). Again, the LDR is grossly erroneous for predicting the fatigue life of turbine disc where creep strain can be accumulated during the service. The metallurgical reason for this phenomenon is that prior creep generates distributed grain boundary cracks first, which would then propagate and coalesce with each other during subsequent fatigue, thus reducing the life significantly as compared to that without prior creep damage. The reverse fatigue-creep interaction may not have the same effect, because creep cracks would preferably propagate along the grain boundaries, which may not interact with the intragranular slip bands induced by prior fatigue. Eq. (8.12) can keep track of accumulation of IDD with deformation under different load sequences, but the LDR does not.

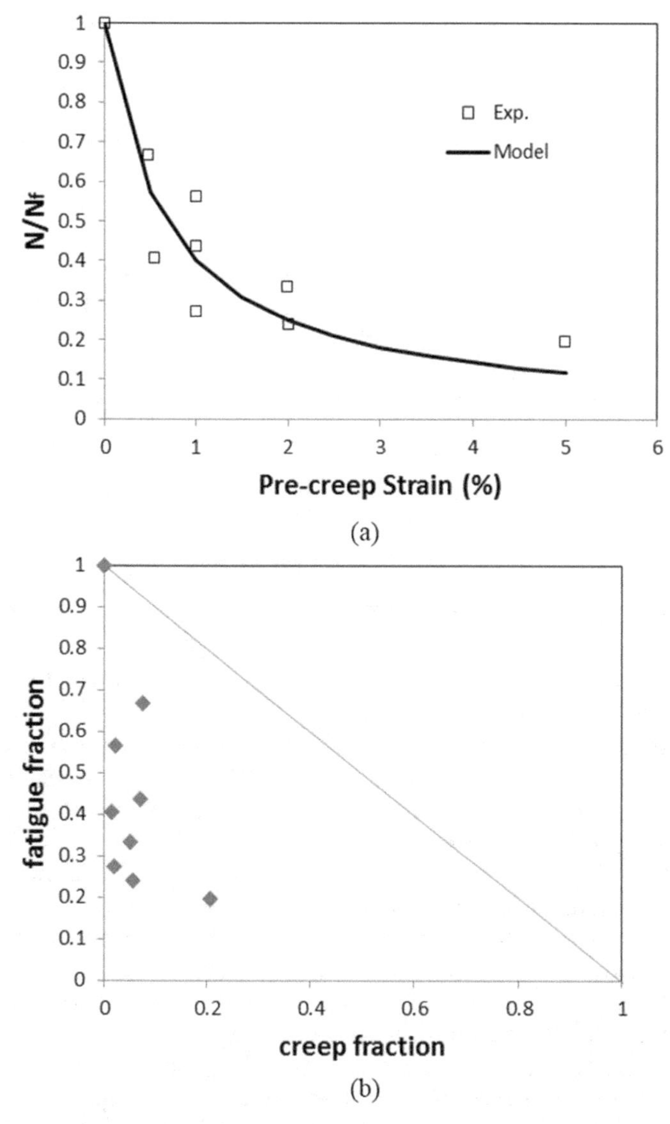

Figure 8.13. Fatigue life ratio as function of (a) prior-creep strain, and (b) creep time fraction in Waspaloy.

Another experimental program was carried out in the author's laboratory to study the effect of prior-creep strain on HCF of IMI 834 (Wu et al. 2018). The creep tests were conducted up to 5% total strain, as shown in Figure 8.14. Under these test conditions, creep deformation was all in the transient (primary plus secondary) stage. However, metallographic examination found that prior creep deformation >3% did cause multiple surface crack nucleation and created internal voids along grain boundaries as well, as shown in

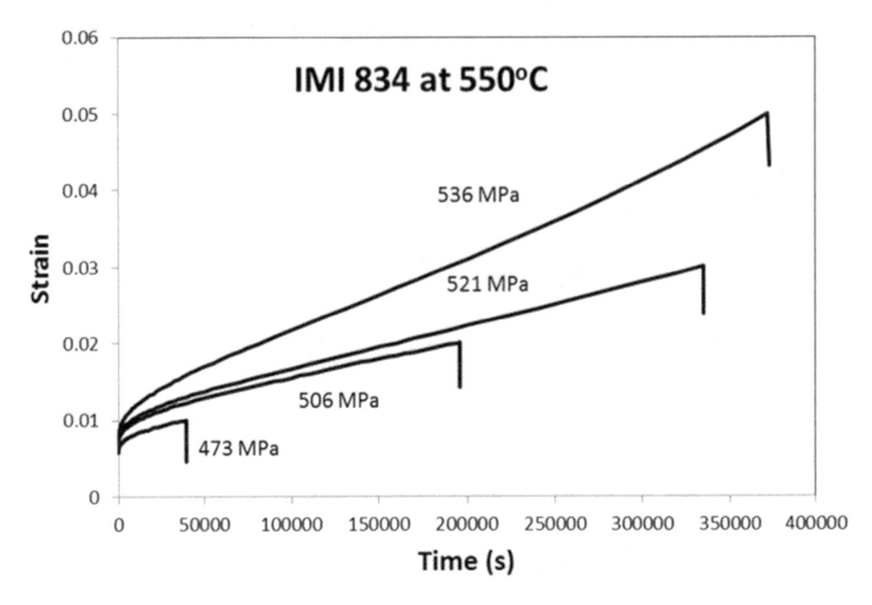

Figure 8.14. Creep curves of IMI 834 under various stresses at 550 °C.

Figure 8.15. The surface crack size distributions in the longitudinal-sections of 5% and 3% crept samples are shown in Figure 8.16. It shows that the size and number of the creep-nucleated surface cracks increases with creep strain. Hardt et al. (1998) studied LCF of IMI 834 with prior-creep strain, but they reported "negligible" void formation up to creep strain of 5% (Hardt et al. 1998, 1999), and therefore they attributed the LCF life debit to degradation of lamellar structure by prior creep. Here, metallurgical evidence exists to indicate physical damage did form during creep, even at the transient stage.

Room-temperature HCF failed specimens with prior-creep strains were also examined in both longitudinal and transverse sections, examples are shown in Figure 8.17. Figure 8.17 (a) shows a surface nucleated fatigue crack propagating into the material, apart from the main crack that failed the specimen. In addition, an internal crack is also seen, growing along the grain boundary. In the transverse section, Figure 8.17 (b), several internal cracks are also found that have not connected to surface or linked themselves yet. Again, metallurgical evidence corroborates the hypothesis of ICFT that the overall damage accumulation consists of surface fatigue crack nucleation and propagation in coalescence with internally distributed damage. The HCF life of IMI 834 with prior-creep strain follows Eq. (8.12), as shown in Figure 8.18. As comparison, the curve representing the ductility exhaustion theory (Eq. (4.13), the dashed line) with creep ductility of 6% is also shown. Certainly the ductility exhaustion theory does not represent the creep-fatigue interaction in this case, which significantly underestimates the effect of prior creep in life reduction.

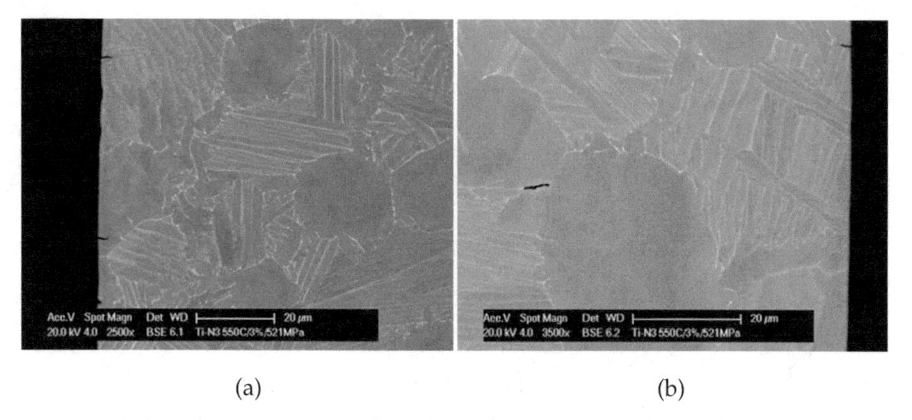

Figure 8.15. Longitudinal section of specimen with 3% creep strain at 550 °C: (a) multiple surface cracks, and (b) internal cracks/voids at grain boundaries.

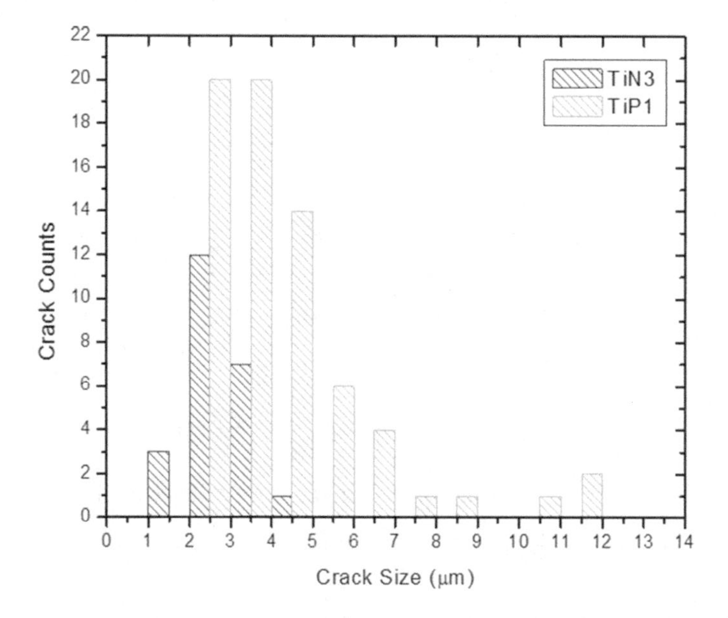

Figure 8.16. Surface crack size distribution in both P1 (5% prior creep strain) and N3 (3% prior creep strain) specimens.

8.5 Summary

This chapter has demonstrated that the basic dislocation-crack nucleation model, Eq. (6.17), is also applicable to HCF in the form of S-N curve (in a case study for Ti-6Al-4V). Because HCF deformation is predominantly elastic at the macroscopic scale, a constant fatigue limit has to be assumed, even though this may be a variable at the microstructural scale. More studies on HCF

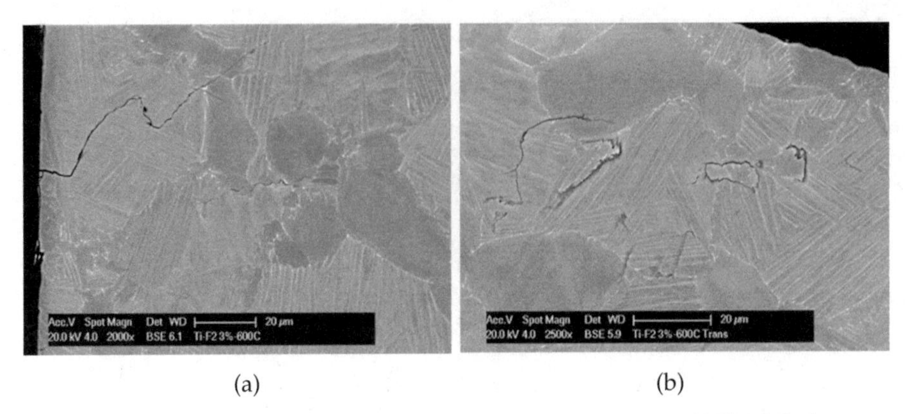

(a) (b)

Figure 8.17. (a) Longitudinal and (b) transverse sections of HCF failed specimen with 3% prior-creep strain.

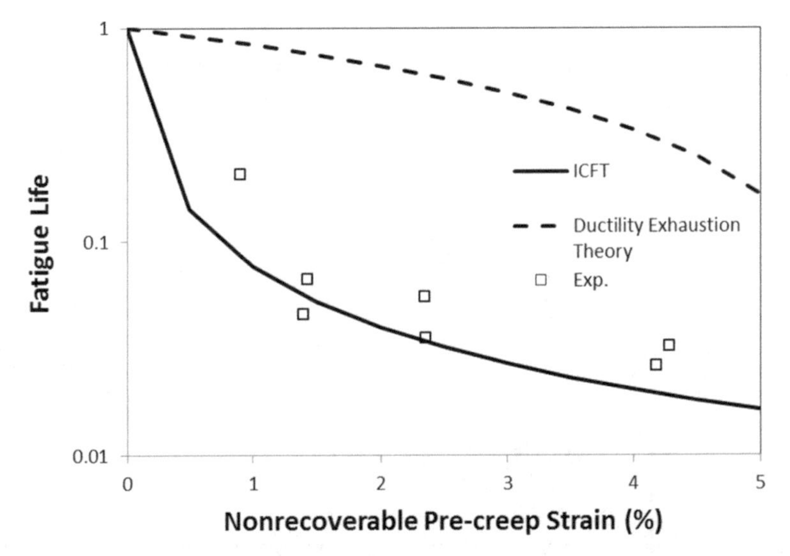

Figure 8.18. Fatigue life ratio as function of prior-creep strain in IMI 834, best-fit with Eq. (8.12).

should be conducted on HCF life and fatigue endurance limit, quantitatively linking to microstructural features.

When other damage is introduced to the material, HCF life can be significantly reduced. The analysis has to be conducted in a mechanism-delineated way. Particularly, when internally distributed damage (IDD) is created, whether through concurrent dwell or prior dwell (creep), Eq. (4.18) or Eq. (8.8), is the right equation to describe life debit due to dwell/creep-fatigue interaction, whereas the LDR is not. As for IDD, different mechanisms produce different types of damage such as ZSK dislocation pile-ups in "cold dwell" and/or creep voids/cracks in "hot creep". Both cold and hot dwell-fatigue life governed by the same holistic damage accumulation equation.

CHAPTER

9

Microscopic Crack Nucleation and Growth

Microscopic crack nucleation and growth is undoubtedly the early stage of material life evolution, which takes a significant portion of the material's total life (as high as 90% in high cycle fatigue, for example). Because of their shear sizes, usually in the micron range, microcracks are difficult to detect. In engineering structural integrity programs, *crack initiation* is often defined as the minimum detectable crack size, but it depends on the non-destructive inspection (NDI) technique used. Henceforth, crack initiation refers to the event when a crack is generated in shape and form identifiable using available inspection methods; whereas crack nucleation refers to crack formation by a particular physical mechanism. It is the focus of this chapter to describe the mechanism(s) and processes of crack formation from embryos to discontinuity at the microstructural level.

Crack nucleation mechanisms have been a subject of extensive discussion, ever since the early 1940s (Orowan 1940). These mechanisms can be categorized as follows (Suresh 1998).

- *Crack nucleation at surface*, which occurs either as a result of intrusion/extrusion due to slip irreversibility (usually in association with persistent slip bands), or oxidation/corrosion, or by fretting.
- *Crack nucleation at subsurface*, which occurs from inclusions, pores, or as dislocation pile-ups.
- *Crack nucleation at grain/inter-phase boundaries,* which occurs in the form of either cavities or wedge cracks.

Among the above mechanisms of crack nucleation, other than vacancy condensation and environmental effects such as oxidation and corrosion, the very basic forms of crack nucleation are dislocation pile-ups, which can accumulate at pores and inclusion interfaces and inter-phase/grain boundaries. Crack formation as distributed dislocation pile-ups has long been treated by the continuously distributed dislocation theory (Stroh 1958, Bilby et al. 1962, Taira et al. 1978, Navarro and los de Rios 1988, Weertman 1996). Dislocation pile-up accumulation with the same sign creates a wedge crack with one tip in tensile mode and the other in compressive mode, whereas dislocation pile-up distribution with opposite signs at the two ends results in a crack with both tips in the tensile mode. The former type is called the Zener-Stroh-Koehler (ZSK) pile-up (Zener 1948, Stroh 1957, Koehler 1952), and the latter called the Bilby-Cottrell-Swinden (BCS) pile-up (Bilby, Cottrell and Swinden 1963). A surface crack is a special case of the latter where only half of the asymmetrical distribution is actually in effect.

In a crystalline material, a crack at its nucleation stage must see the anisotropic effect of the crystalline structure, which may induce coupling between mode-I, II and III fracture on dislocation pile-up formation. Most of the existing dislocation pile-up models were proposed only for isotropic materials (Bilby et al. 1962, Taira et al. 1978, Navarro and los de Rios 1988, Weertman 1996). To understand the complex nature of crack nucleation in crystalline solids, the description of dislocation pile-ups must be given in the context of anisotropic elasticity and plasticity. In this chapter, dislocation pile-ups are described as continuously distributed dislocations in anisotropic solids. Crack nucleation by the energy criteria and microscopic crack interaction with microstructural obstacles such as grain boundaries are discussed. These formulations are useful in bridging between nano-scale discontinuities (dislocations) and macroscopic cracks.

9.1 Theory of Continuously Distributed Dislocations

The continuously distributed dislocation theory is originated from the work by Eshelby, Read and Shockley (1953) and later extended and used by many researchers, among them are Stroh (1958), Bilby et al. (1963), Suo (1990), Asundi and Deng (1995), Ting (1996), Weertman (1996), Fulton and Gao (1997), Wu et al. (2001), Wu (2005), for application to anisotropic materials containing point and line defects.

It has been briefly introduced in Chapter 1 that the stress and displacement fields of a unit dislocation in a homogeneous anisotropic elastic body is given by

$$\mathbf{u} = \{u_1, u_2, u_3\}^T = \mathbf{A}\mathbf{h}(z) + \bar{\mathbf{A}}\overline{\mathbf{h}(z)} \tag{9.1a}$$

$$\mathbf{s} = \{\sigma_{11}, \sigma_{12}, \sigma_{13}\}^T = -\mathbf{LPh}'(z) + \overline{\mathbf{LP}}\overline{\mathbf{h}'(z)} \tag{9.1b}$$

$$\mathbf{t} = \{\sigma_{21}, \sigma_{22}, \sigma_{23}\}^T = \mathbf{Lh}'(z) + \bar{\mathbf{L}}\overline{\mathbf{h}'(z)} \tag{9.1c}$$

where $\mathbf{A}$ is the matrix of equilibrium eigenvectors, $L_{i\alpha} = [C_{i2k1} + p_\alpha C_{i2k2}]A_{k\alpha}$, $\mathbf{P}$ is a diagonal matrix of the three complex eigenvalues p_α, and the complex field potential vector $\mathbf{h}(z)$ is

$$\mathbf{h}(z) = \frac{1}{2\pi i} < \text{In } z > \mathbf{L}^T\mathbf{b} \tag{9.2}$$

where **b** is the Burgers vector. The function $<f(z)>$ denotes a 3 × 3 diagonal matrix of diag. $\{f(z_1), f(z_2)$, and $f(z_3)\}$, $z_\alpha = x_1 + p_\alpha x_2$ is a complex variable with the origin of the coordinate at the dislocation core, p_α is the eigenvalue with a positive imaginary part ($Im\ (p_\alpha) > 0, \alpha = 1, 2, 3$), and $i = \sqrt{-1}$.

The displacement function must satisfy the equilibrium condition:

$$C_{ijkl}u_{k,lj} = 0 \tag{9.3}$$

which leads to the following eigenvalue equation for p_α ($\alpha = 1, 2, 3$)

$$|C_{i1k1} + pC_{i1k2} + pC_{i2k1} + p^2C_{i2k2}| = 0 \tag{9.4}$$

Corresponding to each eigenvalue, its eigenvector is defined by

$$(C_{i1k1} + p_\alpha C_{i1k2} + p_\alpha C_{i2k1} + p_\alpha{}^2 C_{i2k2})A_{k\alpha} = 0 \qquad \alpha = 1, 2, 3 \tag{9.5}$$

For characterization of stress distribution, a material matrix **F** is defined as

$$\mathbf{F} = -2i\mathbf{L}\mathbf{L}^T \tag{9.6}$$

The reference coordinate system for the dislocation pile-up is chosen such that the crack is always lying along the x_1-axis with its center at the zero point. Thus, the stress component **t** on the crack surface can be expressed as

$$t_i(x_1) = \frac{F_{ij}b_j}{2\pi x_1} \tag{9.7}$$

Note that, for isotropic materials, **F** is a diagonal matrix with $F_{11} = F_{22} = \mu/(1 - \nu)$ and $F_{33} = \mu$, where μ is the shear modulus and ν is the Poisson's ratio.

Suppose that dislocations distributed along the x_1-axis are represented by a continuous distribution function

$$\mathbf{B}(x) = \{B_1(x), B_2(x), B_3(x)\}^T \tag{9.8}$$

with the subscript i = 1, 2, 3 designating the edge-glide, edge-climb and screw types, which correspond to Mode II (shearing), I (opening), and III (anti-plane shearing), respectively, as schematically shown in Figure 9.1. Replacing b_j with an infinitesimal element of dislocation $B_j(\xi)\,d\xi$ in Eq. (9.7) and summing up all the contributions from the pile-up distribution, the total stress acting on point x_1 is given by

$$t_i(x_1) = \int_{-c_i}^{c_i} \frac{F_{ij}B_j(\xi)d\xi}{2\pi(x_1 - \xi)} \tag{9.9}$$

where ξ is the coordinate variable over the dislocation distribution range $[-c_i, c_i]$.

In the following sections, we will determine the dislocation distributions for two fundamental types of dislocation pile-ups: (i) the Zener-Stroh-Koehler (ZSK) pile-up, and (ii) Bilby-Cottrell-Swinden (BCS) pile-up.

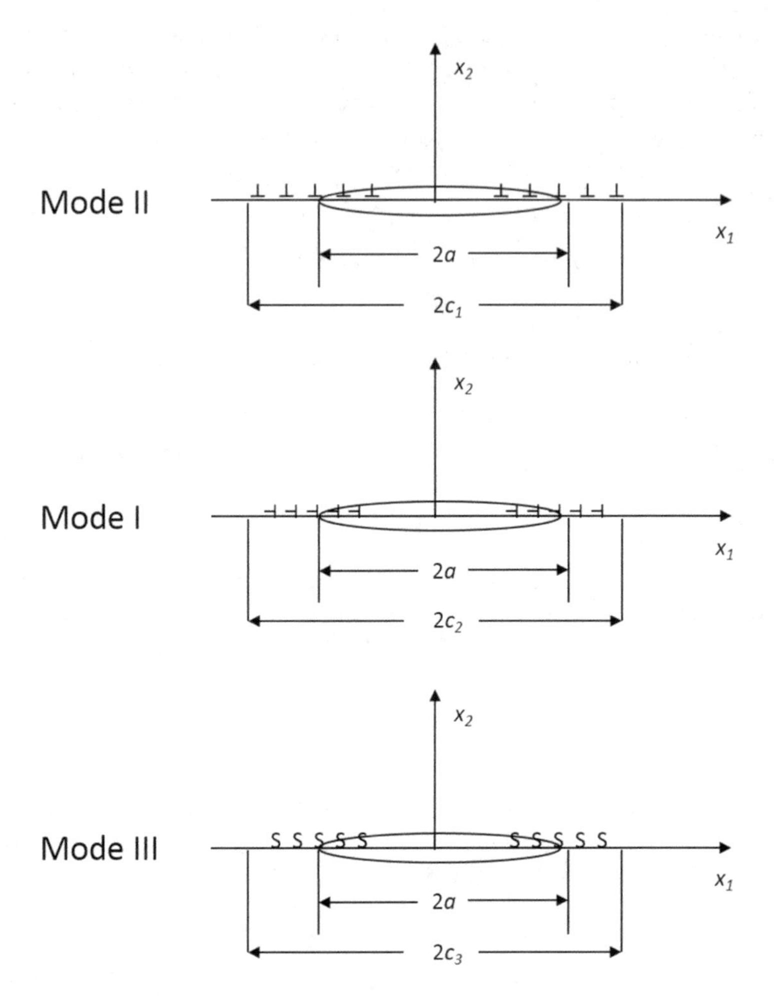

Figure 9.1. Schematic representation of Mode I, II, III cracks with climb, glide and screw dislocations.

9.2 Resolved Shear Stress

In a crystalline solid, dislocation movements and hence slip proceeds on preferred crystallographic planes (defined by the plane normal n_k) and in definite directions (defined by the slip direction b_k). Such movements depend on the resolved shear stresses on those planes. Consider an element of material under stress σ_{ij} $(i, j = 1, 2, 3)$ in a reference coordinate system

$$(\sigma_{ij}) = \begin{pmatrix} \sigma_{11} & \sigma_{12} & \sigma_{13} \\ \sigma_{21} & \sigma_{22} & \sigma_{23} \\ \sigma_{31} & \sigma_{32} & \sigma_{33} \end{pmatrix}$$

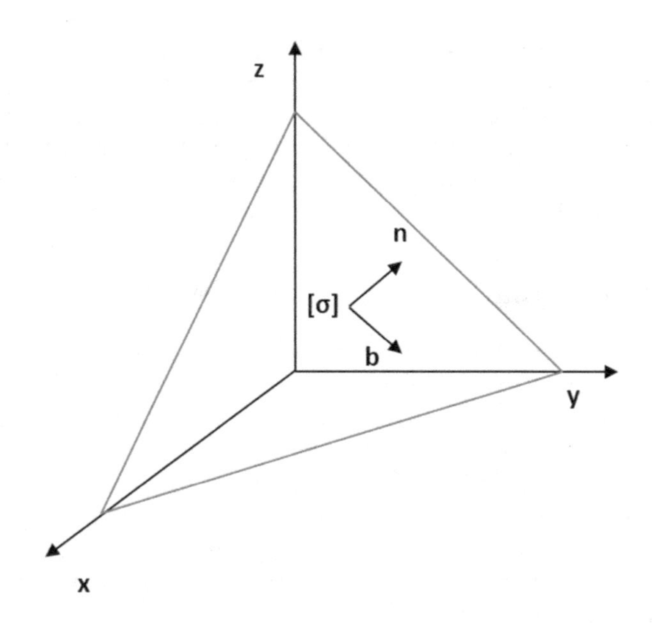

Figure 9.2. Resolved stress on an arbitrary plane with normal **n**.

On a crystallographic plane (n_k), Figure 9.2, the resolved shear stress in a direction (b_k) can be evaluated as

$$\tau = \mathbf{n} \cdot \sigma \cdot \mathbf{b} = \mathbf{b} \cdot \sigma \cdot \mathbf{n} = \frac{1}{2}(b_k n_l + b_l n_k)\sigma_{kl} \tag{9.10}$$

The activation of a definite slip system satisfies the following condition:

$$\tau = \sigma_{kl}\mu_{kl} \geq \tau_{cr} \tag{9.11}$$

where μ_{kl} = ½ ($b_k n_l + b_l n_k$) is the generalized Schmid factor for the active slip system.

The continuously distributed dislocation theory generalizes the mechanics of physical dislocations. In the mathematical sense, dislocation distributions along any given direction can be regarded as the projections of Burgers vectors along that direction. In keeping with the Stroh formalism, in the present treatment, we assume that dislocations are distributed along the x_1-axis. Hence, the friction resistance on the dislocation distribution plane is $t_i^F = \{\sigma_{21}, \sigma_{22}, \sigma_{23}\}^F$, which is related to the critical resolved shear stress τ_{cr} on the active slip plane by

$$t^F = \left\{ \frac{\tau_{cr}}{\mu_{21}} \quad \frac{\tau_{cr}}{\mu_{22}} \quad \frac{\tau_{cr}}{\mu_{23}} \right\} \tag{9.12}$$

In general, t_i^F could be deduced from the yield phenomena for an anisotropic material. Therefore, t_i^F can be regarded as orientation-dependent material constants. The quantity would need to be determined from physical experiments on internal friction or dislocation velocity measurements or deduced from mechanical testing of single crystal material specimens under simple tension or shearing. In theory, they can also be estimated based on the Peierls-Nabarro mechanism for single-phase materials (e.g., pure metals) or its modifications when precipitation of second phase occurs in an alloy as the additional strengthening mechanism (Hirth and Lothe 1998).

9.3 Crack Nucleation by ZSK Distribution

In the early studies on the mechanisms of metal fracture, a concept of crack nucleation was collectively developed by Zener (1948), Koehler (1952) and Stroh (1957). This concept was based on the observation that dislocation piled ups in front of an obstacle obstructing the propagation of slip bands was in fact crack nuclei. Therefore, these types of cracks are referred to as Zener-Stroh-Koehler (ZSK) cracks. Weertman (1996) has given a mathematical treatment to such cracks but only for isotropic materials. Since dislocations mostly reside on definite slip planes in a crystallite, mathematical solutions are due for general anisotropic crystalline materials.

Zener (1948) pointed out that the Griffith type cracks are not likely to form as the first step in the fracture of metals where the propensity of plastic deformation (crystallographic slip) tend to ease the stress build up. It may then be envisaged that Griffith cracks are present in metals as the result of coalescence of finer ZSK cracks inside the material and/or interactions of the metal with the chemistry of the environment at surface. The condition of existence of a ZSK crack is not limited to tension; it may also form even under compression, because plastic deformation occurs in shear mode. The presence of microstructural inhomogeneities certainly aggravates the accumulation of dislocation pile-ups. It is inferred from this mechanism that crack nuclei may form during LCF within the entire plastic deformation range, whether in tension or compression. For crystalline materials with preferred slip systems, it is important to realize that material orientation has an effect on the crack behaviour not only by how much stress is resolved on the slip plane, but also by the anisotropic properties and their coupling effects. In engineering today, single crystal materials are used as components (from electronic devices to gas turbine engines) to carry thermomechanical loads. Besides, in polycrystalline materials, crack nucleation occurs first in individual grains. Therefore, the study of ZSK cracks in anisotropic materials is very important for fully understanding the mechanism and process of crack nucleation in metallic materials.

For ZSK type of dislocation pile-up to form as a crack, it has to satisfy the following conditions: (i) the total force exerted by the distributed dislocations produces a net-zero stress ($t_i = 0$) along the crack surface $[-a, a]$ and, (ii) by the nature of this type of pile-up, which is comprised of dislocations with the same sign, as shown in Figure 9.1, one end, i.e. $[-c_i, -a]$, is in the negative yield ($t_i = -t_i^F$) and the other, i.e. $[a, c_i]$, is in the positive yield ($t_i = t_i^F$). These conditions are summarized in the following equation (Weertman 1996):

$$t_i(x_1) = \begin{cases} t_i^F & a < x_1 < c_i \\ 0 & -a < x_1 < a \\ -t_i^F & -c_i < x_1 < -a \end{cases} \tag{9.13}$$

For mode I, the "positive/negative yield" means yield in tension/compression respectively, while in Mode II and III, the sense is just relative to the reference coordinate. When the crack occurs as a result of the operation of a single slip system, it is in a shear mode (Mode II or III), then t_i^F is the critical resolved shear stress for that particular slip system. Mode I cracks (in metals), however, are more likely the results of multiple slips or duplex slips, then the corresponding "friction resistance", t_2^F, can be determined using the Schmid law, Eq. (9.11), based on the slip systems involved. In practical cases, for example, the slip-band cracking examined by Thompson, Wadsworth and Louat (1956) at the early stage of fatigue was most likely in shearing modes, even the specimen was under uniaxial tension; whereas long cracks in polycrystalline materials and cracks lying on a symmetrical plane of single crystals, such as the (010) plane in face centred cubic materials, under uniaxial tension are mode-I cracks.

Relating Eq. (9.9) to the conditions as described by Eq. (9.13), for stress equilibrium, we have:

$$\int_{-c_i}^{c_i} \frac{F_{ij}B_j(\xi)d\xi}{2\pi(x_1 - \xi)} = \begin{cases} t_i^F & a < x_1 < c_i \\ 0 & -a < x_1 < a \\ -t_i^F & -c_i < x_1 < -a \end{cases} \tag{9.14}$$

Generally, when the crack is under a mixed mode condition, there exists an order of dominance, $a < c_i \leq c_j \leq c_k$ $(i \neq j \neq k)$, and Eq. (9.14) can be solved by that order. As an example, without losing generality, we shall seek the solution for the condition of $a < c_3 \leq c_1 \leq c_2$, which applies to the case of mode-I dominance. Then, each dislocation distribution component can be solved using Muskhelishvili's method (1953) (see Appendix A) (Wu 2005). The solutions are given as follows.

First of all, Eq. (9.14) represents the situation of asymmetrical strip yielding, i.e., the material yields in compression in the region $[-c_i, -a]$, but yields in tension in the region $[a, c_i]$. The distributed dislocation pile-up vanishes at the two ends of $[-c_i, c_i]$, where c_i $(i = 1, 2, 3)$ satisfy:

$$\sqrt{c_2^2 - a^2} = \frac{F_{2j}b_T^{(j)}}{4t_2^F} \qquad (j = 1, 2, 3) \tag{9.15a}$$

$$\sqrt{c_1^2 - a^2} = \frac{b_T^{(1)} + M_{1J}^{-1}F_{J3}b_T^{(3)}}{4M_{1J}^{-1}t_J^F} \qquad (J = 1, 2) \tag{9.15b}$$

$$\sqrt{c_3^2 - a^2} = \frac{b_T^{(3)}}{4F_{3j}^{-1}t_j^F} \qquad (j = 1, 2, 3) \tag{9.15c}$$

where $b_T{}^{(i)}$ are the total Burgers vector in each coordinate direction, and **M** is a 2×2 principal sub-matrix of **F** ($I, J = 1, 2$):

$$\mathrm{M}^{-1} = \begin{bmatrix} F_{11} & F_{12} \\ F_{21} & F_{22} \end{bmatrix}^{-1}$$

The dislocation distribution functions are obtained by solving Eq. (9.14), as:

$$B_2(x_1) = \frac{2F_{22}^{-1}t_2^F}{\pi}\psi(x_1, c_2) - F_{22}^{-1}F_{21}B_1(x_1) - F_{22}^{-1}F_{23}B_3(x_1) \quad |x_1| < c_2 \tag{9.16a}$$

$$B_1(x_1) = \frac{2M_{1J}^{-1}t_J^F}{\pi}\psi(x_1, c_1) - M_{1J}^{-1}F_{J3}B_3(x_1) \qquad |x_1| < c_1 \tag{9.16b}$$

$$B_3(x_1) = \frac{2F_{3j}^{-1}t_j^F}{\pi}\psi(x_1, c_3) \qquad |x_1| < c_3 \tag{9.16c}$$

$$(j = 1, 2, 3; J = 1, 2)$$

where

$$\psi(x, c) = \ln\left|\frac{\sqrt{c^2 - a^2} + \sqrt{c^2 - x^2}}{\sqrt{c^2 - a^2} - \sqrt{c^2 - x^2}}\right| \tag{9.16d}$$

Eq. (9.15) shows that the total length of the dislocation distribution, c, in a particular mode is affected by dislocation pile-ups in other modes, due to anisotropic elastic coupling, and so is true for the plastic yielding zone size, which equals to c–a. The dislocation distribution functions, as given by Eq. (9.16), in general, also consist of multiple pile-ups for modes of higher dominance. The elementary distribution function, i.e., $\psi(x_1, c_i)$, represents one dislocation pile-up, symmetrical about the crack centre, with extreme (to the infinity) density at $x_1 = \pm a$, and vanishing at the endpoints of the plastic zone, $x_1 = \pm c_i$. In general, the magnitudes of these quantities in an anisotropic material depend on the orientation of the materials, since they are related to F_{ij} and t_i^F.

Now, we have obtained the general solution for a ZSK crack with plastic strip-yielding in an anisotropic material under the mixed mode I, II and III conditions showing that $a < c_3 \le c_1 \le c_2$. For the simplest case of isotropic materials, **F** is a diagonal matrix with $F_{11} = F_{22} = \mu/(1 - \nu)$ and $F_{33} = \mu$, where μ is the shear modulus and ν is the Poisson's ratio, it is then easy to show that Eqs (9.15) and (9.16) reduces to the ones obtained by Weertman (1996):

$$B_i(x_1) = \frac{2\alpha_i t_i^F}{\pi\mu}\psi(x_1, c_i) \tag{9.17}$$

$$\sqrt{c_i^2 - a^2} = \frac{\mu b_T^{(i)}}{4\alpha_i t_i^F} \tag{9.18}$$

where $\alpha_{1,2} = (1 - \nu)$, and $\alpha_3 = 1$.

Comparing the present solution with that known for isotropic materials, it is interesting to note that in an anisotropic material, there are generally couplings between I-II-III modes of fracture. Since the "strength" of a pure ZSK crack is controlled by the dislocation pile-up accumulations, the mode dominance is therefore dependent upon the total Burgers vector in the respective direction. For now, we just deal with the situation of mode-I dominance with $a < c_3 \leq c_1 \leq c_2$, implying that $b_T^{(3)} \leq b_T^{(1)} \leq b_T^{(2)}$. Other cases can be solved in a similar manner. In general, the dislocation distribution and the crack-opening displacement of the higher mode of dominance are increasingly affected by dislocations of lesser mode(s), as compared to its counterparts in a similar isotropic material, due to the coupling effect of elastic anisotropy.

In the limiting case when $t_i^F \rightarrow \infty$, then c_i $(i = 1, 2, 3) \rightarrow a$, which means that the dislocation distributions are limited within the crack length in an elastic material. The distribution functions, Eq. (9.14), then reduces to:

$$B_i(x_1) = \frac{b_T^{(i)}}{\pi\sqrt{a^2 - x_1^2}} \tag{9.19}$$

which is identical to that in isotropic elastic materials (Weertman 1996).

A numerical example is given here for a ZSK pile-up (crack) on a (111) plane in a single crystal Ni-base superalloy with the typical elastic compliances as given below (Nye 1957).

$$S_{11} = S_{22} = S_{33} = 0.00799\ (\text{GPa})^{-1};$$
$$S_{12} = S_{23} = S_{31} = -0.00312\ (\text{GPa})^{-1};$$
$$S_{44} = S_{55} = S_{66} = 0.00844\ (\text{GPa})^{-1}.$$

Choosing the Cartesian coordinates x-y-z as: x—$[1\bar{1}0]$, y—$[111]$ and z—$[11\bar{2}]$ (by convention, x is the crack direction, y is the direction perpendicular to the crack, and z the anti-plane axis, the eigen-matrix $\mathbf{F}^{-1}$ can be solved, following the procedure as outlined in section 1.5, as:

$$F^{-1} = \begin{bmatrix} 0.00889 & 0 & -0.00267 \\ 0 & 0.00802 & 0 \\ 0.00889 & 0 & -0.00267 \end{bmatrix}$$

The material friction resistances are estimated, based on the yield properties of SRR99 (Li and Smith 1995), to be:

$$t_{1F} = 530\ \text{MPa},\ t_{2F} = 839\ \text{MPa},\ t_{3F} = 492\ \text{MPa}.$$

Suppose that the total accumulations of Burgers vectors take place in the order of $b_T^{(1)}/a = 0.02$, $b_T^{(2)}/a = 0.05$, and $b_T^{(3)}/a = 0.02$, the respective

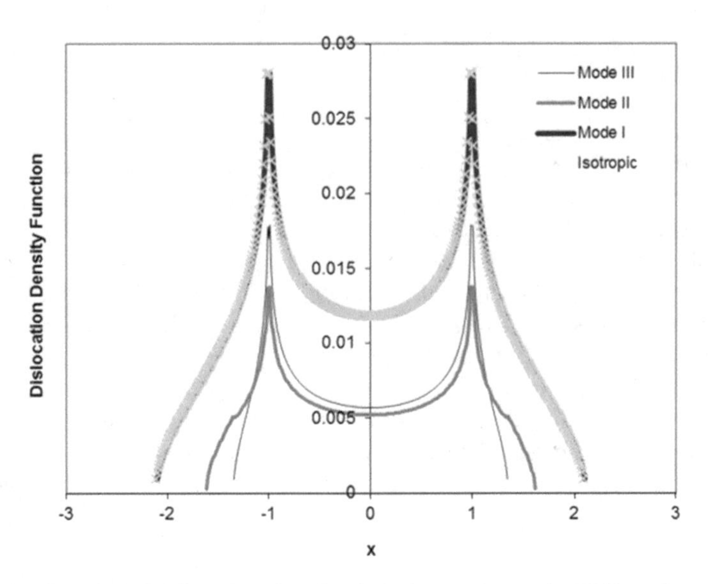

Figure 9.3. Dislocation distributions for a mix-mode ZSK crack.

dislocation density function can be determined by Eq. (9.16), and distribution is shown in Figure 9.3. As one can see, ZSK wedge crack has a symmetrical dislocation pile-up distribution. For comparison, the Mode-I distribution in an isotropic material under the same condition is also shown. It can be seen that there is practically no difference between the dislocation distribution in the isotropic material and the anisotropic material for Mode I, as long as the two materials have the same elastic modulus. However, the Mode II and III are different, even though the total Burgers vectors are equal for the two modes. This is due to the elastic anisotropic coupling present in the $\mathbf{F}^{-1}$ matrix. The coupling, of course, is dependent on the orientation of the crystal with respective to the load. The effect will become important particularly in mixed-mode situations, which are mostly true for crack nucleation conditions. Since the degree of coupling is determined from the matrix $\mathbf{F}^{-1}$ for the particular material/orientation, due considerations should be given to this aspect, when studying crack behaviour in anisotropic materials.

For interests of fracture mechanics, the displacement discontinuity, as an accumulation of the Burgers vectors, can be calculated by the following integral:

$$\Delta u_i = -\int_{-c_i}^{x_1} B_i(x)dx \qquad |x_1| < c_i \tag{9.20}$$

From Eq. (9.20) the crack-opening displacements are given by

$$\Delta u_3 = \frac{2F_{3j}^{-1}t_j^F}{\pi}\{a\varphi(x_1,c_3) - x_1\psi(x_1,c_3) + \chi(x_1,c_3)\} \qquad |x_1| < c_3 \tag{9.21a}$$

$$\Delta u_1 = \frac{2M_{1J}^{-1}t_J^F}{\pi}\{a\varphi(x_1,c_1) - x_1\psi(x_1,c_1) + \chi(x_1,c_1)\} - M_{1J}^{-1}F_{J3}u_3(x_1) \qquad |x_1| < c_1 \quad (9.21b)$$

$$\Delta u_2 = \frac{2F_{22}^{-1}t_2^F}{\pi}\{a\varphi(x_1,c_2) - x_1\psi(x_1,c_2) + \chi(x_1,c_2)\} - F_{22}^{-1}F_{21}u_1(x_1) - F_{22}^{-1}F_{23}u_3(x_1) \qquad |x_1| < c_2 \quad (9.21c)$$

$(j = 1, 2, 3; J = 1, 2)$

where

$$\varphi(x,c) = \ln\left|\frac{a\sqrt{c^2 - x^2} + x\sqrt{c^2 - a^2}}{a\sqrt{c^2 - x^2} - x\sqrt{c^2 - a^2}}\right| \quad (9.21d)$$

$$\chi(x,c) = 2\sqrt{c^2 - a^2}\left(\frac{\pi}{2} - \sin^{-1}\frac{x}{c}\right) \quad (9.21e)$$

The crack-tip opening displacements can be obtained, by taking the values from Eq. (9.21) at $x_1 = a$, as:

$$\Delta u_3(a) = \frac{4F_{3j}^{-1}t_j^F}{\pi}\left[\sqrt{c_3^2 - a^2}\left(\frac{\pi}{2} - \sin^{-1}\frac{a}{c_3}\right) - a\ln\frac{c_3}{a}\right] \quad (9.22a)$$

$$\Delta u_1(a) = \frac{4M_{1J}^{-1}t_J^F}{\pi}\left[\sqrt{c_1^2 - a^2}\left(\frac{\pi}{2} - \sin^{-1}\frac{a}{c_1}\right) - a\ln\frac{c_1}{a}\right] - M_{1J}^{-1}F_{J3}u_3(a) \quad (9.22b)$$

$$\Delta u_2(a) = \frac{4F_{22}^{-1}t_2^F}{\pi}\left[\sqrt{c_2^2 - a^2}\left(\frac{\pi}{2} - \sin^{-1}\frac{a}{c_2}\right) - a\ln\frac{c_2}{a}\right] - \frac{F_{21}}{F_{22}}u_1(a) - \frac{F_{23}}{F_{22}}u_3(a) \quad (9.22c)$$

When the case is purely elastic, i.e., $c_i = a$ $(i = 1, 2, 3)$, the crack-tip opening displacements are equal to zero. Then, the crack is characterized by the stress intensity factor. To derive the stress intensity factor, we first evaluate the elastic stress ahead of the crack-tip, as

$$t_i^A(x_1) = \int_{-a}^{a}\frac{F_{ij}B_j(\xi)d\xi}{2\pi(x_1 - \xi)} = \frac{F_{ij}b_T^{(j)}}{2\pi\sqrt{x_1^2 - a^2}}\left[1 - \frac{2}{\pi}\tan^{-1}\frac{\sqrt{x_1^2 - a^2}}{a}\right] \quad (9.23)$$

From Eq. (9.23) the stress intensity can be obtained as

$$K_i = \lim_{x_1 \to a}\sqrt{2\pi(x_1 - a)}t_i(x_1) = \frac{F_{ij}b_T^{(j)}}{2\sqrt{\pi a}} \quad (9.24)$$

Unlike the stress intensity factor of a Griffith crack, which takes the form: $K = \sigma\sqrt{\pi a}$, and it does not depend on material's properties, the stress intensity factor of a ZSK crack does depend on the material's anisotropic properties through matrix **F**. As such, nucleation and growth of ZSK cracks in crystalline materials are expected to be orientation-dependent and also couples with other modes. This is not totally unexpected, since a ZSK crack is driven by Burgers vectors, which then is orientation-dependent; whereas a Griffith (elastic) crack is totally stress-driven, which is not orientation dependent. The latter point will be further elaborated in section 9.4.

Another interesting property of the crack is the elastic energy release rate, which can generally be obtained from the Irwin closure integral, as follows:

$$G = \lim_{\Delta a \to 0} \frac{1}{2\Delta a} \int_{a}^{a+\Delta a} t_i^A(x) u_i(a + \Delta a - x) dx \tag{9.25}$$

For the elastic case,

$$G = \frac{b_T^{(i)} F_{ij} b_T^{(j)}}{8\pi a} = \frac{1}{2} K_i F_{ij}^{-1} K_j \tag{9.26}$$

For the elastic-plastic case,

$$G = t_i^F u_i(a) \tag{9.27}$$

By the Griffith energy release criterion (Griffith 1921), crack nucleation occurs when

$$G = 2w_s \tag{9.28}$$

where w_s is the fracture energy per unit area.

With the energy release rate given by Eq. (9.26), we have

$$G = \frac{b_T^{(i)} F_{ij} b_T^{(j)}}{8\pi a} = 2w_s \tag{9.29}$$

The above treatment only deals with existing dislocations of population n. In a deformation process, the number of dislocations in a pile-up changes with the deformation rate. Recall that, in section 1.6.2, we have obtained the total number of dislocations during a hold time period, τ, as given by Eq. (1.41):

$$n = \frac{\dot{\gamma}_p}{\kappa}\left(1 - e^{-\kappa\tau}\right) \tag{9.30}$$

As an example, consider that in a titanium alloy dislocations pile up at the boundary of α-grain as shown in Figure 9.4, the crack nucleation condition is:

$$\frac{(F_{11}\cos^2\theta + F_{22}\sin^2\theta)b_T^2}{8\pi a} = 2w_s \tag{9.31}$$

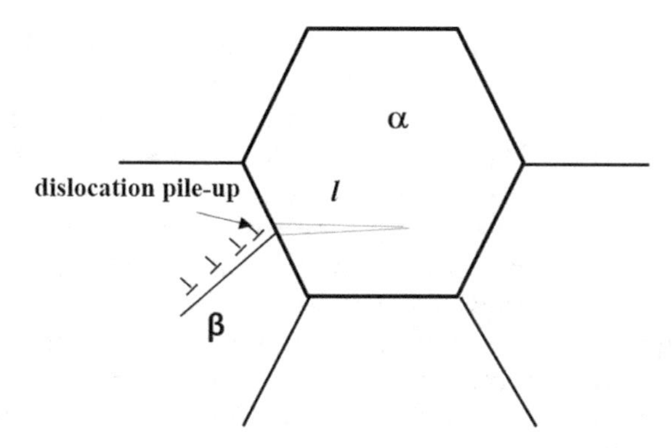

Figure 9.4. Schematics of the formation of a ZSK crack at grain boundary.

from which the crack size ($l = 2a$) can be determined as

$$l = 2a = \frac{(F_{11}\cos^2\theta + F_{22}\sin^2\theta)b_T^2}{8\pi w_s} = \frac{\bar{F}_{22}n^2b^2}{8\pi w_s} \tag{9.32}$$

where $\bar{F}_{22}$ is the average stiffness if θ is about 45°, $b_T = nb$ is the total pile-up Burger's vector.

Substituting Eq. (9.30) into Eq. (9.32), we obtain

$$l = \frac{\bar{F}_{22}b^2}{8\pi w_s}\left(\frac{\dot{\gamma}_p}{\kappa}\right)^2\left(1 - e^{-\kappa\tau}\right)^2 \tag{9.33}$$

The elastic constants of α-titanium are (Nye 1957):

$$C_{11} = 160,\ C_{12} = 90,\ C_{13} = 60,\ C_{33} = 181,\ C_{44} = 47 \text{ GPa}$$

For a crack lying in the basal plane, the orientation stiffness matrix can be calculated as:

$$F = \begin{bmatrix} 68.4 & 0 & 0 \\ 0 & 72.7 & 0 \\ 0 & 0 & 45.9 \end{bmatrix} \tag{9.34}$$

Then, the value of $\bar{F}_{22}$ in Eq. (5.30-5.32) is calculated as 70.55 GPa for θ = 45°. This solution has been used in section 8.3 to describe damage accumulation during cold dwell fatigue. The example emphasises the importance of considering microstructural damage in ICFT, which gives a quantitative description of the dwell sensitivity.

9.4 Crack Nucleation by BCS Distribution

Bilby, Cottrell and Swinden (1963) first developed an elastic-plastic crack model, envisaging the crack and its associated plastic zone (a plastic strip) as an inverted dislocation pile-up. Later, Taira, Tanaka and Nakai (1978), Navarro and de los Rios (1988), and Weertman (1966, 1996) extended the description for various crack conditions, including both bounded and unbounded conditions. The bounded solution usually describes the situation where the dislocation density vanishes at the end of the plastic zone contained within a single crystallite; whereas the unbounded solution describes a microscopic crack that resides in a fully plastically deformed grain where the crack-tip plastic zone extends to the grain boundaries and is blocked by other (neighbouring) grains. The previous solutions were developed for isotropic materials, and therefore their applications were limited to such materials, in a rigorous sense. At the microstructure level, however, when a crack just nucleates within one grain, it immediately sees the effect of crystalline anisotropy on its formation and subsequent growth. Therefore, descriptions of microstructurally small cracks need to be given with consideration of crystalline anisotropy.

Consider an infinite anisotropic medium containing a slit-like crack lying in the plane of x_1 ($|x_1| < a, x_2 = 0$), and subjected to uniform stresses, σ_{ij}^0, with the signs of the dislocations indicating the mode, as shown in Figure 9.1 (note that, for BCS distribution, the dislocation sign on one-half crack plane is opposite to the other half). Dislocations are assumed to be continuously distributed over the range $[-c_i, c_i]$, satisfying the following conditions: (i) the total force exerted by these distributed dislocations produces a net-zero stress along the crack surface $[-a, a]$, and (ii) a uniform stress condition ($t_i = t_i^F$) prevails in the yielded region $a < |x_1| < c$.

According to the Stroh formalism, section 9.1, the resultant forces of the distributed dislocations must satisfy the boundary condition of the crack as follows.

$$t_i(x_1) = \int_{-c_i}^{c_i} \frac{F_{ij}B_j(\xi)d\xi}{2\pi(x_1 - \xi)} = \begin{cases} -t_i^0 & |x_1| < a \\ -t_i^0 + t_i^F & a < |x_1| < c_i \end{cases} \tag{9.35}$$

We will consider two situations: (I) the crack and its associated plastic zone are fully contained within one grain or the entire single crystal, i.e., $c_i < d$ (d is the half grain size) and (II) the crack and its associated plastic zone is blocked by the grain boundary, i.e., $c_1 = c_2 = c_3 = d$. An extreme case to the first situation is that $c_1 = c_2 = c_3 = a < d$, which means that the crack is purely elastic and no dislocation distribution exist beyond the crack length. This case has been solved by Barnett and Asaro (1973). Here, the general elastic-plastic solution for cracks in the aforementioned situations will be given below.

9.4.1 The Unbounded Solution

For mathematical ease, first we solve for the situation (II) where dislocation distributions are completely blocked by grain boundaries such that $c_1 = c_2 = c_3 = d$. Then, Eq. (9.35) turns into:

$$t_i(x_1) = \int_{-d}^{d} \frac{F_{ij}B_j(\xi)d\xi}{2\pi(x_1-\xi)} = \begin{cases} -t_i^0 & |x_1| < a \\ -t_i^0 + t_i^F & a < |x_1| < d \end{cases} \tag{9.36}$$

The general solution to Eq. (9.36), according to Muskhelishvili (1953) and Taira et al. (1978), can be obtained as:

$$\begin{aligned} B_i(x_1) &= -\frac{2F_{ij}^{-1}}{\pi\sqrt{d^2-x_1^2}} \int_{-d}^{d} \frac{\sqrt{d^2-x_1^2}\,t_j(\xi)d\xi}{(x_1-\xi)} \\ &= \frac{2x_1F_{ij}^{-1}}{\sqrt{d^2-x_1^2}}\left[t_j^0 - \frac{2}{\pi}t_j^F \cos^{-1}\frac{a}{d}\right] + \frac{2F_{ij}^{-1}t_j^F}{\pi}\varphi(x_1,d) \end{aligned} \tag{9.37}$$

where

$$\varphi(x,y) = \ln\left|\frac{a\sqrt{y^2-x^2}+x\sqrt{y^2-a^2}}{a\sqrt{y^2-x^2}-x\sqrt{y^2-a^2}}\right| \tag{9.38}$$

Apparently, the dislocation density functions are unbounded at $x_1 = \pm d$.

The displacement discontinuity, as an accumulation of the Burgers vector, can be calculated as:

$$\Delta u_i = \frac{2F_{ij}^{-1}t_j^F}{\pi}\{a\psi(x_1,d) - x_1\varphi(x_1,d)\} + 2\sqrt{d^2-x_1^2}\,F_{ij}^{-1}\left[t_j^0 - \frac{2}{\pi}t_j^F \cos^{-1}\frac{a}{d}\right] \tag{9.39}$$

The crack-tip opening displacement (CTOD) can be obtained, letting $x_1 \to a$, as

$$\Delta u_i(a) = \frac{4aF_{ij}^{-1}t_j^F}{\pi}\ln\frac{d}{a} + 2\sqrt{d^2-a^2}\,F_{ij}^{-1}\left[t_j^0 - \frac{2}{\pi}t_j^F \cos^{-1}\frac{a}{d}\right] \tag{9.40}$$

9.4.2 The Bounded Solution

The crack and its associated plastic zone is entirely contained within one grain, only if $t_i^0 < t_i^F$, which yields a bounded solution for the dislocation distribution within the grain. Under a general mixed mode condition, there should exist an order of dominance, $a < c_i \le c_j \le c_k\,(i \ne j \ne k) < d$, and Eq. (9.35) can be solved by that order. Without losing generality, we assume for the following treatment that $a < c_3 \le c_1 \le c_2$ (mode-I dominance). Then, Eq. (9.35) can be written as:

$$\int_{-c_2}^{c_2} \frac{F_{ij}B_j(\xi)d\xi}{2\pi(x_1-\xi)} = t_i(x_1), \qquad a < |x_1| < c_i \tag{9.41}$$

Unification of the integral bounds is done by considering the fact that $B_i(x_1) = 0$ when $|x_1| > c_i$ and c_2 is the maximum. Again, Eq. (9.35) can be solved from the smallest to the largest (dominant) distribution. Details are given in Appendix B. The distribution sizes c_i satisfies the following condition:

$$\frac{a}{c_3} = \cos\frac{\pi F_{3j}^{-1} t_j^0}{2F_{3j}^{-1} t_j^F} \tag{9.42a}$$

$$\frac{a}{c_1} = \cos\frac{\pi M_{1J}^{-1} t_J^0}{2M_{1J}^{-1} t_J^F} \qquad (J = 1, 2) \tag{9.42b}$$

$$\frac{a}{c_2} = \cos\frac{\pi t_2^0}{2t_2^F} \tag{9.42c}$$

where matrix $\mathbf{M}^{-1}$ is the inverse of a principal submatrix of $\mathbf{F}$, as given by

$$\mathrm{M}^{-1} = \begin{bmatrix} F_{11} & F_{12} \\ F_{21} & F_{22} \end{bmatrix}^{-1}$$

The dislocation distributions are solved for Mode I, II, and III respectively, as:

$$B_3(x_1) = \frac{2F_{3j}^{-1} t_j^F}{\pi}\varphi(x_1, c_3) \qquad |x_1| < c_3 \tag{9.43a}$$

$$B_1(x_1) = \frac{2M_{1J}^{-1} t_J^F}{\pi}\varphi(x_1, c_1) - M_{1J}^{-1} F_{J3} B_3(x_1) \qquad |x_1| < c_1 \tag{9.43b}$$

$$B_2(x_1) = \frac{2F_{22}^{-1} t_2^F}{\pi}\varphi(x_1, c_2) - F_{22}^{-1} F_{21} B_1(x_1) - F_{22}^{-1} F_{23} B_3(x_1) \qquad |x_1| < c_2 \tag{9.43c}$$

Following the same integral as Eq. (9.20), the mode-I, II and III crack opening displacements can be obtained respectively, as

$$\Delta u_2 = \frac{2F_{22}^{-1} t_2^F}{\pi}\{a\psi(x_1, c_2) - x_1\varphi(x_1, c_2)\} - F_{22}^{-1} F_{21} u_1(x_1) - F_{22}^{-1} F_{23} u_3(x_1) \quad |x_1| < c_2 \tag{9.44a}$$

$$\Delta u_1 = \frac{2M_{1J}^{-1} t_J^F}{\pi}\{a\psi(x_1, c_1) - x_1\varphi(x_1, c_1)\} - M_{1J}^{-1} F_{J3} u_3(x_1) \quad |x_1| < c_1 \tag{9.44b}$$

$$\Delta u_3 = \frac{2F_{3j}^{-1} t_j^F}{\pi}\{a\psi(x_1, c_3) - x_1\varphi(x_1, c_3)\} \qquad |x_1| < c_3 \tag{9.44c}$$

and the CTOD (when $x_1 = a$) are:

$$\Delta u_3(a) = \frac{4a}{\pi} F_{3j}^{-1} t_j^F \ln \frac{c_3}{a} \tag{9.45a}$$

$$\Delta u_1(a) = \frac{4a}{\pi} \left(F_{1j}^{-1} t_j^F \ln \frac{c_3}{a} + M_{1J}^{-1} t_J^F \ln \frac{c_1}{c_3} \right) \tag{9.45b}$$

$$\Delta u_2(a) = \frac{4a}{\pi} \left(F_{2j}^{-1} t_j^F \ln \frac{c_3}{a} + M_{2J}^{-1} t_J^F \ln \frac{c_1}{c_3} + F_{22}^{-1} t_2^F \ln \frac{c_2}{c_1} \right) \tag{9.45c}$$

$$(j = 1, 2, 3; J = 1, 2)$$

Using the Irwin closure integral, Eq. (9.25), the energy release rates can be calculated as

$$G = t_j^F \, \Delta u_j \Big|_{x_1 = a} \tag{9.46}$$

In this section, the bounded solution for a crack fully contained within one crystallite is presented for the conditions that $a < c_3 \le c_1 \le c_2 < d$, i.e., mode-I dominance. But, the solution procedure is not limited to this case, the applications of any mode dominance can be similarly addressed by switching the subscripts, to the corresponding order of dominance.

It is interesting to note that the dominant plastic zone size $(c_2 - a)$ does not depend on the material's elastic properties, but only on the crack size and the ratio of the applied stress to the yield stress, which is similar to that in an isotropic material. The dislocation distribution functions and crack opening displacements, however, are shown to be generally dependent on the elastic coupling between modes I, II and III due to the effect of crystalline anisotropy, since they are related to the material eigenvalue matrix, F_{ij}, and the friction resistance, t_i^F.

For each mode, the dislocation distribution function is a linear combination of the $\varphi(x_1, c_i)$ function, which also has the same characteristics as that for the isotropic materials: it represents an inverted dislocation pile-up, anti-symmetrical about the crack centre, with extreme (to infinity) density at $x_1 = \pm a$, and vanishing at the endpoints of the plastic zone, $x_1 = \pm c_i$. Due to elastic coupling, the dislocation distribution function for the mode of higher dominance may consist of multiple pile-ups, as seen in Eq. (9.43a). An example is given for a Ni-base single crystal superalloy with elastic constants (for the cubic axes) as:

$$C_{1111} = 250.5 \text{ GPa}, C_{1122} = 160.5 \text{ GPa}, C_{1212} = 118.5 \text{GPa}$$

Here, we consider a (010)[010] crack with the (lmn) indicating the crack plane normal—x_2 and $[gkh]$ indicating the crack direction—x_1. The F^{-1} matrix and the friction resistances for this orientation is given as follows.

$$F^{-1} = \begin{bmatrix} 9.66 & 0 & 0 \\ 0 & 9.66 & 0 \\ 0 & 0 & 8.44 \end{bmatrix} 10^{-12} (\text{Pa})^{-1}$$

The F_{ij}^{-1} values are obtained by solving the eigenvalue problem as described in section 1.4, and t_i^F values are the minimum values estimated from Eq. (9.11) for the octahedral and cubic slip systems with a constant critical resolved shear stress $\tau_{cr} = 464$ MPa.

The dislocation density for a crack of unit size (a = 1) in the (100)[010] orientation under unidirectional loading, $t_2 = \sigma_{22} = 610$ (MPa), is shown in Figure 9.5. It can be seen that a BCS crack has an asymmetric dislocation pile-up distribution. For this case, because of its crystallographic symmetry, coupling does not occur in the $\mathbf{F}^{-1}$ matrix. The dislocation distribution for the (010)[100] crack is therefore purely mode I as in an isotropic material. Examples for other orientations such as $(111)[\bar{1}10]$ where slip bands most likely lie and hence natural cracks are likely to occur, will be discussed later.

9.5 Fatigue Crack Nucleation in Anisotropic Materials

Tanaka and Mura (1981) gave a theoretical treatment for fatigue crack nucleation in terms of BCS type dislocation pile-ups in isotropic materials. But, they obtained the pile-up strain by integrating the displacement function, which led to a dimensional error in the solution, as discussed in section 6.1. For easy comparison with the original derivation, Wu (2018a) first corrected this problem for the isotropic formulation of dislocation pile-

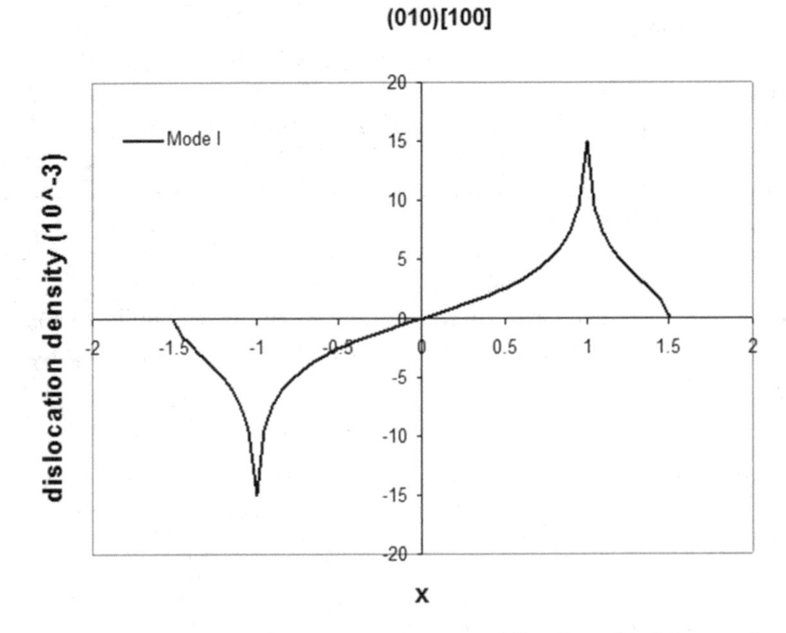

Figure 9.5. Dislocation distribution for the (010)[010] crack of $a = 1$ under stress $\sigma_{22} = 610$ MPa.

up (section 6.1). Here, we will give a treatment to fatigue crack nucleation using the Stroh formalism for anisotropic crystalline materials (Wu 2018b). Without repetition, we only consider the case where crack nucleation occurs in a surface grain, Figure 6.1, as mostly observed during fatigue. In this case, the necessary condition for crack nucleation by BCS pile-ups is that the inverted pile-ups exist at least in two adjacent layers in the form of (a) vacancy dipole, which leads to intrusion; (b) interstitial dipole which leads to extrusion, or c) tripole that corresponds to an intrusion-extrusion pair.

Under fully reversed cyclic loading conditions, the fatigue process proceeds as in the following.

In the first cycle, a pileup occur in layer I, satisfying

$$\int_{-d}^{d} \frac{F_{ij}B_j^{(1)}(\xi)d\xi}{2\pi(x_1-\xi)} + t_i^0 - t_i^F = 0 \tag{9.47}$$

Note that integration over the negative range corresponds to the imaginary part of pileup beyond the free surface. Eq. (9.47) is the same as Eq. (9.35) for the condition prior to crack formation ($a = 0$), and therefore the solution is

$$B_i^{(1)}(x_1) = \frac{2x_1F_{ij}^{-1}(t_j^0 - t_j^F)}{\sqrt{d^2 - x_1^2}} \quad \text{(when } t^0{}_j > t^F{}_j\text{)} \tag{9.48}$$

The plastic strain associated with the pile-up is given by

$$\gamma_i^{(1)} = \frac{1}{d}\int_0^d B_i^{(1)}(x)dx = 2F_{ij}^{-1}(t_j^0 - t_j^F) \tag{9.49}$$

The stored energy associated with the dislocation pile-up is given by

$$U^{(1)} = \frac{1}{2}(t_i^0 - t_i^F)\gamma_i^{(1)} = (t_i^0 - t_i^F)F_{ij}^{-1}(t_j^0 - t_j^F) \tag{9.50}$$

On loading reversal, another pile-up occurs in layer II, satisfying

$$\int_{-d}^{d} \frac{F_{ij}B_j^{(2)}(\xi)d\xi}{2\pi(x_1-\xi)} + \int_{-d}^{d} \frac{F_{ij}B_j^{(1)}(\xi)d\xi}{2\pi(x_1-\xi)} - t_i^0 + t_i^F = 0 \tag{9.51}$$

The distribution function for the pileup in layer II is thus given by

$$B_i^{(2)}(x_1) = -\frac{2x_1F_{ij}^{-1}(\Delta t_j^0 - 2t_j^F)}{\sqrt{d^2 - x_1^2}} \quad \text{(when } t^0{}_j > t^F{}_j\text{)} \tag{9.52}$$

where $\Delta t_j^0 = 2t_j^0$ is the fully reversed stress range.

The plastic strain associated with the pile-up in layer II is given by

$$\gamma_i^{(2)} = \frac{1}{d}\int_0^d B_i^{(2)}(x)dx = -2F_{ij}^{-1}(\Delta t_j^0 - 2t_j^F) \tag{9.53}$$

And hence, the stored energy associated with the dislocation pile-up in layer II is given by

$$U^{(2)} = -\frac{1}{2}(\Delta t_i^0 - 2t_i^F)\gamma_i^{(2)} \tag{9.54}$$

On the k-th reversal, the increment of dislocation $\Delta B_i^{(k)}(x_1)$, the strain $\Delta\gamma_i^{(k)}$, and the energy $\Delta U_i^{(k)}$ are obtained in a similar manner:

$$\Delta B_i^{(k)}(x_1) = (-1)^{k+1}\frac{2x_1 F_{ij}^{-1}(\Delta t_j^0 - 2t_j^F)}{\sqrt{d^2 - x_1^2}} \quad (\text{when } t_j^0 > t_j^F) \tag{9.55}$$

$$\Delta\gamma_i^{(k)} = (-1)^{k+1}\Delta\gamma_i = (-1)^{k+1}2F_{ij}^{-1}(\Delta t_j^0 - 2t_j^F) \tag{9.56}$$

$$\Delta U^{(k)} = \frac{1}{2}(\Delta t_i^0 - 2t_i^F)\Delta\gamma_i \tag{9.57}$$

The index k takes $2N$ at the minimum and $2N$+1 at the maximum stress after N cycles.

The number of dislocations in the pile-up is given by

$$N_i = \frac{1}{b_i}\int_0^d \Delta B_i(x)dx = 2F_{ij}^{-1}(\Delta t_j^0 - 2t_j^F)\frac{d}{b_i} \tag{9.58}$$

The entire pile-up may burst into a crack once the stored energy in the strip of dislocation pile-up (bd) becomes equal to the energy to form new crack surfaces. The latter condition can be described by the energy balance as:

$$2N\Delta Ubd = 4dw_s \tag{9.59}$$

where w_s is the surface energy per unit area.

Here, it is emphasized that even though dislocation pile-ups may appear as discontinuities under metallurgical examination, the distribution planes remain to be continuum planes that transmit stress in full magnitude before crack nucleation satisfying Eq. (9.59). Then, the cycle to crack nucleation is given by

$$N_c = \frac{2w_s}{(\Delta t_i^0 - 2t_i^F)F_{ij}^{-1}(\Delta t_j^0 - 2t_j^F)b} \tag{9.60}$$

or in terms of strain:

$$N_c = \frac{8w_s}{\Delta\gamma_i F_{ij}\Delta\gamma_j b} \tag{9.61}$$

Eq. (9.61) reduces to Eq. (6.16) for a single mode-II crack nucleation in an isotropic material, when $F_{11}^{-1} = (1 - \nu)/\mu$. A fatigue case study for single crystal Ni-base superalloys will be discussed later in Chapter 11.

As implied by Eq. (9.48), even application of a single stress component may induce dislocation distributions of all modes, depending on the coupling terms in the **F** matrix. As cyclic loading continues, the accumulation of these mixed mode pile-ups will eventually burst into a crack when the stored strain energy is enough to create a pair of crack surfaces with the life as given by Eq. (9.60) or Eq. (9.61). In this sense, metal fatigue is almost unavoidable under repeated loading of any kind exceeding the lattice frictional resistance t_i^F. In materials with sufficient number of active slip systems such as f.c.c. and b.c.c. materials under large plastic deformation, dislocation pile-ups would exist in all possible slip systems such that the material appears to be almost "isotropic". In the previous chapter, section 6.1, it is shown that the isotropic formulation can satisfactorily describe low-cycle fatigue failure of polycrystalline materials with appreciable plastic deformation. However, it can be expected that this is not the case when the load amplitude is relatively small. Then, only a few dislocation pile-ups exist in the most favorable slip systems, depending on the microstructure, while the rest of the body remains elastic. This situation corresponds to high-cycle fatigue. Computational microstructual fatigue study is currently underway in the author's laboratory.

9.6 Microscopic Crack Growth in Anisotropic Materials

The early stage of crack growth most inevitably occurs at the microstructural level. This problem has been dealt with by Tanaka et al. and Navarro and los de Rios using the continuously distributed dislocation theory for isotropic materials. In Tanaka et al.'s approach, the microscopic crack growth is envisaged to consist of (i) extended slip bands (ESB), (ii) blocked slip bands (BSB), and (iii) propagated slip bands (PSB). At the ESB stage, the crack and its associated plastic zone are contained in one grain. As the crack propagates, the extension of the plastic zone will be first blocked by grain boundaries, and the crack propagation turns into the BSB situation. Eventually, the plastic zone will break through the blocking grain boundary and extend to the next grain, and the crack propagation is at the PSB stage. This process repeats necessarily as the crack advance through subsequent grains. In Navarro and los de Rios' model, crack propagation is envisaged as successive BSB processes. These models consider that crack growth rate is controlled by the crack opening/sliding displacement.

In this section, microscopic crack is treated realistically as propagating in an anisotropic solid. Actually, the solutions to the BSB and ESB have been given in sections 9.4.1 and 9.4.2, except that for microscopic crack propagation, we should consider Mode II dominance, since microscopic crack growth mainly proceeds along slip bands, which is classified as Stage I crack growth.

In most practical cases, slips prefer to occur on the most favourably oriented slip systems, where the resolved shear stress is at the maximum, and the local mode I (relative to the crack coordinate) is mainly an elastic response. Therefore, for in-plane ESB problems, it is reasonable to assume that only glide and screw types of dislocations exist on the crystallographic plane such that $a < c_2 \leq c_3 \leq c_1$, i.e., mode-II dominance. Omitting the intermediate steps, we just give the final solution as follows.

For the ESB case, switching the subscripts in accordance to the above order of mode dominance, the dislocation distributions can be obtained as

$$B_1(x_1) = \frac{2F_{11}^{-1}t_1^F}{\pi}\varphi(x_1, c_1) - F_{11}^{-1}F_{12}B_2(x_1) - F_{11}^{-1}F_{13}B_3(x_1) \quad |x_1| < c_1 \tag{9.62a}$$

$$B_2(x_1) = \frac{2F_{2j}^{-1}t_j^F}{\pi}\varphi(x_1, c_2) \quad |x_1| < c_2 \tag{9.62b}$$

$$B_3(x_1) = \frac{2M_{3J}^{-1}t_J^F}{\pi}\varphi(x_1, c_3) - M_{3J}^{-1}F_{J2}B_2(x_1) \quad |x_1| < c_3 \tag{9.62c}$$

with the plastic zones defined as

$$\frac{a}{c_2} = \cos\frac{\pi F_{2j}^{-1}t_j^0}{2F_{2j}^{-1}t_j^F} \tag{9.63a}$$

$$\frac{a}{c_3} = \cos\frac{\pi M_{3J}^{-1}t_J^0}{2M_{3J}^{-1}t_J^F} \quad (J = 1, 3) \tag{9.63b}$$

$$\frac{a}{c_1} = \cos\frac{\pi t_1^0}{2t_1^F} \tag{9.63c}$$

where matrix $\mathbf{M}^{-1}$ is the inverse of a principal submatrix of $\mathbf{F}$, as defined by

$$\mathbf{M}^{-1} = \begin{bmatrix} F_{11} & F_{13} \\ F_{13} & F_{33} \end{bmatrix}^{-1} \tag{9.64}$$

The CTOD (when $x_1 = a$) are expressed as

$$\Delta u_2(a) = \frac{4a}{\pi}F_{2j}^{-1}t_j^F \ln\frac{c_2}{a} \tag{9.65a}$$

$$\Delta u_3(a) = \frac{4a}{\pi}\left(F_{3j}^{-1}t_j^F \ln\frac{c_2}{a} + M_{3J}^{-1}t_J^F \ln\frac{c_3}{c_2}\right) \tag{9.65b}$$

$$\Delta u_1(a) = \frac{4a}{\pi}\left(F_{1j}^{-1}t_j^F \ln\frac{c_2}{a} + M_{1J}^{-1}t_J^F \ln\frac{c_3}{c_2} + F_{11}^{-1}t_1^F \ln\frac{c_1}{c_3}\right) \tag{9.65c}$$

$(j = 1, 2, 3; J = 1, 3)$

For BSB problems, assuming the same order of mode dominance, the blocking will first occur to mode II, and secondly to mode III, and finally all will be blocked, as the crack approaches the grain boundary.

When only mode II is blocked, the other two modes are still at the ESB stage and their solutions are the same as given before. Then, the blocked mode II will be described by

$$B_1(x_1) = \frac{2x_1 F_{11}^{-1}}{\sqrt{d^2 - x_1^2}}\left[t_1^0 - \frac{2}{\pi} t_1^F \cos^{-1}\frac{a}{d}\right] + \frac{2F_{11}^{-1} t_1^F}{\pi}\varphi(x_1, d) - F_{11}^{-1}F_{12}B_2(x_1) - F_{11}^{-1}F_{13}B_3(x_1) \tag{9.66}$$

where c_2 and c_3 are given by Eqs (9.63a-b), and $B_2(x_1)$ and $B_3(x_1)$ are given by Eqs (9.62a-b), respectively.

The crack-tip sliding displacement is

$$\Delta u_1(a) = \frac{4a}{\pi}\left(F_{1j}^{-1} t_j^F \ln\frac{c_2}{a} + M_{1J}^{-1} t_J^F \ln\frac{c_3}{c_2} + F_{11}^{-1} t_1^F \ln\frac{c_1}{c_3}\right) + 2\sqrt{c_1^2 - a^2}\, F_{11}^{-1}\left[t_1^0 - \frac{2}{\pi} t_1^F \cos^{-1}\frac{a}{c_1}\right] \tag{9.67}$$

When both mode II and III are blocked,

$$B_I(x_1) = \frac{2x_1 F_{IJ}^{-1}}{\sqrt{d^2 - x_1^2}}\left[t_J^0 - \frac{2}{\pi} t_J^F \cos^{-1}\frac{a}{d}\right] + \frac{2F_{IJ}^{-1} t_J^F}{\pi}\varphi(x_1, d) - M_{IJ}^{-1}F_{J2}B_2(x_1) \tag{9.68}$$

and

$$\Delta u_I(a) = \frac{4a}{\pi}\left(F_{IJ}^{-1} t_J^F \ln\frac{c_2}{a} + M_{IJ}^{-1} t_J^F \ln\frac{d}{c_2}\right) + 2\sqrt{d^2 - a^2}\, F_{IJ}^{-1}\left[t_J^0 - \frac{2}{\pi} t_J^F \cos^{-1}\frac{a}{d}\right] \quad (I, J = 1, 3) \tag{9.69}$$

When all modes are blocked, the solution is just given by Eqs (9.37)–(9.40).

For the case of PSB, a general treatment is very complex. Here, we assume only the dominant mode penetrates the grain boundary into the next grain such that the equilibrium equation can be written as in the following.

$$t_1(x_1) = \int_{-c}^{c} \frac{F_{1j}B_j(\xi)d\xi}{2\pi(x_1 - \xi)} = \begin{cases} -t_1^{(1)} & |x_1| < a \\ -t_1^{(1)} + \tau_{cr} & a < |x_1| < d \\ -t_1^{(2)} + \tau_{cr} & d < |x_1| < c \end{cases} \tag{9.70}$$

where τ_{cr} is the critical resolved shear stress for the active slip system, and τ_1 and τ_2 are the resolved shear stress on the active slip system in the adjacent grains, 1 denoting the grain containing the crack, and 2 denoting the grain to be penetrated. Here, it is recognized that slip bands propagates along slip systems with a constant friction resistance τ_{cr}, but the resolved shear stress in different grains may be different due to their orientations.

By the similar criteria as proposed by Tanaka et al. (1986), when

$$K_{II} = \left[t_1^0 - \frac{2}{\pi} t_1^F \cos^{-1}\frac{a}{d} \right] \sqrt{\pi d} \geq K_{cr}^m \tag{9.71}$$

where $K_{cr}{}^m$ is the critical microstructural stress intensity, the propagated slip band is defined by

$$\cos^{-1}\left(\frac{a}{c}\right) + (k_1 - k_2)\cos^{-1}\left(\frac{d}{c}\right) = k_1 \frac{\pi}{2} \tag{9.72}$$

where $k_1 = \tau_1 / \tau_{cr}$ and $k_2 = \tau_2 / \tau_{cr}$.

Then, the dislocation density can be obtained as

$$B_1(x_1) = \frac{2F_{11}^{-1}\tau_{cr}}{\pi}[\varphi(x_1, a, c) + (k_1 - k_2)\varphi(x_1, d, c)] - F_{11}^{-1}F_{12}B_2(x_1) - F_{11}^{-1}F_{13}B_3(x_1) \tag{9.73}$$

The crack-tip opening (sliding) displacement of Mode II PSB can be obtained as

$$\Delta u_1(a) = \frac{2F_{11}^{-1}\tau_{cr}}{\pi}\left[2a\ln\frac{c}{a} + (k_1 - k_2)\{a\varphi(a,d,c) - d\psi(a,d,c)\}\right] - F_{11}^{-1}F_{12}\Delta u_2(a) - F_{11}^{-1}F_{13}\Delta u_3(a) \tag{9.74}$$

where

$$\varphi(a,d,c) = \ln\left|\frac{d\sqrt{c^2 - a^2} + a\sqrt{c^2 - d^2}}{d\sqrt{c^2 - a^2} - a\sqrt{c^2 - d^2}}\right|$$

$$\psi(a,d,c) = \ln\left|\frac{\sqrt{c^2 - d^2} + \sqrt{c^2 - a^2}}{\sqrt{c^2 - d^2} - \sqrt{c^2 - a^2}}\right|$$

and Δu_2 and Δu_3 are given by Eq. (9.65a-b), respectively.

Take, again, a typical Ni-base superalloy for example. In this case, the slip band form on a {111} plane in a {110} direction. The $\mathbf{F}^{-1}$ matrix and the friction resistances for the $\{\bar{1}11\} < 101 >$ orientation is given by Eq. (9.62)

$$\mathbf{F}^{-1} = \begin{bmatrix} 8.89 & 0 & -2.67 \\ 0 & 8.03 & 0 \\ -2.67 & 0 & 13.97 \end{bmatrix} 10^{-12}\,(\mathrm{Pa})^{-1} \quad \mathrm{t}^{\mathrm{F}} = \{464,\ 985,\ 464\}\ \mathrm{MPa} \tag{9.75}$$

Marx et al. (2006) studied microcrack propagation in a directionally solidified Ni-base superalloy. In one specimen (DSVV1), a crack of a = 80 μm was initiated by a focused ion beam in one grain (Euler angles: $\varphi_1 = 91.7°$, $\Phi = 98.7°$, $\varphi_2 = 0.3°$), which propagated along the above slip system into another grain (Euler angles: $\varphi_1 = 166.4°$, $\Phi = 21.1°$, $\varphi_2 = 289.6°$, a horizontal distance of 60 μm away from the initial crack-tip) under a stress amplitude of 300 MPa at $R = -0.1$. The crack propagation was actually in mixed modes, since the crack direction was approximately 45° inclined to the loading axis, and the crystal anisotropy would induce Mode II-III coupling, as shown in Eq. (9.76). The initial crack condition can be modelled by dislocation pile-ups, assuming $t^F = \{464, 985, 464\}$, as shown in Figure 9.6. Crack opening (sliding) displacements are calculated, using Eqs (9.65), (9.67) and (9.74), for the crack under ESB, BSB and PSB conditions, respectively, as the crack approaches to the grain boundary. The mode-II sliding displacement obtained from the present model is shown in Figure 9.7a in comparison with the isotropic model proposed by Tanaka et al. (1986). One can see that crystal anisotropy induces a coupling effect that increase the crack-tip sliding displacement, which becomes stronger as the crack approaches to the grain boundary until the blocking effect takes over. Since we only consider that the mode II crack-tip plastic zone penetrates into the next grain, the CTOD profile appears as if in an isotropic material when the crack is in the PSB mode. The crack growth rate behaviour observed by Marx et al. (2006) is shown in Figure 9.7b. It is seen that the crack growth rate follows a similar trend of CTOD.

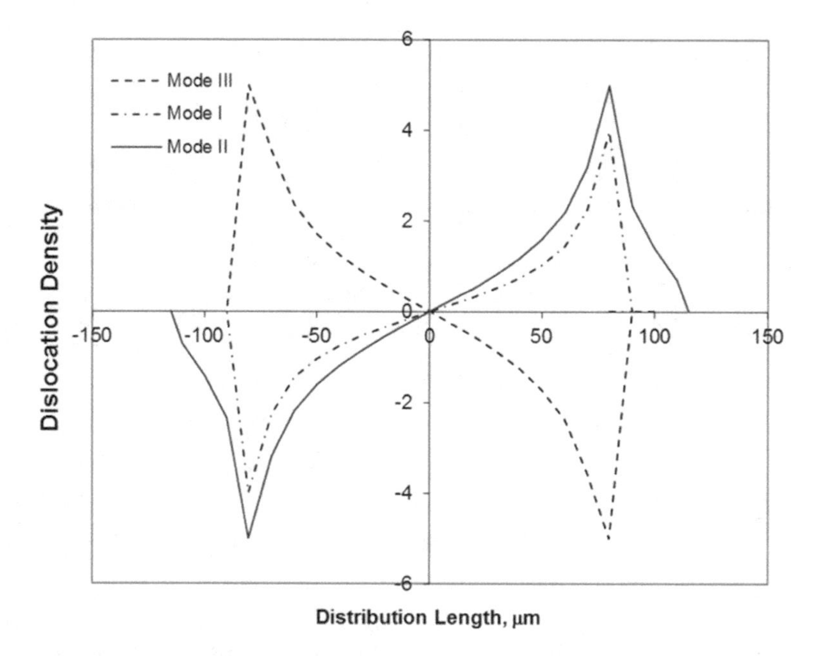

Figure 9.6. Dislocation pile-ups representing the $\{\bar{1}11\} < 101 >$ crack in specimen DSVV1.

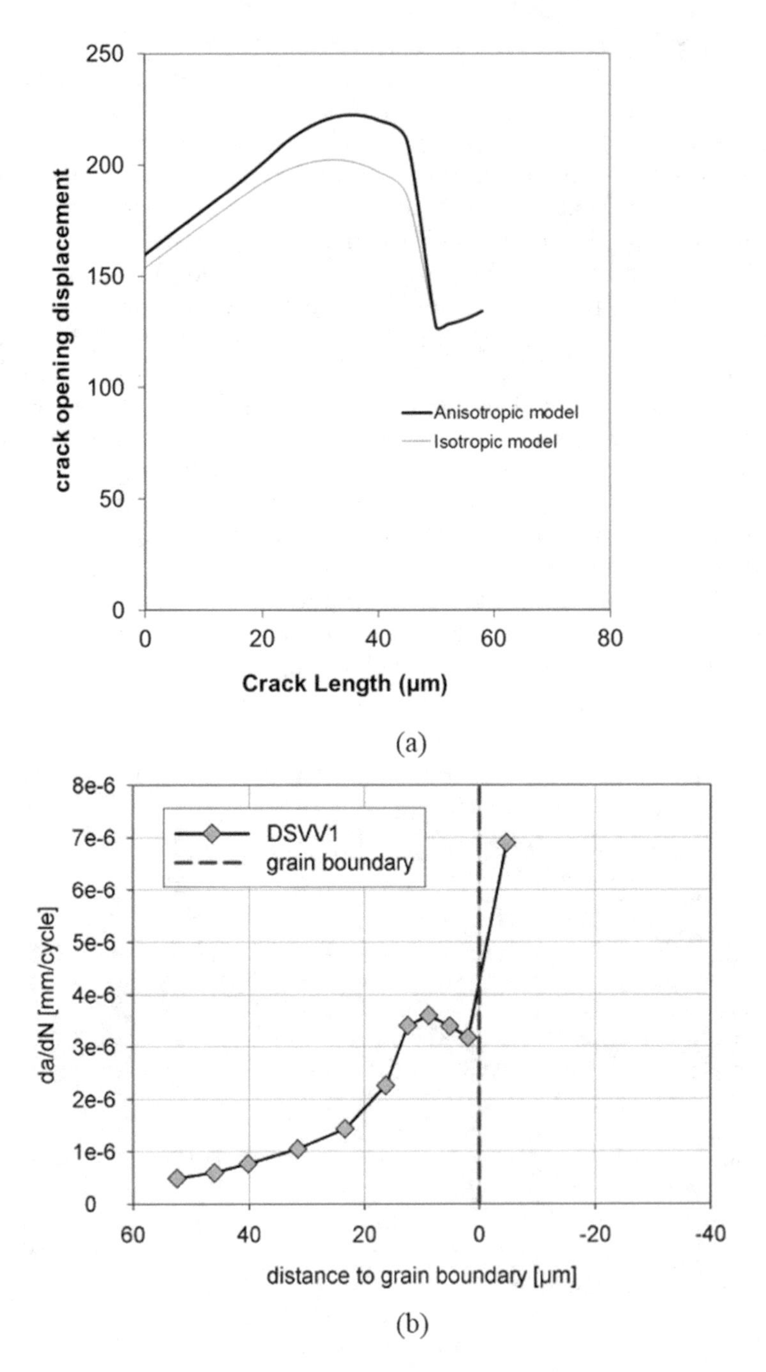

Figure 9.7. (a) Crack opening (sliding) displacement and (b) crack growth rate in specimen DSVV1 (Marx et al. 2006).

9.7 Growth of Small Fatigue Cracks

The phenomenon of small fatigue crack growth (in the range from a few to hundreds of microns) was first reported by Pearson (1975), who found that the growth rate of small fatigue cracks in aluminum alloys were much higher than the "long" fatigue cracks (usually tens of millimeters in length) under the same driving force as defined by linear elastic fracture mechanics (see Chapter 10). Subsequently, intensive research has been devoted to the study of small fatigue crack growth behaviours in various metallic materials, which revealed an "anomalous" behaviour as compared to the long cracks (Suresh 1998). As summarized by Liao (2009, 2010), the "anomalous" behaviours of small-cracks are referred to the following observations:

- they grow faster than predicted by linear elastic fracture mechanics (LEFM) using large crack data (see Chapter 10);
- they can grow at stress intensity factor range ΔK levels well below the large crack threshold ΔK_{th} (see Chapter 10 for definition of ΔK and ΔK_{th});
- their growth can be decelerated or accelerated, arrested or coalesced, depending on the microstructure and stress level; and
- their growth rates have significantly greater scatter than those of large cracks.

In the length scale of a natural crack growth process, small cracks occur first. Therefore, the observed small crack behaviours are actually normal or intrinsic to the material, whereas the perception from LEFM (for long crack growth behaviour) with a driving force of ΔK and threshold ΔK_{th} is not suitable for small cracks (Miller 1982). The fact that the small fatigue crack regime is defined in terms of the long-crack ΔK is the result of LEFM characterization in reverse order. Readers of this book have to wait until Chapter 10 to see the LEFM treatment of long cracks, as the cracks have not grown there yet.

As suggested by the correlation shown in Figure 9.7, small fatigue crack growth rate is proportional to the crack-tip sliding displacement (actually, in a cyclic process slip reversal should also be considered, but it is omitted here for simplification). As shown in Eq. (9.65), (9.67) and (9.74), the crack-tip sliding displacement under various grain boundary blocking conditions is all linearly dependent on the crack length, but the stress dependence follows a non-linear relationship with the ratio of the applied stress to the lattice friction stress, t^0/t^F. It is easy to show that under small-scale yielding conditions, i.e., when $t^0/t^F << 1$, the stress dependence reduces to the power of two (2). Navarro and los de Rios (1988) has demonstrated this stress-dependence transition using the continuously distributed dislocation formulation for isotropic materials. In this case, the crack length multiplied by the square of stress happens to be equal to the square of stress intensity factor K. But, at high t^0/t^F ratios, the small fatigue crack growth rate can be represented by the following relationship on average:

$$\frac{da}{dN} \propto \Delta u_1 \propto \left(\frac{t^0}{t^F}\right)^n a \tag{9.76}$$

where n corresponds to the dominant term in the Taylor series expansion of Eq. (9.65), (9.67) or (9.74) as t^0/t^F approaches 1. In this case, especially when small cracks form in stress concentration regions such as at the notch root, the small-scale yielding condition is violated and the K-similitude will break down.

In section 9.6, we have demonstrated that when a small crack grows inside a single grain, it can be affected by the crystallographic orientation. Hence, for a polycrystalline material, small crack growth is a stochastic process. In complex engineering alloys, inclusion also adds as an additional stochastic variable. For example, inclusions in aluminum alloys vary in size and shape, and furthermore, some may be cracked during the manufacturing process (Merati and Awatta 2005). This has been characterized as the initial discontinuity state (IDS) with certain probability of distribution (Liao 2009). Liao (2010) further extended Tanaka et al. (1986)'s isotropic CTOD formulation considering global plasticity such as in the case of LCF for isotropic materials, which can be formulated as:

$$CTOD = CTOD_{elastic} + CTOD_{plastic} \tag{9.77}$$

In this process, the crack grows consecutively passing the grains from 1 to k, as shown in Figure 9.8. Liao (2010) performed a Monte Carlo simulation of small-crack growth within three grains in 2024-T351 aluminum alloy sheet, using the isotropic CTOD model, in comparison with the AGARD data (Newman, Jr. and Edward 1988) generated from single edge notch tension (SENT) specimens, as shown in Figure 9.9. In the simulation, the frictional stress, grain size and critical microscopic stress intensity factor K_c^m were

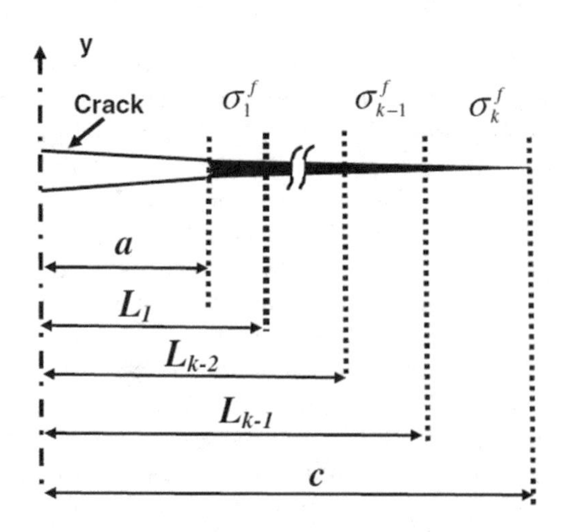

Figure 9.8. Crack and slip band tips in multiple grains.

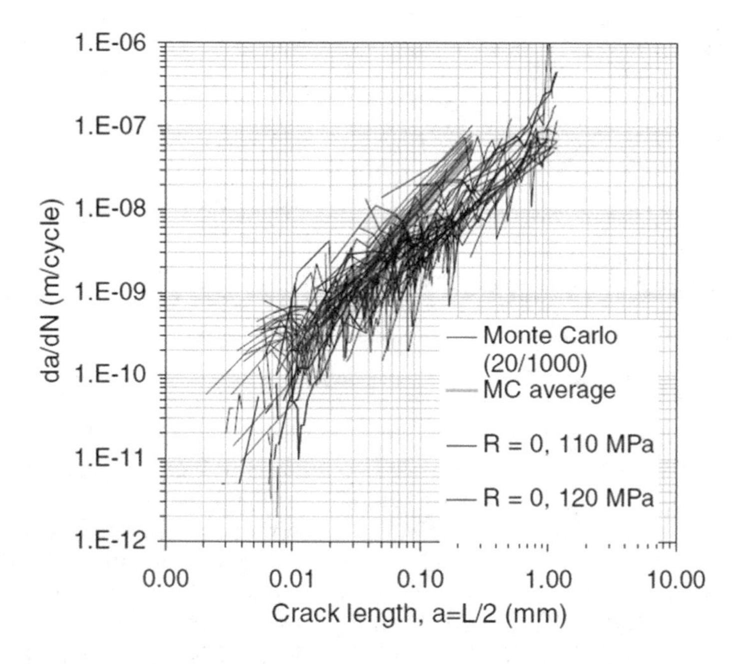

Figure 9.9. Relationship between da/dN and crack length: Monte Carlo simulation and test results (AGARD).

considered as random variables, but not the grain orientation. The Monte Carlo simulation results (only 20 of 1000 are shown in the figure) generally reproduced the highly scattered short-crack growth rate behaviour with grain boundary blocking, which indicate that short-crack growth is indeed a stochastic process. In future studies, the effect of crystalline anisotropy (grain orientation) on short-crack growth behaviours should be described with appropriate ESB, BSB and PSB formulations as given in the previous sections.

The subject of this chapter constitutes the first two phases of the holistic structural integrity process (HOLSIP) that consists of crack nucleation, short crack, long crack and final instability (Hoeppner 1986, Komorowski 2003, 2016). Currently, at the National Research Council Canada, the HOLSIP is implemented for aircraft life cycle management. Effort is underway to link the modeling of the above four phases to provide a true holistic treatment of fracture.

CHAPTER

10

Macroscopic Crack Growth

Macroscopic crack growth is the last stage of damage accumulation leading to catastrophic fracture (instability). The propagation of macroscopic cracks is an immediate danger to the structural integrity. From the early catastrophic fracture of the Liberty ship in USA (1946) to more recent findings of small cracks on the A380 wings, which resulted in a global recall of the newest super jumbo jet fleet (Daily Mail, 2 March, 2012), fatigue cracks still bother structural engineers. As discussed in the previous chapters, fatigue under various thermal-mechanical conditions will inevitably lead to formation of cracks. Crack nucleation and microscopic crack growth are particularly insidious, because their sheer sizes are difficult to detect on a large structure. With the maturation of fracture mechanics, people have developed methods to deal with macroscopic crack growth, and apply the damage tolerance philosophy to quantify the crack growth period, to safe-guard the structure from catastrophic failure. In a structural integrity program, five basic questions still remain to be fully understood:

1. When to expect cracks to form?
2. What is the strength of the component as a function of crack size?
3. How long does it take for a crack to grow from a detectable crack size to the maximum permissible crack size?
4. Does fatigue crack growth threshold condition exist under service conditions?
5. How the load profile (including dwell), load sequence and environment (including temperature) may affect crack growth?

These issues are at a high stake, because component/structural failure can have a huge impact on economic operation and safety of engineering systems such as gas turbine engines, aircraft, pipelines, offshore platforms, and ships etc. Therefore, a mechanism-based understanding of the crack growth process is needed to ensure holistic structural integrity.

10.1 Fracture Mechanics

Modern fracture mechanics stems from the Griffith theory that crack formation occurs by transfer of the stored strain energy to the energy forming new surfaces (Griffith 1921). We have applied this criterion to crack nucleation in previous chapters. Once the presence of crack is found, fracture mechanics seeks to find the driving force for crack growth. This is most often accomplished through experiments to characterize the material's resistance to crack growth and fracture. Traditional solid mechanics treats materials as homogenous continuums and uses analytical methods to calculate the crack-driving forces. Nowadays, most fracture mechanics solutions for engineering components are obtained using FEM satisfying the stress equilibrium and strain compatibility conditions with the appropriate material constitutive laws. Depending on the material's constitutive behaviour and the applied condition, whether Hooke's law for elasticity or power-law for plasticity applies, fracture mechanics are categorized into: i) linear elastic fracture mechanics; and ii) non-linear fracture mechanics, which will be briefly introduced in the following sections.

10.1.1 Linear Elastic Fracture Mechanics

Linear-elastic fracture mechanics addresses crack problems on the premise of linear elasticity with Hooke's law. Griffith (1921) used the stress solution of Inglis (1913) for an elliptical hole in an infinite elastic plate to deduce the net change in potential energy required for the crack (the sharp elliptical hole) to advance, and he obtained the following relation:

$$\sigma_f \sqrt{\pi a} = \sqrt{2Ew_s} \tag{10.1}$$

where σ_f is the fracture stress to cause the creation of two new crack surfaces.

Initially, the Griffith formula only infers unstable fracture of brittle materials such as glass. Eq. (10.1) does not hold true for most ductile materials, so its utility in characterizing crack growth in metallic materials is very limited.

Williams (1957) considered a semi-infinite crack in an isotropic infinite, Figure 10.1, and obtained the crack-tip field solutions for plane strain and generalized plane stress conditions, as:

$$\begin{Bmatrix} \sigma_{xx} \\ \sigma_{yy} \\ \sigma_{xy} \end{Bmatrix} = \frac{K_I}{\sqrt{2\pi r}} \begin{Bmatrix} 1 - \sin\frac{\theta}{2}\sin\frac{3\theta}{2} \\ 1 + \sin\frac{\theta}{2}\sin\frac{3\theta}{2} \\ \sin\frac{\theta}{2}\cos\frac{3\theta}{2} \end{Bmatrix} \quad \text{(for mode I)} \tag{10.2}$$

$$\begin{Bmatrix} \sigma_{xx} \\ \sigma_{yy} \\ \sigma_{xy} \end{Bmatrix} = \frac{K_{II}}{\sqrt{2\pi r}} \begin{Bmatrix} -\sin\frac{\theta}{2}\left(2 + \cos\frac{\theta}{2}\cos\frac{3\theta}{2}\right) \\ \sin\frac{\theta}{2}\cos\frac{\theta}{2}\sin\frac{3\theta}{2} \\ \cos\frac{\theta}{2}\left(1 - \sin\frac{\theta}{2}\sin\frac{3\theta}{2}\right) \end{Bmatrix} \quad \text{(for mode II)} \tag{10.3}$$

$$\begin{Bmatrix} \sigma_{xz} \\ \sigma_{yz} \end{Bmatrix} = \frac{K_{III}}{\sqrt{2\pi r}} \begin{Bmatrix} -\sin\frac{\theta}{2} \\ \cos\frac{\theta}{2} \end{Bmatrix} \quad \text{(for mode III)} \tag{10.4}$$

The near-tip stress field is characterized by the stress intensity factor K for each fracture mode (I—tensile, II—in-plane shear or III—anti-plane shear). Standard K-solutions have been obtained for simple geometrical configurations, which can be found in fracture mechanics handbooks (e.g. Sih 1973). Generally, the stress intensity factor can be expressed, as

$$K_I = Y\sigma\sqrt{\pi a} \tag{10.5}$$

where Y is the shape and boundary correction factor for the component containing the crack.

For most engineering components, no standard solutions are available, and the stress intensity factors have to be deduced numerically from FEM analyses. Most advanced commercial FEM software provides such function.

By LEFM, the stress at the crack tip has a singularity of $r^{-1/2}$ and it approaches infinity, as $r \to 0$, which is not true in real materials. In actuality, the stress near the crack tip within real materials will yield first, because of plastic deformation. George Irwin estimated the extent of the crack-tip plastic zone by equating the yield strength of the material to the far-field mode-I fracture stress along the crack (x direction) and solved for the effective radius, as

$$r_p = \frac{1}{2\pi}\left(\frac{K_I}{\sigma_y}\right)^2 \tag{10.6}$$

This equation gives the approximate ideal radius of the plastic zone ahead of the crack tip under static loading. In reality, because the mode-I crack-tip is in a strong hydrostatic stress state along the x-direction and plastic deformation is prevalent under shear stresses, the real plastic zone

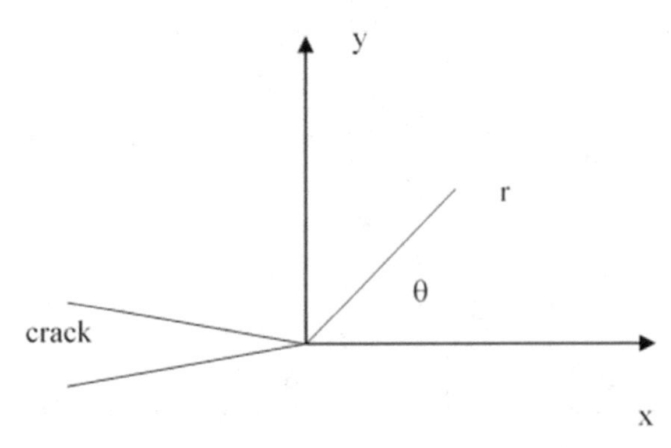

Figure 10.1. The coordinate system at a 2D crack-tip.

shape is not circular as Eq. (10.6) may imply. Nonetheless, it gives a good estimation of the extent of crack-tip plasticity. Eq. (10.6) is often used to define the so called *small-scale yielding* condition, i.e. $r_p/a << 1$, which usually defines the domain of validity of LFEM.

As the crack-tip stress field is characterized by the stress intensity factor K, under small scale yielding conditions, a material property is thus defined as the fracture toughness, K_{IC}, such that when $K_I < K_{IC}$ the crack remains to be stable, and when $K_I > K_{IC}$ the crack will cause sudden fracture of the material. Note that the fracture toughness criterion is similar to Eq. (10.1), but its value is often much greater than that evaluated from the metal surface energy. Hence, fracture toughness tests are conducted to determine this material property (ASTM E1820 – Standard test method for measurement of fracture toughness plane stress/plane strain, 2000), and it is often used as a structural design criterion.

10.1.2 Non-Linear Fracture Mechanics

Nonlinear fracture mechanics generally deals with crack problems in materials exhibiting non-linear constitutive behaviours. Instead of Hooke's law, the nonlinear material behaviour of monotonic plasticity can be described by

$$\varepsilon_{ij} = p\frac{s_{ij}}{\sigma_e} \tag{10.7a}$$

where p is the plastic multiplier:

$$p = \frac{3}{2}\varepsilon_0\left(\frac{\sigma_e}{\sigma_0}\right)^n \tag{10.7b}$$

where σ_0 and ε_0 are the reference stress and strain values, and n is the strain-hardening exponent, similar to Eq. (2.1), s_{ij} is the deviatoric stress tensor, σ_e is the equivalent stress, as defined in section 1.3.

The crack-tip stress field in the power-law plastic material has been obtained by Hutchinson (1968) and Rice and Rosengren (1968), as

$$\sigma_{ij} = \sigma_0\left(\frac{J}{\sigma_0\varepsilon_0 I_n r}\right)^{1/(n+1)} \tilde{\sigma}_{ij}(\theta, n) \tag{10.8}$$

which is known as the HRR field.

As one may notice, the singularity of the crack-tip stress field in a power-law plastic material depends on the material's strain hardening exponent. Therefore, the way to extract the stress intensity factor in a linear elastic material is not valid from the mechanics point of view, since it is not universally valid to all materials. Rice proposed a contour-independent energy integral as the crack-tip parameter for nonlinear materials, as defined by (Rice 1968)

$$J = \int_{\Gamma} \left(W\,dy - \mathbf{T} \cdot \frac{\partial \mathbf{u}}{\partial \mathbf{x}} ds \right) \tag{10.9a}$$

where **u** is the displacement vector, **T** is the surface traction vector, W is the strain energy density, and s is the arc length along the contour Γ. Eq. (10.9) is called the J-integral.

For elastic materials, the J-integral reduces to

$$J = G = \begin{cases} \dfrac{K^2}{E} & \text{for plane stress} \\ \dfrac{(1-v^2)K^2}{E} & \text{for plane strain} \end{cases} \tag{10.9b}$$

If Eq. (10.9) is used in formality as the fracture criterion, it reduces to the Griffith equation for brittle materials (no plasticity, $G = 2w_s$), but the "fracture toughness" will be much greater when the J-integral involves crack-tip plasticity.

When the material at the crack-tip exhibits power-law creep behaviour, the crack-tip stress can be obtained (Riedel and Rice 1980), similar to the HRR field, as:

$$\sigma_{ij} = \sigma_0 \left(\frac{C(t)}{\sigma_0 \dot{\varepsilon}_0 I_n r} \right)^{\frac{1}{n+1}} \tilde{\sigma}_{ij}(\theta, n) \tag{10.10}$$

where $C(t)$ is a path-independent energy integral as defined, similar to the J-integral, as (Landes and Begley 1976)

$$C(t) = \int_{\Gamma} \dot{W}\,dy - \mathbf{T} \cdot \frac{\partial \mathbf{u}}{\partial x} ds \tag{10.11}$$

The path-independent integral $C(t)$ will reach a steady-state value, C^*, when creeping is in the steady-state. This parameter has been used to correlate creep crack growth rates in engineering alloys. As creep occurs by a number of mechanisms, as elucidated in Chapter 5, the relative contribution of each mechanism in the crack-tip zone must be understood, which will be discussed in detail later.

10.2 Fatigue Crack Growth

Fracture mechanics finds its engineering applications for fracture control when subcritical crack growth is characterized by fracture mechanics parameters. Fatigue crack growth rate (FCGR) was first described by Paris (1963) in terms of the cyclic stress intensity factor range, ΔK, as

$$\frac{da}{dN} = C(\Delta K)^n \tag{10.12}$$

where C and n are material constants.

Even though Eq. (10.12) was originally established purely on an empirical basis, it has found wide applications for characterization of almost every engineering alloy to date. The applicability of the Paris equation is warranted by the *small-scale yielding condition*, where the crack-tip plastic zone size is much smaller than the crack size.

In fatigue crack growth testing, a phenomenon called *fatigue threshold* was observed at the lower end of the Paris regime, usually below 8 MPa $\sqrt{m}$, under K-decreasing conditions where FCGR would drop steeply to a level ~10^{-9} m/cycle. The fatigue threshold condition was of great interest to structural integrity engineers, since it implies that if the fatigue design is below the threshold, one would see no crack propagation. Many empirical FCGR relations were proposed, trying to closely depict the "near-threshold" behaviour (Kanninen and Popelar 1987). However, small fatigue cracks have been found to grow at rates well above the long crack "near-threshold" behaviour, under constant-amplitude *K*-increasing conditions (Miller 1989). Then, a question is raised why fatigue threshold exist for long cracks but not for short cracks? And more seriously, does it really exist at all under service loading conditions?

To gain fundamental understanding of the Paris equation and to elucidate its applicability under various service conditions, particularly to better answer the above questions, a mechanism-based fatigue crack growth model has been derived based on the physical process of restricted slip reversibility (Wu et al. 1993), as shown in the section below.

10.2.1 The Restricted Slip Reversal Model

The development of fatigue crack growth models has come a long way. In 1960s, many metallurgical studies were conducted to examine the role of slip processes in inducing fatigue damage, i.e., crack nucleation and propagation (Wadsworth 1963, Snowden 1963, Forsyth 1963, Neumann 1974). With regards to crack propagation, particularly, alternating forward slip and slip reversal processes were found to cause the formation of striations during fatigue crack propagation, the slip reversal was responsible for sharpening the vertex of the crack (McClintock 1971, Neumann 1974, Laird and Smith 1982). Fatigue striations have been recognized as the basic feature of fatigue fracture, for example, as shown in Figure 10.2 for a Ni-base superalloy.

Early models of fatigue crack growth were mainly developed on the premise of continuum mechanics in which the analysis of crack-tip plasticity was coupled with critical damage criteria to formulate crack growth rate equations (McClintock 1963, Rice 1967, Tomkins 1968, Mura and Lin 1974). These continuum damage criteria were defined in terms of either accumulated plastic work (hysteresis energy), or accumulated plastic deformation (fracture strain) or a balance of total energy input into the system. Later, some dislocation-based models were proposed with consideration of dislocation activities ahead of the crack tip. Yokobori et al. (1974, 1975) used dislocation

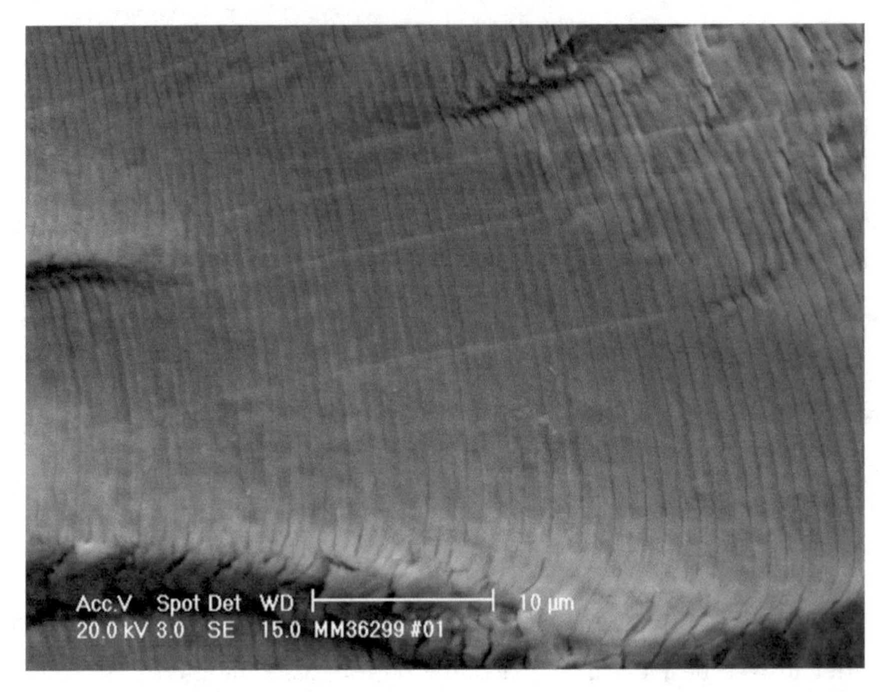

Figure 10.2. Fatigue striations on the fracture surface of a Ni-base superalloy, IN901.

dynamics to determine the number of dislocations emitted from the crack tip which provide the required crack tip opening displacement. Weertman (1981) considered the plastic blunting process and related crack growth rate to the crack tip opening displacement by considering geometrical factors ahead of the crack-tip. Taylor and Knott (1984) and Gerberich et al. (1987) also used dislocation dynamics concepts to assume that the crack growth rate is proportional to the near-crack-tip strain range or dislocation group velocity. However, these dislocation models did not consider the boundary of slip activities, i.e., the size of the crack-tip plastic zone should have an effect.

Fong and Thomas (1988a, 1988b) proposed a restricted-slip-reversibility (RSR) concept to depict the crack propagation process at the crack tip, as shown schematically in Figure 10.3 (a). The model explains transgranular crack propagation events as follows.

(i) Upon loading, slip systems on two favourably oriented slip planes S1 and S2 are activated.
(ii) Forward slip occurs solely on S1 during the rising load cycle, producing a slip step of length l_f.
(iii) During the decreasing load cycle, an increment of slip reversal l_r occurs on S1.

(iv) A final slip reversal occurs on S2, producing a sharp crack tip. This process may be repeated over several load cycles (N) to produce a final crack length (a) increment.

(v) A similar process occur on another favourably oriented slip system variant along another pair of parallel slip planes, S3 and S4, as shown in the schematics (v) to (viii) of Figure 10.3a.

The proceedings of RSR on alternative slip planes will create striations on the crack plane. Based on the above RSR process, Wu et al. (1993) developed a transgranular fatigue crack growth model, considering the deformation kinetics of slip processes.

Slip occurs as a result of dislocation glide, which is a thermally activated process. It proceeds by the dual occurrences of forward and reverse activation steps over the rate-controlling potential barrier caused by lattice resistance or defect obstacles, as discussed in section 1.6. In a plastically deforming crystalline material, plastic strain accumulates whenever slip occurs in stress concentration regions. In the case of a growing crack, it is well established that plastic strain accumulates in the form of a plastic zone ahead of the crack tip and the slip activity is limited to this plastic zone. Figure 10.3 (b) shows a schematic of slip activity within the plastic zone in conformity with the RSR model for crack growth.

By the RSR process (Figure 10.3a), the fatigue crack growth rate can be formulated as:

$$\frac{da}{dN} = (l_f - l_r)\cos\theta \tag{10.13}$$

where θ represents the favourably oriented slip direction within the plastic zone in Figure 10.3 (b).

Each slip step occurs by dislocation glide along a segment of material of length dx and width b on the preferred slip plane. The slip distance dl will

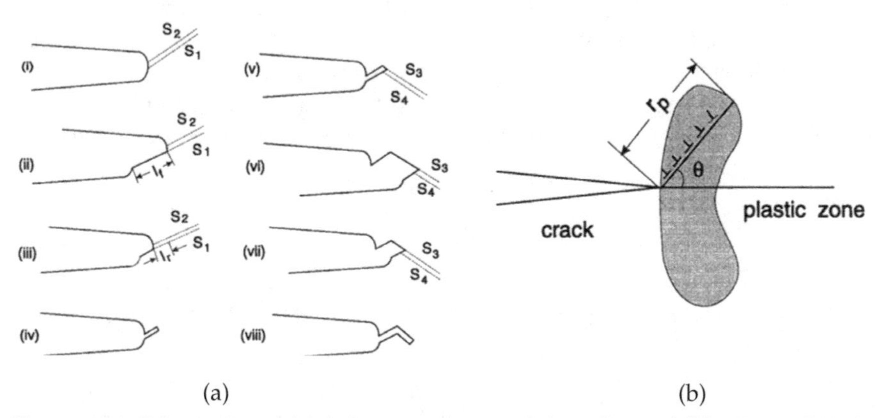

Figure 10.3. Schematics of (a) fatigue crack growth by RSR, and (b) RSR within the confine of the crack-tip plastic zone.

be equal to the number of mobile dislocations contained in this segment, $n = \rho_m bdx$, times the average glide distanceλ, as given by

$$dl = \bar{\lambda}\rho_m bdx = \gamma dx \tag{10.14}$$

where ρ_m is the mobile dislocation density and b is the Burgers vector. Note that the average slip distance can be translated into shear strain (γ) by using the relation, following Gilman (1940),

$$\gamma = b\rho_m\bar{\lambda}. \tag{10.15}$$

The total slip step length l can be obtained by integrating equation (10.14) from the dislocation free zone size (at present we consider the smallest one, b) to the plastic zone size r_p:

$$l = \int_b^{r_p} \gamma dx = \bar{\gamma} r_p \tag{10.16}$$

where

$$\bar{\gamma} \equiv \frac{1}{r_p}\int_b^{r_p} \gamma dx \tag{10.17}$$

is defined as the average plastic strain in the plastic zone. In the following, for simplicity of writing, the symbol γ is used here to depict the average plastic strain as defined in the context of equation (10.17) unless otherwise stated.

Substituting equation (10.16) into equation (10.13), we obtain:

$$\frac{da}{dN} = (\gamma_f - \gamma_r) r_p \cos\theta. \tag{10.18}$$

where, the subscripts "f" and "r" refer to forward slip and slip reversal respectively. Equation (10.18) expresses the fatigue response of a growing crack in terms of the net strain ahead of the crack tip. This equation agrees, in principle, with Laird and Smith's (1982) conclusion that plastic deformation is responsible for crack propagation in ductile metals. Similar expression has also been obtained by Tomkins (1968), based on continuum mechanics. When dealing with crack growth, r_p represents the size of a localized yield zone. In the case of fatigue crack growth under small scale yielding condition and from the analysis of linear fracture mechanics, the cyclic plastic zone r_p can be expressed as (Rice 1967):

$$r_{px} = r_p \cos\theta = \frac{1}{3\pi}\left(\frac{\Delta K}{2\sigma_y}\right)^2. \tag{10.19}$$

Using equation (10.19) for $r_p \cos\theta$ in equation (10.18), crack growth rate can be expressed by:

$$\frac{da}{dN} = \frac{1}{12\pi}(\gamma_f - \gamma_r)\left(\frac{\Delta K}{\sigma_y}\right)^2. \tag{10.20}$$

The description of plastic deformation at the crack tip follows the theory of deformation kinetics, as detailed in section 1.6, and by the mechanism of dislocation-glide, the shear strain rate at high stresses can be expressed as:

$$\dot{\gamma} = \dot{\gamma}_0 \exp\left(-\frac{\Delta G^{\neq} - V\tau_{eff}}{kT} \right) \tag{10.21}$$

where $\dot{\gamma}_0$ is a pre-exponential factor; $\Delta G^{\neq}$ is the activation energy, V is the activation volume, k is the Boltzmann constant, T is the absolute temperature; τ_{eff} is the effective stress (applied stress minus the internal stress) experienced by the dislocation at the activation sites. Eq. (10.21) is actually the reduced form of Eq. (1.38), at high stress near the crack tip. The effective stress is equal to the applied stress minus the back stress. According to Eq. (1.46-1.47), neglecting dislocation climb, the effective stress can be expressed as

$$\tau_{eff} = \tau_a - H\gamma - \tau_o \tag{10.22}$$

where τ_a is the applied stress, H is the strain hardening coefficient, γ is the plastic shear strain, and τ_o is the initial internal stress.

Under small-scale-yielding conditions, which are appropriate for fatigue crack growth conditions, we substitute the elastic crack-tip stress function, as given by Eq. (10.2), in Eq. (10.22). Inserting Eq. (10.22) into Eq. (10.21) and carrying on the integration for the average plastic strain, as expressed in Eq. (10.17), we have

$$\dot{\gamma} = \frac{1}{r_p} \int_b^{r_p} \dot{\gamma}(r,\theta)\, dr \tag{10.23a}$$

and hence

$$\dot{\gamma} = \dot{\gamma}_0 \exp\left(-\frac{\Delta G^{\neq} - (\alpha K - VH\gamma - V\tau_0)}{kT} \right) \tag{10.23b}$$

where the fracture work factor α can be expressed as

$$\alpha = \frac{V}{\sqrt{2\pi\xi_m}}$$

and ξ_m is the mean-value distance.

Lin and Thomson (1987) have argued that the presence of dislocations in the vicinity of the crack tip may induce a shielding effect that weakens the singularity of the stress field at the crack tip. Dislocation shielding only results in a modified field that controls the dislocation free zone within the order of 1.5-30 nm ahead of the crack tip (Lii et al. 1990), and the dislocation free zone makes no contribution to plastic deformation. Therefore, the assumption of the K-controlled field is appropriate. Since the stress decreases rapidly away from the crack tip, contribution to the plastic strain will mostly come from activations of dislocations glide within a short distance ahead of the crack tip. Therefore, the mean-value distance ξ_m depends largely on the microstructure and defect density in the material at the crack tip, but is less relevant to the plastic zone size. Therefore, α can be considered to be a material constant as a first approximation.

During cyclic loading, forward slip occurs in the load-rising period, and slip reversal occurs in the load-decreasing period. The mechanisms controlling the forward slip and the slip reversal processes are the same in mechanical fatigue, but these mechanisms will be different in corrosion fatigue where for example the oxide film formation on fresh metal surface may affect the slip reversal. For generality, we first consider that two separate slip mechanisms operate in alternating slip processes.

The forward strain γ_f can be obtained by integrating Eq. (10.23) over the load-rising period,

$$\int_0^{\gamma_f} \exp\left(\frac{V_f H\gamma}{kT}\right) d\gamma = \dot{\gamma}_0 \exp\left(-\frac{\Delta G_f^{\neq} + V_f \tau_{0f}}{kT}\right) \int_0^{t_R} \exp\left(\frac{\alpha_f K}{kT}\right) dt \quad (10.24a)$$

which leads to

$$\frac{kT}{V_f H}\left[\exp\left(\frac{V_f H\gamma_f}{kT}\right) - 1\right] = \frac{\dot{\gamma}_0 kT}{2f\alpha_f \Delta K} \exp\left(-\frac{\Delta G_f^{\neq} + V_f \tau_{0f}}{kT}\right)$$

$$\exp\left(\frac{\alpha_f R\Delta K}{(1-R)kT}\right)\left[\exp\left(\frac{\alpha_f \Delta K}{kT}\right) - 1\right] \quad (10.24b)$$

where f is the loading frequency

Often at a low temperature, $T < 0.3\ T_m$ (T_m is the absolute melting temperature):

$$\exp\left(\frac{V_f H\gamma_f}{kT}\right) >> 1, \exp\left(\frac{\alpha_f \Delta K}{kT}\right) >> 1 \quad (10.25)$$

For example, for a typical value of the activation volume of 80 b^3, $b = 2.5\times10^{-10}$ m, and a work hardening coefficient of 1000 MPa, if the ductility at room temperature is taken as 10%, the magnitude of the term $\exp(VH_{\gamma f}/kT)$ approximates e^{30} which is much greater than unity. For mechanically assisted thermal activation, the activation work, $\alpha K - VH\gamma - \tau_o$, is positive. Hence $\alpha_f \Delta K > VH_{\lambda f}$ and $\exp(\alpha_f \Delta K/kT) >> 1$. Therefore, Eq. (10.25) is valid for most fatigue conditions. Then the forward plastic strain in Eq. (10.24) can be approximated by

$$\gamma_f = \frac{kT}{V_f H} \ln\left[\frac{V_f H\dot{\gamma}_{0f}}{2f\alpha_f \Delta K} \exp\left(-\frac{\Delta G_f^{\neq} + V_f \tau_{0f}}{kT}\right)\right] + \frac{\alpha_f \Delta K}{(1-R)V_f H} \quad (10.26)$$

Similarly, over the load-decreasing cycle, the reverse slip plastic strain can be depicted by

$$\gamma_r = \frac{kT}{V_r H}\ln\left[\frac{V_r H\dot{\gamma}_{0r}}{2f\alpha_r \Delta K}\exp\left(-\frac{\Delta G_r^{\neq} + V_r\tau_{0r}}{kT}\right)\right] + \frac{\alpha_r \Delta K}{(1-R)V_r H} \tag{10.27}$$

The net cyclic plastic strain range ahead of the crack tip can be calculated as:

$$\gamma_f - \gamma_r = \ln\left[\frac{\left(\frac{V_f H\dot{\gamma}_{0f}}{2f\alpha_f \Delta K}\right)^{\frac{kT}{V_f H}}\exp\left(-\frac{\Delta G_f^{\neq} + V_f\tau_{0f}}{V_f H}\right)}{\left(\frac{V_r H\dot{\gamma}_{0r}}{2f\alpha_r \Delta K}\right)^{\frac{kT}{V_r H}}\exp\left(-\frac{\Delta G_r^{\neq} + V_r\tau_{0r}}{V_r H}\right)}\right] + \left(\frac{\alpha_f}{V_f H} - \frac{\alpha_r}{V_r H}\right)\frac{\Delta K}{1-R} \tag{10.28}$$

Then, the crack growth rate according to equation (10.20) can be expressed as:

$$\frac{da}{dN} = \frac{(\alpha_f V_r - \alpha_r V_f)}{12\pi(1-R)V_f V_r H}(\Delta K - \Delta K_{th})\left(\frac{\Delta K}{\sigma_y}\right)^2 \tag{10.29}$$

where

$$\Delta K_{th} = \frac{(1-R)HV_f V_r}{\alpha_f V_r - \alpha_r V_f}\left[\frac{1}{H}\left(\tau_{0f} - \tau_{0r} + \frac{\Delta G_f^{\neq}}{V_f} - \frac{\Delta G_r^{\neq}}{V_r}\right)\ln\left[\frac{\left(\frac{V_f H\dot{\gamma}_{0f}}{2f\alpha_f \Delta K}\right)^{\frac{kT}{V_f H}}}{\left(\frac{V_r H\dot{\gamma}_{0r}}{2f\alpha_r \Delta K}\right)^{\frac{kT}{V_r H}}}\right]\right] \tag{10.30}$$

Equation (10.30) defines the fatigue threshold condition in terms of a complex function of temperature, frequency, and stress ratio. The ΔK_{th} value should be microstructure/environment sensitive, because it is controlled by activation energies, activation volumes and work factors.

In mechanical fatigue, where the microstructure (grain size, precipitate size and shape) is stable and environmental effects are absent, α_f and α_r are of the same order of magnitude, and both forward and reverse slips occur by the same mechanism so that $\Delta G_f = \Delta G_r$, $V_f = V_r = V$, $\tau_{of} = \tau_{or}$ and $\dot{\gamma}_{0f} = \dot{\gamma}_{0r}$, and hence ΔK_{th} approaches zero, as $kT/VH << 1$. Therefore, for pure mechanical fatigue:

$$\frac{da}{dN} = \frac{(\alpha_f - \alpha_r)}{12\pi(1-R)VH\sigma_y^2}(\Delta K)^3. \tag{10.31}$$

Three main conclusions can be deduced from Equation (10.31) as follows. First, FCGR follows the Paris relationship with a physically defined proportional factor and a power law exponent of 3. This was indeed observed for fatigue crack growth rate data of a variety of alloys tested at room temperature in vacuum (Speidel 1977), as shown in Figure 10.4, where crack growth rate vs. ΔK relationships in a log-log plot exhibit a slope of 3 in the Paris regime. In addition, it was reported that $n = 3$ represents fatigue crack growth in ductile steels (Miline et al. 1988) and $n = 2.92$ in offshore steels (Smith and Cooper 1989). Numerous experiments have also shown that the power law exponent of the Paris law falls within the range of 2.7~3.4 within the Paris regime. The power law exponent values found in these experiments are closer to 3 than 2 or 4, the values predicted by other models. Therefore, it may be fairly concluded that the RSR model is suitable for conventional materials where deformation is promoted by homogeneous slips.

Second, the RSR model shows that the fatigue crack growth rate depends on the materials' yield strength inversely to a power of 2, and it is also inversely proportional to the work-hardening coefficient. This implies the dependence of mechanical fatigue on microstructure and temperature. For example, the yield strength depends on the grain size through the Hall-Petch relationship, and it can be inferred that a coarse-grained material would exhibit a faster fatigue crack growth rate than its fine-grained counterpart. The material's yield strength is also dependent on temperature, which give rise to temperature-dependence in mechanical fatigue. But, this temperature dependence is rather moderate as compared to crack growth rates under cyclic conditions at high temperatures or in aggressive environments. High temperature fatigue and corrosion fatigue phenomena usually include additional rate processes that have an exponential temperature-dependence, as reflected in the elementary rate constant, Eq. (1.32). Pure fatigue is essentially a "low temperature" phenomena occurring by the mechanism of dislocation glide, as depicted by the RSR model. The dependence of fatigue crack growth rate on yield strength was demonstrated by Benson and Edmonds (1978) on 0.5Cr-0.5 Mo-0.25 V steel. A range of yield strength values, varying between 466 MPa and 834 MPa, were obtained using different heat treatment procedures. The work hardening coefficient, H, for each heat-treated material is estimated by dividing the difference between the ultimate tensile strength and the yield strength with the elongation, typically 13%. Then, the H-compensated crack growth rates at $\Delta K = 15.69\ \text{MPa}\sqrt{m}$, which is far above the fatigue threshold of this material, are plotted against σ_y^2 as shown in Figure 10.5.

Third, interestingly, Eq. (10.29)–(10.31) further predict that the threshold of fatigue crack growth does not occur in the absence of microstructural change and environmental effects ahead of the crack tip such as in the case of pure mechanical fatigue. Some apparent thresholds ΔK_{th} have been observed in fatigue crack growth rate testing using load-decreasing procedures. These apparent "thresholds" are often attributed to external shielding viz. crack closure, which will be discussed in a separate section.

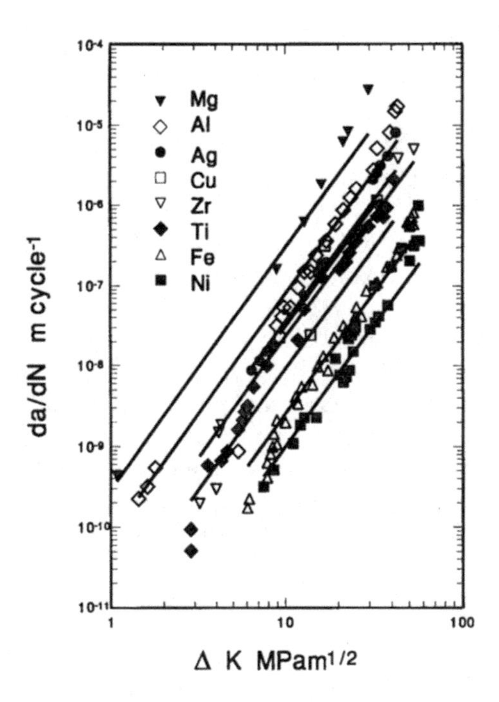

Figure 10.4. Fatigue crack growth behaviours of a number of metals at room temperature in vacuum.

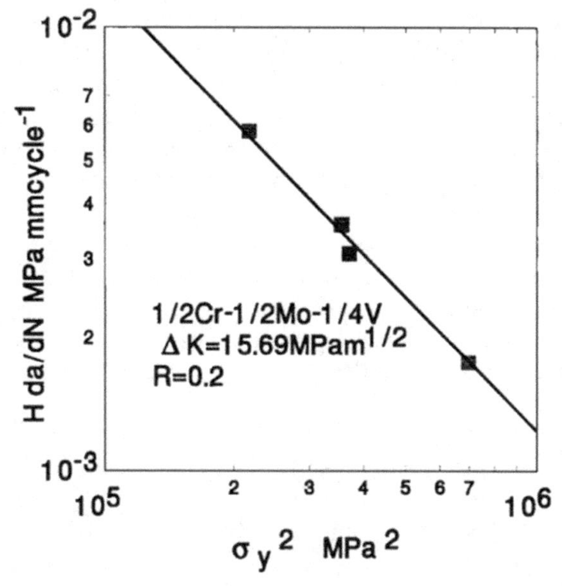

Figure 10.5. H-compensated fatigue crack growth rate vs. σ_y^2 for 0.5 Cr-0.5 Mo-0.25 V steel.

10.2.2 Fatigue Crack Growth in Textured Material

Depending on their thermomechanical processing, engineered materials are often produced with crystallites in certain preferred crystallographic orientations, which is called texture. Common textures are: (1) cube component: (001)[100]; (2) brass component: (110)[$\bar{1}$12]; (3) copper component: (112)[11$\bar{1}$]; and (4) S component: (123)[63$\bar{4}$]. The degree is dependent on the percentage of crystals having the preferred orientation such that a polycrystalline material can have a weak, moderate or strong texture. Texture can have a great influence on materials properties such as strength, deformation behaviour, weldability, chemical reactivity and stress corrosion cracking resistance. Failures can correlate with the crystalline textures formed during fabrication or use of that component. Therefore, understanding the effects of texture are critical in material/process selection and component design. In this section, we only concern fatigue crack growth in textured materials.

Aluminium-lithium (Al-Li) alloys can be made with a strong brass-type texture. They were developed in the early 1980s with 7 to 10 pct of increase in specific strength as compared to traditional high strength aluminum alloys, aiming for next-generation aerospace structural applications. These alloys often exhibit outstanding fatigue crack propagation resistance but with extremely marked anisotropy, which were well documented, e.g., in *Aluminum-Lithium Alloys-I to V* conference proceedings from 1981 to 1989. Such performance has been traced to the intense inhomogeneous slip along certain preferred planes and in certain directions resulting from the presence of coherent shearable δ' precipitates and strong textures in these alloys. Localized slip planarity promotes macroscopic crack deflection and leads to an unusual fracture mode with an extraordinary surface roughness comprised of well-defined slip band facets, e.g. Figure 10.6 (a) for 8090 Al-Li alloy. It was thought that the superior fatigue crack growth resistance of Al-Li alloys was due to roughness-induced crack closure (Venkateswara Rao et al. 1988). It is this distinguished fracture mode that has some general implications on fatigue crack growth in materials with textures, which is worth discussing.

8090 Al-Li alloy plate has a strong {110} <112> (the "brass" type) texture, which results in the slip plane normal having an angle of ϕ with respect to the loading direction in β-orientation (β is the angle of the loading direction with respect to the rolling direction), Figure 10.6 (b). The relationship between ϕ and β can be derived based on the existence of {110} <112> texture (Wu et al. 1994), as:

$$\cos\phi = \frac{1}{\sqrt{3}}\sin(\beta + 54.74) \tag{10.32}$$

By stress transformation, the stress intensity factor with the inclined crack surface is given by

$$K_{\phi} = K\cos^2\phi \tag{10.33}$$

where K is the stress intensity factor with the crack plane normal to the loading direction.

Substituting Eq. (10.33) into Eq. (10.31), but also substituting $\sigma_y \cos^2\phi$ for σ_y, we obtain

$$\frac{da}{dN} = \frac{(\alpha_f - \alpha_r)\sin^2(\beta + 54.74)}{36\pi VH\sigma_y^2(1-R)}(\Delta K)^3. \tag{10.34}$$

Figure 10.7 shows the comparison of FCGRs for a number of conventional aluminum alloys and aluminum-lithium alloys, 2090-T81 and 8090-T8771 (Wu et al. 1994). Compensated by $(1-R)\,H\sigma_y^2$, the FCGR data of conventional aluminum alloys are represented by the upper solid line and Al-Li alloys are represented by the lower solid line, both with a slope of 3 on the log-log scale. In the case of conventional aluminum alloys, $(\alpha_f - \alpha_r)/V = 0.697\ \text{m}^{-1/2}$. Using this $(\alpha_f - \alpha_r)/V$ value for substitution in Eq. (10.34), the predicted FCGR curve (the dashed line) falls very close to the experimentally observed FCGR in the case of 2090-T81 and 8090-T8771 L-T specimen ($\beta = 0$), Figure 10.7. These calculations clearly indicate that relative to conventional aluminum alloys, 90 percent of the total reduction in compensated FCGR for aluminum-lithium alloys is accounted for merely by the texture dependent geometric factor $\cos^2\phi$. The same equation also describes well FCGRs in other orientations of $\beta = 15°$, $30°$ and $45°$ (Wu et al. 1994). Crack closure was monitored to be insignificant under constant-amplitude K-increasing conditions during the FCGR testing on 8090-T8771 for all orientations (Wu et al. 1994). Therefore, it shows that the superior FCGR resistance of Al-Li alloys is attributed to the intrinsic texture-effect, but not extrinsic shielding mechanism viz. crack closure.

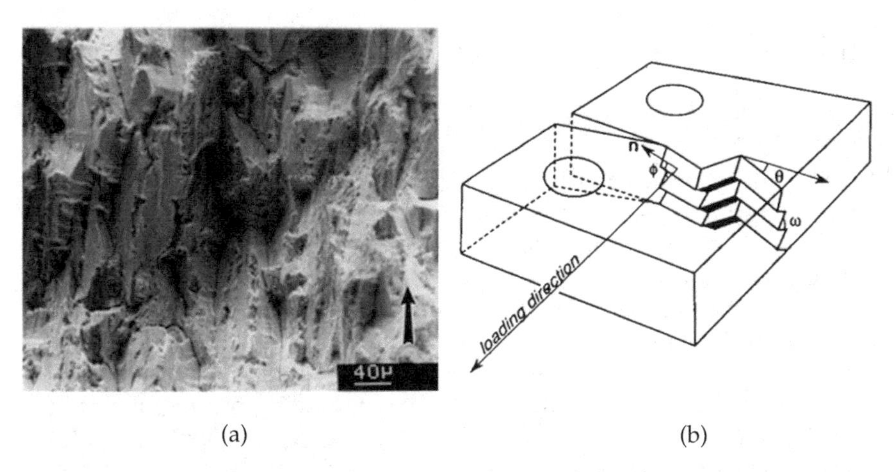

Figure 10.6. (a) Fatigue fracture mode in 8090-T8771 Al-Li alloy plate; (b) the slip band facet normal with respect to the loading direction.

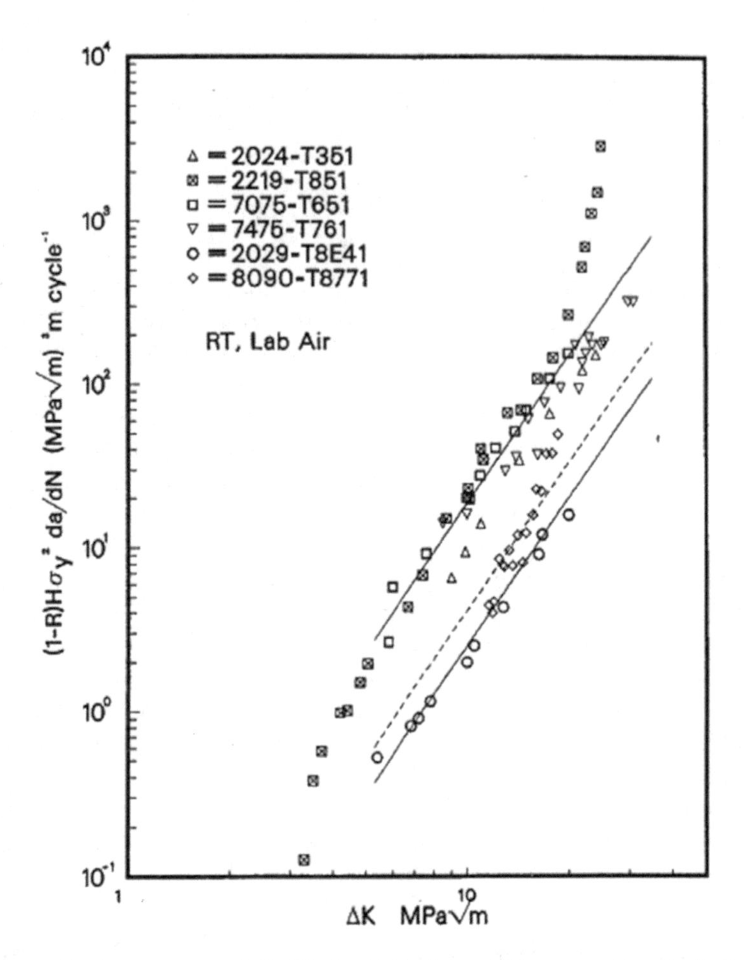

Figure 10.7. Comparison of FCGRs in Al-Li alloys with conventional Al alloys.

10.2.3 *K*-Similitude and Crack Closure

The LEFM description of fatigue crack growth is predicated upon the concept of *K*-similitude, that is, identical near-tip conditions will prevail in specimens of different sizes containing different crack geometries if the magnitude of the stress intensity factor *K* is the same. By the RSR model, it is understood that the *K*-similitude is ensured by the crack-tip slip activity and slip boundary. However, this similitude breaks down when crack closure occurs.

Crack closure refers to premature contact of the two crack surfaces at a level above the minimum load during the unloading portion of a fatigue cycle. The contact could be attributed to various reasons including plastic deformation left along the crack wake, surface roughness or corrosion deposits, etc. (Suresh and Ritchie 1984). It was thought to be the main factor responsible for the fatigue crack growth threshold. The concept of crack closure was first proposed by Elber (1970) based on the observation that the crack opening displacement (COD) exhibited a non-linear relationship versus the load (P) during a tensile fatigue cycle, as shown in Figure 10.8. The deflection point was taken as the crack opening load, P_{op}, corresponding to the crack opening stress intensity, K_{op}. Below K_{op}, the crack could not be fully opened. It was proposed that the effective stress intensity range, ΔK_{eff}, at the fatigue crack is given by

$$\Delta K_{eff} = \begin{cases} K_{\max} - K_{op} & (K_{op} > K_{\min}) \\ K_{\max} - K_{\min} & (K_{op} < K_{\min}) \end{cases} \tag{10.35}$$

By Eq. (10.35), it is simply assumed that crack closure completely shields any mechanical effect at stress intensities below K_{op}. This concept is rather intuitive and has been widely accepted for interpretation of the fatigue threshold phenomena and the stress ratio effect (Suresh and Ritchie 1984, Suresh 1998). But, it does not provide the exact similitude condition, since the closure extent below P_{op} can be different, even at the same ΔK_{eff}. Louat et al. (1993), using the continuously distributed dislocation theory, analysed the crack-tip stress field in the presence of crack closure. They obtained approximate closed-form solutions for two limiting conditions: (i) crack closure by contact of two thin asperities very near to the crack tip, and (ii) a fully closed crack where every point along the crack wake remains in contact below K_{op}. Wu (1995) examined the energy of crack closure, using the Irwin-Kies's approach (1954) based on the COD vs. P response. As it is discussed below, only the Irwin-Kies's approach provides the K-similitude with crack closure, satisfying the energy conservation principle.

Within the regime of linear elastic fracture mechanics, i.e., under small scale yielding conditions, the ideal relationship between the load and the crack opening displacement should be perfectly linear. With the interference

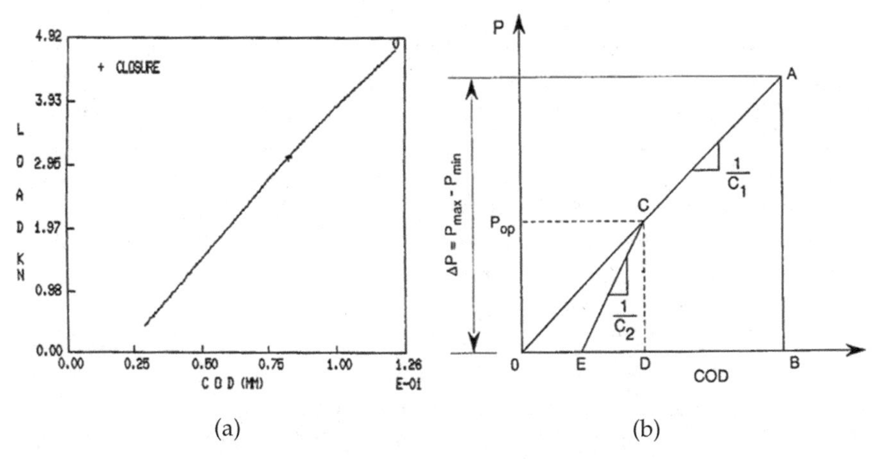

Figure 10.8. (a) Typical COD vs. P curve for a CT specimen; and (b) the idealized bi-linear behaviour as represented by line EC and CA.

of crack closure, the observed COD vs. P response exhibits a non-linear relationship. Figure 10.8 (a) shows a typical COD vs. P response of a compact-tension (CT) specimen when crack closure occurs, where the COD is monitored at the crack mouth using a clip-on gauge. Since COD measured at any fixed distance from the load line can be translated to load-line COD, without losing generality, emphasis will only be placed on the load-line COD in the following discussion. Figure 10.8 (b) shows an idealization of the observed COD vs. P behaviour which can be represented by two segments of linear relationships, line AC and line CE with compliance C_1 and C_2 respectively (compliance is defined as the increment of displacement divided by the increment of load, $C = \Delta u/\Delta P$). These two lines intersect at point C which corresponds to the conventionally defined P_{op}. In the coordinate system of Figure 10.8 (b), the origin is chosen at the point of minimum load and minimum displacement without crack closure, and therefore load (P) and displacement (u) have their relative values above their minima when taken in calculations, but their meaning can be extended to the absolute quantities without using separate symbols. This convention also applies to other associated mechanical quantities such as stress intensity factor.

In the system represented by Figure 10.8 (b), the mechanical work is

$$\Pi = \int P du \tag{10.36}$$

If crack closure did not occur, the mechanical work would be equal to the area of triangle OAB, denoted as Π_{OAB}. Due to the interference of crack closure, the input energy is equal to the area of the quadrilateral $ECAB$, denoted as Π_{ECAB}. According to the principle of energy conservation, the amount of energy shielded by crack closure, Π_{sh} (the shaded area), must be equal to the energy difference $\Pi_{OAB} - \Pi_{ECAB} = \Pi_{OCE}$. Therefore,

$$\Pi_{sh} = \Pi_{OAB} - \Pi_{ECAB} = \Pi_{OCD} - \Pi_{ECD} = \frac{1}{2} C_1 P_{op}^2 - \frac{1}{2} C_2 P_{op}^2 \tag{10.37}$$

The idealized physical picture, Figure 10.8(b), depicts the crack closure event as follows. The crack remains partially closed to a length of a_2 at loads below P_{op} (a_2 can be calculated from the inverse function of compliance at the value of C_2). At P_{op}, the crack is opened up to its full length a_1. This opening-up process behaves like a "brittle fracture" process, therefore, the crack-opening force against crack closure can be represented by the energy release rate. Following Irwin and Kies's definition, the equivalent shielding stress intensity range, ΔK_{sh}, can be defined as

$$\frac{\Delta K_{sh}^2}{E} = \frac{1}{B}\frac{d\Pi_{sh}}{da} = \frac{1}{2B}\frac{d(C_1 P_{op}^2)}{da} - \frac{1}{2B}\frac{d(C_2 P_{op}^2)}{da} \tag{10.38}$$

where E is the elastic modulus and B is the specimen thickness.

Finite element analysis by Newman (1977) has shown that $P_{op}/P_{\max}$ is constant during steady-state crack growth under constant amplitude loading conditions. In load shedding tests, crack growth consists of a series of constant-amplitude crack growth steps with the load amplitude decreased following a prescribed gradient, $(dK/da)/K$. Therefore, under most FCGR testing conditions, it can be assumed that P_{op} is constant during each step of steady-state crack growth, and the derivatives in Eq. (10.38) are only taken to the compliance.

In terms of Irwin-Kies's definition:

$$K = \frac{P}{B}\left[\frac{1}{2}\frac{d(CEB)}{da}\right]^{1/2} \quad (10.39)$$

Eq. (10.38) can be written as

$$\frac{\Delta K_{sh}^2}{E} = \frac{K_{op}^2}{E} - \frac{K_{cl}^2}{E} \quad (10.40)$$

where K_{op} and K_{cl} are given by

$$K_{op} = \frac{P_{op}}{B}\left[\frac{1}{2}\frac{d(C_1EB)}{da}\right]^{1/2} \quad (10.41)$$

and

$$K_{cl} = \frac{P_{op}}{B}\left[\frac{1}{2}\frac{d(C_2EB)}{da}\right]^{1/2} \quad (10.42)$$

K_{op} is the crack-opening stress intensity, i.e., the critical stress intensity at which a crack becomes fully open to its current length. Here, K_{cl} is defined as crack closure stress intensity at which a crack remains closed to the extent in relation with the change in the slope of COD vs. P curve.

From Eq. (10.40),

$$\Delta K_{sh} = \sqrt{K_{op}^2 - K_{cl}^2} = \chi K_{op} \quad (10.43)$$

where

$$\chi = \sqrt{1 - \frac{K_{cl}^2}{K_{op}^2}} \quad (10.44)$$

χ shall be called the *shielding factor*.

From the energy conservation standpoint, the effective stress intensity range should be defined by

$$\Delta K_{eff} = K_{\max} - \Delta K_{sh} = K_{\max} - \chi K_{op} \quad (10.45)$$

There are two limiting cases for Eq. (10.45): (i) if the closed crack is rigid, i.e., COD vs. P behaviour follows line CD in Figure 10.8 (b), χ will be equal to 1, then Eq. (10.45) becomes equivalent to Eq. (10.35); (ii) if crack closure occurs very close to the crack tip, i.e., $a_1 \approx a_2$, K_{op} and K_{cl} will be very close and χ will be very small. These inferences agree with the conclusion obtained by Loaut et al. (1993). According to their results, $\chi \approx 0.2$ for crack closure that occurs very closely behind the crack tip. For crack closure occurring under real testing conditions, the shielding stress intensity range, ΔK_{sh}, has to be evaluated from experimentally observed COD vs P response, according to Eq. (10.45).

It should be emphasised that originally the concept of stress intensity factor was experimentally established by Irwin and Kies using the compliance energy method (K is so named after Kies), which laid the foundation of the linear elastic fracture mechanics. The crack closure process illustrated above consists of sudden crack length change from a_2 (corresponding C_2) to a_1 (corresponding C_1), like "brittle" fracture. Therefore, the shielding stress intensity factor of this process has to be defined in terms of the same energy criterion of Irwin-Kies, to be consistent with the original K definition. In previous FCGR test analyses, no attention has been paid to COD behaviour below K_{op}, and C_2 data are rarely given in the literature. According to the above analysis, three quantities, C_1, C_2, and P_{op}, determine how much mechanical energy is shielded by crack closure, therefore, C_2 should not be ignored in analysing the effect of crack closure. In the subsequent FCGR data analysis, Eqs. (10. 43) to (10.45) have to be built into consideration.

10.2.4 Does Fatigue-Threshold Exist?

The existence of fatigue thresholds below which cracks do not propagate remains to be the biggest question in debate in the FCGR study. It is important for structural integrity assessment as to whether or when to use such a condition to calculate the crack growth life for safe inspection intervals. For long cracks, the ASTM Standard E-647 prescribes two testing procedures to obtain FCGR: (i) the constant-amplitude procedure, which creates ΔK-increasing conditions as crack propagates, and (ii) the load-shedding procedure, following a constant normalized K-gradient, $[1/K][dK/da]$, which creates ΔK-decreasing conditions as crack propagates. Particularly, load shedding procedures are used to investigate the occurrence of fatigue thresholds, because the pre-cracking ΔK is usually high, ~10 MPa. In ASTM E-647 load shedding tests, crack opening displacement is also forced to decrease with increasing crack length. Crack closure is mostly observed under such ΔK-decreasing conditions, which has certainly brought uncertainties in the interpretation of the observed fatigue threshold phenomenon. Since crack closure has rarely been reported to occur along with naturally-generated small cracks or under plane-strain ΔK-increasing conditions, correction of the crack closure effect to deduce the intrinsic fatigue crack growth properties from the load-shedding tests becomes extremely important for structural integrity designs with true material properties.

Realizing that the ASTM E-647 load shedding procedure induces COD-decreasing conditions, which may aggravate crack closure, Wu et al. (1995a) developed a power-law load shedding procedure, to maintain constant COD, while decreasing ΔK to interrogate fatigue threshold, with reduced crack closure at a given stress ratio R.

First, fit the measured COD with a power-law function, as

$$COD = \delta_0 + A\frac{K}{E}\left(\frac{a}{W}\right)^m \sqrt{W} \tag{10.46}$$

where δ_0, A and m are parameters determined from fitting to the compliance-curve of the CT specimen.

Using Eq. (10.46), the constant COD condition is defined by

$$\frac{dCOD}{da} = \frac{A}{E}\left(a^m \frac{dK}{da} + mKa^{m-1}\right)W^{1/2-m} = 0 \tag{10.47}$$

from which it yields

$$\frac{1}{K}\frac{dK}{da} = -\frac{m}{a} \tag{10.48}$$

Integration of Eq. (10.48) leads to

$$\frac{K}{K_0} = \left(\frac{a_0}{a}\right)^m \tag{10.49}$$

Thus, a power-law load shedding schedule is developed to keep COD constant. Ideally, at a constant R-ratio, and if crack closure does not occur, COD_{max} and COD_{min} as well as ΔCOD will all be constant in this procedure. If crack closure occurs, which is likely, COD_{min} will be affected. Therefore, only COD_{max} is controlled in testing.

Having obtained the energy-based correction for crack closure in section 10.2.3, we now proceed to examine how exactly crack closure affects the FCGR behaviour in the above three testing procedures (Wu et al. 1995a, 1995b): (i) constant amplitude (CA), (ii) constant K-gradient load-shedding, and (iii) constant COD_{max}. In the ΔK-increasing test, the pre-crack was prepared under a ΔK-decreasing condition with a normalized K-gradient of –0.18 mm^{-1}, then material of 1 mm in width along the crack wake was removed, and the subsequent testing was conducted in load control using constant-amplitude sine-wave loading. In the ΔK-increasing (CA) test in the ΔK range of 3.75 ~12 MPa$\sqrt{m}$, the COD vs. P response was observed to be linear above P_{min} (no crack closure). In the ASTM ΔK-decreasing test, the pre-crack was prepared in CT specimen at constant ΔK = 10 MPa$\sqrt{m}$ up to 13 mm, then a constant normalized K-gradient procedure was carried out with $[1/K][dK/da]$ = –0.05 mm^{-1}. In this test, crack closure level was observed to be the highest, ~0.8 K_{max}, as shown in Figure 10.9. In the power-law procedures, the normalized K-gradient, $1/K[dK/da] = m/a$ (m = 0.9), which varied from –0.08 to –0.02 mm^{-1} as crack length increases during the test, satisfying the ASTM E-647 specification (>–0.08 mm^{-1}). In this test, crack closure was moderate, ~0.4 K_{max}, at first, then gradually reduced to the level comparable to K_{min}, as shown in Figure 10.9. These tests clearly show that the level of crack closure varies with the test procedure, in other words, the amount of crack closure depends on prior loading history. These experimental findings suggest that crack closure is an artefact of load-shedding procedure.

The FCGRs generated using these three testing procedures on 8090-T8771 L-T are shown in Figure 10.10. Apparent fatigue thresholds can be seen in the two load-shedding tests with that in the ASTM-E647 test the highest at ΔK = 5 MPa$\sqrt{m}$, and that in the power-law procedure ~3 MPa $\sqrt{m}$. FCGR obtained under the ΔK-increasing condition are regarded as the intrinsic FCGR of 8090-T8771 L-T at R = 0.1, free of closure-interference. It is observed that the ΔK-decreasing FCGRs are all lower than the ΔK-increasing FCGRs. Under the same loading conditions (ΔK, R, and f) and in the same environment, the retardation of FCGR is apparently caused by crack closure. Then, FCGRs are plotted as function of the ΔK_{eff} defined by Eq. (10.35) and that defined by Eq. (10.45), as shown in Figure 10.11(a-b), respectively. It is apparent that the ΔK_{eff} given by Eq. (10.35) does not consolidate the data, whereas the ΔK_{eff} given by Eq. (10.45) does within a reasonable scatter band (a factor of 2), as shown in Figure 10.11(b). Therefore, the ΔK_{eff} defined by Eq. (10.45) represents the true crack growth driving force.

The K-similitude of the linear elastic fracture mechanics (LEFM) ensures the applicability of the Paris law, regardless of specimen shape, which has been found to be true under ΔK-increasing conditions. However, as shown in the aforementioned experiments, this similitude breaks down in ΔK-decreasing crack growth processes, due to different crack wake interferences generated using different load-shedding procedures (initial K_{max}, and K-gradient) (Wu et al. 1995). Eq. (10.35) was initially proposed for correction of the situation, but it does not follow the Irwin-Kies definition of K derived from the crack compliance (Wu 1995).

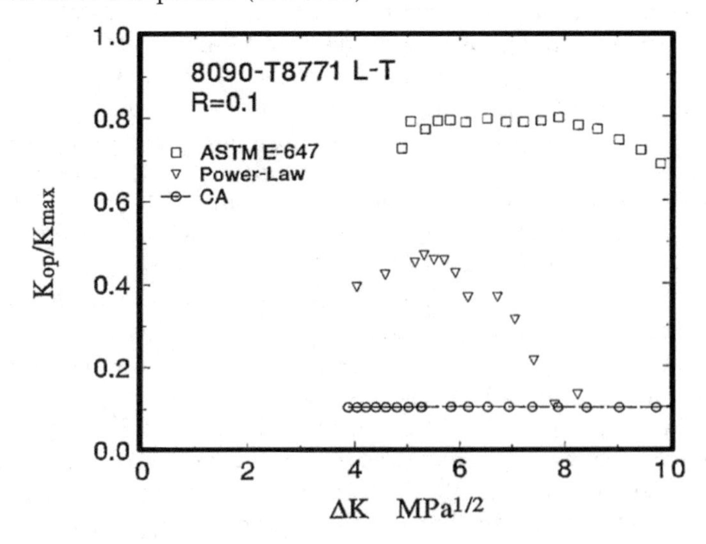

Figure 10.9. Crack closure levels in Al-Li 8090-T8771 under both ΔK-decreasing and ΔK-increasing conditions.

Another cause of breaking down the similitude could lie in the history-dependence of cyclic plasticity ahead of the crack-tip (Zhao and Tong 2008). With kinematic/isotropic hardening (see section 2.1), load drop would result in crack-tip cyclic deformation to occur below the yield surface established by the previous load, before crack comes out of the previous plastic zone, which means that the material in the plastic zone undergoes a different cyclic deformation process under *K*-decreasing conditions as compared to that under *K*-increasing conditions. If the load shedding is continuous, sooner or later, the crack-tip cyclic deformation would retrieve to an "elastic condition" that terminates the crack propagation. This aspect is worth further investigation.

The above experiments and analysis do show that crack closure is mostly an artefact of load-shedding procedure and its effect should be corrected with the energy argument given by (10.45). Under *K*-increasing conditions, crack closure does not exist above K_{min} under plane-strain conditions. This approach has been demonstrated on 8090-T8711 in other orientations (Wu et al. 1995b), and on other alloys including 7075-T651, Type 316SS and IN718 (Wu et al. 1995a). On the other hand, crack closure was found mostly to exist under plane-stress conditions (Elber 1971, Tanaka and Nakai 1983,

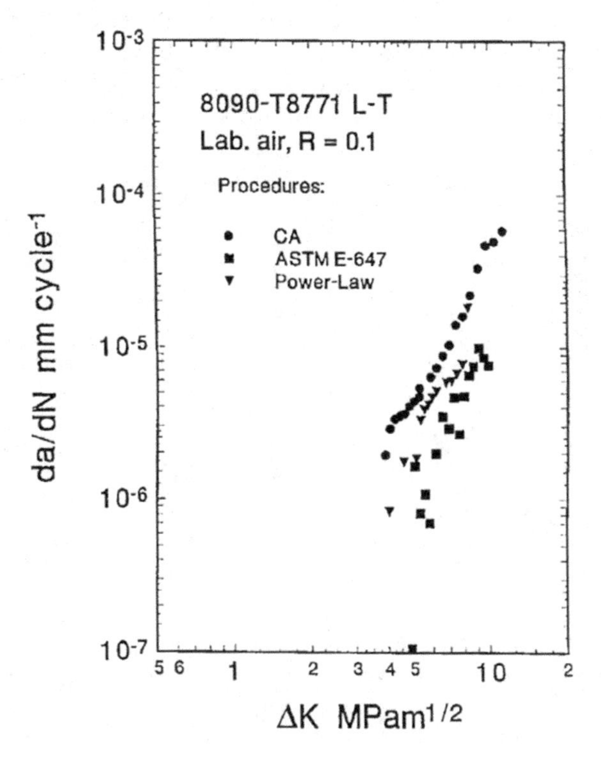

Figure 10.10. FCGRs in 8090-T8771 Al-Li alloy generated using (i) constant amplitude (CA), (ii) ASTM-E647 load-shedding; and power-law load-shedding procedures.

Lee and Sharpe 1986, Larsen et al. 1986, Davidson and Lankford 1986). Despite that Newman (1983) developed a FEM model for plasticity-induced crack closure, which would be only suitable to address fully-closed crack, by virtue of the idealization of continuum mechanics, there is no general crack closure model to delineate different crack closure mechanisms under service conditions.

In light of the above experimental findings and discussions, it is questionable whether intrinsic fatigue threshold exist in K-increasing fatigue crack growth processes as in real engineering cases, where fatigue cracks nucleate at very small size (in the order of a few Burgers vectors) from intrusion and extrusion pairs ($\Delta K < 0.1$ MPa $\sqrt{m}$), grow initially as small cracks within the confinement of a few grains ($\Delta K < 5$ MPa $\sqrt{m}$), and later become "long" cracks. If true fatigue threshold existed, say above 0.1 MPa $\sqrt{m}$, one would not see long cracks from such a natural fatigue process.

10.3 Environment-Assisted Fatigue Crack Growth

As discussed in section 10.2.1, the RSR model implies that for stable microstructure the fatigue threshold occurs solely as a result of environmental interactions, which is extrinsic to the material in nature. The model describes this effect by assuming that the environment influences the characteristic activation quantities ($\Delta G^{\neq}$, α, V) in both forward and reverse slip processes. Quantitatively, ΔK_{th} will become significant in equation (10.30) only if t_{of} is

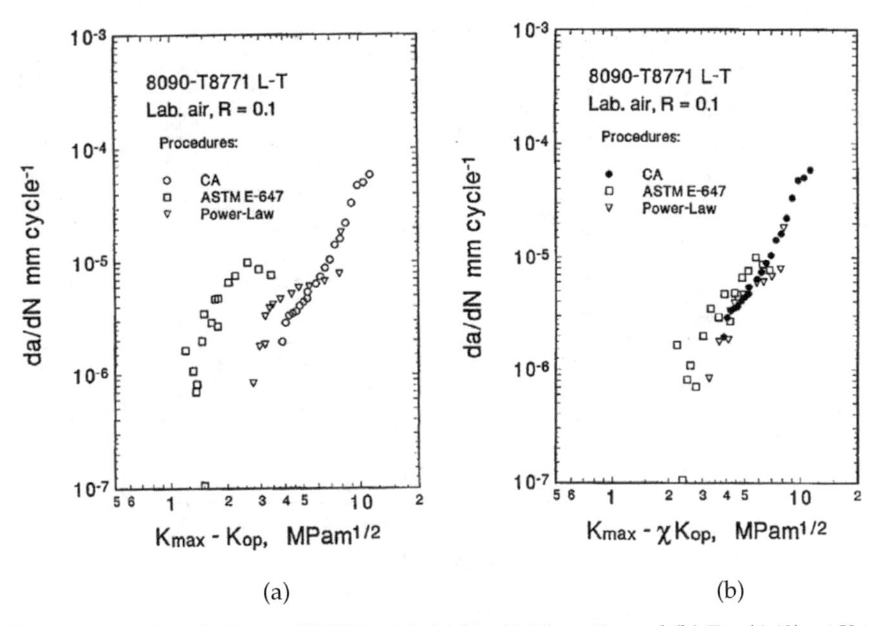

Figure 10.11. Correlations of FCGR with (a) Eq. (6.32) - ΔK_{eff} and (b) Eq. (6.42) - ΔK_{eff}, after Wu (1995).

greater than t_{or}, ΔG_f is smaller than ΔG_r and/or V_r is considerably greater than V_f. In practice, this may happen when environmental effects weaken the point obstacles after a new crack surface is exposed, which will result in a decrease in both ΔG_r and t_{or}, or when some weak point obstacles are cut by dislocations during the forward slip and do not act as obstacles during the reverse slip process, which will result in an increase of V_r. Indeed, corrosion fatigue is a complex process and details of the kinetics for processes such as corrosion debris formation need to be considered. Without alluding to a specific corrosion process or mechanism, corrosion fatigue is left as an open subject in this book. In the following, we only formulate a generalized corrosion fatigue process, as discussed by Wu and Wallace (1994).

With the presence of a reactive environment, corrosion will attack the crack-tip, viz. oxidation film rupture, hydrogen embrittlement, stress-corrosion cracking, active path dissolution, etc. Regardless of the mechanism(s) by which corrosion damage accumulates, it is the rupture of the environment-affected layer that contributes to crack growth, additional to the component promoted by RSR. Therefore, the characteristic dimension of the corrosion event, l (which can be the oxide film thickness, the hydrogen embrittlement zone size, and the depth of dissolution), is the quantity adding to the rate of crack growth in a deleterious environment. Upon incorporating this corrosion damage zone into the RSR model, as schematically shown in Figure 10.12, a physical description of the general environment-assisted fatigue crack growth can be derived as follows.

Assume that the corrosion damage zone is a circular area of radius l, and it ruptures (or dissolution occurs) along the main crack plane (In reality, the rupture may occur along some crystallographic planes or grain boundaries, but it is assumed that l represents the projection onto the main crack plane to the first order approximation). Then, the forward crack increment is given by

$$(\Delta a)_f = \gamma_f(r_p - l)\cos\theta + l \tag{10.50a}$$

and the slip reversal is given by

$$(\Delta a)_r = \gamma_r(r_p - l)\cos\theta \tag{10.50b}$$

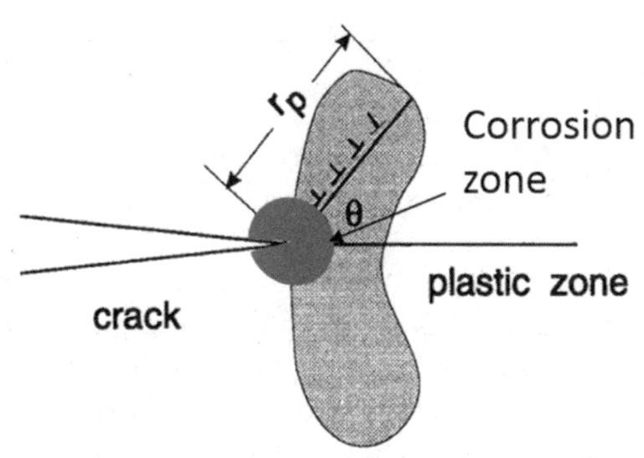

Figure 10.12. Schematic of a corrosion damage zone incorporated in the RSR model.

Hence, the crack growth rate in a deleterious environment can be formulated as:

$$\frac{da}{dN} = (\Delta a)_f - (\Delta a)_r = (\gamma_f - \gamma_r) r_p \cos\theta + [1 - (\gamma_f - \gamma_r)\cos\theta] l \qquad (10.51)$$

According to this slip-dissolution/rupture model, environment-assisted crack growth consists of (i) an RSR component given by Eq. (10.29), and ii) an environmental contribution, $(da/dN)_{\text{env}} = l$ weighted by a factor of $\Theta = 1 - (\gamma_f - \gamma_r)\cos\theta$. When l is small, RSR dominates the fatigue crack process. But when l approach r_p, crack growth is totally dominated by the environmental component, $(da/dN)_{env}$. In the cases where it can be simplified that $\Theta \approx 1$, Eq. (10.36) reduces to the linear superposition rule, as many researchers assumed (Wei and Shim 1984).

By the ICFT, if we further incorporate internally distributed damage under combined cyclic and dwell loading, we can write the general fatigue crack growth equation as

$$\frac{da}{dN} = D\left[\left(\frac{da}{dN}\right)_{RSR} + \Theta\left(\frac{da}{dN}\right)_{env}\right] \qquad (10.52)$$

where D is given by Eq. (4.17). Eq. (10.52) is essentially identical to Eq. (4.15).

10.4 Creep Crack Growth

Creep crack growth resistance is one of the basic damage-tolerance properties required for large structures such as nuclear reactor pressure vessels and land-based gas turbines operated for long durations at high temperatures ($T \geq 0.3\, T_M$, where T_M is the absolute melting temperature). Hence, creep crack growth has been a long-time subject of study. Creep crack growth occurs under constant load at high temperature in air, which involves creep-oxidation interaction. It has been reported that failure by creep crack growth can occur under temperature and stress conditions which normally do not cause significant creep deformation in the material (Gooch 1984). Some alloys, such as Alloy 718 and cold-worked stainless steels, exhibit poor creep crack growth resistance even though they possess good creep rupture strengths. Therefore, it is important to understand creep crack growth behaviours and their dependence on microstructure and environment in a unified sense, which has practical significance for design of damage tolerant components operating at high temperatures.

As discussed in section 10.1, the stress field ahead of the crack tip in a power-law creeping material is controlled by the path-independent energy integral $C(t)$ or its steady-state value C^*. Good correlations were found between creep crack growth rates (CCGR) and C^* for creep-ductile materials such as Discaloy (Landes and Begley 1976) and steels (Liu and Hsu 1985), where CCGR was found to be almost linearly proportional to C^*.

On the other hand, the stress intensity factor (K) has also been used to correlate CCGR under small-scale creep deformation conditions and in the presence of environmental embrittlement effects. Particularly, power-law of CCGR vs. K relations is commonly used to characterize the creep crack growth behaviour for high strength heat resistant alloys such as superalloys (Floreen 1975, Floreen and Kane 1976, Larson and Floreen 1977, Sadananda and Shahinian 1977, Diboine et al. 1987). The general observation is that the power-law exponent of crack growth falls in the range of 2 to 8, but it is not equal to the power law creep stress exponent for the same material, and the proportional constant is a temperature-dependent term but also dependent on the material's microstructure, especially the grain size and grain boundary morphology.

Many CCGR models have been developed based on either power-law creep deformation or cavitation mechanisms, trying to explain the various aspects of creep crack growth phenomena. Barnby (1975) proposed a deformation-based model, simply assuming that CCGR is proportional to the steady state creep rate. Dimelfi and Nix (1977) considered cavity growth by power-law creep in the elastic crack-tip stress field, which leads to CCGR being proportional to K^n, where n is equal to the power-law creep exponent; Cocks and Ashby (1982) considered creep crack propagating in the Riedel-Rice field and obtained an expression of $da/dt \sim C^{*n/n+1}$ (the power exponent, $n/(n+1) \sim 1$); Miller and Pilkington (1980) considered cavity growth by power-law creep in the surrounding elastic matrix, and included a grain-size dependence giving $da/dt \sim K^n d^{-n/2+2}$. Vitek (1978) proposed a diffusion-based model considering removal of atoms from the crack tip via stress induced grain boundary diffusion and their deposition at grain boundaries, which predicts that $da/dt \sim K^4$. Sadanada (1978) proposed a creep crack growth model that combines two competing mechanisms: cavity coalescence via grain boundary diffusion and plastic flow providing additional sinks of vacancies via bulk diffusion. Some general fracture mechanics treatments have also been given, purely from a continuum point of view (Saxena 1986, Riedel 1989, Riedel 1990). None of the above models considered the contribution of grain boundary sliding, and particularly few included the effect of microstructure. Xu et al. (1999) developed a CCGR model considering the multi-mechanism zones ahead of the crack under steady-state creep conditions and derived a unified CCGR expression that covers both K-dependence and C^*-dependence, depending on the microstructure.

10.4.1 The Crack-Tip Stress Distribution

It has been understood that creep deformation and damage can proceed by multiple mechanisms, as discussed in Chapters 2-5. The mechanism-based strain decomposition is given by Eq. (4.14). In general, because deformation in the crack-tip region is restrained by the surrounding elastic deformation, the crack-tip stress field really builds up as the result of stress relaxation by operation of rate-dependent mechanisms, including intragranular deformation (ID) and GBS. Since it is always the fastest deformation process that plays the predominant role in creep and stress relaxation process, the creep crack-tip stress field can therefore be divided into three regions: A) power-law zone by ID; B) GBS zone; and C) the elastic region, as shown schematically in Figure 10.13. In the following, we will give a detailed treatment for the crack-tip stress field in the three zones and the associated CCGR law.

Power-law Zone

The stress field of a creep crack has been obtained by Riedel and Rice (1980) for materials obeying power-law creep. For mathematical simplicity, whether by intragranular dislocation glide or climb, we can formulate the creep rate as power-law (Chapter 2):

$$\dot{\varepsilon}_g = A\sigma^n \tag{10.53}$$

where A is a material constant and n is the power-law exponent.

Then, according to the RR field, the maximum stress in the power-law zone is given by

$$\sigma = \left(\frac{C(t)}{AI_n r}\right)^{1/(n+1)} \tag{10.54}$$

GBS Zone

As the stress decreases away from the crack tip, intragranular dislocation motion may become limited, and GBS assumes the dominant role in the stress relaxation process. From the solution of GBS in the presence of grain boundary precipitates (Wu and Koul 1995), we have

$$\dot{\varepsilon}_{gbs} = A_0\left(\frac{b}{d}\right)^q\left(\frac{\lambda + r}{b}\right)^{q-1}\left(\frac{\sigma}{\mu}\right)^p = B\sigma^p \tag{10.55}$$

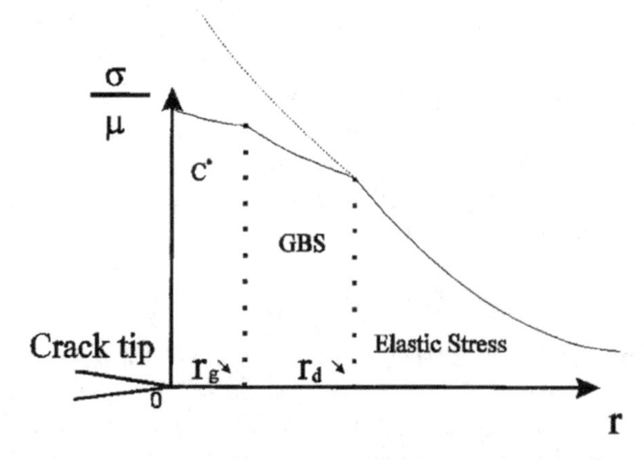

Figure 10.13. A schematic of stress distribution in the ID power-law zone, GBS zone, and elastic region ahead of a creep crack tip.

with the constants and microstructural parameters as defined in section 2.4.

Following the RR-field distribution, the stress in the GBS-dominated region is given by

$$\sigma = \left(\frac{C(t)}{BI_p r}\right)^{1/(p+1)} \tag{10.56}$$

The boundary between the intragranular deformation zone and the GBS zone can be approximately defined by equating the two stresses given by Eqs. (10.54) and (10.56), as

$$\left(\frac{C(t)}{AI_n r_g}\right)^{1/(n+1)} = \left(\frac{C(t)}{BI_p r_g}\right)^{1/(p+1)} \tag{10.57}$$

from which the boundary between the ID zone and GBS zone r_g is determined as

$$r_g = \frac{(I_n A)^{\frac{p+1}{n-p}}}{(I_p B)^{\frac{n+1}{n-p}}} C(t) \tag{10.58}$$

Elastic Region

The vast region further away from the GBS zone can be considered as the elastic region. The boundary between the GBS zone and the elastic region can be defined by equating the GBS-controlled stress to the elastic stress, as

$$\left(\frac{C(t)}{BI_p r_d}\right)^{1/(p+1)} = \frac{K}{\sqrt{2\pi r_d}} \tag{10.59}$$

From Eq. (10.59), r_d can be determined as

$$r_d = \left(\frac{K}{\sqrt{2\pi}}\right)^{\frac{2(p+1)}{p-1}} \left(\frac{I_p B}{C(t)}\right)^{\frac{2}{p-1}} \tag{10.60}$$

10.4.2 The Creep Crack Growth Model

As discussed above, the crack tip field can be divided into A) intragranular deformation zone, B) GBS zone and C) elastic region. In the A zone ($0 < r < r_g$), while intragranular deformation relaxes the elastic stress near the crack-tip, this stress drives GBS responsible to create intergranular cracking over the region $A+B$ ($0 < r < r_g + r_d$).

Let time $t = 0$ when an element of material at a distance $L = r_g + r_d$ ahead of the crack tip begins to experience GBS. After time t, this distance reduces to $r = L - vt$, where v (= da/dt) is the crack growth rate. Suppose that, during this time, GBS displacement (S) increases to

$$S = \varepsilon_{gbs}\boldsymbol{d} \tag{10.61}$$

From integration of the GBS strain rate, we find that:

$$\varepsilon_{gbs} = \int_0^{L/v} \dot{\varepsilon}_{gbs}(L - vt)\,dt = \frac{1}{v}\int_0^L \dot{\varepsilon}_{gbs}(r)\,dr \tag{10.62}$$

Assuming that rupture occurs when the GBS displacement S becomes equal to a critical distance λ, combining equations (10.61) and (10.62), we have

$$\lambda = S_{cr} = \varepsilon_{gbs}d = \frac{d}{v}\int_0^L \dot{\varepsilon}_{gbs}(r)\,dr \tag{10.63}$$

Rearranging Eq. (10.63), we obtain the creep crack growth rate as:

$$v = \left(\frac{d}{\lambda}\right)\overline{\dot{\varepsilon}_{gbs}}L \tag{10.64}$$

where

$$\overline{\dot{\varepsilon}_{gbs}} = \frac{1}{L}\int_0^L \dot{\varepsilon}_{gbs}(r)\,dr \tag{10.65}$$

is the average GBS strain rate over the entire damage zone.

It is interesting to note that, similar to the transgranular fatigue crack growth rate expression, Eq. (10.18), creep crack growth rate is also found to be proportional to the strain rate times the total damage zone size L ahead of the crack-tip. In the case of intergranular cracking, λ can be approximately equal to the grain size, if the grain boundary is perfectly planar and clean (no precipitates and cavities). In engineering materials (non-perfect grain boundaries, λ may be equal to a) the distance between two adjacent tension-type ledges, b) the carbide spacing, c) the cavity spacing, and d) the wave length of a serrated grain boundary.

The average strain rate can be calculated by the integral of Eq. (10.65), as

$$\overline{\dot{\varepsilon}_{gbs}} = \frac{1}{L}\int_0^L \dot{\varepsilon}_{gbs}(r)\,dr$$

$$= \frac{1}{L}\left[\int_0^{r_g} B\left(\frac{C(t)}{I_n A r}\right)^{\frac{n}{n+1}} dr + \int_0^{r_d} B\left(\frac{C(t)}{I_p B r}\right)^{\frac{p}{p+1}} dr\right]$$

$$= \frac{B}{L}\left[\frac{n+1}{n-p+1}\left(\frac{C(t)}{I_n A}\right)^{\frac{p}{n+1}} r_g^{\frac{n-p+1}{n+1}} + (p+1)\left(\frac{C(t)}{I_p B}\right)^{\frac{p}{p+1}} r_d^{\frac{1}{p+1}}\right] \tag{10.66}$$

Substituting Eqs. (10.58) and (10.60) for r_g and r_d into Eq. (10.66), we obtain

$$\overline{\dot{\varepsilon}_{gbs}} = \frac{B}{L}\left[\frac{n+1}{n-p+1}\frac{(I_nA)^{\frac{1}{n-p}}}{(I_pB)^{\frac{n-p+1}{n-p}}}C(t)+(p+1)\left(\frac{K}{\sqrt{2\pi}}\right)^{\frac{2}{p-1}}\left(\frac{C(t)}{I_pB}\right)^{\frac{p(p-1)-2}{(p+1)(p-1)}}\right] \tag{10.67}$$

Then, substituting Eq. (10.67) into Eq. (10.64), we obtain the CCGR as

$$\frac{da}{dt} = \frac{d}{\lambda}\left[\frac{n+1}{n-p+1}\left(\frac{I_nA}{I_pB}\right)^{\frac{1}{n-p}}C(t)+(p+1)B\left(\frac{K}{\sqrt{2\pi}}\right)^{\frac{2}{p-1}}\left(\frac{C(t)}{I_pB}\right)^{\frac{p(p-1)-2}{(p-1)(p+1)}}\right] \tag{10.68}$$

10.4.2.1 Creep-Ductile Case

From Eq. (10.68), it can be seen that intergranular creep crack growth rate may consist of two components: one is contributed by GBS in the stress field relaxed by intragranular deformation and the other is contributed from the GBS mechanisms alone. In creep ductile materials with small grain size, the creep zone may contain many grains, and therefore intragranular deformation plays a significant role in stress relaxation, while the contribution of GBS to

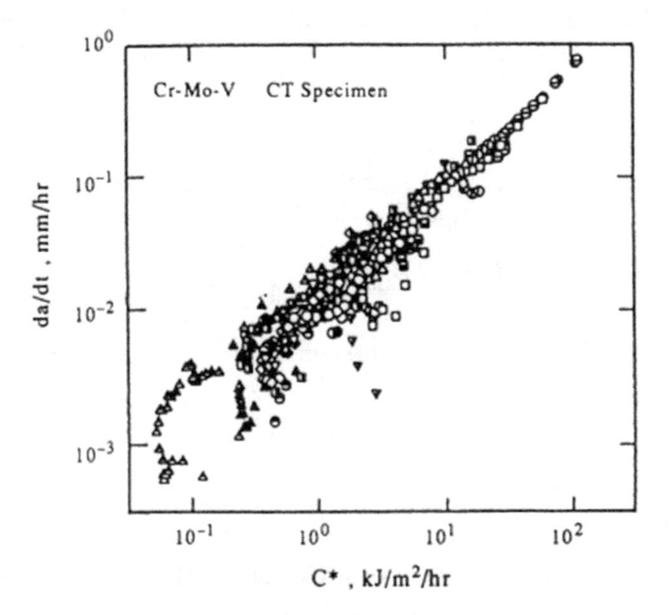

Figure 10.14. Creep crack growth rate plotted against C* for Cr-Mo-V steel (Toshimitsu and Yokobori, Jr. 1999).

stress relaxation may be relatively small. For these cases, the first term in Eq. (10.68) dominates such that

$$\frac{da}{dt} = \frac{d}{\lambda}\frac{n+1}{n-p+1}\left(\frac{AI_n}{BI_p}\right)^{\frac{1}{n-1}} C(t) \tag{10.69}$$

This relationship has been observed in many engineering alloys such as Discaloy (Landes and Begley 1976), Cr-Mo-V steel (Nikbin et al. 1977, Toshimitsu and Yokobori, Jr. 1999), Type 304 steel (Taira et al. 1979) and steels (Liu and Hsu 1985). Figure 10.14 shows the vs. C^* correlation for Cr-Mo-V steel.

10.4.2.2 Creep-Brittle Case

In creep brittle materials (e.g., cast Ni-base superalloys) containing large grains, the role of intragranular deformation in crack-tip stress relaxation and creep crack growth will be limited. In that case, crack-tip deformation will mostly be concentrated in the grain boundary region, and therefore GBS plays a dominant role in the crack growth process. Then, the second term of Eq. (10.68) dominates the rate process:

$$\frac{da}{dt} = \frac{(p+1)d}{\lambda}\left(\frac{1}{\sqrt{2\pi}}\right)^{\frac{2p}{p-1}} B^{\frac{1}{p-1}} K_I^{\frac{2}{p-1}} \left(\frac{C(t)}{I_n}\right)^{\frac{p^2-p-2}{(p-1)(p+1)}} \tag{10.70}$$

At first, it seems that the creep crack growth rate depends on both K and $C(t)$. But, since the theoretical value of p for GBS is 2 (see section 2.4), Eq. (10.70) further reduces to

$$\frac{da}{dt} = \frac{3}{4\pi^2}\frac{d}{\lambda}BK^2 \tag{10.71}$$

The proportional constant B combines the effects of grain boundary diffusion and microstructural features as given by

$$B = \frac{D_{gb}\mu b}{kT}\left(\frac{b}{d}\right)^q \left(\frac{\lambda + r}{b}\right)^{q-1} \propto d^{-q} \tag{10.72}$$

From Eq. (10.72), we can predict the grain size dependence of CCGR as follows.

$$\frac{da}{dt} \propto \begin{cases} d^0 & \text{(no } GB \text{ precipitates)} \\ \dfrac{1}{d} & \text{(discrete } GB \text{ precipitates)} \\ \dfrac{1}{d^2} & \text{(continuous } GB \text{ precipitates)} \end{cases} \tag{10.73}$$

Creep crack growth tests were conducted by Xu et al. (1999) on Udimet 520 at 540 °C in both argon atmosphere and laboratory air, in order to verify the above model. The materials were heat treated to obtain different grain sizes, all containing a certain amount of intragranular γ', but without any grain boundary $M_{23}C_6$ precipitates. The test temperature was selected to ensure that no grain boundary precipitates would form during testing. Fractographic studies revealed that predominantly intergranular fracture and extensive crack tip branching occurred both in air and in argon, as shown in Figure 10.15 (a) and (b) respectively. The metallurgical evidence suggests that GBS-controlled intergranular cracking and oxidation were the dominant damage mechanisms. No evidence of cavitation was found in any of the specimens. The test and material conditions satisfy the model's assumptions of the dominance of GBS.

The CCGR in the argon environment, plotted in a double logarithm scale against the stress intensity factor K, exhibits a linear behaviour with the slope of 2, as shown in Figure 10.16. For Udimet 520 with planar grain boundaries and no grain boundary precipitates, the stress index of GBS rate is 2 and hence, as predicted by Eq. (10.71), the creep crack growth rate exhibits a dependence on the stress intensity factor (K) to a power of 2. All the deformation-based creep models fail to predict the K^2 dependence for creep crack growth in an inert or vacuum environment, because they relate creep crack growth rate to the power-law ($n > 2$) for intragranular deformation, even though the creep fracture always occurs in an intergranular mode. Other diffusion-based models also do not describe this phenomenon either. The diffusional mechanisms in the above tests are of secondary importance, because large intergranular cracks and secondary cracking are not likely to be caused by the removal of atoms via grain boundary diffusion.

With regards to microstructure-dependence, the present CCGR model predicts that the grain-size dependence of CCGR increases with the grain boundary precipitate distribution condition, Eq. (10.73). For materials with planar and clean grain boundaries (no grain boundary precipitates), it predicts

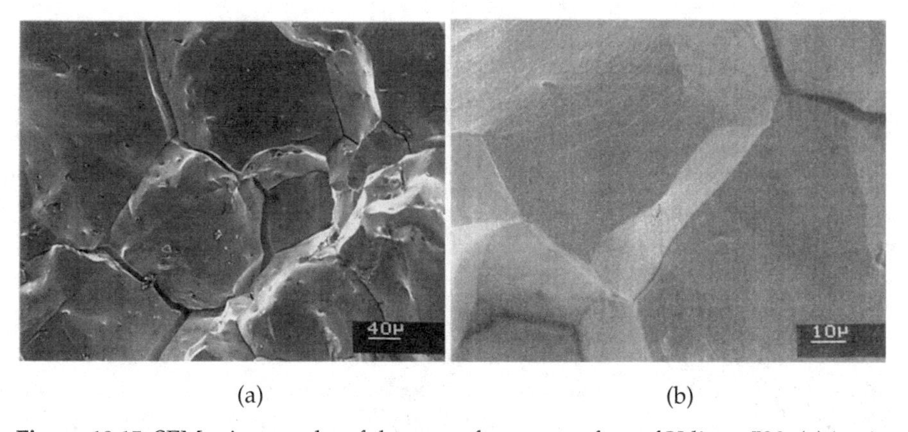

(a) (b)

Figure 10.15. SEM micrographs of the creep fracture surface of Udimet 520, (a) in air and (b) in argon, after Xu et al. (1999).

no grain-size dependency. The data on Udimet 520 in Xu et al.'s study (1999) showed that the CCGR data were not sensitive to grain size variation over a grain size range of 235-464 μm, which corroborated the model prediction. An inverse grain size dependence of CCGR is predicted for materials with a discrete grain boundary precipitate distribution, Eq. (10.73). In superalloys that have received standard heat-treatments, a discrete distribution of grain boundary precipitates (e.g. $M_{23}C_6$ or δ phase) is often present and its effect is to increase creep-rupture resistance by suppressing GBS. In the literature, coarse grain size superalloys have always been reported to have lower CCGRs after standard heat treatments (Floreen 1975, Floreen and Kane 1976, Liu et al. 1991). For example, an inverse relation between CCGRs and grain size has been reported for Alloy 718 at 650 °C (Liu et al. 1991). Therefore, the model prediction in the presence of discrete grain boundary carbides is generally supported by the results reported in the literature. A stronger grain-size dependence is predicted by the model for materials with a continuous grain boundary precipitate network, Eq. (10.73). Such network may result from overaging or service exposure. No experimental results are presently available on materials containing continuous carbide network in the open literature with which these theoretical predictions could be compared. This is a serious issue concerning aging gas turbine components.

10.4.2.3 Environmental Effect

As shown in Figure 10.16, the creep crack growth rate in Udimet 520 tested in air exhibits an increasing power-dependence on K from 2 to a high value of 6.5 as K increases. This change in the K-dependence of CCGR with the environment, i.e. argon versus air, suggests that oxidation has a strong influence on the CCG process.

The oxidation effects can be attributed to the reduction of the critical GBS distance λ to advance the crack in Eq. (10.71). An effective λ in the presence of oxide penetration may be equaled to λ_0-X, where λ_0 is the critical GBS distance in the absence of oxidation effects, and X is the penetrating oxide depth measured from the external oxide surface. As time t increases, oxides will penetrate deeper into the material and λ will be significantly reduced, and therefore the CCGR will increase. Taking the effect of oxidation into consideration, Eq. (10.71) takes the form

$$\frac{da}{dt} = \frac{3}{4\pi^2} \frac{d}{\lambda_0 - X} BK^2 \tag{10.74}$$

The curve for the CCGR in air was fitted using Eq. (10.74) to the test data by Xu et al. (1999). Higher K-dependence indices have also been observed in other superalloys. For example, the CCGRs of Alloy 718 were reported to be very sensitive to environmental effects, and an m value of 8 was even observed at 540 °C (Floreen and Kane 1976, Sadananda and Shahinian 1977). Eq. (10.74) provides a means to delineate oxidation from material-intrinsic creep crack growth behaviour.

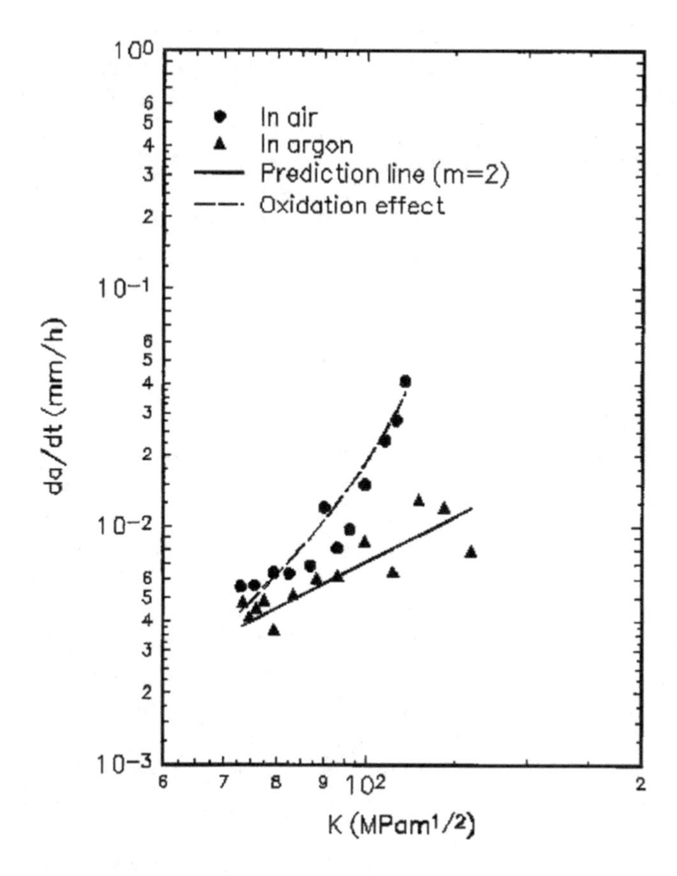

Figure 10.16. Creep crack growth rate in Udimet 520 in argon and air at 540 °C (Xu et al. 1999).

10.5 Creep-Fatigue Crack Growth

Crack growth under creep-fatigue interaction conditions is another important phenomenon affecting the structural integrity of gas turbine components, since evidently most of them operate under such conditions during the takeoff-cruise-landing cycle of the aircraft. The subject has been studied by many researchers for many years, but mostly experimental results were reported without quantitative model analysis (e.g., Saxena et al. 1981, Bain et al. 1988, Telesman et al. 2008). It has been found that fatigue crack growth rate with hold times are sensitive to microstructure and environment as well. However, the actual interplay of various deformation and damage mechanisms in this process remains to be a "mystery".

Zhao and Tong (2008) performed an FEM simulation of crack-tip deformation and crack growth behaviour in a CT specimen of Ni-based superalloy at elevated temperature, using Chaboche's unified constitutive law of viscoplasticity. Their simulation shows that, first for stationary cracks, cyclic deformation ahead of the crack-tip ratchets positively in the loading direction with the peak cyclic stress gradually relaxed and the local hysteresis loops evolve towards a balanced position with the local cyclic stress ratio of $R = -1$, regardless of the remote stress (load) ratio. Figure 10.17 shows the crack-tip cyclic deformation for the first 30 cycles under triangular loading waves with $P_{max} = 7$ kN, $R = 0$; and with $P_{max} = 10$ kN, $R = 0.3$. The higher loading profile with high stress ratio results in a longer stress relaxation process with larger ratchet strain for a given number of cycles. Obviously, crack-tip ratcheting displacement leads to no crack closure at positive R ratios under K-increasing conditions. Futher ponder on the crack-tip stress-strain loop evolution process as shown in Figure 10.17, one can perceive that if a load-shedding procedure is followed, the crack-tip cyclic deformation will fall below the previously established yield surface, step by step. This, even without crack closure interference, will eventually terminate the cyclic crack-tip plastic deformation process, leading to "crack arrest", as reasoned by the RSR model. Hence, crack arrest by load-shedding should not be defined as the "fatigue threshold" which means the condition to start crack growth.

Secondly, the finite element simulation results also show that low frequencies and super-imposed hold periods at peak loads significantly enhance strain accumulation at crack tip, as shown in Figure 10.18. The holding period certainly enhances the strain accumulation, and hence creep damage, ahead of the crack-tip. The load-dwell also induces stress relaxation. This is consistent with the observed dependency of crack-growth rate on frequency and dwell period from the experimental results (Dalby

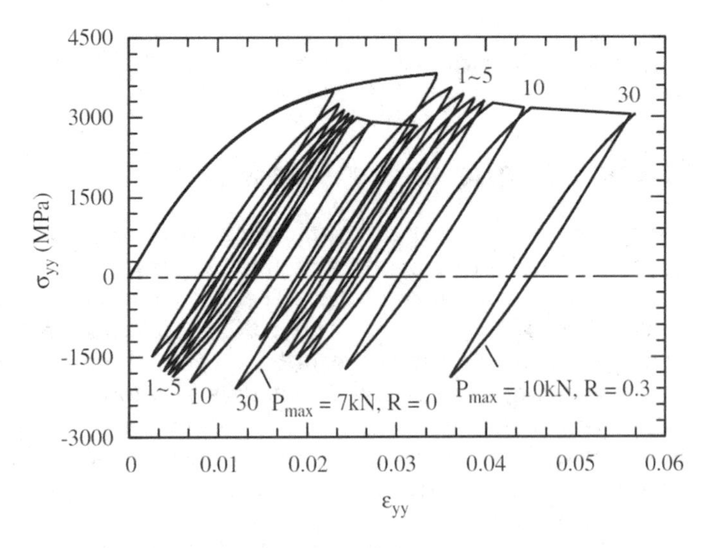

Figure 10.17. Evolution of stress-strain loops over 30 cycles at a Gauss integration point just ahead of the crack-tip for load ratios $R = 0$ and 0.3 and with a constant load range $\Delta P = 7$ kN. After Zhao and Tong (2008).

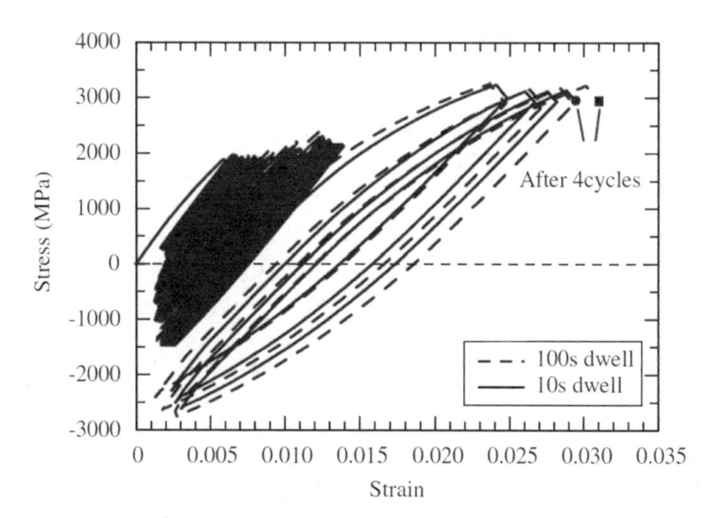

Figure 10.18. Evolution of stress-strain loops at a monitored element ahead of a growing crack for two different holding times. Results for 4 cycles are presented after the crack grows into the monitored element and the crack is assumed stationary momentarily. After Zhao and Tong (2008).

and Tong 2005), i.e., faster crack-growth rate at lower frequency and longer dwell period.

Ratcheting refers to the progressive accumulation of inelastic strain under cyclic stress. In this case, the stress–strain loops would progressively shift in the direction of the mean load. Using the ICFT, creep-fatigue crack growth rate can be described as the result of creep-fatigue interaction ahead of of the crack-tip.

Recall that, in section 4.3, we have obtained the total crack growth rate under general thermomechanical loading conditions, as

$$\frac{da}{dN} = \left(1 + \frac{l_c + l_z}{\lambda}\right)\left\{\left(\frac{da}{dN}\right)_f + \left(\frac{da}{dN}\right)_{env}\right\} \tag{10.75}$$

Under creep-fatigue crack growth conditions at high temperature (here we consider the cold-dwell damage is nil, i.e. $l_z = 0$), and the crack-tip stress is relaxed by creep, according to the Norton law, $\dot{\varepsilon}_v = A\sigma^n$, such at

$$\dot{\sigma} = -E\dot{\varepsilon}_v = -EA\sigma^n \tag{10.76}$$

Then, stress relaxation is solved from Eq. (10.76) as

$$\frac{1}{\sigma^{n-1}} - \frac{1}{\sigma_0{}^{n-1}} = (n-1)EAt_h \tag{10.77}$$

and

$$\Delta\varepsilon_v = \frac{\sigma_0 - \sigma}{E} \tag{10.78}$$

where σ_0 is the yield stress and t_h is the dwell time.

Then, the creep damage size can be estimated, according to Eq. (3.7), as

$$l_c = d\Delta\varepsilon_v \tag{10.79}$$

When the dwell period is short, it can be shown that

$$l_c = A\sigma_0^n t_h d \tag{10.80}$$

Here, Eq. (10.75-80) is a general representation of the dwell fatigue crack growth. Detailed analysis should be given to phenomena that depend on the dominant stress-relaxation (creep) mechanism in particular engineering alloys, as discussed below.

Telesman et al. (2008) observed strong microstructural dependence of hold-time fatigue crack growth rate in the LSHR P/M disk superalloy. They used heat treatments to maintain a constant grain size and used different cooling rate and subsequent aging treatments to control the γ' size, thus obtained different versions of the material with different primary, secondary and tertiary γ' size and distribution. They found that the dwell-fatigue crack growth rates could be attributed to the stress relaxation potential of each microstructure, which is controlled by the mean tertiary γ' size. Their experimental results showed that slow cooling rate and prolonged aging treatment produced larger γ' size in the Ni-base superalloy, which resulted in lower dwell-fatigue crack growth rates. From Eq. (10.80), one can infer the same conclusion, because the dwell damage l_c is very sensitive to the material's yield strength, which is controlled by the γ'size and distribution. They compared the dwell-fatigue crack growth resistance of different γ' microstructures with rather uniform grain size varying from ASTM 6.6 to 7.1.

Tsang et al. (2011) studied the dwell-fatigue crack growth behaviour of a new superalloy, ATI 718Plus alloy (718Plus), with different microstructures obtained through different heat treatments: (1) standard heat treatment (HT1); (2) standard heat treatment plus thermal exposure at 732 °C for 1000 h (HT2); (3) fine-grained condition with modified δ phase (HT3); and (4) fine-grain overaged through a modified heat treatment (HT4). The microstructural characteristics are summarized in Table 10.1, in comparison with Waspaloy in standard heat treated condition (W1) and W1 plus additional thermal exposure (W2). The fatigue crack growth rates with and without 100 sec. dwell at 649 °C and 704 °C are shown in Figure 10.19.

The results show that the steady state fatigue crack growth rates of each microstructural variation in both alloys are identical. The 100 sec. dwell period induced increased crack growth rates, but the extent varied with microstructure. For ATI 718Plus, the best dwell fatigue crack growth resistance lies in HT2 with the largest grain size and secondary γ' size (there is no tertiary γ' in this material). Again, from Eq. (10.80) and Hall-Petch relationship, it can be inferred that the HT2 microstructure has a greater stress relaxation potential than the other microstructures. The trend for Waspaloy, even though with limited data, seems to follow Telesman et al.'s conclusion, as controlled by the tertiary γ' size. The dwell-fatigue crack growth was observed to proceed in a mixed transgranular and intergranular mode, shifting toward predominantly intergranular mode at higher temperature (704 °C).

Table 10.1. Microstructural characteristics of ATI 718Plus and Waspaloy (Tsang et al. 2011)

Alloy	ID	Grain size (μm)			δ fraction		γ′ size (nm)	
		mean	std. dev.	ASTM#	(wt.%)	std. dev.	secondary	tertiary
718Plus	HT1	24	±2.1	7.5	6.5	±1.0	27.6	n/a
	HT2	61	±2.4	4.8	5.9	±0.9	59.6	n/a
	HT3	58	±3.5	5	7.6	±2.9	31.4	n/a
	HT4	31	±2.1	6.8	11.9	±2.7	38.4	n/a
Waspaloy	W1	20	±1.9	8	n/a	n/a	144.4	40.5
	W2	24	±1.9	7.5	n/a	n/a	135.4	44.3

Creep-fatigue crack growth in a P/M Ni-base superalloy—Astroloy with straight and serrated grain boundaries was studied by Danflou et al. (1992). The conventional microstructure (straight g.b.) was obtained using a fast cooling rate of 100 °C/min., and the serrated g.b. material was obtained using a slow cooling rate ~1 °C/min between 1100 °C and 1050 °C, after homonization above the γ′ solvus at 1150 °C. Then a double aging treatment (700 °C/24h + 800 °C/4h) was given to all the specimens. The grain size of the two materials is in the range of 60-90 μm, and the morphology of the serrated g. b. is close to sinusoidal wave with average amplitude of 2 μm and wave length of 5 μm (Loyer Danflou et al. 1992). Their experimental crack growth rate data generated using a 10 (ramping) - 300 (hold) - 10 (unloading) sec. cycle profile are plotted in Figure 10.20. The creep fatigue fracture mode is mostly intergranular, the fatigue crack growth rate with straight g. b. is represented by

$$\frac{da}{dN} = 6.0 \times 10^{-8} \Delta K^3 \tag{10.81}$$

and the fatigue crack growth rate with serrated g. b. is represented by

$$\frac{da}{dN} = 1.7 \times 10^{-8} \Delta K^3 \tag{10.82}$$

The observed creep-fatigue crack growth behaviour can be explained by the ICFT crack growth model, Eq. (10.75). First of all, the fatigue crack growth rate exponent is 3, as predicted by the RSR model. Second, when stress relaxation mainly occur by GBS, according to Eq. (2.27) and (10.80), the difference between the fatigue crack growth rates with internally distributed damage in the two microstructures is the g. b. serration factor. Then, according to Eq. (2.23), ϕ is calculated to be 0.245, whereas the experimental best-fit coefficients in Eq. (10.81) and (10.82) have a ratio of 0.283 between the straight and serrated g. b.

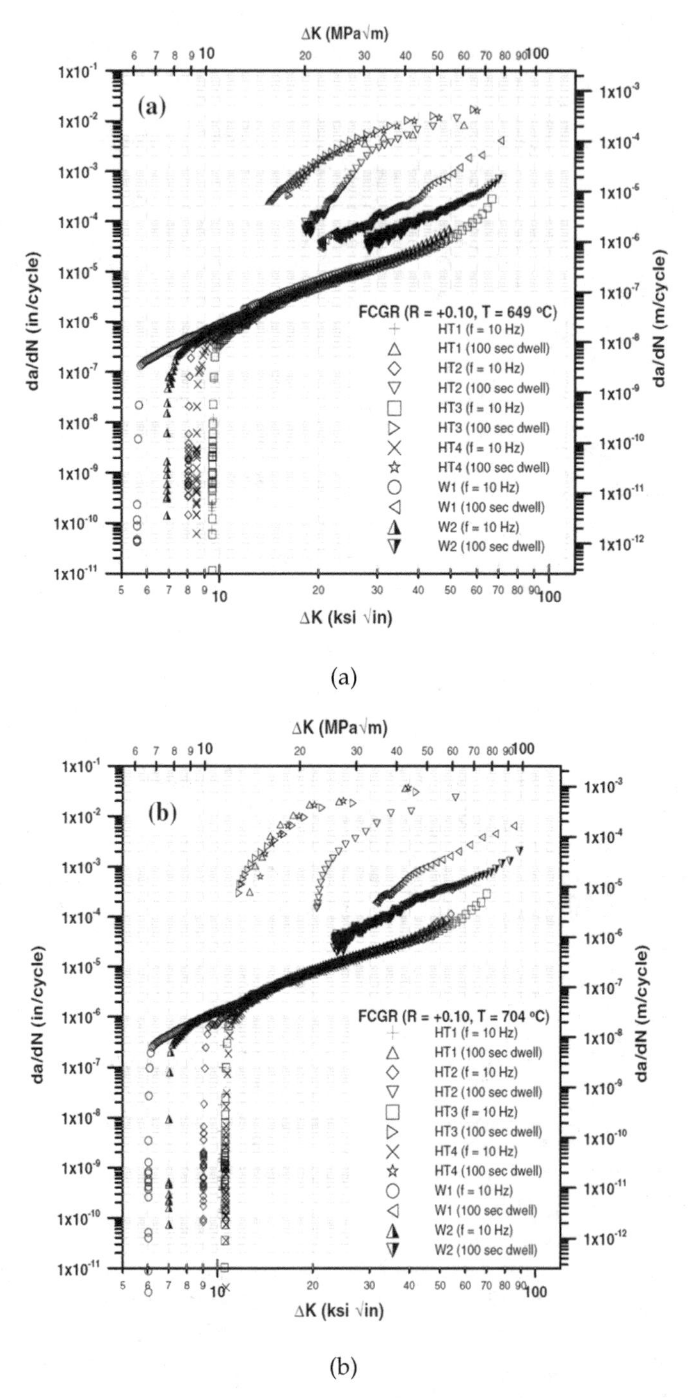

(a)

(b)

Figure 10.19. The fatigue crack growth rates in ATI 718Plus and Waspaloy with and without 100 sec. dwell at (a) 649 °C and (b) 704 °C, after Tsang et al. (2011).

In conclusion, the ICFT crack growth model offers an in-depth explanation of creep-fatigue interactions in crack growth. This interaction depends on intragranular strengthening mechanisms as well as grain boundary sliding. Intragranular strengthening mechanisms would set out the initial condition of crack-tip stress relaxation with higher σ_0 in highly strengthened materials, and therefore, more sensitive to dwell in creep-fatigue crack growth processes. Particular, in the case of intergranular creep-fatigue fracture, where the crack-tip stress relaxation is controlled by GBS, the dwell-fatigue crack growth rate also depends strongly on grain size and grain boundary morphology, including grain boundary serration and g. b. precipitate distribution. No other existing models could offer such explanations. Therefore, more studies of creep-fatigue crack growth behaviour should be conducted with ICFT.

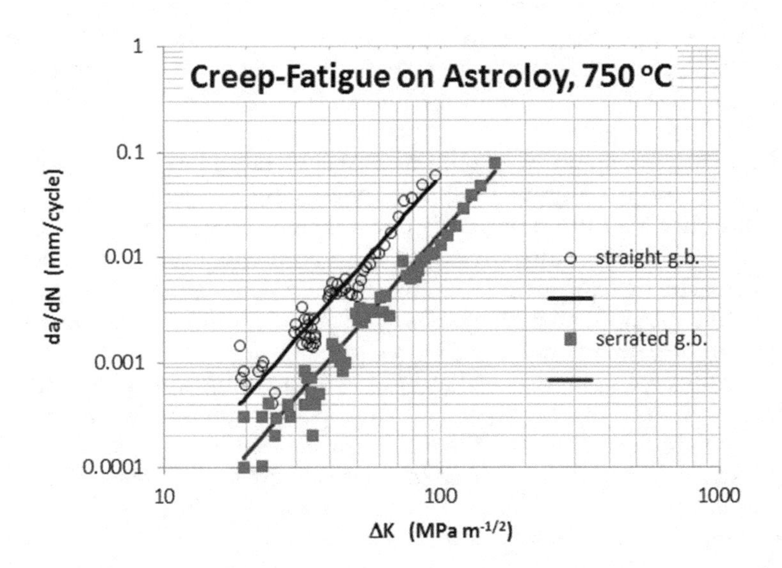

Figure 10.20. Creep-fatigue crack growth in Astroloy at 750 °C under 10 - 300 -10 s cycle profile, data after Danflou et al. (1992).

CHAPTER

11

Single Crystal Ni-Base Superalloys

Single crystal Ni-base superalloys are characterized with a γ-γ' microstructure, typically as shown in Figure 1.6. The γ' is an intermetallic Ni_3 (Al, Ti) phase that has an ordered $L1_2$ (f.c.c.) crystal structure coherent with the γ matrix. The volume fraction of γ' can be as high as 70%. The first generation single crystal Ni-base superalloys were introduced in the 1970s. The removal of grain boundaries apparently eliminates grain boundary sliding. Also, it removes the need of grain boundary enhancing elements thus allowing more latitude for addition of matrix enhancing elements. For every increase of 25 °C in the temperature capability of creep resistance, a new generation is marked. The development of single crystal Ni-base superalloys has gone through five generations with increased addition of refractory elements such as rhenium (Re) and ruthenium (Ru). Addition of Re really helps the alloy's creep resistance, but above certain limit, Ru has to be added to reduce the propensity of forming topologically close-packed phases (σ phase and μ phase) that are detrimental to the alloy's properties (Reed 2006). Since both Re and Ru are heavy and expensive elements, there seems to be a limit to the alloying development, because of the material's solubility, apart from cost and weight considerations.

Owing to its high temperature capability, single crystal Ni-base superalloys are used as turbine blade material in advanced gas turbine engines. However, for accurate blade design analysis and life prediction, single crystal anisotropy must be considered. Single crystal turbine blades are mainly cast in the [001] direction (parallel with the centrifugal force), which has the maximum compliance for low-cycle fatigue and creep resistance. But, a gas turbine blade generally has complicated geometrical features such as serrated roots, internal cooling channels and cooling holes, which induce a complex 3D stress field in the blade. This chapter is dedicated to deal with 3D anisotropic deformation and crack growth in single crystal Ni-base superalloys.

11.1 Anisotropic Deformation Behaviour

The anisotropic elasticity and elastic-perfectly plastic behaviour of single crystal Ni-base superalloys have been discussed in Chapter 1 and 9. The classical theory of anisotropic plasticity has also long been developed (Hill 1998). In the unified constitutive theory, time-dependent inelastic deformation is addressed mainly using two approaches: (1) the continuum approach and (2) the crystallographic slip approach. The continuum approach basically follows the Hill theory but with incorporation of internal state variables in a unified viscoplastic formulation (Nouailhas and Chaboche 1991). The crystallographic slip approach employs slip models that are based on octahedral slip $\{111\} < \bar{1}10 >$ and cube slip {100}<011> systems, as proposed by Ghosh et al. (1990), Li et al. (1998), Han et al. (2001), Meric and Cailletaud (1991). However, there have been debates on what slip systems are activated to provide the necessary deformation. For example, MacLachlan et al. (2001) proposed a continuum-damage-mechanics-based crystallographic slip model based on the $\{111\} < \bar{1}10 >$ and $\{111\} < \bar{1}\bar{1}2 >$ slip systems, excluding cubic slip. The crystallographic slip models certainly provide a deeper insight on the anisotropic behaviour of single crystals, but they are more complicated than the Hill formulation. From the mechanism delineation point of view, it is more convenient to formulate the rate-independent plasticity and time-dependent creep of single crystal materials in the Hill's fashion than on slip-system basis. Mathematically, the Hill formulation only employs the necessary Hill's constants for the crystalline anisotropy and ensures the tensorial properties of the resulting stress and strain. The crystallographic slip models, on the other hand, need to employ a set of slip property constants for each slip system possibly to be activated, which are mathematically redundant. The Hill's formulation is already implemented in most commercial FEM for macroscopic component analysis.

As discussed in Chapter 4, the ICFT emphasizes deformation mechanism delineation, in order to describe the damage accumulation in relation to specific controlling mechanisms, for holistic life prediction. Here, we adopt the Hill theory for general formulation of anisotropic plasticity and creep, without losing the sight on possible slip systems involved.

The generalized Hill theory for anisotropic plasticity/creep is formulated as (Dassault Systèmes 2008):

$$\begin{Bmatrix} \dot{\varepsilon}_{11} \\ \dot{\varepsilon}_{22} \\ \dot{\varepsilon}_{33} \\ \dot{\varepsilon}_{23} \\ \dot{\varepsilon}_{13} \\ \dot{\varepsilon}_{12} \end{Bmatrix} = \frac{\dot{p}}{\bar{\sigma}} \begin{bmatrix} G+H & -H & -G & & & \\ -H & F+H & -F & & 0 & \\ -G & -F & F+G & & & \\ & & & 2L & & \\ & 0 & & & 2M & \\ & & & & & 2N \end{bmatrix} \begin{Bmatrix} \sigma_{11} \\ \sigma_{22} \\ \sigma_{33} \\ \sigma_{23} \\ \sigma_{13} \\ \sigma_{12} \end{Bmatrix} \tag{11.1}$$

where

$$\bar{\sigma} = \sqrt{F(\sigma_{22}-\sigma_{33})^2 + G(\sigma_{11}-\sigma_{33})^2 + H(\sigma_{22}-\sigma_{11})^2 + 2L\sigma_{23}{}^2 + 2M\sigma_{13}{}^2 + 2N\sigma_{12}{}^2}$$

and $\dot{p} = f(\sigma_s)$ is the strain rate multiplier, σ_s is the von Mises stress. For rate-independent plasticity, the strain rate quantities are replaced by the corresponding strain increments, to conform with the incremental plasticity formulation.

Naturally, Eq. (11.1) results in no volume expansion, i.e. $\dot{\varepsilon}_{ii} = 0$.

Considering the crystallographic symmetry of f.c.c. structure, we can assume that $F = G = H$, and $L = M = N$, so that the anisotropic material matrix reduces with two independent variables, which means that plasticity/creep experiments only need to be carried out along two directions for calibration of the constitutive law.

In the following, we shall proceed with an example for anisotropic creep analysis.

In analyzing single crystal deformation, a relationship between the loading coordinates **x**′ and the crystal axes **x** has to be established first. The coordinate transformation relations are as given in Table 11.1.

Table 11.1. Directional relations between coordinate x and x'

Coordinates	x	y	z
x'	α_1	β_1	γ_1
y'	α_2	β_2	γ_2
z'	α_3	β_3	γ_3

Following the tensor transformation rule as discussed in Chapter 1, we have:

$$\dot{\varepsilon}' = \frac{\dot{p}}{\bar{\sigma}} Q_\varepsilon^T C Q_\sigma \sigma' \tag{11.2}$$

where C is the compliance matrix as given in Eq. (12.1), and

$$Q_\varepsilon = \begin{bmatrix} \alpha_1^2 & \alpha_2^2 & \alpha_3^2 & \alpha_2\alpha_3 & \alpha_1\alpha_3 & \alpha_2\alpha_1 \\ \beta_1^2 & \beta_2^2 & \beta_3^2 & \beta_2\beta_3 & \beta_1\beta_3 & \beta_2\beta_1 \\ \gamma_1^2 & \gamma_2^2 & \gamma_3^2 & \gamma_2\gamma_3 & \gamma_1\gamma_3 & \gamma_2\gamma_1 \\ 2\beta_1\gamma_1 & 2\beta_2\gamma_2 & 2\beta_3\gamma_3 & \beta_2\gamma_3+\gamma_2\beta_3 & \beta_1\gamma_3+\gamma_1\beta_3 & \beta_2\gamma_1+\gamma_2\beta_1 \\ 2\alpha_1\gamma_1 & 2\alpha_2\gamma_2 & 2\alpha_3\gamma_3 & \alpha_2\gamma_3+\gamma_2\alpha_3 & \alpha_1\gamma_3+\gamma_1\alpha_3 & \alpha_2\gamma_1+\gamma_2\alpha_1 \\ 2\alpha_1\beta_1 & 2\alpha_2\beta_2 & 2\alpha_3\beta_3 & \beta_2\alpha_3+\alpha_2\beta_3 & \beta_1\alpha_3+\alpha_1\beta_3 & \beta_2\alpha_1+\alpha_2\beta_1 \end{bmatrix} \tag{11.3}$$

$$Q_\sigma = \begin{bmatrix} \alpha_1^2 & \alpha_2^2 & \alpha_3^2 & 2\alpha_2\alpha_3 & 2\alpha_1\alpha_3 & 2\alpha_2\alpha_1 \\ \beta_1^2 & \beta_2^2 & \beta_3^2 & 2\beta_2\beta_3 & 2\beta_1\beta_3 & 2\beta_2\beta_1 \\ \gamma_1^2 & \gamma_2^2 & \gamma_3^2 & 2\gamma_2\gamma_3 & 2\gamma_1\gamma_3 & 2\gamma_2\gamma_1 \\ \beta_1\gamma_1 & \beta_2\gamma_2 & \beta_3\gamma_3 & \beta_2\gamma_3+\gamma_2\beta_3 & \beta_1\gamma_3+\gamma_1\beta_3 & \beta_2\gamma_1+\gamma_2\beta_1 \\ \alpha_1\gamma_1 & \alpha_2\gamma_2 & \alpha_3\gamma_3 & \alpha_2\gamma_3+\gamma_2\alpha_3 & \alpha_1\gamma_3+\gamma_1\alpha_3 & \alpha_2\gamma_1+\gamma_2\alpha_1 \\ \alpha_1\beta_1 & \alpha_2\beta_2 & \alpha_3\beta_3 & \beta_2\alpha_3+\alpha_2\beta_3 & \beta_1\alpha_3+\alpha_1\beta_3 & \beta_2\alpha_1+\alpha_2\beta_1 \end{bmatrix} \tag{11.4}$$

For Ni-base single crystal superalloys with f.c.c structure, because of crystallographic symmetry, $F = G = H$, and $L = M = N$. Therefore, there are only two independent rate variables controlling the anisotropic creep behaviour of Ni-base single crystal superalloys: one can be determined from creep tests along the <100> direction and the other from the <111> direction, because it happens that:

$$\dot{\varepsilon}_{\langle 100\rangle} = \sqrt{2G}\dot{p} \tag{11.5}$$

and

$$\dot{\varepsilon}_{\langle 111\rangle} = \sqrt{\frac{2L}{3}}\dot{p} \tag{11.6}$$

The creep rates in these two orientations can be delineated using the mechanism-based approach, as described in Chapters 4 and 5. Taking a 2nd generation Ni-base single crystal superalloy, CMSX-4, for example, based on the creep test data obtained by MacLachlan et al. (2001), the creep-mechanism strain functions for the <100> and <111> directions at 950 °C can be expressed as (note that GBS is eliminated in Eq. (5.8))

$$\varepsilon = \frac{1}{M^*}\left[\exp\left(M^*\dot{\varepsilon}_i t\right) - 1\right] \tag{11.7}$$

where is the minimum creep rate in each direction, given by

$$\dot{\varepsilon}_i = A_i \exp\left(\frac{V\sigma_s}{kT}\right) \tag{11.8}$$

The creep model parameters for <100> and <111> orientations are given in Table 11.2.

Table 11.2. Creep model parameters for CMSX-4 at 950 °C

Orientation	A	V/kT	M*
<100>	1.18E-7	0.018	550
<111>	9.71E-9	0.022	900

Then, the uniaxial creep strain can be written as:

$$\varepsilon = p\sqrt{G[(\beta_1^2 - \gamma_1^2)^2 + (\alpha_1^2 - \gamma_1^2)^2 + (\beta_1^2 - \alpha_1^2)^2] + 2L[(\beta_1\gamma_1)^2 + (\alpha_1\gamma_1)^2 + (\beta_1\alpha_1)^2]} \tag{11.9}$$

The parameter Gp^2 and Lp^2 can be solved from Eq. (11.5) and (11.6), respectively. The model description of CMSX-4 creep behaviours at 950 °C are shown in Figure 11.1 with (a) <100> and (b) <111> as the calibration, and (c) <110> and d) <123> as predictions, in comparison with the experimental data. MacLachlan et al. used the crystallographic damage-mechanics model to describe the same experimental behaviours, but their model deviation was large for the predicted curves other than the calibrated <100> and <111> orientations. For example, their model description for the <123> orientation is shown in Figure 11.1d, which results in three times over-estimation of the creep life. This does not necessarily deny crystallographic slip as the controlling deformation mechanism, but may be merely a model fitting issue with redundant constants. The comparative analyses simply show that Hill's theory, satisfying all the macroscopic tensorial relationships, is more effective dealing with anisotropic creep behaviour. The advantage of this formulation is that it is compatible with the Hill's theory for anisotropic plasticity and can be easily implemented into commercial FEM codes for component analysis.

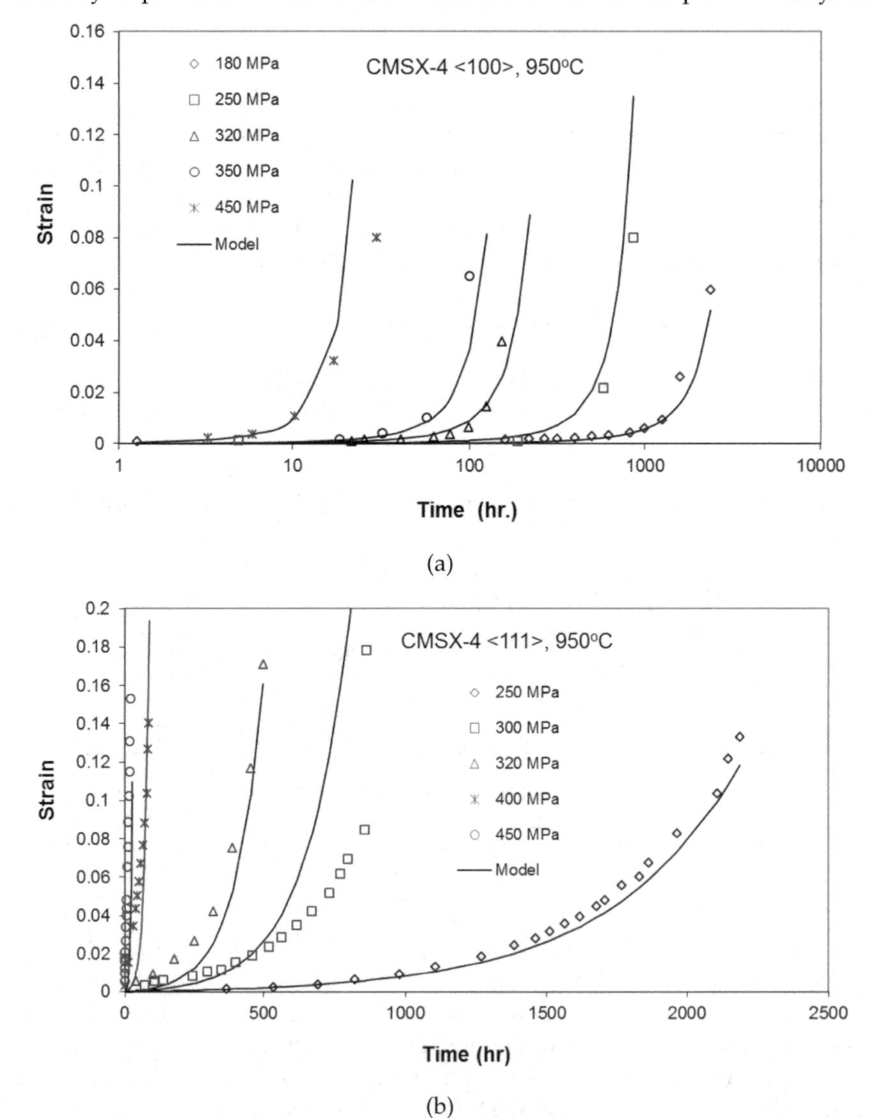

(a)

(b)

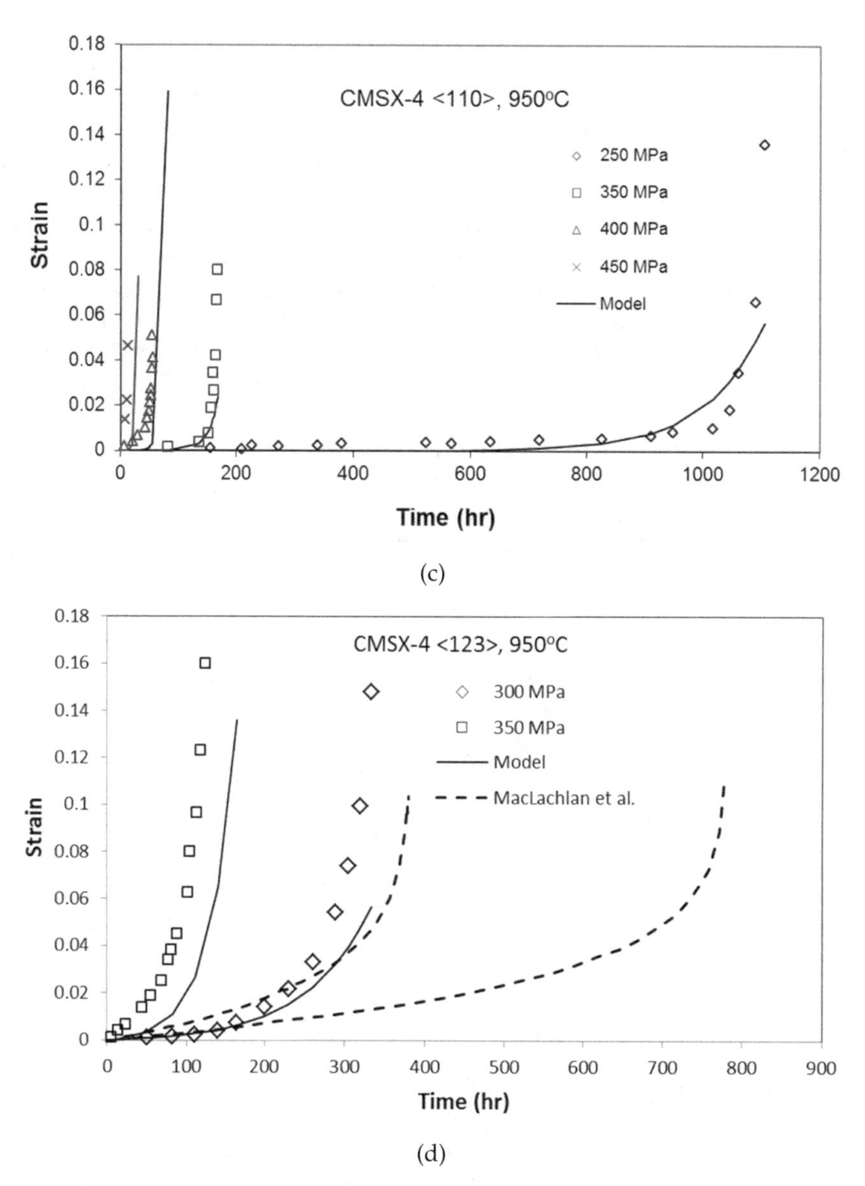

Figure 11.1. CMSX-4 creep behaviours at 950 °C in (a) <100>, (b) <111>, (c) <110>, and (d) <123> orientations.

11.2 Anisotropic Fatigue

As discussed in Chapters 6 and 9, fatigue involves dislocation pile-up in a crystalline material, in this case, single crystal Ni-base superalloy. Therefore, characterization of single crystal fatigue must be related to specific slip systems. Arakere and Orozco (2001) analyzed low-cycle fatigue of single crystal Ni-base turbine blade superalloy PWA 1493 tested at 650 °C. They applied various multi-axial fatigue failure theories on the critical slip systems in this f.c.c. single crystal material, including i) the *critical plane theory* where shear and normal strain on a critical plane are combined to define the failure equation (Brown and Miller 1973, Kandil et al. 1982), ii) the *modified critical plane theory* where mean stress is also added to the critical plane failure equation (Socie et al. 1985), iii) the *alternate shear* model (Fatemi and Socie 1988), and iv) Smith-Watson-Topper model (1970), because the critical slip systems in an f.c.c. material do operate under multi-axial stresses even when the material is under uniaxial loading. However, none of the above theories were satisfactory in correlating the experimental data; neither did the uniaxial Coffin-Manson strain equation. Instead, the maximum shear stress amplitude on the critical slip system appeared to provide a better correlation.

To provide the rationale for the above correlation, here we use the dislocation pile-up model developed in Chapter 9, i.e., Eq. (9.60), to re-analyze the experimental data from Arakere and Orozco for the <001>, <111> and <011> orientations. The dislocation pile-ups are assumed to reside on the {111}<110> octahedral slip system in the <001> and <011> orientations, and on the {100}<110> cube slip system in the <111> orientations, to be consistent with Arakere and Orozco's analysis. The $\mathbf{F}^{-1}$ matrices for the octahedral and cube slip systems are obtained, respectively, based on the elastic properties of nickel, using the Stroh formalism, as

$$F^{-1}_{\{111\}<110>} = \begin{bmatrix} 8.89 & 0 & -2.67 \\ 0 & 8.03 & 0 \\ -2.67 & 0 & 13.97 \end{bmatrix} 10^{-12}\,(\text{Pa})^{-1} \tag{11.10}$$

and

$$F^{-1}_{\{100\}<110>} = \begin{bmatrix} 5.45 & 0 & 0 \\ 0 & 5.45 & 0 \\ 0 & 0 & 8.44 \end{bmatrix} 10^{-12}\,(\text{Pa})^{-1} \tag{11.11}$$

Suppose that the dislocation dipole pile-ups are formed only by edge-glide dislocations on these slip systems. Using the surface energy value of Ni, $w_s = 2.38\ \text{J/m}^2$, and assume that the machining surface roughness introduces a factor of $R_s = 1/3$, from Eq. (9.60), we obtain:

$$\Delta\tau = \tau^F + \frac{1}{2}\sqrt{\frac{2w_s R_s}{F_{11}^{-1} N b}} \tag{11.12}$$

Note here the stress amplitude $\Delta\tau$ is half of the fully reversed stress range in Eq. (11.12). Then the slip system shear stresses are calculated using Eq. (11.12), based on the observed life given in Table 11.3, assuming the lattice resistance $\tau^F = 345$ MPa and for the <001> orientation, and $\tau^F = 550$ MPa for <111> and <011> orientations. The difference in the lattice resistances is perhaps due to orientation dependence of γ-γ'interfacial properties that affect the dislocation pile-up as opposed to that in a single-phase material. The model calculated results are shown Figure 11.2 in comparison with Arakere and Orozco's data.

It is seen that the model description agrees well with the experimental observation. In particular, there is a strong dependence of fatigue life on the material's anisotropic elastic matrix $\mathbf{F}^{-1}$. Such effect would not be accounted for by an "isotropic" model. For further refinement of the model, the material's surface properties can be calculated from the first principles of physics via *ab initio* density functional theory. Atomistic calculation of Peierls-Nabarro stress for single phase materials can also be attempted. These computational methods can advance for evaluation of complex engineering alloys. Then, we may proceed to purely computational fatigue design using the anisotropic fatigue crack nucleation model developed in this book.

Table 11.3. LCF of PWA 1493 (Arakere and Orozco 2001) and prediction of Eq. (11.12)

Orientation	ε_{max}	ε_{min}	$\Delta\gamma$	$\Delta\tau$ (MPa)	$\Delta\tau$ (predicted)	Life
<001>oct	0.01509	0.00014	0.0198	745.2	713.3531	1326
<001>oct	0.0174	0.0027	0.0194	731.4	681.0688	1593
<001>oct	0.0112	0.0002	0.014728	550.62	546.8924	4414
<001>oct	0.01202	0.00008	0.016	602.37	523.086	5673
<001>oct	0.00891	0.00018	0.012	446.43	423.0742	29516
<111> cube	0.01219	–0.006	0.02106	2311.5	3909.712	26
<111> cube	0.0096	0.0015	0.00924	1028.1	1140.031	843
<111> cube	0.00809	0.00008	0.009406	1021.2	1087.455	1016
<111> cube	0.006	0	0.0076	759	843.3672	3410
<111> cube	0.00291	0.00284	0.0067	738.3	753.2961	7101
<111> cube	0.00591	0.00015	0.006724	731.4	749.7413	7356
<111> cube	0.01205	0.00625	0.007	793.5	742.6927	7904
<011>oct	0.0092	0.0004	0.01435	814.2	809.4884	2672
<011>oct	0.00896	0.00013	0.015	848.7	704.5543	7532
<011>oct	0.00695	0.00019	0.01069	624.45	627.1594	30220

11.3 Anisotropic Fatigue Crack Growth

Single crystal Ni-base superalloys are the materials of choice for gas turbine blades because of their superior creep resistance at high temperature. Fatigue crack growth resistance and its orientation dependence is also an important consideration. The primary orientation of single crystal blade is cast in <001> direction, in parallel to the centrifugal force direction, but its

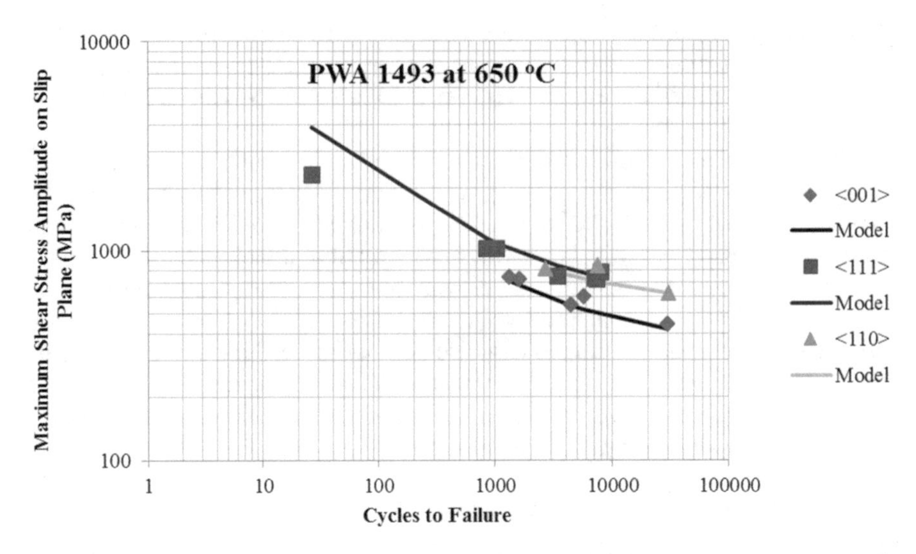

Figure 11.2. LCF life of PWA 1493 as function of the critical shear stress range at 650 °C.

secondary orientation (i.e. the orientation of the blade root serration notch) is not controlled. Therefore, to ensure blade structural integrity, fatigue crack growth resistance in all directions need to be known. Fatigue crack growth rates in various single crystal Ni-base superalloys have been studied (Chan et al. 1986, Antolovich et al. 1997, Henderson and Martin 1992, Reed et al. 2000, Wu et al. 2001, Joyce et al. 2003). Here, CMSX-4 is chosen as an example, particularly with both <100> and <110> crack growth directions in consideration: orientation A (OA) with the crack growth direction in <110> direction, which is represented by $<\bar{1}\bar{1}0>/(001)$; and orientation B (OB) with crack growth direction in the <100> direction, which is represented by $<0\bar{1}0>/(001)$, as shown in Figure 11.3.

The elastic constants of CMSX-4 at 650 °C are given by Sieborger et al. (2001), with respect to its cubic axes, as:

$$C_{1111} = 221 \text{ GPa}, C_{1122} = 145 \text{ GPa}, C_{1212} = 110 \text{ GPa}$$

Then, the $\mathbf{F}^{-1}$ matrix for a crack in [100] direction is evaluated to be as follows:

$$F_{[100]}^{-1} = \begin{bmatrix} 0.0117 & 0 & 0 \\ 0 & 0.0117 & 0 \\ 0 & 0 & 0.0091 \end{bmatrix} (\text{GPa}^{-1}) \tag{11.13}$$

The $\mathbf{F}^{-1}$ matrix for a crack in [110] direction is:

$$F_{[110]}^{-1} = \begin{bmatrix} 0.00844 & 0 & 0 \\ 0 & 0.00972 & 0 \\ 0 & 0 & 0.01547 \end{bmatrix} (\text{GPa}^{-1}) \tag{11.14}$$

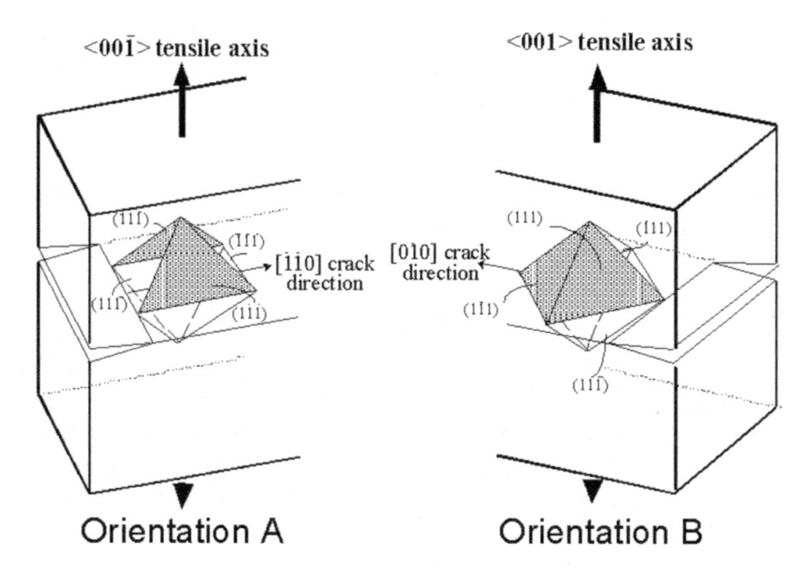

Figure 11.3. Crack-crystal orientations (after Reed et al. 2000).

Here, in the crack/loading coordinate system, the index 1 indicates the crack direction, and the index 2 indicates the mode-I loading direction. The two matrices are all diagonal matrices, which means that there is no mixed-mode elastic coupling in these two orientations. Under uniaxial loading in the <001> direction, according to Eq. (9.42), $c_3 = c_1 = a$, $c_2 = a \sec[\pi\sigma/(2\sigma^F)]$ ($t_2 = \sigma$, $t_2^F = \sigma^F$). Therefore, for Mode-I crack under small-scale yielding condition, according to Eq. (9.45), the CTOD takes the form:

$$\Delta u_2 = \frac{K^2}{2F_{22}\sigma^F} \tag{11.15}$$

where K is the stress intensity factor. Eq. (11.15) is very similar to the CTOD expression for an isotropic material, except that the elastic constant F_{22} and the yield strength σ^F are orientation dependent.

In vacuum (without environmental effects), fatigue crack growth proceeds primarily by octahedral slip and slip reversal, as discussed in Chapter 10. As shown in Figure 11.3, OA has four slip systems that can operate to advance the crack, whereas OB has only two. For crack opening, suppose that each slip system contributes equally, all having the same Schmid factor, the following relation should hold true:

$$m \cdot \frac{l}{\sqrt{2}} = \delta \tag{11.16}$$

where m is the number of contributing slip systems, l is the slip distance ($\delta = \Delta u_2$) is the crack opening displacement given by Eq. (11.15). By the RSR model, FCGR is directly related to the slip distance l, and hence it can be used to correlate FCGR in vacuum.

Fatigue crack growth rates (FCGR) in CMSX-4 tested in vacuum and air at 650 °C are shown in Figure 11.4. The experimental data are taken from Joyce et al. (2003). A marked orientation dependence is observed in vacuum, but the orientation dependence tends to diminish at higher ΔK in air. Using Eq. (11.16) as the correlation parameter, the FCGR in *OA* and *OB* are represented by the lines in comparison with the experimental data. In this correlation, the yield strength is assumed to be $\sigma_F = 900$ MPa.

$$\frac{da}{dN} = 5.66\left(\frac{\Delta K^2}{\sqrt{2}mF_{22}\sigma_F}\right)^{1.5} \tag{11.17}$$

Joyce et al. 2003) examined the fracture surfaces, where a transition from fine-scale cooperative slip along alternating {111} slip planes at low-to-mid ΔK levels to far coarser extended slip band cracking along these planes above a critical ΔK level was observed. Orientation A generally has slower FCGRs overall than *B*. The model shows that the difference can be mainly attributed to the orientation dependence in the **F** matrix, as reflected by Eq. (11.17). Interestingly, overall, the FCGR depends on the stress intensity factor range to a power of 3, as the RSR model indicates.

With regards to crack growth in air at 650 °C, oxidation has strong effects on FCGR, depending on the ΔK level. At low ΔK, FCGR in *OB*-air are

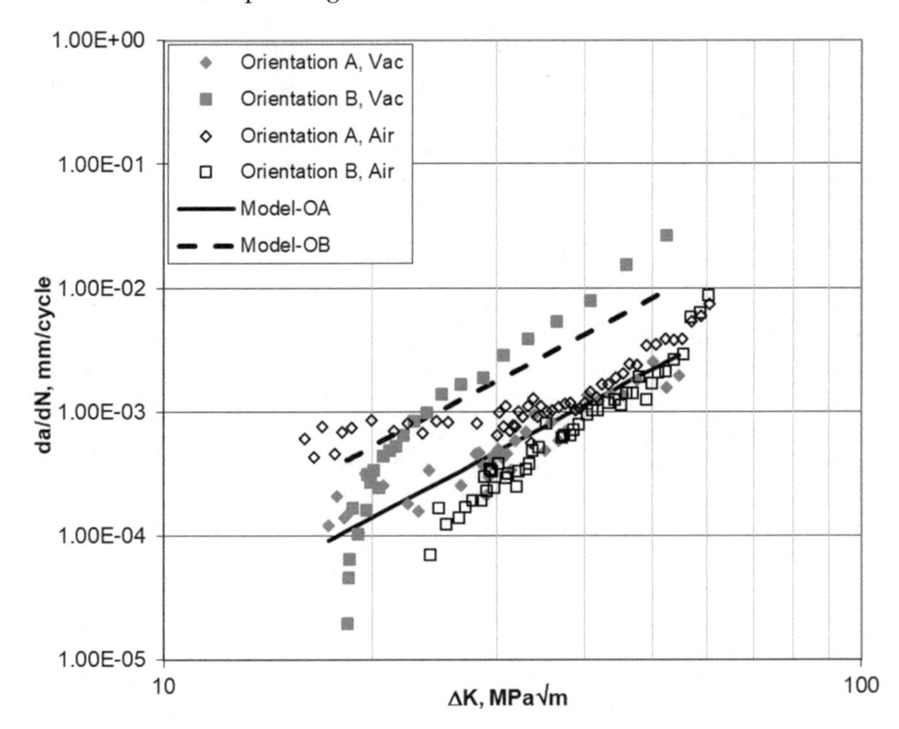

Figure 11.4. (a) FCGR as function of ΔK, and (b) as function of CTOD in orientation *A* and *B* of single crystal Udimet 720 Ni-base superalloy.

higher than that in vacuum, but gradually reduced to no marked orientation dependence at $\Delta K > 30$ MPa$\sqrt{m}$. This could be attributed to effects of oxidation. On the other hand, oxidation does not seem to have a significant effect on FCGR in OA. Effectively, fatigue crack growth in air appears to be isotropic, perhaps because of slip homogenization by oxidation.

11.4 Notch Fatigue

A study on notch fatigue behaviour of CMSX-4 was carried out by Miller et al. (2007). Fatigue testing was conducted on single notch specimen (50×8×8 mm) under a load of 6.2 kN in three point bending with a load ratio of 0.1 at 650 °C. A trapezoidal 1-1-1-1 waveform was employed, comprising a 1 second dwell at maximum and minimum loads, with ramp times of 1 second, hence giving an overall frequency of 0.25 Hz. Due to the high stress concentration at the notch, crack initiation almost all occurred at the internal cast pores. Figure 11.5 shows a crack initiation site in CMSX-4 tested at 650 °C. The following assumptions are made for the fatigue failure process.

1. Elliptical cracks initiate instantaneously at the inter-dendrite pores upon loading. Both the major and minor axes of these cracks are in [100] directions, due to the nature of inter-dendrite pore formation.
2. The growth of internal (subsurface) cracks is limited by the crack growth rate in the [100] direction (along the major/minor axes) in a vacuum-like environment, such that the expansion of the crack remains elliptical. The growth of subsurface cracks is considered to be under a constant stress field, corresponding to the stress at the pore location, just below the notch root.
3. Once the crack breaks through to the specimen surface, it is assumed to take a circular shape with a radius equal to the diameter of the internal crack, and the subsequently the crack grows in the gradient stress field of the notch exposed to air.

Based on the above assumptions 1-3, the total crack growth life of notch fatigue process can be calculated using Eq. (11.17). Table 1 lists the size and depth of the crack-initiating pores and gives the results of the fracture mechanics calculation in comparison with the experimental life. Specimen failure is considered to occur whenever the maximum stress intensity factor reached the critical value of 60 MPa$\sqrt{m}$. In most of the cases, crack growth starts at an internal pore, and growth of such subsurface cracks takes more than half of the total life, as shown by the number in the bracket. The overall agreement of the predicted life with the experimental observation is well within the scatter band of a factor of 2. Most importantly, the above approach provides a good explanation of the notch fatigue data scatter on CMSX-4 under seemingly identical mechanical loading condition. For example, the *OA* and *OB* specimens having a similar initial crack-initiating pore but located in different depth had fatigue lives by 4 times of difference (25,500 cycles vs.

6,500 cycles), just because the subsurface crack in the *OB* specimen was closer to the surface and internal crack growth in *OB* was much faster than that in the *OA* specimen. In the same orientation *OA*, a deeper embedded pore (specimen 4) caused a significant life reduction than a surface pore of similar size (specimen 3). This was also because the internal crack grew faster as controlled by [100] FCGR in vacuum (assumption 2) and, when it broke into the surface, it formed a larger surface crack, while the smaller pore (crack) at the surface grew much slower in air.

Through the above analysis, the large scatter of notch fatigue is rationalized using the fracture mechanics description for single crystal Ni-base superalloys. This may help better control of the structural integrity of turbine blades with statistical information of pore distribution and applicable non-destructive inspection methods.

Table 11.4. Notch fatigue life of CMSX-4

Orientation	Major axis (μm)	Minor axis (μm)	Depth (μm)	Test life (cycle)	Predicted life (cycle)
OB	90	10	130	6,500	11,100 (4,600*)
OA	100	25	250	25,500	49,000 (30,000)
OA	40	15	0	21,661	32,000
OA	50	30	200	5,270	6,500 (4,000)
OB	50	5	350	13,717	14,900 (12,800)

*The number in the bracket indicates the life of internal crack growth.

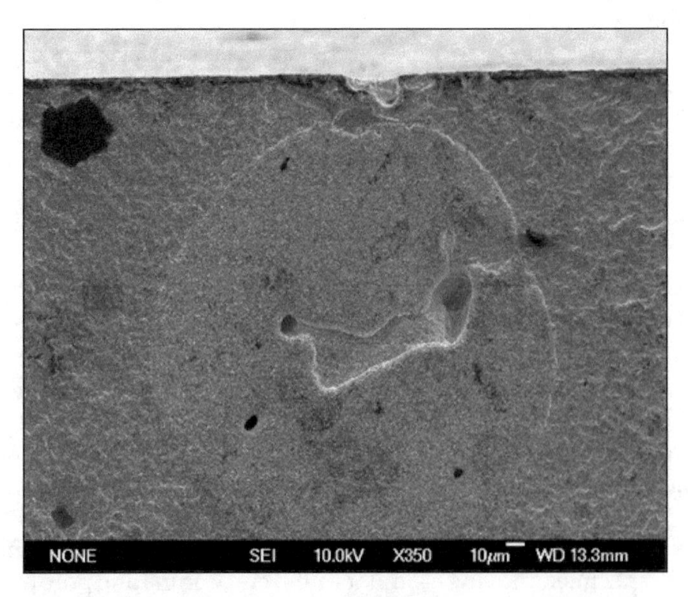

Figure 11.5. Sub-surface initiation in CMSX-4 at 650 °C after 6,500 cycles.

CHAPTER

12

Thermal Barrier Coatings

With increasing firing temperature exceeding 1400 °C, to achieve high thermal efficiency and reduce greenhouse gas emissions, gas turbine engine hot-section gas path components such as combustors, ducts, vanes and blades all need to be protected from the extreme high temperature, which otherwise would cause the metal component to melt. Thermal barrier coating (TBC) is effectively a thermal insulation layer for these components. A typical TBC system consists of a yttria-stabilized zirconia (YSZ) top coat, a metallic bond coat (BC) on top of the substrate material/component, as shown schematically in Figure 12.1. The TBC topcoat is usually deposited using either a) thermal spray techniques or b) electron beam physical vapor deposition (EB-PVD) process. Thermal spray TBC generally has a laminar splat microstructure, the advantage of which is lower thermal conductivity, because more interfaces are in the path of heat conduction. EBPVD-TBC typically has a columnar microstructure with gaps/porosity in between, which possesses strain tolerance. Because of its strain-tolerant characteristics (with a lower effective modulus), EB-PVD TBC is often applied on rotating parts such as turbine blades of aero-engines, whereas thermal spray TBCs are often applied on static parts such as combustors, vanes and transition ducts for lower costs. Industrial gas turbine blades may also have thermal spray TBC for the same reason. The bond coat of TBC can be various types ranging from MCrAlY (M = Ni and/or Co) to PtAl diffusion coating, depending on the application and cost considerations. The bond coat plays a dual role of enhancing adhesion between the ceramic topcoat and the metallic substrate and also acting as a protective layer to the substrate from oxidation/corrosion. Failure of either TBC or bond coat can severely harm the substrate component in the gas turbine environment. Therefore, understanding the failure mechanisms of TBC becomes a subject of intensive research in recent years.

It has been understood that various factors such as thermal expansion mismatch between the ceramic topcoat and metallic bond coat and TGO contribute to TBC failure (Miller 1984, Bennett 1986, Rabiei and Evans 2000, Nuiser et al. 2000, Vaßen et al. 2001, Busso et al. 2001). Particularly, damage accumulation in TBC proceeds by collective nucleation and growth

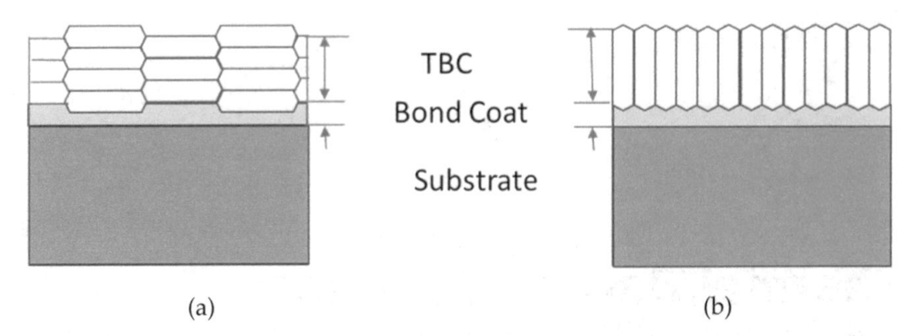

Figure 12.1. Schematic of typical TBC systems: (a) thermal-spray, (b) EBPVD.

of numerous microcracks (Wu et al. 2003, Chen et al. 2005, 2006, 2008). Stress and damage mechanics analyses have been performed using finite element models of TBC systems comprised of topcoat/TGO/ bond-coat/substrate structure (Nuiser et al. 2000, Vaßen et al. 2001, Busso et al. 2001), trying to understand the stress distribution within TBC and establish the life prediction algorithms. For example, Vaßen et al. (2001) proposed a crack growth model considering TBC failure as the result of fatigue crack propagation through the internal stress field in TBC. Busso et al. (2001) developed a damage mechanics formulation for TBC failure. These models have provided some mechanistic pictures of TBC failure, but both models are deterministic by fitting. Further understanding the stochastic nature of the crack evolution process in TBC is necessary for accurate assessment of TBC durability. Obviously, the largest crack poses as the imminent threat to TBC structural integrity, and the probability of its occurrence determines the TBC life.

12.1 Microstructural Evolution of APS-TBC during Cyclic Oxidation

Microstructural evolution in air-plasma spayed (APS) TBC, experiencing a cyclic oxidation profile of 24 hrs. per cycle from 25 °C to 1200 °C, have been examined. The TBC samples were manufactured by air-plasma spray, which consisted of yttria partially stabilized zirconia (ZrO_2-8%Y_2O_3) top coat, Ni-22Cr-10Al-1Y (wt.%) bond coat and Inconel 625 substrate. The sample size was 12.5 mm in diameter. The thickness of the ceramic coat was 250-310 μm, and that of the bond coat was 160-180 μm. At each time interval of 24 hrs, 48 hrs, 72 hrs, 120 hrs, and 144 hrs, one sample was retrieved. The last sample failed after 161 hrs after more than 30% of the topcoat spalled. All samples were then sectioned and examined under a Philips XL30S field emission gun scanning electron microscope (SEM). The SEM was operated at an accelerating voltage between 5 kV and 20 kV.

The evolution of TBC microstructures from the as-sprayed condition to the subsequently exposed conditions are shown in Figure 12.2. The

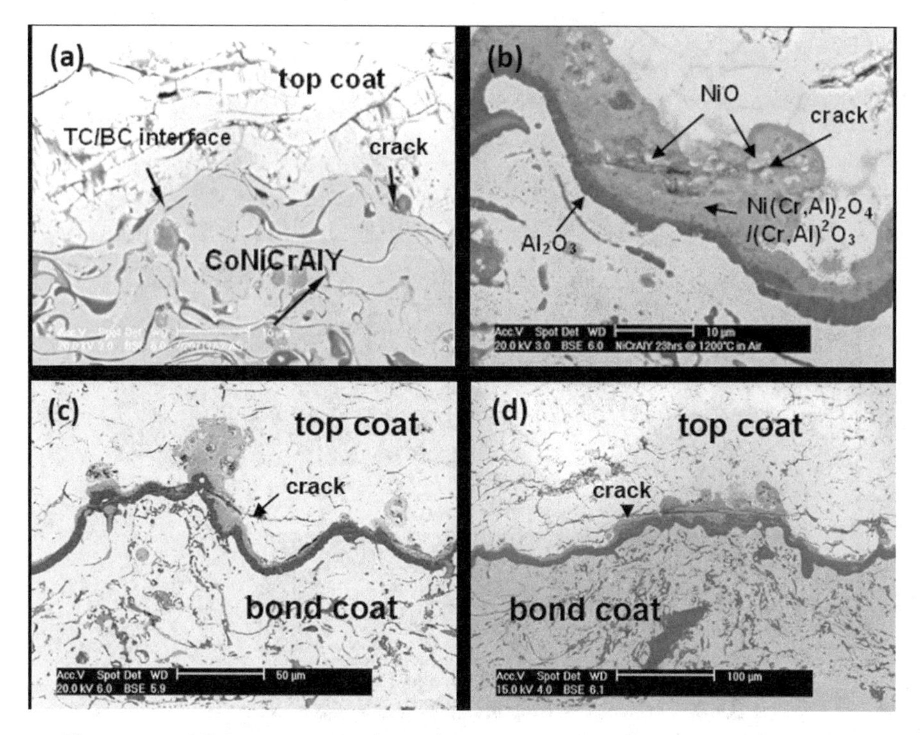

Figure 12.2. Microstructures of air-plasma sprayed TBC: (a) pristine, and after (b) 24 hrs, (c) 48 hrs, and (d) 144 hrs of exposure.

as-sprayed microstructure is a typical laminar microstructure consisting of thermal spray splats, Figure 12.2a. The topcoat contains discontinuities such as pores and splat boundaries; the bond coat is relatively dense but has been partially oxidized during the plasma spraying process.

Upon thermal exposure in air at 1200 °C, mixed oxides containing $(Cr, Al)_2O_3 \cdot Ni(Cr, Al)_2O_4 NiO$ grew rapidly along with the continuous formation of an Al_2O_3 layer in the TBC/bond coat interface region, as shown in Figures 12.2b to 12.2d. Cracking was mostly seen to be associated with the formation of $(Cr, Al)_2O_3$, spinel ($Ni(Cr, Al)_2O_4$) and nickel oxide (NiO). Formation of the mixed oxides was quite heterogeneous, which also would introduce local volume change. The volumes of the crystalline cells of the basic phases in the bond coat and TGO are shown in Figure 12.3 and Table 12.1. Cracks usually nucleated within or near the mixed oxides (Figures 12.2b and 12.2c) and grew into the ceramic topcoat. At a later stage, the microcracks would coalesce to form a long dominant crack in the ceramic topcoat near the interface region (Figure 12.2d), which would lead to TBC spallation. Apparently, the crack length increases with the TGO thickness and exposure time. The TGO curve in this case (with an oxidation constant k_{ox} = 0.8 μm^2/hr) is shown in Figure 12.4.

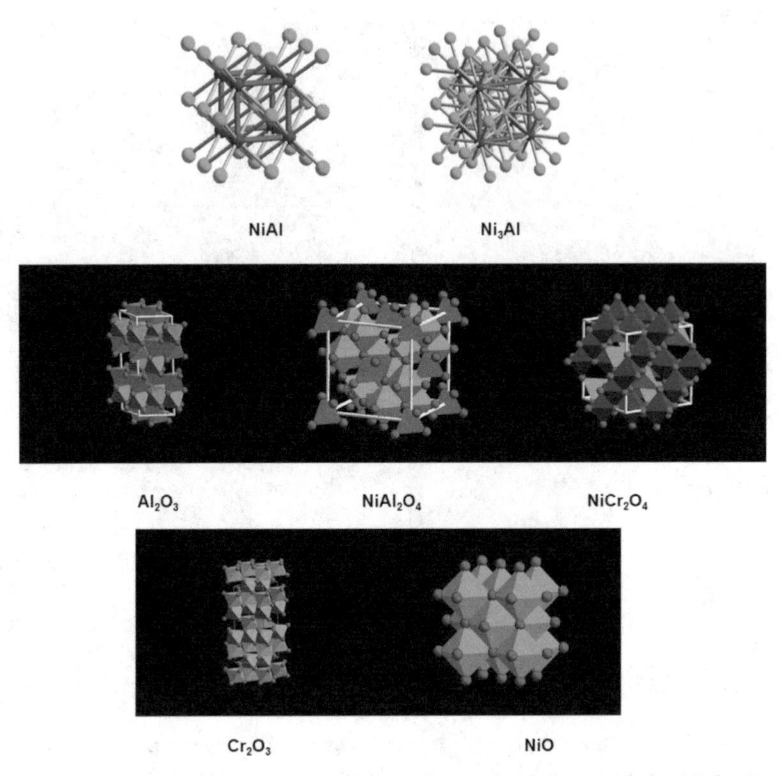

Figure 12.3. Crystalline structures of NiAl, Ni_3Al, Al_2O_3, $NiAl_2O_4$, $NiCr_2O_4$, Cr_2O_3 and NiO. (Courtesy of Toth Information Systems, Ottawa, ON, Canada).

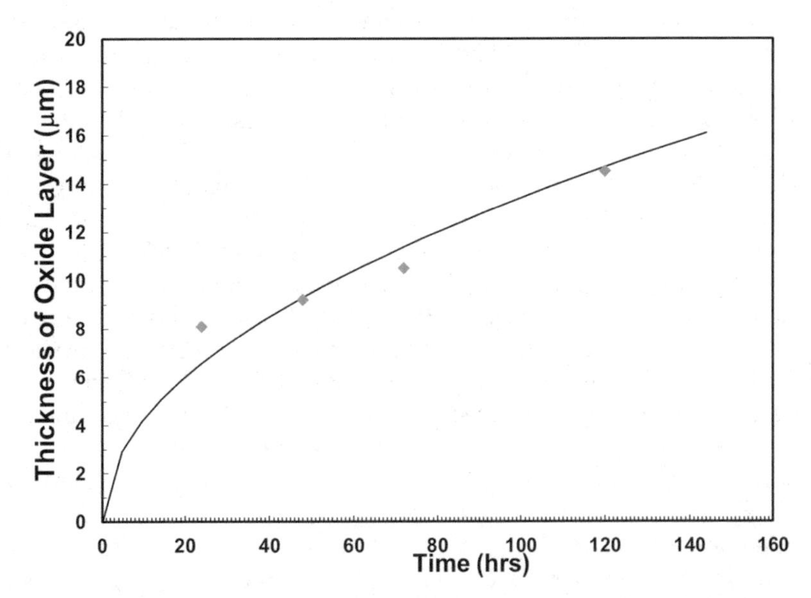

Figure 12.4. TGO thickness as function of exposure time at 1200 °C.

Table 12.1. Phase volumes

	Cryst.	Units/Cell	Vol. Å^3
NiAl	cubic	1	24.06
Ni_3Al	cubic	1	45.38
Al_2O_3	h.c.p	6	254
$NiAl_2O_4$	cubic	8	522
$NiCr_2O_4$	cubic	8	575
Cr_2O_3	h.c.p.	6	288.47
NiO	cubic	4	72.90

12.2 The Crack Number Density Theory

By the nature of plasma spray deposition process, there are always variations in chemical composition, grain structures and interface morphology in commercially produced TBC, which would affect the coating life. As discussed in the above section, local variations in aluminum and nickel contents could result in early formation of spinel ($Ni(Cr, Al)_2O_4$), instead of alumina (Al_2O_3), which would lead to early crack nucleation (Wu et al. 2003, Chen et al. 2005), because it would induce a large local volume strain, in addition to the thermal expansion mismatch strain. The combination of all these microstructural variations would result in a wide scatter in the durability of TBC. Previously, Wu et al. (2003) characterized the crack size distribution in an air plasma-sprayed TBC at different stages of thermal exposure, in terms of lognormal distribution, but the physical relationship between the TGO growth and crack evolution processes have not been elucidated. In this section, the crack evolution process is described using the crack number density theory originally proposed by Fang et al. (1995) as follows.

The crack-number-density (CND) theory describes the evolution of crack number by a differential equation as (Fang et al. 1995):

$$\frac{\partial n}{\partial t}+\frac{\partial}{\partial c}[An]=n_N \tag{12.1}$$

where n is the number of cracks at time t, c is the crack length, $A = \dot{c}$ is the crack growth rate, and n_N is the crack nucleation rate.

Eq. (12.1) states that the incremental crack number is the sum of contributions from crack nucleation and growth into the size c at any time t. The theory itself does not specify how cracks nucleate and how fast they will grow. It has to be combined with the relevant physical theories for the specific crack growth process.

The general solution of Eq. (12.1), with the initial value of n = 0, can be written in the following form (Fang et al. 1995):

$$n(c,t) = \begin{cases} n_N(c)t, & c < c_0 \\ \frac{1}{A(c)}\int_\eta^c n_N(u)du, & c \geq c_0 \end{cases} \tag{12.2a}$$

where c_0 is a threshold value below which the crack does not grow, and η is an intermediate crack size variable that satisfies

$$t = \int_\eta^c \frac{du}{A(u)} \tag{12.2b}$$

In TBC, the growth of TGO thickness (h) generally follows the parabolic oxidation law, as

$$h = \sqrt{2kt} \tag{12.3}$$

where k is the oxidation constant.

A fracture mechanics model of TBC cracking has been proposed by Evans et al. (2001), which depicts crack growth as wedge-opened by a growing sphere corresponding to TGO formation along the undulating interface between the topcoat and bond coat. The stress intensity factor K for such a crack configuration is expressed as

$$K = \frac{3}{2(1+v)\sqrt{\pi}}\left(\frac{R}{c}\right)^{3/2}\frac{(m-1)E}{3(1-v)m}\left(\frac{h}{R}\right)\sqrt{R} \tag{12.4}$$

where R is the radius of undulation at crack initiation site, m is the ratio of volume change by oxidation, E is the Young's modulus and v is Poisson's ratio of TBC.

As one can see from Eq. (12.4), the stress intensity factor increases with the TGO thickness, but inversely proportional to $c^{3/2}$. As TGO thickens, the stress intensity factor will reach a critical value, i.e. the fracture toughness K_{IC}, to cause crack growth. But, crack extension immediately lowers the stress intensity factor such that the crack growth remains stable in TBC. By re-arrangement of Eq. (12.4), we have:

$$c = \left[\frac{(m-1)ERh}{2m\left(1-v^2\right)\sqrt{\pi}K_{IC}}\right]^{\frac{2}{3}} = Ch^{2/3} \tag{12.5a}$$

where

$$C = \left[\frac{(m-1)ER}{2m\left(1-v^2\right)\sqrt{\pi}K_{IC}}\right]^{\frac{2}{3}} \tag{12.5b}$$

From Eq. (12.5), we can derive the crack growth rate as function of crack length, c, as

$$A = \dot{c} = \frac{2}{3}Ch^{-1/3}\dot{h} = \frac{2}{3}k_{ox}Ch^{-4/3} = \frac{2}{3}k_{ox}C^3c^{-2} \tag{12.6}$$

We assume that $c_0 = 0$ and crack nucleation in TBC follows a Gamma distribution, $\Gamma(\alpha, \beta)$ with $\alpha = 2$ and $\beta = 1/(\phi h)$, where ϕ is a scaling factor, such that

$$n_N = B\frac{\beta^{\alpha}}{\Gamma(\alpha)}c^{\alpha-1}e^{-\beta c} \tag{12.7}$$

where B is a constant.

Substituting Eq. (12.6) and (12.7) into Eq. (12.2) through integration, we obtain the crack number density function as

$$n(c,t) = \frac{B}{A\Gamma(\alpha)}\left[\beta\left(\eta e^{-\beta\eta} - ce^{-\beta c}\right) + \left(e^{-\beta\eta} - e^{-\beta c}\right)\right] \tag{12.8a}$$

where, from Eq. (12.2b),

$$\eta = \sqrt[3]{c^3 - 2k_{ox}C^3 t} \tag{12.8b}$$

Using Eq. (12.8), the total number of cracks at any given exposure time can be evaluated as

$$N(t) = \int_0^{\infty} n(c,t)\,dc \tag{12.9}$$

Hence, the CND probability is given by

$$\rho(c,t) = \frac{n(c,t)}{N(t)} = \frac{\frac{1}{A}\left[\beta\left(\eta e^{-\beta\eta} - ce^{-\beta c}\right) + \left(e^{-\beta\eta} - e^{-\beta c}\right)\right]}{\int_0^{\infty}\frac{1}{A}\left[\beta\left(\eta e^{-\beta\eta} - ce^{-\beta c}\right) + \left(e^{-\beta\eta} - e^{-\beta c}\right)\right]dc} \tag{12.10}$$

The experimental crack size distributions are shown in Figure 12.5 (a-d), for different exposure times. When normalized with the total number of cracks for the particular exposure time, the data represent the probability density of the crack size distribution in the TBC. By the heterogeneous nature of $(Cr, Al)_2O_3 \cdot Ni(Cr, Al)_2O_4 \cdot NiO$ formation, due to local compositional inhomogeneity, crack nucleation is certainly stochastic in nature. Also, due to compositional and microstructural variations, local conditions of crack nucleation and growth in the APS-TBC are not identical, as idealized by Eq. (12.5). The Young's modulus E and the fracture toughness K_{IC}, may also vary from batch to batch (Busso et al. 2001, Tang and Schoenung 2006, Marinis et al. 2013). Therefore, to describe the crack evolution process using Eq. (12.10), the crack growth constant C and the distribution shape factor ϕ need to be calibrated in the statistical sense. Based on the 24-hour crack size data, the best-fit values are: $C = 0.394\ \mu m^{3/2}$ and $\phi = 0.24$. The subsequent distribution curves are calculated using Eq. (12.10), as shown in Figure 12.5 (a-d) for each exposure stage (Wu 2019).

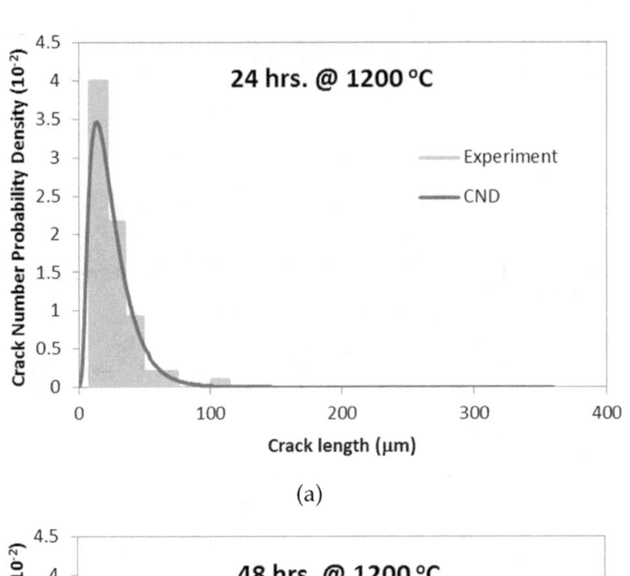

24 hrs. @ 1200 °C
Experiment
CND
Crack Number Probability Density (10⁻²)
Crack length (μm)
0
0.5
1
1.5
2
2.5
3
3.5
4
4.5
0
100
200
300
400

(a)

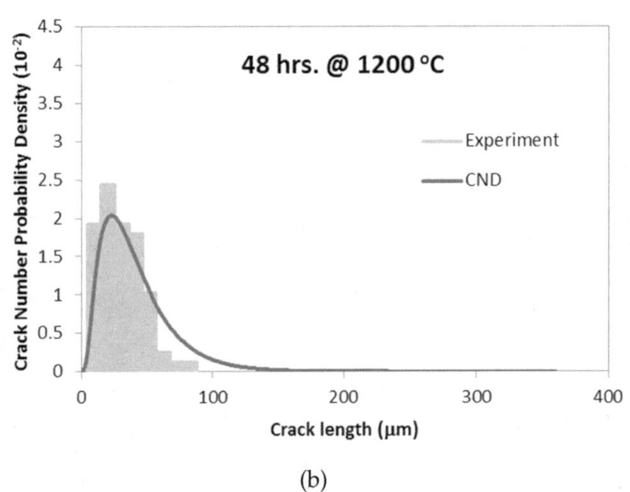

48 hrs. @ 1200 °C
Experiment
CND
Crack Number Probability Density (10⁻²)
Crack length (μm)
0
0.5
1
1.5
2
2.5
3
3.5
4
4.5
0
100
200
300
400

(b)

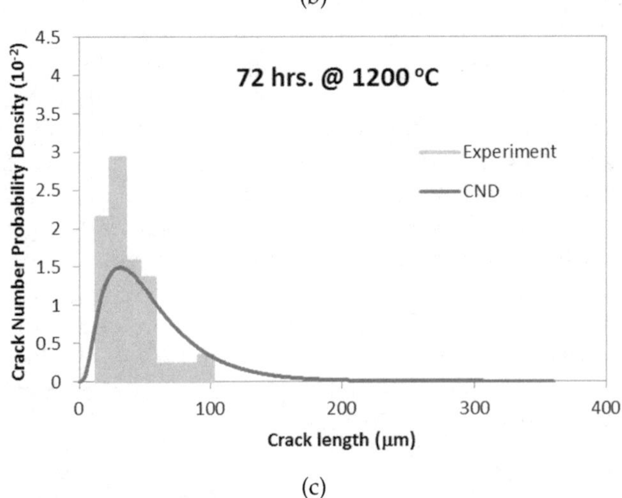

72 hrs. @ 1200 °C
Experiment
CND
Crack Number Probability Density (10⁻²)
Crack length (μm)
0
0.5
1
1.5
2
2.5
3
3.5
4
4.5
0
100
200
300
400

(c)

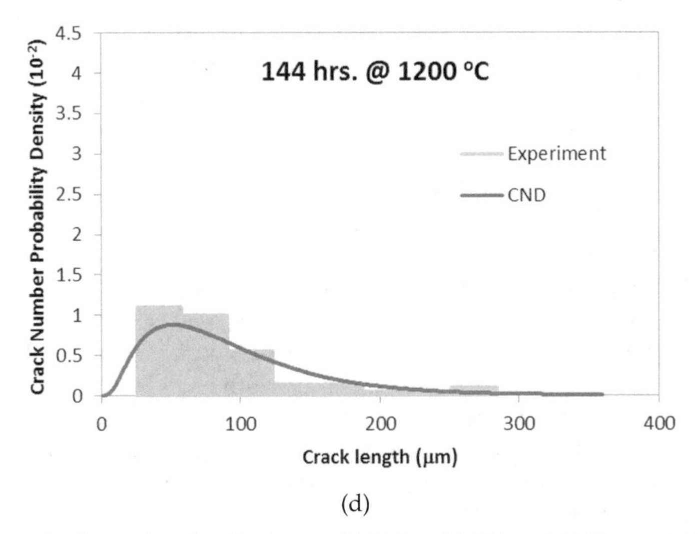

(d)

Figure 12.5. Crack number distributions at (a) 24 hrs, (b) 48 hrs, (c) 72 hrs and (d) 144 hrs.

A general trend can be observed that both the mean crack size and distribution width increases with the exposure time. The CND theory combined with the Evans-He-Hutchinson model provides a physics-based description of crack nucleation and growth phenomena in relation to TGO growth. Particularly, crack nucleation is observed to be related to the heterogeneous formation of $(Cr,Al)_2O_3$· $Ni(Cr,Al)_2O_4$·NiO with local volume change (strain), and crack growth is thought to be driven by further TGO growth.

Under the test conditions of this study, $(Cr, Al)_2O_3 \cdot Ni(Cr, Al)_2O_4 \cdot NiO$ oxides formed in the very beginning of the oxidation process. This occurred at sites where the local aluminum concentration was low and nickel had segregated as a result of compositional inhomogeneity in the air-plasma-sprayed TBC system (Chen et al. 2005). These zones were situated along the original ceramic/bond coat interface region, and it is believed that the mixed oxides were produced because, locally, there was insufficient aluminum reacting to form an initial alumina layer to serve effectively as a diffusion barrier to nickel. Lower concentration of aluminum in the nickel segregated areas, particularly in air-plasma sprayed NiCrAlY, may change the thermodynamics of oxide formation such that it favors $Ni(Cr, Al)_2O_4$/NiO rather than Al_2O_3, as opposed to the opposite in the ideal case (Hindam and Whittle 1982). These observations have provided the physical basis of relating crack nucleation with TGO formation in a statistical manner.

Assuming crack nucleation follows the Gamma distribution, it is shown that the CND theoretical description of crack evolution agree rather well with the actual crack-size measurement data, as a function of exposure time. Apparently, the widening of the distribution scatter can be attributed to TGO thickening, as it is controlled by the distribution shape parameter

β, which is inversely proportional to h. The mechanism of crack growth is thought to be due to TGO growth as a wedge to drive the crack open. From the fracture mechanics point of view, crack nucleation and growth in TBC occurs by the same wedge-opening mechanism, the former is more related to heterogeneous formation of (Cr, $Al)_2O_3 \cdot Ni(Cr, Al)_2O_4 \cdot NiO$ clusters, while the latter is driven by the entire TGO growth. It is envisaged that when the local stored elastic energy due to volumetric misfit of the oxides is equal to the decohesion energy, crack nucleation occurs, and the crack advances in association with the thickening of the TGO layer to keep the energy in balance, as implied by Eqs (12.4) and (12.5). The CND model can be used for probabilistic TBC life prediction with an allowable maximum crack size, i.e., in terms of the probability to reach the maximum crack size at any exposure time. In practice, the distribution parameters C and ϕ can be determined (calibrated) with experimental observation of crack-size distribution only once at a given exposure time. This obviously has an advantage over deterministic model-fitting approaches.

12.3 Effects of Heat Treatment on TBC Microstructural Evolution

As discussed in the above section, TBC durability is limited by the early heterogeneous formation of (Cr, $Al)_2O_3 \cdot Ni(Cr, Al)_2O_4 \cdot NiO$ clusters. It would be of great interest to explore ways to extend the TBC life through heat treatment. An experimental program was carried out by Chen et al. (2005, 2006, 2008) to study the microstructural evolution and oxidation behaviour of TBCs including APS TBC, APS plus a vacuum heat treatment (VHT $<10^{-5}$ torr) at 1080 °C for 4 hours, and APS plus a low-pressure oxygen heat treatment (LOPT ~ 2×10^{-3} torr) at 1080 °C for 24 hours, all with a CoNiCrAlY bond coat on ϕ16 mm × 10 mm Inconel 625 disks.

Cyclic oxidation was conducted with a thermal cycling profile consisting of 10-min ramping to 1050 °C, holding for 45 minutes, followed by 30~40 min cooling to the ambient temperature (25 °C). Samples were retrieved from the cyclic furnace at predetermined time (cycle) intervals and then sectioned for metallurgical examination of the microstructure evolution in terms of oxidation and cracking. The as-sprayed APS microstructure is shown in Figure 12.6 (a). The bond coat contained segmented Al_2O_3 veins as a result of partial oxidation of the bond coat during coating deposition. SEM/EDS analysis showed that the aluminum concentration was highly non-uniform in the as-sprayed bond coat microstructure, with some regions having very low Al content compared to that in the CoNiCrAlY powder (≈16 at.%) (Chen et al. 2008). Compared to the as-sprayed APS-TBC, the APS-VHT-TBC contained increased amount of Al_2O_3, but mostly distributed as a discontinuous layer at the ceramic/bond coat interface, the as shown in Figure 12.6 (b). In the APS-LPOT-TBC, a rather uniform and nearly continuous Al_2O_3 layer formed at the ceramic/bond coat interface as shown in Figure 12.6 (c). A small amount of $Al_2O_3 \cdot (Co, Ni)(Cr, Al)_2O_4 \cdot NiO$ (ASN) also formed within the ceramic/bond coat interface region. The aluminum concentrations in the VHT and LOPT microstructures were rather uniform, but became less than 11 at.%

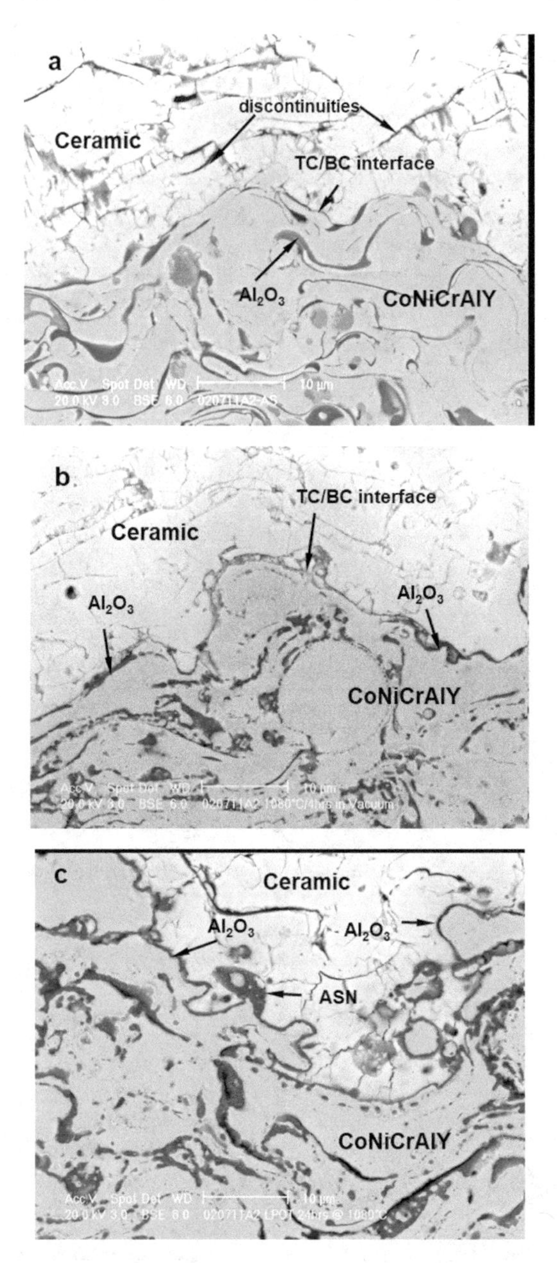

Figure 12.6. Microstructures of (a) APS-TBC, (b) VHT-TBC and (c) LPOT-TBCs with CoNiCrAlY bond coat.

(Chen et al. 2008). The different Al_2O_3 morphology at the ceramic/bond coat interface in the initial microstructures of these TBCs would influence the microstructural evolution later during cyclic oxidation exposure.

Upon thermal exposure (after one cycle), a layer of predominantly chromium oxide and spinel (CS: $(Cr, Al)_2O_3$+ (Co, Ni) $(Cr, Al)_2O_4$) formed along the interface between the ceramic and the bond coat in APS-TBC, as shown in Figure 12.7 (a). The Al_2O_3 portion in the oxide layer was less than 20%. At the same time, chromium oxide-spinel-nickel oxide (CSN) clusters also formed at the ceramic/bond coat interface. These mixed oxides are generally called thermally grown oxides (TGO). As such, the TGO formed in the APS-TBC was comprised of a CS layer and CSN clusters. As thermal cycling continued after 50 cycles, the TGO layer grew thicker, a layer of Al_2O_3 started to form underneath the CS layer and CSN clusters, as shown in Figure 12.7 (b). By comparison, in APS-LOPT-TBC after 50 cycles a layer of 90 pct Al_2O_3 had grown; and in APS-VHT-TBC a layer of 70 pct Al_2O_3 had grown, mixed with CS, as shown in Figure 12.7 (c)-(d), respectively.

In the last stage towards TBC spallation, heterogeneous growth of mixed oxides, mostly $(Cr, Al)_2O_3$, (Co, Ni)$(Cr, Al)_2O_4$ and NiO, formed underneath the previously formed CS layer in APS-TBC, as shown in Figure 12.8 (a). In addition, large cracks could be seen, which originated from of the pre-existing

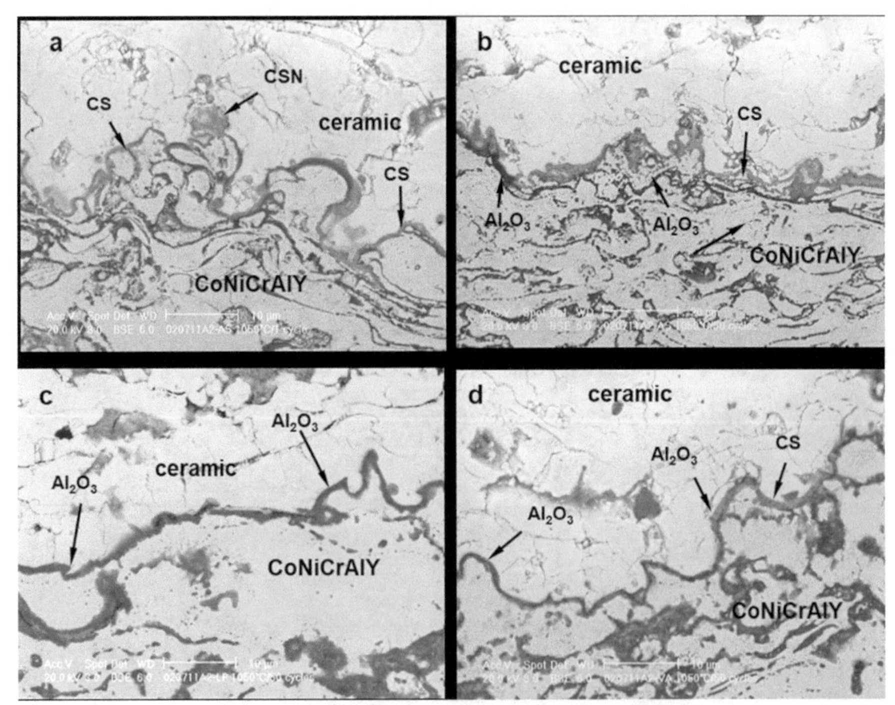

Figure 12.7. (a) Microstructure of APS-TBC after 1 cycle, (b) APS-TBC after 50 cycles, (c) APS-LOPT-TBC after 50 cycles, and (d) APS-VHT-TBC after 50 cycles of thermal cycling with 45 min hold at 1050 °C.

discontinuities (splat boundaries) near the ceramic/bond coat interface. This resulted APS-TBC failure after about 500 cycles. A similar failure scenario could be observed in APS-VHT/LOPT-TBCs, albeit it occurred after many more cycles. In the last stage of failure, the previously formed Al_2O_3 layer disappeared, which had transformed to CS and CSN. A close-up look of the VHT microstructure after 2000 cycles is shown in Figure 12.8 (b). The heterogeneous growth of CSN could cause large local volume expansion that would open up the pre-existing discontinuities in the ceramic, leading to crack propagation and coalescence and eventually TBC spallation.

The TGO formation can be qualitatively explained by the Ni-Cr-Al phase diagram, according to Hindam and Whittle (1982), as shown in Figure 12.9. In a CoNiCrAlY coating of standard composition that contains 16 at.% Al,

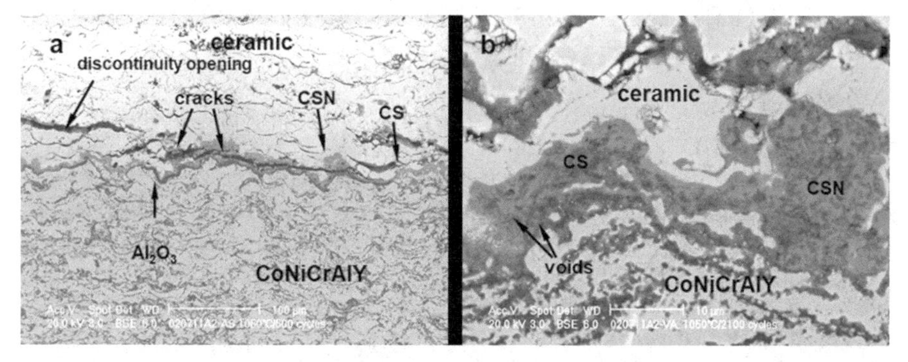

Figure 12.8. (a) Crack formation within the ceramic topcoat as a result of discontinuity opening associated with the CSN in APS-TBC after 500 cycles, (b) heterogeneous growth of CSN in VHT-TBC after 2100 cycles.

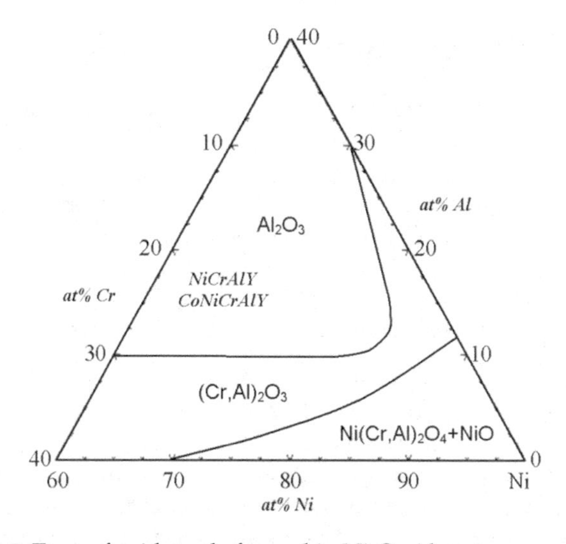

Figure 12.9. Type of oxide scale formed in Ni-Cr-Al system upon thermal exposure in air, after Hindam and Whittle (1982).

Al_2O_3 would form first during thermal exposure in air. But in reality, a layer of chromia + spinel plus mixed oxide clusters formed first in the as-sprayed APS-TBC. This is because the APS deposition process could have consumed some aluminum via partial oxidation, resulting in lower aluminum content in the APS-CoNiCrAlY bond coat. Lower aluminum concentration in the as-sprayed APS samples leads to movement from the Al_2O_3 region into the (Cr, Al)$_2$O$_3$ region and then eventually to the Ni(Cr, Al)$_2$O$_4$/NiO region (Figure 12.9).

The VHT and LPOT treatments, on the other hand, imposed a low oxygen pressure environment that promotes the formation of Al_2O_3. It has been suggested that the formation of $NiAl_2O_4$ and NiO could be avoided in the Ni-Al system at 940 °C with an oxygen pressure $< 4.9 \times 10^{-14}$ atm ($P_{O2} < 5\times10^{-9}$ Pa) (Kuznetsov 1993). This study also shows that the spinel and nickel oxide formation can be almost avoided in the CoNiCrAlY at a much higher oxygen pressure ($P_{O2} \approx 0.056$ Pa). In the VHT treatment, however, the oxygen pressure is too low ($P_{O2} < 2.8\times10^{-4}$ Pa) to develop a continuous Al_2O_3 layer at 1050 °C.

When a continuous oxide layer formed at the ceramic/bond coat interface, the oxygen partial pressure at the oxide/metal interface would decrease to the formation pressure of Al_2O_3 which is the lowest value among the various species present, making Al_2O_3 the most favorable oxide product. However, as Al became depleted, (Cr, Al)$_2$O$_3$ starts to form at the Al_2O_3/ bond coat interface after extended thermal cycles. As such, the transformation of the initially formed Al_2O_3 layer into a chromia + spinel layer in the VHT- and LPOT-treated samples can therefore be attributed to the reaction of (Cr, Al)$_2$O$_3$ and/or (Co, Ni)O with Al_2O_3.

12.4 TGO Behaviour

As shown in the micrographs in the above section, TGO formation is heterogeneous along a rough ceramic-metallic interface, especially with CSN clusters, the thickness at individual locations does not provide an overall picture of the phenomenon. Therefore, to take into account the entire TGO, an equivalent TGO thickness, δ_{eq}, is defined as:

$$\delta_{eq} = \frac{\sum(\text{cross sectional } TGO \text{ area})}{\sum(\text{cross sectional length of } TC/BC \text{ interface})} \tag{12.11}$$

Figure 12.10 shows the TGO growth behaviours of 1) APS-TBC, 2) ASP-VHT-TBC, and 3) ASP-LOPT-TBC during cyclic oxidation. The relationship of δ^2_{eq} vs. number of thermal cycles initially exhibits a linear relationship, obeying the parabolic law of oxidation; however, at the later stage, TGO growth accelerates. A similar observation was made in a FeCrAl alloy in terms of the weight change under oxidation conditions (Meier et al. 1995). These observations clearly show that TGO growth exhibits a two-stage

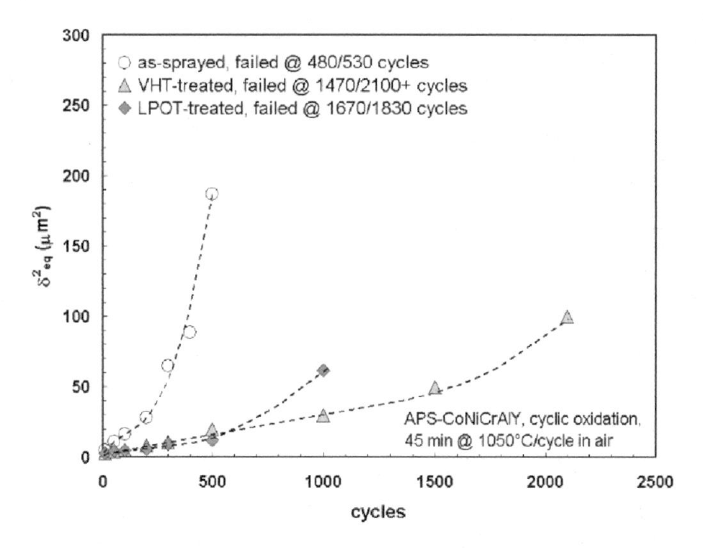

Figure 12.10. Growth of TGO in APS, APS-VHT and APS-LOPT TBCs during cyclic oxidation.

behaviour: (i) the parabolic stage, involving growth of an Al_2O_3/CS layer; and (ii) an acceleration stage, involving growth of CS and CSN. When an Al_2O_3 layer is promoted through VHT and/or LPOT treatments, TGO has an extended parabolic stage, compared to that in the as-sprayed APS-TBC.

The growth of TGO started to accelerate very early in APS-TBC at ~300 cycles, but much later in LPOT and VHT TBCs, at ~1000 and 2000 cycles, respectively. The onset of this oxidation acceleration stage was clearly associated with heterogeneous oxidation (Figure 12.8). In the as-sprayed APS-TBC, it occurred underneath the previously formed CS layer, but in the VHT- and LPOT-TBCs, it occurred by transformation from the Al_2O_3 layer to CS +CSN and subsequent heterogeneous oxidation underneath, after extended thermal exposure. The higher TGO growth rate in the as-sprayed TBC can be attributed to the CS layer initially formed at the ceramic/bond coat interface, since diffusion through chromia is faster than through Al_2O_3 (Wallwork and Hed 1971, Pettit and Meier 1984). As aluminum is further depleted from the bond coat, heterogeneous oxidation will occur underneath the TGO, leading to the onset of accelerated TGO growth. On the other hand, TGO growth in the VHT- and LPOT-TBCs would remain slow due to slow diffusion through the Al_2O_3 before the initially formed Al_2O_3 completely transformed to chromia plus spinel, which prolonged the parabolic TGO growth. As aluminum was further depleted in the bond coat, the initially formed Al_2O_3 layer would completely transform into a CS layer at the ceramic/bond coat interface, and consequently, heterogeneous oxidation would occur, leading to an accelerated TGO growth. The growth behaviours of Cr_2O_3 and Al_2O_3 in relation to the microstructural evolution are schematically shown in Figure 12.11.

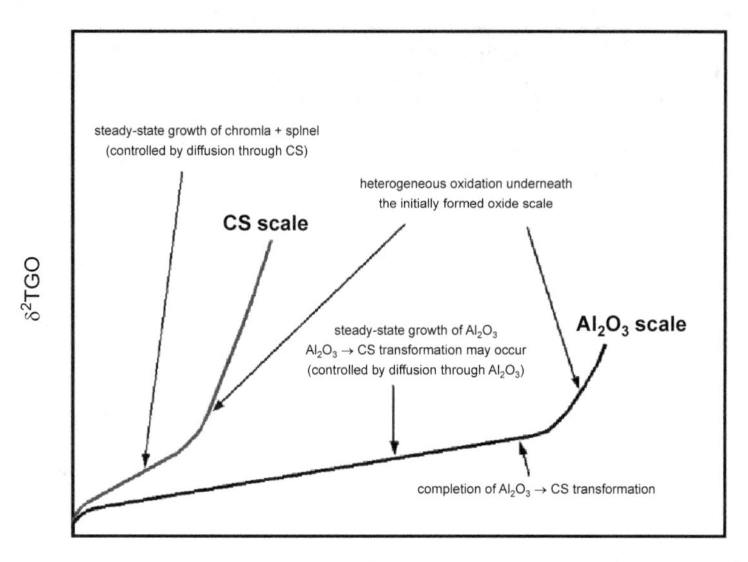

Figure 12.11. A systematic description of TGO thickening process.

The frequency effect on TBC life has also been studied by Chen et al. (2012). Figure 12.12 shows the comparison of TGO growth behaviours in APS-TBC with different hold times. It is evident that the higher frequency (the shorter holding time) resulted in a higher TGO growth rates, which may be attributed to faster oxygen transportation through the small cracks

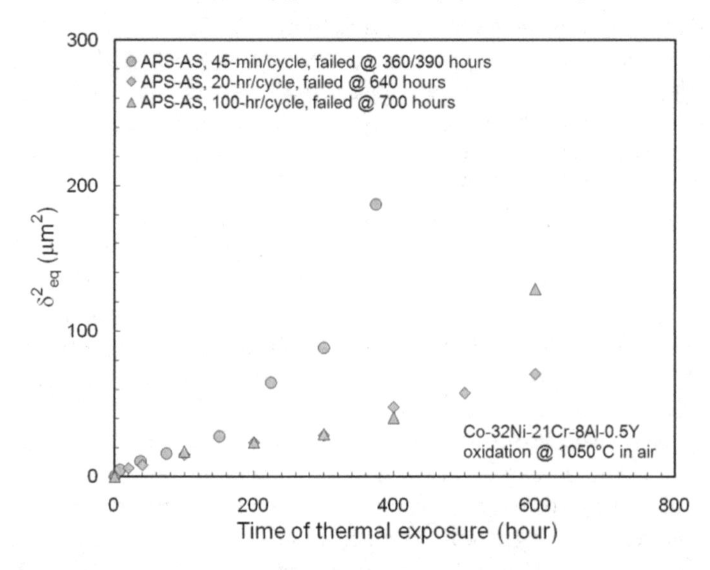

Figure 12.12. TGO growth in APS-TBC system under thermal cycling profiles with different holding times.

developed in the oxide scale due to the more frequent thermal cycling. The TGO growth curves for the 20-hour/cycle and 100-hour/cycle were nearly identical to 600 hours, implying that thermal cycling has a minimal effect on TGO growth when the holding time is long enough (e.g., longer than 20 hours).

12.5 Crack Growth Behaviour

In a thermal spray TBC with heterogeneous microstructure and TGO growth along the ceramic-metallic interface, as shown in Figures 12.2, 12.7-8, numerous microcracks would form during extended thermal exposure. It is impossible to keep tracking the growth of individual cracks, but use a statistical approach. Therefore, the CND theory is advantageous, as discussed in section 12.2. In order to establish the failure criteria with the CND theory, the maximum crack length seems to be the critical indication of the damage-state in TBC, since spallation would occur with the longest crack when it reached a critical condition to buckle, by the extreme value theory. The maximum crack length measured in the above three types of TBCs are shown in Figure 12.13 (a), as functions of TGO thickness. It is interesting to see that despite the different TGO growth behaviours in the as-sprayed APS, VHT- and LPOT TBCs, the maximum crack length follows a unique power-law relationship with the TGO thickness, as

$$a_{\max} = k\delta_{eq}^{n} \tag{12.12}$$

Regression analysis shows that the overall best-fit value of n is 0.597. Actually, the theoretical exponent of 2/3 in Eq. (12.5) gives a better fit for long cracks. As Eq. (12.5) predicted and Figure 12.13 (a) shows, the maximum crack length increases with TGO thickness.

In APS-TBC, CSN clusters and the CS layer are more susceptible to crack nucleation and propagation (Chen et al. 2005, 2006, 2008). Crack propagation in thermal spray TBCs appears to proceed via opening and coalescence of discontinuities in the ceramic in the later stage of thermal cycling (Figure 12.8), which is assisted by TGO growth. On the other hand, VHT and LPOT promote the formation of an Al_2O_3 layer at the ceramic/bond coat interface that, albeit not continuous, provides a barrier for further oxidation such that the TGO growth rate is lower in these two types of TBCs. As reasoned by the CND model, crack number and size development would be lower with a thinner TGO layer. Eventually, as thermal exposure further proceeds, the Al_2O_3 would transform to chromia plus spinel, resulting in increased cracking associated with the TGO, and consequently longer cracks within the ceramic/bond coat interface region.

APS-TBC samples experiencing thermal cycles with different hold-times were also examined. When plotting the maximum crack length as a function of TGO thickness for the APS-TBC subjected to thermal cycling with different holding times, it appears that the $a_{\max}$ vs. δ_{eq}

relationship can also be represented by a similar master curve, as shown in Figure 12.13(b). Regression analysis shows that the power-law relationship has a best-fit exponent ~0.7149, which is very close to the theoretical value of the Evan-He-Hutchinson model, Eq. (12.5). To further verify the relationship, an isothermal oxidation test was conducted to examine the validity of Eq. (12.12) for the extreme case. This additional test was performed at 1050 °C for 375 hours (not failed), which is equal to the total time of five hundred 45-min thermal cycles, and the datum also falls closely on the master curve within the experimental scatter, as shown in Figure 12.13(b).

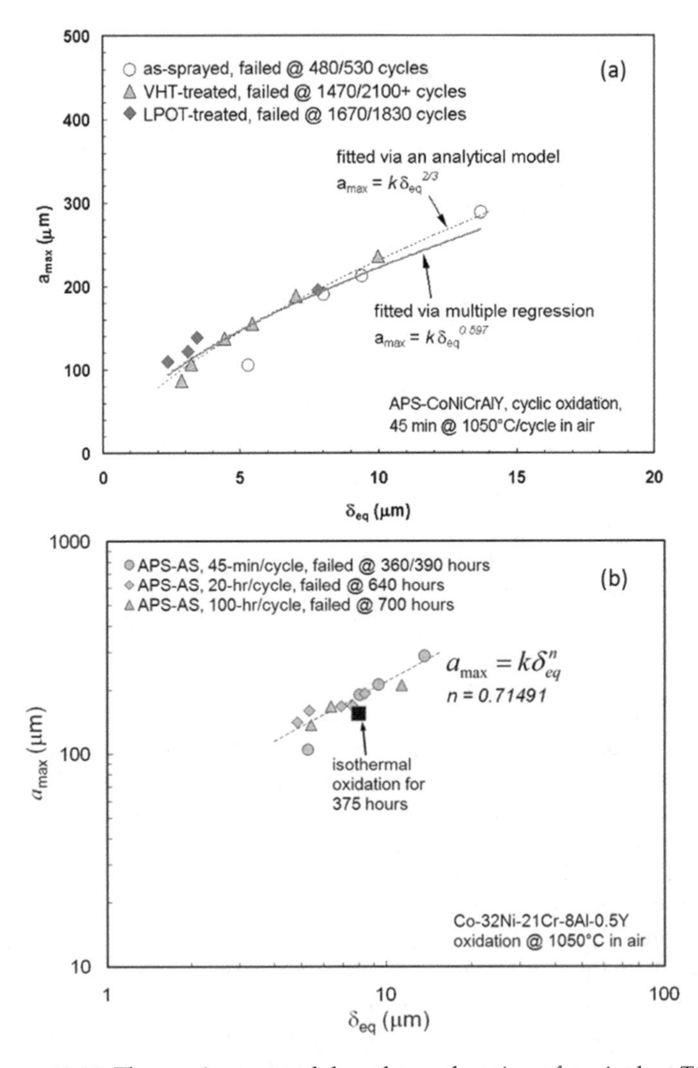

Figure 12.13. The maximum crack length as a function of equivalent TGO thickness, (a) in APS, VHT and LPOT TBC, and (b) in APS but subjected to different thermal cycles.

This correlation suggests that the a_{max} vs. δ_{eq} relationship is thermal profile-independent, which means it can be used for TBC life prediction.

12.6 TBC Life Prediction

The ultimate goal of understanding crack evolution in TBC is to predict TBC durability. The CND model is developed just for this purpose; combined with the maximum crack length criteria, it will formulate a TBC life prediction method. Before this endeavour, a significant effort has been spent on developing TBC life prediction models, which are also reviewed below.

12.6.1 The NASA Model

The NASA model was the earliest effort for TBC life prediction, which takes the form of Coffin-Manson type equation, as (DeMasi et al. 1989):

$$N = \left(\frac{\Delta\varepsilon_{in}}{\Delta\varepsilon_f} \right)^b \tag{12.13}$$

where N is cycles to failure, $\Delta\varepsilon_{in}$ is the inelastic strain range, $\Delta\varepsilon_f$ is the inelastic strain range that causes failure in one cycle and b is a constant of negative value. The effect of high temperature exposure on life is included in $\Delta\varepsilon_f$ which is a function of TGO growth and the adhesion strength of the TBC. The growth of the TGO influences life through the expression

$$\Delta\varepsilon_f = \Delta\varepsilon_{f0} \left(1 - \frac{\delta}{\delta_c} \right) + \Delta\varepsilon_{in} \left(\frac{\delta}{\delta_c} \right) \tag{12.14}$$

where $\Delta\varepsilon_{f0}$ is the inelastic failure strain range for an unoxidized coating system, δ is the oxide scale thickness, δ_c is the critical oxide layer thickness that causes the coating to fail in one cycle.

This model covers the effects of both mechanical strain and oxidation in ceramic TBC on a component. The model requires a constitutive model to evaluate the inelastic strain range in the ceramics for TBC life prediction. However, the residual stress in TBC by deposition is an unknown factor in the constitutive behaviour.

12.6.2 The Al-Depletion Model

Microstructural evolution in terms of oxidation, inter-diffusion and β phase dissolution in MCrAlY coatings were studied by a number of researchers and the corresponding models were proposed (Lee et al. 1987; Chan et al. 1998, 1999), where Al depletion in the coating was considered as the critical failure criterion. This failure criterion is adopted in COATLIFE—a computer software tool for combustion turbine hot-section coating life management (EPRI/DOE Report 1011593, 2005). Similarly, Renusch et al. (2008) analyzed aluminum depletion in TBC with CoNiCrAlY bond coat. The postulation is that as the aluminium content reaches a critical value, the coating stops acting as an Al reservoir; then the protective interface oxide layer can no longer be maintained and chemical failure occurs. Since removal of Al from the coating causes dissolution of the β phase, either the complete dissolution of β phase or Al depletion to a critical Al content may be used as a suitable criterion, which can be measured from micrographs of cross-sectioned specimens. For example, Renusch et al. (2008) suggested 3 wt.% (~12 at.%) as a critical Al content. The COATLIFE Al-depletion model predicted that below 12 at.%, the Al content would be decelerated to zero in GT33 (an equivalent of CoNICrAlY) on GT-111 at 1066 °C. The aluminum depletion criterion is strongly dependent on the bond coat composition. As governed by the Fick's law, the aluminum content after long-time exposure approaches a constant level asymptotically, as shown in Figure 3.9. A large difference in life can be obtained if the terminal level changes slightly.

In section 3.3, the multi-element diffusion equation is solved. The equation describes Al-depletion in a high velocity oxy-fuel (HVOF) spray plus VHT TBC, as shown in Figure 3.9. The general trend is that Al-depletion first occur at fast rates (within the first 1000 hours) in the temperature range of 1050-1150 °C and it tails-off, approaching a constant level. But, with the error of measurement, it is difficult to obtain a single critical level at failure for the tested coupons. A small difference in the critical aluminum content value can result in a predicted life which differ by a couple of thousand hours. Therefore, this criterion has to be used with careful calibration.

12.6.3 The Constitutive-Damage Model

Because of the particular failure mode of TBC—spallation along the ceramic/bond coat, many researchers focused on the out-of-plane stress near the interface, which has to be evaluated using a microstructural model of TBC. Then, the TBC life is calculated in terms of such stress component. A representative model was proposed by Busso et al. (2001), which takes the following form for APS TBCs:

$$\frac{dD}{dN} = D^m \left(\frac{\sigma_{max}}{F} \right)^p \tag{12.15}$$

where D is a fatigue damage parameter such that $D = 1$ at failure, σ_{max} is the maximum out-of-plane interfacial stress, N is number of cycles and m and F are given by

$$m = 1 - C \left(\frac{\sigma_{max}}{\sigma_{c0}} \right)^{0.818p} \tag{12.16}$$

and

$$F = F_0 \left(1 - F_1 \sigma_{max} \right) \tag{12.17}$$

where σ_{c0} is the initial strength of the TBC and p, C, F_0 and F_1 are material parameters that need to be determined by fitting to the experimental data.

In this model, the fatigue damage is driven by the maximum out-of-plane stress, σ_{max}, which is obtained from finite element analysis of a representative segment of the TBC system. In general, the σ_{max} is contributed by constrained thermal expansion, oxide growth and sintering of the top coat:

$$\sigma_{max} = \sigma_{therm.} + \sigma_{ox} + \sigma_{sintr.} \tag{12.18}$$

This approach requires the establishment of constitutive models for each constituent layer in TBC for complete thermal elastic and viscoplastic FEM analysis, in addition to TGO and sintering models.

12.6.4 The Fracture Mechanics Approach

As it is recognized that spallation is caused by crack development in the ceramic top coat, a micromechanical fracture-mechanics model has been proposed by Vaβen et al. (2001), where the TBC spallation life is calculated by integration of the Paris-law over the out-of-plane stress field, as:

$$N = \int_{c_0}^{c_f} \frac{1}{A}\left(\frac{K_{IC}}{\Delta K}\right)^n dc \tag{12.19}$$

where A is the Paris-law constant, K_{IC} is the coating fracture toughness, c_0 and c_f are the initial and final crack length. The stress intensity factor has to be evaluated using a microstructural FEM model of the TBC. Since it is impossible to keep tracking individual fatigue crack growth in the TBC, all the constants are determined by fitting the integration result to the observed TBC failure cycle or time.

12.6.5 The CND and Maximum Crack Length Model

As discussed through sections 12.2-5, the crack number density theory combined with the maximum crack length criterion can be used to predict the cyclic oxidation life for air-plasma sprayed TBC. First of all, the CND model in combination with the Evans-He-Hutchinson sphere-wedging crack model recognizes the nature of crack nucleation in association with the volumetric misfit as induced by heterogeneous formation of $(Cr, Al)_2O_3 \cdot Ni(Cr, Al)_2O_4 \cdot NiO$ and considers the crack growth mechanism as driven by TGO growth. The model describes the evolution of crack size distribution as a function of time. Particularly, the width of distribution is directly related to TGO thickness, so the maximum crack size. This correlation has been demonstrated for different TBC types subjected to different thermal cycling profiles. Therefore, in principle, the method can be extended to other isothermal and cyclic oxidation conditions, as long as the Arrhenius relation of the oxidation constant is determined. This may lend a powerful tool for probabilistic TBC life prediction.

The sphere-wedging crack model insinuates circular cracks in TBC. When the maximum crack grows to its critical length at the TBC/TGO interface, the top coat will buckle under bi-axial compressive stresses σ induced by the thermal mismatch within the TBC system, leading to TBC spallation. According to the elastic theory for a clamped circular plate, the relationship between the maximum crack size and stress is given by (Timoshenko and Gere 1961):

$$\frac{\sigma}{E} = \frac{\kappa}{3(1-\nu^2)}\left(\frac{H}{a}\right)^2 \tag{12.20}$$

where H is the topcoat thickness, and κ is a constant (to be calibrated by experimental data). It has been shown that the critical crack size for TBC spallation indeed depends on the topcoat thickness, and so does the residual stress in the topcoat by deposition (Zhu et al. 2004).

The method to implement this model is to conduct thermal-cyclic tests of TBC, calibrate the CND model for a given exposure time before failure and then using the calibrated CND model and the maximum crack length criteria to predict the failure at an acceptable probability.

It can be seen from Figures 12.13, the various APS-TBC failed at a critical crack length of ~200-300 μm, which corresponds to a critical TGO thickness of about 11 μm. As such, the VHT-treated samples can be expected to have a lifetime of about 2250 cycles, while the LPOT-treated samples may last 1300 cycles, based on the TGO growth curves (Figure 12.11). Using the same model (the $a_{\max}$ vs. δ_{eq} relationship), the APS-TBC life under thermal cycling with hold times of 45 min, 20 hrs, and 100 hrs is predicted to be 360, 600, and 760 cycles, respectively. Once CND model and Eqs. (12.10) and (12.11) are calibrated based on coupon testing data, TBC failure on a gas turbine component can be predicted with an acceptable confidence.

CHAPTER

13

Ceramics Matrix Composites

Modern aircraft engines present some of the most challenging working environments for metallic materials, today, the temperature of combustion gas entering the turbine exceeds the melting point of the white hot rotating blade alloy! Figure 13.1 shows the trend of engine firing temperature and turbine blade materials developed over the past 50 years. Notably, the development of single crystal Ni-base superalloys, in combination with thermal barrier coating and cooling techniques, has allowed the turbine blade to withstand higher and higher temperatures, but the trend will be replaced by ceramic matrix composites (CMCs).

Aero-engines are desirable to be light weight and high performance with low emission, noise and life cycle cost. Ceramic matrix composites (CMCs) are among the recently advanced materials that have been identified

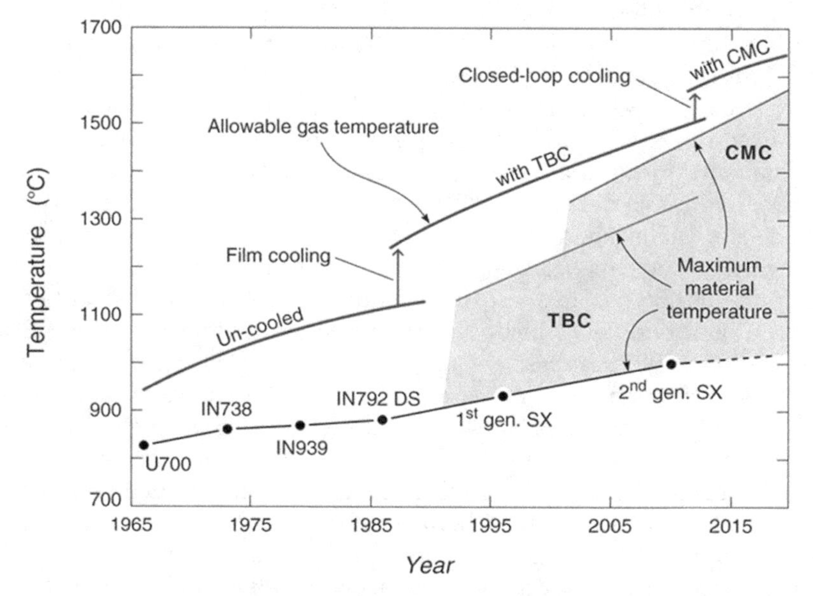

Figure 13.1. The trend of firing temperature with material capability http://www.virginia.edu /ms/research/wadley/high-temp.html.

as a key material system for improving the thrust-to-weight ratio of high-performance aircraft engines with increased turbine inlet temperature (TIT) (Hurst 2018). As the operation profiles of aircraft engines typically consists of takeoff–cruise–landing (shutdown), which cause creep-fatigue damage in combination with environmental effect, the synergistic degradation mechanisms in CMCs have not been well understood, relative to that in metals. The durability of CMCs is so critical that their failures may cause meltdown of the adjacent metallic parts, resulting in catastrophic failure of the engine. Therefore, it is important to thoroughly understand the failure modes and mechanisms CMCs.

13.1 CMC Failure Mechanism

CMCs are typical manufactured in two steps: (1) fibre lay-up, shaped as the desired component; and (2) infiltration of the matrix material. An advantage of continuously woven fiber-reinforced CMCs over monolithic ceramics is the higher fracture resistance via fiber-bridging mechanism (Marshall and Cox 1985, Marshall and Evans 1985, Budiansky et al. 1986, Evans 1990, Warren 1992, Evans et al. 1994).

Creep mechanism indicators have been summarized by Chermant et al. (2002), based on optical and/or scanning electron microscopy (SEM) and transmission electron microscopy (TEM) observations on crept CMC specimens, which include fiber/matrix debonding, matrix cracking, yarn/yarn debonding, fiber and yarn bridging, fiber pull-out, and fiber rupture. The creep behaviour of SiC/SiC in argon at temperatures of 1000-1300°C have been studied by Zhu et al. (1997, 1999) who observed that at low stresses the fracture surface consisted of two regions: (i) the slow crack growth region where fiber fracture flushed with the matrix, and (ii) the fast fracture region where the fibers were pulled out. The matrix creep was presumed to be predominant during the transient and steady-stage until the tertiary stage where progressive debonding of interfaces and fiber pull-out and rupture would occur. The fiber bridging could make out the increased ductility during the tertiary creep. At high stresses, the fracture surface was also rough with pull-out fibers. In this case, upon loading, cracks initiated at large pores between fiber bundles, which were then bridged by intact 0° fibers. Crack propagation in 90° bundles was along fiber/matrix interfaces or connected by pores in the bundles. Crack propagation in 0° bundles at high stresses was similar to that widely observed in unidirectional fiber-reinforced CMCs.

Extensive matrix cracking can cause increase in the specimen compliance. Thus, the reduction of the effective modulus is often regarded as a damage parameter (Rospars et al. 1998, Chermant et al. 2002). Rospars et al. found that the modulus-based damage parameter for the tested SiC/SiC is a function of inelastic strain, irrespective of the applied stress, and they adopted a power-law viscoplastic potential to compute the viscoplastic strain. Zhu et al. (1999) also observed gradual reduction of the elastic modulus of SiC/SiC during fatigue and creep. However, if this damage parameter were adopted in the creep rate equation in a way as Kachanov and Rabotnov proposed, a tertiary stage would gradually appear. But, this is often not observed during creep of CMC (if a tertiary stage indeed occurred in CMC, it occurred rather steeply). On the contrary, creep of CMC exhibits a predominantly transient behaviour (primary plus secondary), and in some cases fracture occurs before the steady-state. Chermant et al. (2002) and Grujicic et al. (2016) used empirical time-power-laws for description of creep strain accumulation in CMCs, the mechanism basis of which were not clearly related to any damage mechanisms mentioned above.

As discussed in Chapter 4 and 5, intragranular dislocation multiplication is responsible for the gradual accretion of tertiary creep, whereas GBS is responsible for the transient behaviour. Because of the atomic bonding nature of the ceramics, intragranular inelastic deformation is negligible in ceramics. Therefore, it can be postulated that creep in CMCs occurs predominantly by GBS. Generally speaking, GBS in different constituents of CMC is highly non-uniform, because of the different grain sizes in fibers and matrix, and hence there will be load transfer between the matrix and fibers during creep. The propensity of load transfer depends on the creep mismatch ratio (CMR), defined as a ratio of fiber creep rate to matrix creep rate (Holmes and Chermant 1994). When CMR < 1, rapid stress relaxation in the matrix transfers the load to the fibers; and vice versa when CMR > 1. The load transfer can therefore be a nonlinear relationship due to the strain compatibility requirement, and it will change in the presence of interfacial debonding and matrix cracking, the details of which need finite element microstructural analysis of the whole composite. For macroscopic modeling (to be used as the constitutive model for component analysis), it is simply assumed that the overall creep behaviour of CMC can be divided into two mechanism regimes: i) overall matrix creep, where fibers and matrix deform compatibly; and ii) fiber creep and bridging (including interfacial sliding that leads to fiber pull-out).

13.2 The Tensile Model of CMC

The basic structure of a CMC can be represented by a unit cell as shown in Figure 13.2. Suppose matrix cracks first exist either as incipient defects introduced during matrix infiltration or occurring as a result of deformation. Upon loading, a plastic zone develops ahead of the crack with fiber-bridging. According to the Dugdale (1960) model as well as the BCS dislocation pile-up model (Chapter 9), the crack tip opening displacement can be formulated as:

$$u = \frac{2(\sigma - \sigma_0)^2 a}{E\sigma_Y} \tag{13.1}$$

where σ is the applied stress, σ_0 is the fiber-bridging stress, a is the crack length, E is the elastic modulus, and σ_Y is the yield strength of the matrix.

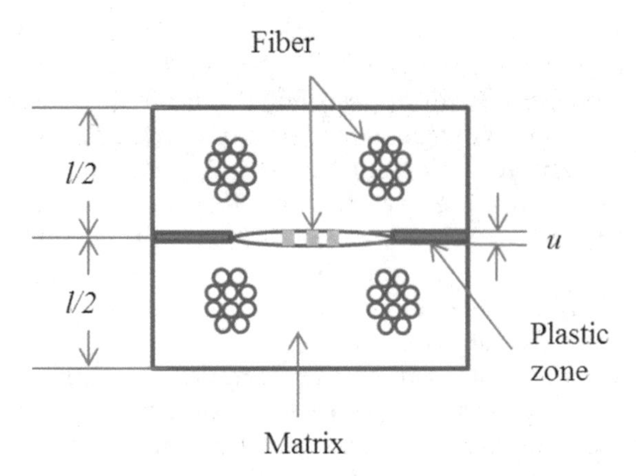

Figure 13.2. Schematic of matrix cracking with fiber bridging.

When the crack develops with crack-tip plasticity, the total strain of the unit cell is equal to the elastic strain of the matrix plus the plastic strain, u/l, as:

$$\varepsilon = \frac{\sigma}{E} + \frac{2(\sigma - \sigma_0)^2 a}{E\sigma_Y l} \tag{13.2}$$

Note that Eq. (13.2) is similar to the Ramberg-Osgood equation for metals, so the total strain of CMC can be rewritten, as:

$$\varepsilon = \frac{\sigma}{E} + \left(\frac{\sigma - \sigma_0}{K}\right)^2 \tag{13.3}$$

where $K = \sqrt{\frac{E\sigma_Y l}{2a}}$.

Figure 13.3 shows the tensile curve of standard SiC/SiC at 1300 °C, where the data are taken from the literature (Zhu et al. 1999), and the curve is fitted using Eq. (13.3) with $E = 204$ GPa, $\sigma_0 = 100$ MPa, and $K = 2750$ MPa. Excellent agreement is found between the model and the experiment. During the tensile loading to a small strain at a fast strain rate, matrix-crack formation is instantaneous, so that K can be assumed as a constant. But, it is expected that crack growth may occur under repeated tensile loading, i.e., fatigue, such that cyclic softening may occur as matrix crack extends.

13.3 The Model of CMC Creep

Matrix creep can occur below a certain stress level, σ_0, without instantaneous matrix cracking. The overall deformation is assumed to be a volume-fraction-weighted average of the components from each constituents. Thus, adopting Wu-Koul's GBS model (in this case, without grain boundary precipitates), we can write the creep strain equation as:

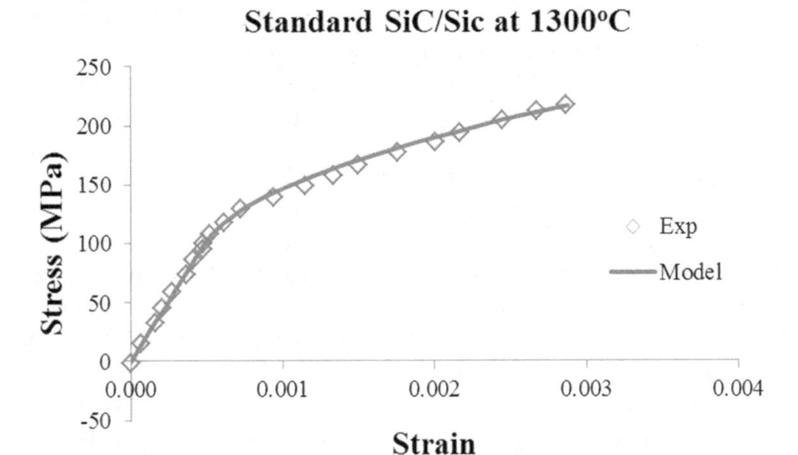

Figure 13.3. Tensile curve of standard SiC/SiC at 1300 °C.

$$\varepsilon_1 = \varepsilon_0 + \dot{\varepsilon}_{ss}t + \frac{\sigma}{H\beta^2}\left[1 - e^{-\frac{\beta^2 H\dot{\varepsilon}_{ss}t}{\sigma(\beta-1)}}\right] \tag{13.4a}$$

$$\dot{\varepsilon}_{ss} = \left[V_m A_m\left(\frac{b}{d_m}\right) + V_f A_f\left(\frac{b}{d_f}\right)\right]\sigma^p = A_1\sigma^p \qquad (\sigma < \sigma_0) \tag{13.4b}$$

where ε_0 is the initial strain, partly elastic-plastic, partly due to compliance of the experimental set up; H is the work hardening coefficient; β is a microstructural parameter; b is the Burgers vector; V_m and V_f are volume fractions, d_m and d_f are the grain sizes, A_m and A_f are creep rate constants of matrix and fibers, respectively, and $\dot{\varepsilon}_{ss}$ is the volume-fraction-averaged steady-state creep rate.

Above the critical stress σ_0, some matrix cracking will occur instantaneously, and fiber-bridging will immediately take effect. Then, creep deformation is controlled by fiber creep, as

$$\varepsilon_2 = \varepsilon_0 + \dot{\varepsilon}_{ss}t + \frac{\sigma}{H\beta^2}\left[1 - e^{-\frac{\beta^2 H\dot{\varepsilon}_{ss}t}{\sigma(\beta-1)}}\right] \tag{13.5a}$$

$$\dot{\varepsilon}_{ss} = \left[V_m' A_m\left(\frac{b}{d_m}\right) + V_f A_f\left(\frac{b}{d_f}\right)\right]\left(\frac{\sigma}{V_f + V_m'}\right)^p = A_2\sigma^p \tag{13.5b}$$

where V'_m is the reduced volume (area) fraction of uncracked matrix.

The above formulation explains the role of fiber-bridging during creep, but the reduced fraction of uncracked matrix is difficult to quantify.

Therefore, as a measure of the macroscopic response, constant A_2 will be estimated from the experimentally observed behaviour.

Zhu et al. (1997, 1999) have studied the creep mechanisms of SiC/SiC in argon at temperatures of 1000-1300 °C. Typically, at high stresses, fast creep fracture occurs with abundant fibers being pulled out; while at low stresses, slow crack growth with fiber fracture flushed with the matrix. The fracture modes of these two mechanisms are shown in Figure 13.4 (a) and (b), respectively. By the nature of ceramics, matrix creep is presumably to occur by grain boundary sliding (GBS), as the ceramics grains are hard to deform. Therefore, matrix creep mostly proceed in the transient stage that consists of primary and secondary creep. Tertiary creep of CMC is expected to occur only when progressive debonding of interfaces and pullout/rupture of fibers occur. Fiber bridging can make out the increased ductility during the tertiary creep. Figure 13.4 (a) shows that at high stresses, the fracture surface is rough with pullout fibers. It can be reasoned that cracks initiate at large pores between fiber bundles, which then propagate along fiber/matrix interfaces of 90°-bundles and/or coalesce with other pores/cracks in the bundles during creep. These cracks are bridged by 0°-fibers. Since crack growth is fast at high stresses, load transfer to fiber occurs in a short time such that fracture appears to be "sudden" with pullout fibers. At low stresses, matrix cracking can be extensive during longer creep time such that fiber pullout is minimal at final fracture.

Below σ_0 without instantaneous matrix cracking, the CMC can be considered as a continuum, where matrix creep proceeds by GBS. Figure 13.5 shows the description of Eq. (13.4) in comparison with the experimental

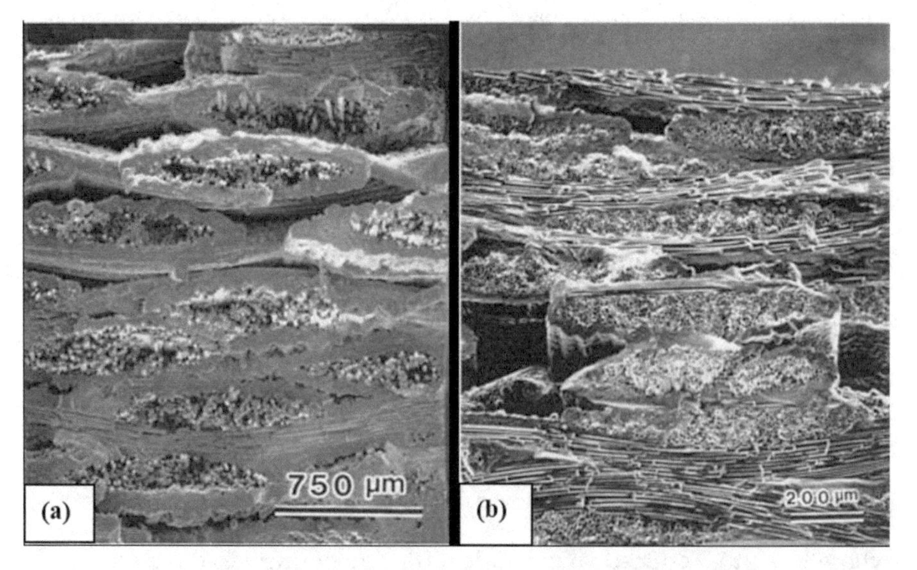

Figure 13.4. Fracture surfaces of SiC/SiC: (a) at 180 MPa; and (b) 45 MPa at 1300 °C in argon, picture taken from Zhu et al. (1997).

behaviour at low stresses as observed by Zhu et al. (1999). The parameters for matrix creep ($\dot{\varepsilon}_{matrix}$) of SiC/SiC at 1300°C in argon are given in Table 13.1. The model is found in good agreement with the experimental observation up to the onset of tertiary creep, at which point extensive matrix cracking and fiber-pullout start to occur. Since a lot of fiber bundles have been broken during the slow crack growth period in association with the transient creep, load transfer to the remaining fibers will cause a sudden increase in specimen elongation, leading to the tertiary creep. At this stage, the remaining fibers will fracture very quickly, because they bear very high load when matrix cracking becomes extensive. The mechanistic interplay of fiber pullout in conjunction with matrix cracking is very complicated and it happens so fast that its duration can be ignored from the lifing point of view.

Table 13.1. Parameters of Eq. (13.4) for SiC/SiC matrix creep at 1300 °C in argon

A_1	p	β	H (GPa)
3.15×10^{-17}	5.04	1.02	576.7

At high stress, it can be assumed that matrix cracking occurs instantaneously, and the subsequent creep behaviour will be dominated by fiber creep and fiber/matrix interfacial sliding. Then, Eq. (13.5) applies to fiber creep, which again is operated by GBS. Figure 13.6 shows the description of Eq. (13.5) for the creep behaviour of SiC/SiC in the stress range from 90 MPa to 180 MPa at 1300 °C in argon. At 90-120 MPa, the creep behaviour is still dominated by matrix creep, but above 120 MPa, fiber creep becomes predominant. The model describes the creep behaviour at 150 -180 MPa with the parameters for $\dot{\varepsilon}_{fiber}$ given in Table 13.2. The instantaneous strain ε_0 includes elastic-plastic strains as evaluated by Eq. (13.3). Again,

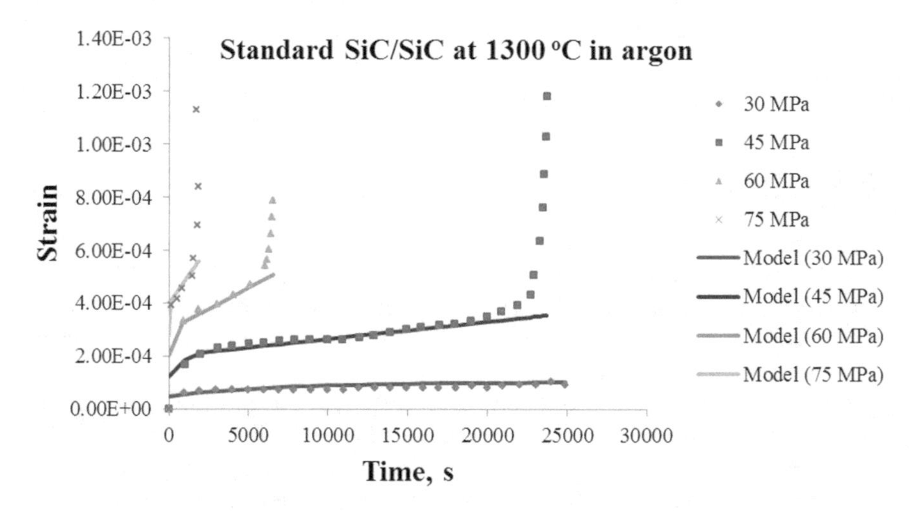

Figure 13.5. Tensile creep strain versus time at low stresses in argon 1300 °C.

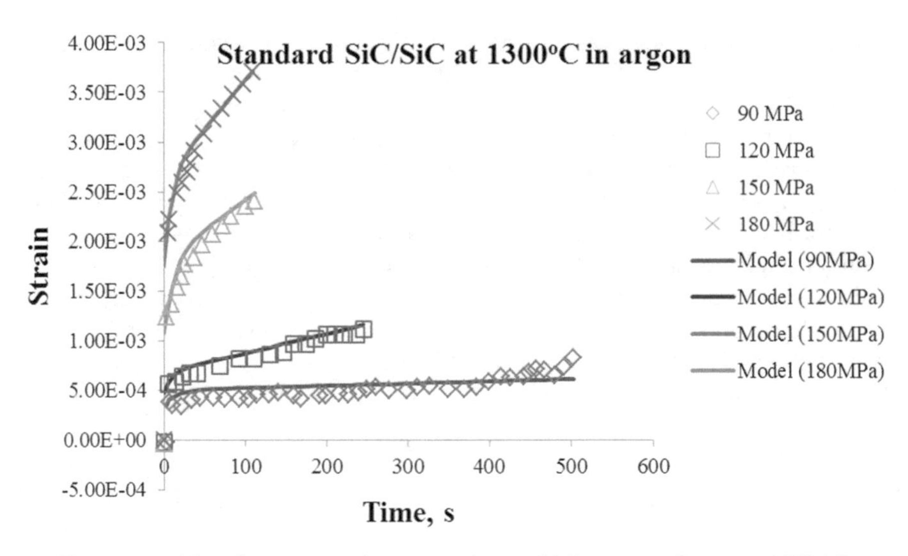

Figure 13.6. Tensile creep strain versus time at high stresses in argon 1300 °C.

the model is found in good agreement with the experimental observation. Particularly, at high stresses, there is only a transient creep stage due to fiber creep. The model infers that at high stress levels with matrix cracking occurring instantaneously upon loading, creep proceeds by GBS in the fibers. It has been shown by Wu and Koul that GBS is responsible for transient creep phenomena, typically with no appreciable tertiary stage. Therefore, when the last load-bearing constituents—fibers—are broken, the CMS ruptures without the apparent tertiary stage.

Table 13.2. Parameters of Eq. (13.4) for SiC/SiC fiber creep at 1300 °C in argon

A_2	p	β	H (GPa)
1.0×10^{-14}	3.95	1.1	165.3

13.4 CMC Fatigue

Fatigue behaviour of SiC/SiC composites has also been investigated by Zhu et al. (1999). The cyclic fatigue lives at room temperature (RT) and 1000 °C in argon are shown in Figure 13.7. The stress-life curve at RT can be described by a Basquin-type equation, as indicated by the best-fit line shown in Figure 13.7; however, the behaviour at 1000 °C deviates away from this relationship at high cycles ($>10^4$) at stresses lower than 180 MPa. The fatigue failure mode at RT is similar to tensile failure with abundant pullout fibers. Cross-sectional examination revealed that fatigue crack nucleation and propagation at RT were mainly along the 90° fiber/matrix interface, whereas at 1000 °C many cracks initiate at the pores (Zhu et al. 1999). Under stress-controlled cycling at high temperature, materials ratchet with accumulative creep strain. The pore-initiated cracks can be classified as internally distributed damage (IDD), according to the ICFT (Chapter 4). Then, the fatigue strength is reduced by a factor of D related to the size of IDD, which is proportional to the ratcheting strain. The ratcheting strain can be calculated using Eq. (13.4) under a given cyclic loading profile. Ignoring the transient ascending period, the matrix ratchet strain under a triangular wave form can be evaluated as:

$$\varepsilon = \frac{\sigma_{max}}{\beta^2 H} + \left(\frac{N\tau}{(p+1)\Delta\sigma}\right)\left(\frac{Ab}{E^p d}\right)\left(\sigma_{max}^{p+1} - \sigma_{min}^{p+1}\right) \tag{13.6}$$

where N is the cycle number and τ is the cycle period.

Eq. (13.6) indicates that the ratchet strain increases with cycle number. Indeed, as the stress cycle extends at high temperature, the fatigue strength reduces significantly, in a fashion similar to the behaviour of HCF with prior creep strain, as discussed in Chapter 8. This failure mechanism does not happen at room temperature.

Obviously, at a given stress level, the mechanism with the shortest crack nucleation life dominates the fatigue process in CMCs. It can be reasoned that at high stresses, cracks may initiate rather quickly along the weak 90 °C fiber/matrix interfaces, but at low stresses when the interface can hold tight, cracks may initiate alternatively at pores. Using Eq. (4.18) with no oxidation term, both the RT and 1000 °C-argon fatigue behaviours are described, which are in agreement with the experimental observations, as shown in Figure 13.7.

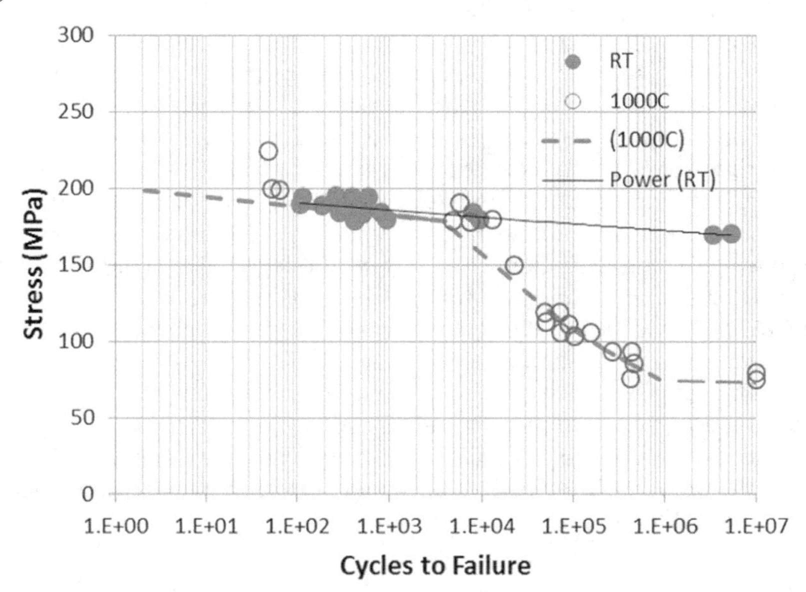

Figure 13.7. Maximum tensile stress versus cycles to failure in SiC/SiC composite at room temperature and 1000 °C.

13.5 CMC Component Design Considerations

Under harsh operating conditions of gas turbine engines, all CMC materials exhibit property degradation due to environmental exposure such as temperature and moisture, and salt, etc., as well as time-at-load effects. Therefore, CMC components must be designed to the "end-of-life" allowables. In stress analysis, CMCs are generally treated as an orthotropic continuum material, which can be dealt with using the classical composite theory. However, there is still lack of a universal model to address life prediction of CMCs under operational conditions, because the synergistic interactions between loading and environmental effects are not well understood yet. Currently, because of lacking the "end-of-life" strength data, components can only be designed to a limited life based on the existing short-term test data, and design improvements is an ongoing exercises as experience is gained through the service (Miller 2000). However, this "look-as-you-go" approach is bound to encounter surprises that may be either costly or catastrophic.

Through the analyses of tensile, creep and fatigue test data for a SiC/SiC as illustrated in sections 13.2-4, we have shown that the mechanism-based ICFT offers an insight into understanding the CMC behaviours. The agreement of the analysis with experimental observations strongly supports the mechanism-delineation of fiber/matrix creep as controlled by GBS. At high stresses, crack initiation may occur instantaneously from defects at fiber/matrix interface. Matrix GBS creep is responsible for matrix cracking during creep and crack initiation from pores during fatigue at lower stress amplitudes. As Eq. (13.3) implies, both the compliance of CMC (the total strain over stress) will increase with either cycles or time-at-load, as cracks develop within the CMC, which has been observed in both creep and fatigue tests. Thus, assuming that the physical deformation and damage mechanisms continue to operate as long as the loading is sustained cyclically or at a constant level, the evolution of material properties and damage can be traced with the mechanism-based model for life prediction. Even for CMC, the holistic damage evolution rule, Eq. (4.18) as derived based on the process of fatigue crack nucleation and propagation in coalescence with internally distributed damage, still applies, as demonstrated in the previous section. For service applications, environmental effect assisting surface crack nucleation and growth should also be included. The particular environmental degradation mechanisms need to be first identified and understood, then appropriate formulation may be developed, which is beyond the scope of this book for now. In short, ICFT is promising to provide a mechanism-based life prediction framework for CMC, but more extensive research and analysis need to be conducted.

CHAPTER

14

Component—Level Life Cycle Management

In industries, product life cycle management (PLCM) is the process of managing the entire life cycle of a product from inception through engineering design and manufacture to service, maintenance and disposal. Life prediction is an integral part of this process. It provides the basis for warranty, and also for inspection and overhaul intervals to ensure safe operation of high-value equipment. Especially, with integrated sensor technology, the health conditions may be monitored to allow diagnosis and prognosis for each individual system based on its actual usage. This is the current trend shifting from reactive maintenance by pre-determined schedule to condition-based proactive maintenance. The ultimate goal of PLCM of large engineering systems such as aircraft, nuclear reactors, pipelines, bridges, ships and offshore platforms etc. is to shorten the lead-time of product development and prevent costly/catastrophic failure in service.

This chapter focuses on the gas turbine component design life and health management philosophies with case study examples. Design and maintenance of other engineering components may follow the same approaches, depending on the criticality of failure.

14.1 The FAA Lifing Requirements for Gas Turbine Engine Components

The Federal Aviation Agency (FAA) of the United States issued an Airworthiness Circular AC33.70-1, specifically addressing lifing issues in design of gas turbine components. It requires a systematic engineering plan that includes flight (mission) profile analysis, performance analysis, fluid dynamics and heat transfer analysis, and stress analysis, with all the results feeding to the life prediction algorithm, as summarized in Figure 14.1. Of course, it also requires validation of the analytical results and maintaining the structural integrity attributes with cyclic rig tests and cyclic engine tests.

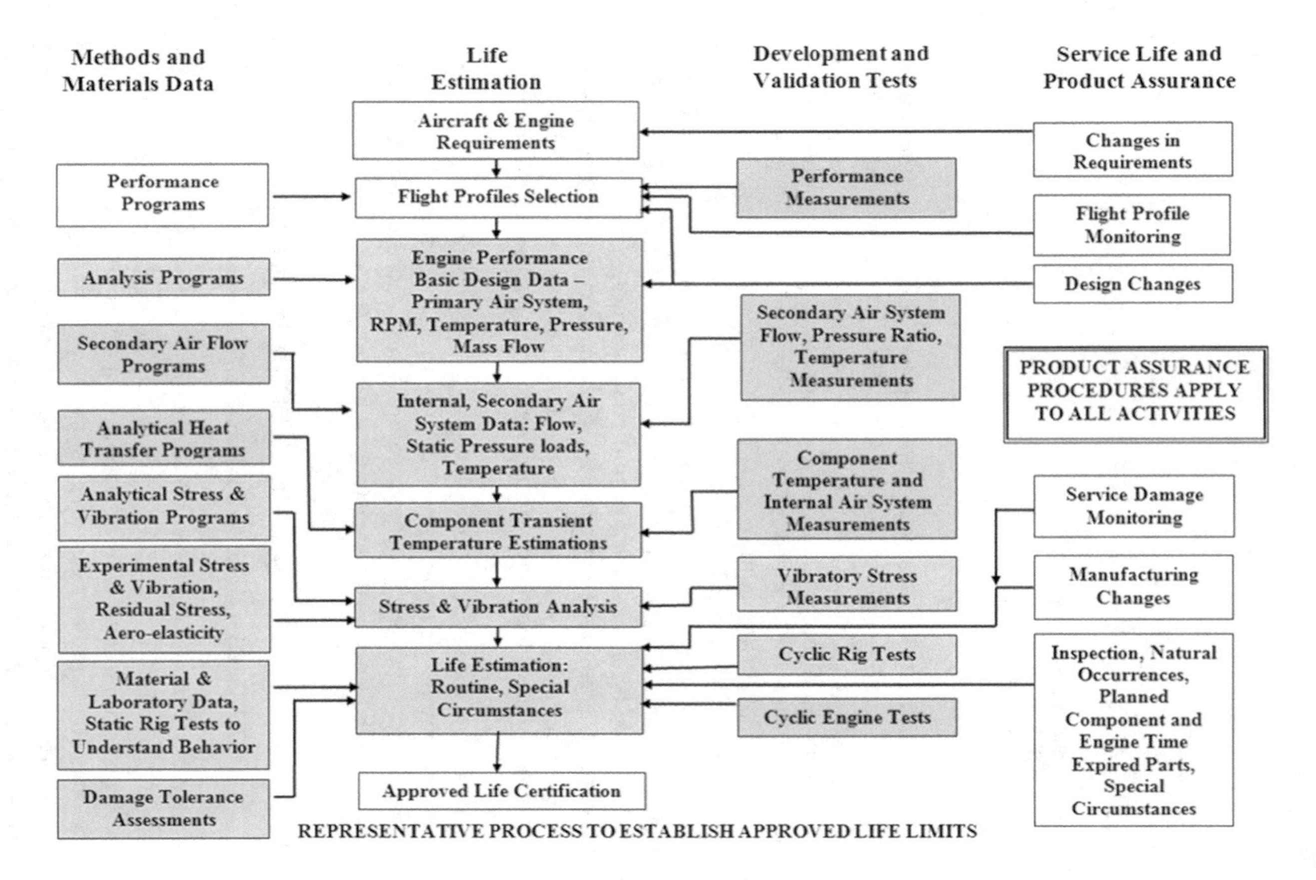

Figure 14.1. FAA-AC33.70-1 lifing requirements.

The FAA-AC basically adopts the safe-life approach for gas turbine engine component lifing. In this approach, materials are assumed to be perfect continuum before failure, and components reaching the safe-life limit must be retired. Actually, this AC only describes the lifing process to the point of approval of entering into service with planned inspection intervals. The past experience of operating gas turbine engines tells us that safe-life parts are not always safe, because defects and discontinuities are inherently associated with manufacturing processes and damage also occurs in the component during service, which leads to unexpected failures. Therefore, FAA-AC33.70-1 also requires that the damage tolerance approach must be used as a supplement to safe-guard the safe life of critical gas turbine components.

14.2 Aircraft/Engine Component Damage Modes

Aircraft/engine component damage induced by the operating conditions varies in a wide range. The damage may be external affecting dimensions and surface finish of a component, as a result, for instance, of erosion, foreign object damage (FOD), wear, corrosion or oxidation. These forms of damage affect the aerodynamic performance and load bearing capacity of gas path components. Conversely, surface cracks and notches induced by low cycle fatigue (LCF), fretting-wear or FOD may lead to high cycle fatigue (HCF) failures under vibration. The damage may also be internal, affecting microstructure of the hot and highly stressed parts, as a result of metallurgical aging reactions, creep and/or fatigue. This form of damage may reduce component strength and lead to component distortion. Its accumulation may cause the initiation of flaws, which may lead to cracking and component failure. Table 14.1 summarizes the generic forms of damage known to affect engine components by type (Immarigeon et al. 2001). These damage modes can be classified into two groups: (I) surface morphology damage; and (II) material internal damage. The implications of these damages on component life are briefly discussed below.

Corrosion can cause various forms of surface damage including pitting and intergranular delamination. Pitting is most notorious because it acts as crack initiators, which can subsequently lead to fatigue crack growth. Steels, aluminum and titanium alloys are particularly susceptible to this form of damage. Corrosion can attack both airframe as well as engine components when salts are ingested into the engine. Foreign object damage (FOD) is another form of surface damage: it can cause nicks and dents on fan blades, acting as crack initiators. Erosion, on the other hand, causes more gradual change in the shape and surface finish of compressor airfoils. Such changes, in addition to reducing compressor efficiency, may lead to resonant excitation and HCF failures of airfoils. In the lifing process, the stochastic nature of FOD, and the progression of surface damage by corrosion and oxidation should be evaluated based on the time of exposure and environmental chemistry.

As discussed in previous chapters, material deformation damage occurs by plasticity and creep. Plasticity is particularly responsible for surface/subsurface crack nucleation under fatigue loading, whereas creep tends to promote internally distributed damage such as grain boundary cavities or cracks. When engines are in operation, metallurgical aging reactions can occur in hot parts, such as turbine blades and vanes. For example, rafting of the γ-γ' structure may occur in single crystal turbine blades during service exposure, as shown in Figure 14.2. Creep strength may also degrade, as shown in Figure 14.3 for nickel base superalloy IN713 (Immarigeon et al. 2001). Loss of creep strength may lead to excessive distortion of hot parts. Vane airfoils may bow while blades may lengthen or untwist. In aircraft engines, in order to protect the components from direct environmental and thermal attacks, turbine blades and vanes are often coated with protective coatings such as environmental or thermal barrier coatings. But due to differences in chemistry, reactions may occur between coating and bond coat/substrate, forming thermally grown oxide layers, precipitate free zone or inter-diffusion zone. The formation of these layers or zones may alter the material properties and thus change the crack nucleation mechanism as opposed to the original substrate material.

Table 14.1. Life-limiting damage modes of turbine engine components

Section	Component	Failure mode
Fan	Blades	FOD, HCF
Compressor	Blades	FOD, ER, COR, HCF
	Vanes	FOD, ER, COR, HCF
	Discs	LCF, C, HCF
	Spacer	LCF, C, HCF
Turbine	Blades	TMF, C, HC, LCF, HCF
	Vanes	TF, HC, C, HCF
	Discs	LCF, C, HCF
	Torque Ring	LCF
Combustor		LCF, TF, C, HC
Shaft		LCF, WR
Compressor discharge case		LCF, COR
Rotating seal		LCF, C, HCF

Abbreviations: FOD: Foreign Object Damage | HCF: High Cycle Fatigue | LCF: Low Cycle Fatigue | TMF: Thermomechanical Fatigue | TF: Thermal Fatigue | ER: Erosion | COR: Corrosion | C: Creep | HC: Hot Corrosion | WR: Wear |

Material internal damage can be regarded as the result of all above metallurgical reactions as well as plastic and creep strain accumulation. It is an insidious form of damage because, in contrast to surface damage, it cannot be readily detected by non-destructive inspection (NDI) techniques. Therefore,

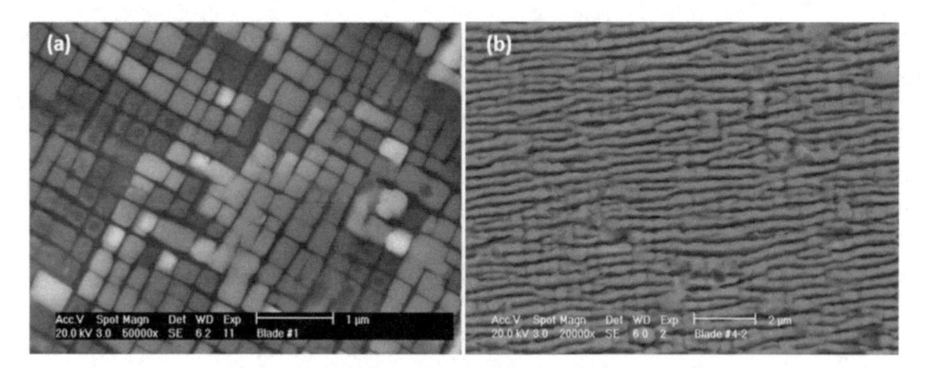

Figure 14.2. Microstructure of single crystal turbine blades: (a) new blade microstructure and (b) rafted microstructure of service-expose blade.

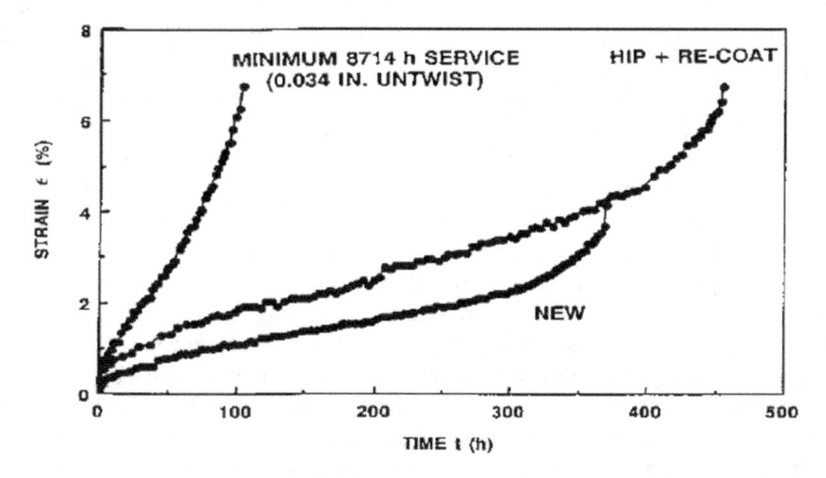

Figure 14.3. Creep curves of new, service-exposed and HIP rejuvenated IN713LC blades.

it needs physics-based models to predict microstructural evolution, lifetime and failure mode of the component under service conditions. The ICFT elucidated in this book is suitable to cover the roles of various mechanisms played in material degradation and failure.

14.3 Safe-Life Approach

Traditional maintenance approach is based on the "safe life", or the life to crack initiation. The part retirement/replacement is prescribed with fixed cycle or time intervals based on the duty-cycle usage. By "safe-life", component failures often occur as surprises. Traditionally, the cycle count is derived from the Coffin-Manson type equation and the time-at-temperature is derived from the Larson-Miller method. The linear damage rule (LDR), Eq. (4.12), is often used to count the total damage when both cyclic and dwell loadings are involved. In Chapter 6 to 8, many cases have been shown where the LDR does not satisfactorily represent the true creep-fatigue interaction. Inaccuracy of LDR has also been widely reported in the open literature (Lloyd and Wareing 1981, Rees 1987, Inoue et al. 1991, Spindler 2007). The problem complicates itself in actual service when engines in different fleets are operated by different users, variations in mission profiles often result in life consumption that deviates from the design duty cycle, which can lead to either premature failure or wasteful replacement of parts. Therefore, the safe-life approach has to be upgraded to deal with life consumption by actual usage.

14.4 The Damage Tolerance Approach

The damage-tolerance analysis (DTA) philosophy assumes that materials or components entering into service have defects in their initial conditions. Then the component life is basically the life of crack propagation starting from an initial flaw to a dysfunction size (the critical fracture crack size divided by a safety factor), according to ENSIP (2002). In a structural integrity program, this means that critical components have to be periodically inspected, to check for the existence of cracks. The period of such inspection is determined based on crack growth analysis by fracture mechanics concepts. The damage tolerance life is usually obtained by integration of the crack growth rate equation such as the Paris law from the initial crack to the critical fracture condition. In such a calculation, whether to assume the existence of fatigue threshold would have a huge impact on the total fatigue life. If fatigue threshold existed, how could cracks develop from the original continuum condition (zero crack size, and hence zero stress intensity factor) to a finite size (crack initiation) remains to be a question.

In life cycle management programs, damage tolerance assessment is used to complement the safe life approach to avoid catastrophic failures due to the presence of "anomalies". On the other hand, the damage tolerance approach, if rigorously implemented, can be seen as a way of extending the life of safe-life expired parts, especially for overly-designed components with conservative safety factors. The damage-tolerance-based life extension philosophy is represented in Figure 14.4. Safe-life expired parts are first inspected. If cracks are found, the part is retired from service. If no actual crack is found, an initial crack is assumed to exist in a size equal to the non-destructive inspection (NDI) limit, and hence the crack propagation life is evaluated to determine the next safe inspection interval (SII) on the maintenance schedule. This way, by virtue of damage tolerance, safe-life expired parts can repeatedly enter into service, until cracks are found on the part. This is called *retire for cause* (RFC). Since crack propagation life is sensitive to the initial crack size, implementation of a damage-tolerance-based life extension scheme relies on advanced NDI techniques with high accuracy and low detection limits.

In DTA, evaluation of stress intensity factor (SIF) is critical. Many tools have been developed for this task. First of all, for simplified cases, there exist

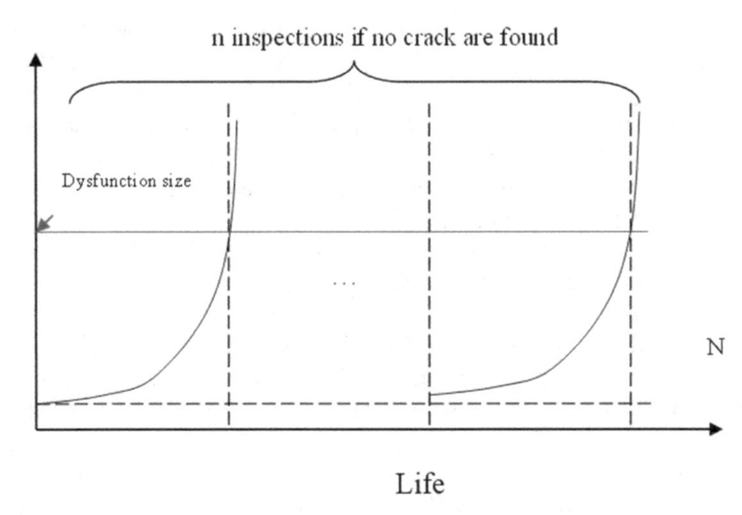

Figure 14.4. Schematic of damage–tolerance life management.

many analytical solutions such as included in MSC. Fatigue, NASGRO and AFGROW libraries with appropriate correction factors. For 3D complicated-shape components, no analytical solutions are generally available; then fracture mechanics analyses must be performed using FEM, which is often cumbersome in meshing the cracked body. To date, some auto-meshing codes such as Zencrack and FRANC3D have been developed to facilitate DTA. Also, the weight function method is an interesting choice for crack growth analysis, which offers the expedience without the need to mesh the crack, e.g., DARWIN©. However, DARWIN only performs stress integration along the major and minor axes of an elliptical crack, which is not a full 3D consideration. A weight function method formula that takes the 3D stress distribution in the integration over the crack surface will be given in a later section.

14.5 The Holistic Structural Integrity Approach

The holistic structural integrity process (HOLSIP) includes the four stages of life evolution: i) crack nucleation, ii) small crack growth, iii) long crack growth, leading to iv) unstable fracture. The holistic life approach is not simply a hybrid approach combining safe-life and damage-tolerance, because there is a missing bridge in between. Currently, the safe-life is defined as the life of "crack initiation" to an arbitrary size, e.g. 0.03 inch (0.8 mm); whereas damage tolerance considers the existence of an initial flaw size as defined by the probability of detection (POD) of the chosen NDI technique, and the fracture mechanics analyses of DTA are mostly based on the long-crack data. In the holistic life process, the size and shape of physical damage evolves with deformation history, as described in the previous chapters of this book. Particularly, surface crack occurs by fatigue and environmental mechanisms, whereas internally distributed damage is mostly promoted by creep or other thermally activated processes. All need to be considered under the influence of component thermal and stress fields in the service history. In HOLSIP, microstructure also plays an important role, which may change with the service exposure. Therefore, before launching into the holistic life cycle management scheme, the physics of failure has to be clearly understood. In the previous chapters, various basic damage processes in association with deformation mechanisms have been formulated. Detailed characterization of microstructural evolution and its effects on deformation and damage accumulation mechanisms are needed for HOLSIP.

In recent years, the concept of integrated vehicle health management (IVHM) has emerged, which integrates sensing, diagnosis and prognosis capabilities to assess the current health state and predict the future health state of an engineering system. It calls for the HOLSIP methodology. As sensors enable us to monitor the operating condition of an engineering system, it becomes eminently needed to update part life consumptions based on the historical usage profile and calculate the remaining useful life for mission planning and maintenance decision-making. In service, gas turbine engines are often operated with mission profiles that deviate from the design duty cycles, because of the specific needs of the operators, which will have an impact on actual life consumption. The safe-life based fixed-schedule maintenance strategy cannot meet this requirement, especially when the "safe-life" is estimated by empirical relations. The HOLSIP-based life prediction methods are needed in shifting the paradigms from reactive maintenance to proactive maintenance.

Among all issues of IVHM, understanding the physics of failure and predicting the lifetimes of critical components are at the core. For example, to monitor the structural integrity, one should know when, where and by what mechanism(s) cracks may start to develop under mission loading profiles that are usually combination of cyclic, dwell and vibrational loading. Gas turbine engine components usually operate under extreme thermal-mechanical loading in a corrosive environment. The combined effects of mechanical loads and temperature, compounded with microstructural degradation and environmental effects, often induce multiple deformation and damage modes i.e. corrosion/oxidation, resulting in mix-mode failures, as listed in Table 14.1. The challenge is to predict the failure mechanism(s), mode(s) and lifetime where it would happen on the component. Therefore, one needs a mechanism-delineated theory to capture the evolution of life in gas turbine components.

The ICFT, as given in the context of Eq. (4.16), describes the holistic damage accumulation process consisting of surface/subsurface crack nucleation and its propagation in coalescence with internally distributed damage/discontinuities, which naturally undergoes the aforementioned four stages. Its simple integration form, Eq. (4.18), has been used to describe material coupon life, as demonstrated in the previous chapters. For a structural component, the rate form for each stage needs to be formulated with particular consideration of the scale effect, i.e., with regards to cracks at microscopic and macroscopic scales, as dealt with in Chapters 6 to 10. The total component life is therefore comprised of material "point" life at crack nucleation location, and subsequent crack propagation life (from microstructurally short crack to macroscopic long crack) in the component thermal-stress field, because the crack growth driving force, e.g., K, depends on the component geometry and stress distribution. The safe-life approach and the long-crack damage-tolerance approach address the two issues separately without a common definition of crack initiation. The gap between the point failure in a continuum and crack growth with a macroscopic size needs to be filled based on the physics of damage.

In the proceeding chapters, we have dealt with the four stages of damage accumulation separately, based on the rate theory and kinetics of dislocation mobility and micro-mechanisms considering interactions of dislocation movement with microstructures. Particularly at high temperatures and in the presence of an aggressive environment, the deformation and damage processes often occur not by one but multiple mechanisms, which are compounded with microstructural degradation to induce multiple modes of damage accumulation and fracture. A complete description should include:

- Definition and description of crack nucleation process, as a result of competition of multiple deformation and damage mechanisms
- Material microstructural degradation and nucleation and growth of internally distributed damage
- Description of microscopic crack growth in coalescence with internally distributed damage occurring by multiple deformation and damage mechanisms under local stress distribution
- Description of macroscopic crack growth in coalescence with internally distributed damage occurring by multiple deformation and damage mechanisms in the component thermal and stress field.

Therefore, for IVHM, it is necessary to incorporate ICFT into the structure integrity engineering plan (Figure 14.1), which applies the aforementioned analysis and validation methods to individual or fleet of vehicles with consideration of particular mission (flight) profiles, to evaluate life consumption and the potential failure mode. With physics-based models and mechanism genome database, a virtual system—the digital twins—may be established as the ultimate tool of platform maintenance and mission planning.

14.6 Case Study—Nozzle Guide Vane

14.6.1 Stress-Strain Analysis

A gas turbine nozzle guide vane (NGV) is selected as an example to demonstrate the crack nucleation analysis in HOLSIP. What is different from

Figure 14.5. Temperature distribution in the NGV.

Color version at the end of the book

the traditional safe-life analysis is that here we use the mechanism-based ICFT to describe the material behaviour and life prediction such that the mechanism-strain components and failure modes are identified at the failure location(s). Material selection choices are also discussed for improved design and maintenance.

A FEM model of the component was built with hexahedron solid elements. The thermal fluid dynamics and heat transfer problem of the NGV was solved using Fluent, and the aerothermal loads were applied as the boundary condition of the solid model (Wu et al. 2017). The temperature distribution in the airfoil is shown in Figure 14.5. Two materials are considered for comparison: material A is Mar-M 509, which is described by the constitutive model given in Table 14.2; material B has the same tensile and fatigue properties but twice the creep rates. These material models are implemented into MSC. Marc for the FEM analyses.

Stress-strain analyses were performed for the NGV under take-off and cruise conditions. The von Mises stress distribution at the maximum power during take-off is shown in Figure 14.6. The accumulation of plastic and creep strains in the NGV are shown in Figure 14.7 and 14.8, respectively. It is evident that the maxima of both occur at the corners of the upper and lower gas outlets, which are deemed to be critical fracture locations of low-cycle thermal fatigue. The accumulation of the equivalent plastic and creep

Table 14.2. Mechanism-based constitutive equations of material A

Mechanism	Equation (σ in MPa)	Parameter (T in Kelvin)
Plasticity	$\sigma = \sigma_0 + K\varepsilon_p^{1/n}$ ($\sigma_0 = 0$ in this case)	$K = 1.3261\times10^{-3}T^2 - 1.8479T + 1.4696 \times 10^3$ (MPa) when $T < 973$ K $K = -1.8T + 2.678 \times 10^3$ (MPa) when $T > 973$ K $n = 8.77$
Dislocation glide	$\dot{\varepsilon}_g = B\sigma^n$	$B = 8.5 \times 10^{-21}$ *exp(−73866.5/T) (1/s) $n = 18.32$
Dislocation climb	$\dot{\varepsilon}_c = (1 + M\varepsilon_c)C\sigma^n$	$C = 8.258 \times 10^7$ *exp(−101454/T) (1/s) $m = 8.961$ $M = 4000$
Grain boundary sliding	$\Delta\varepsilon_{gbs} = \dot{\varepsilon}_s\Delta t + \frac{\dot{\varepsilon}_s\Delta t}{(\beta - 1)}\exp\left(-\frac{\beta^2 H\dot{\varepsilon}_s t}{\sigma(\beta - 1)}\right)$ $\dot{\varepsilon}_s = A\sigma^p$	$A = 8.33 \times 10^{-21}$ exp(−13034/T) (1/s) $p = 2$ $\beta = 1.05$ $H = 0.24$*E (MPa)

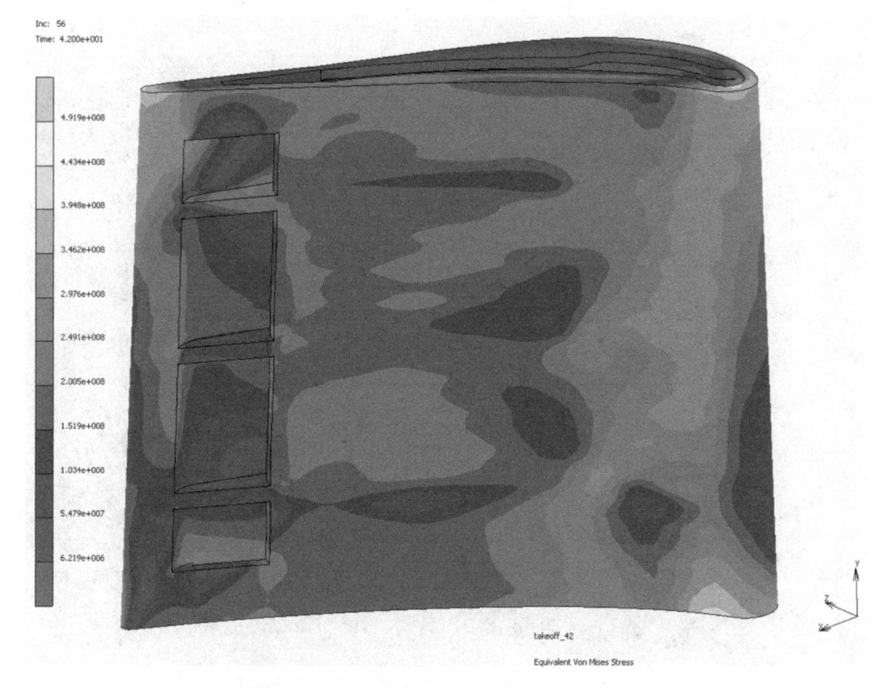

Figure 14.6. Stress distribution in the NGV (material A).

Color version at the end of the book

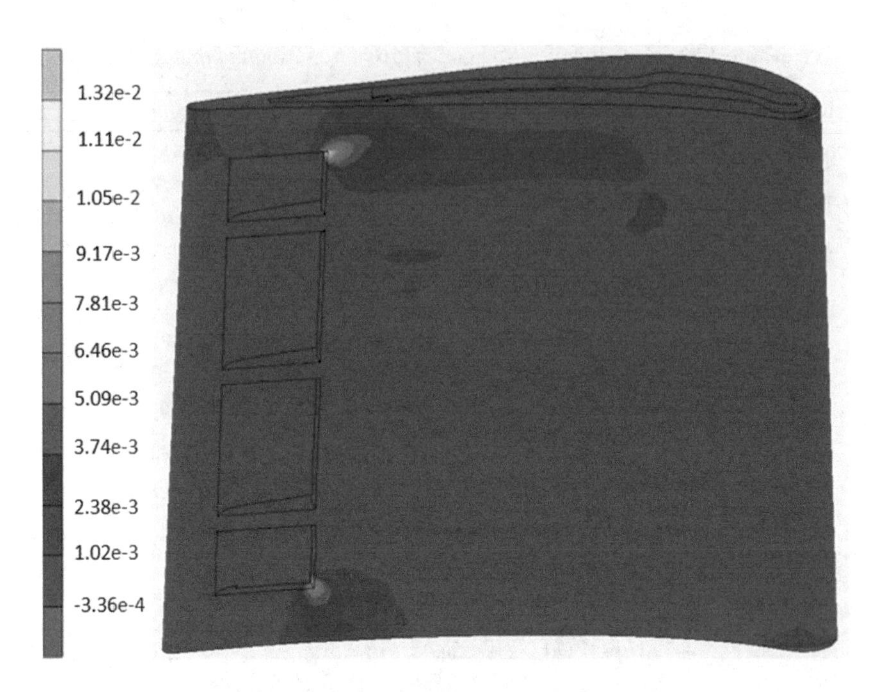

Figure 14.7. Locations of plastic strain accumulation (material A).

Color version at the end of the book

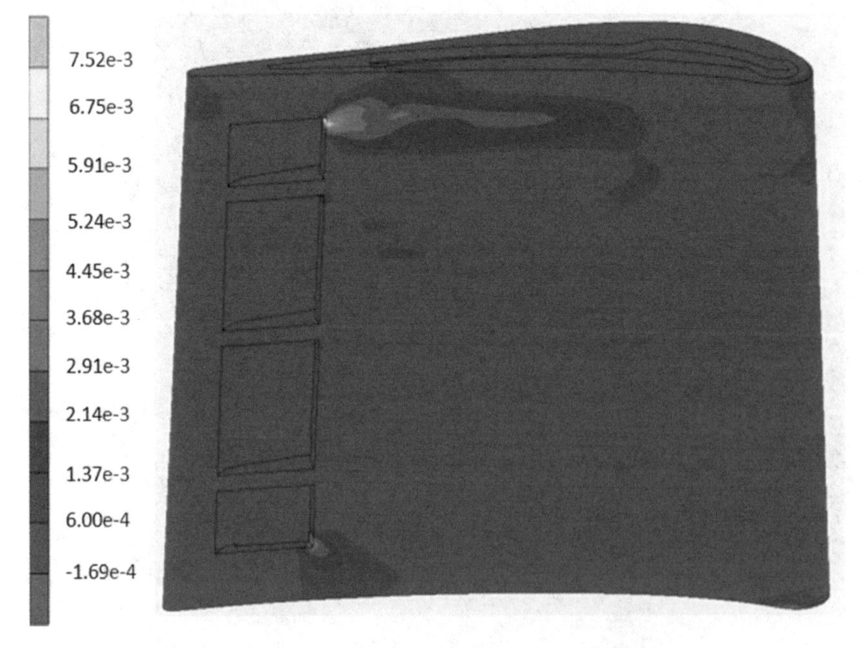

Figure 14.8. Locations of creep strain accumulation (material A).

Color version at the end of the book

strains, and von Mises stress with time at the corner node during take-off are shown in Figure 14.9, where the time is normalized by the standard take-off time ~42 sec. as defined by the International Civil Aviation Organization. The maximum von Mises stress occurs almost immediately at engine start-up, which induced maximum plasticity. However, as soon as creep strain starts to accumulate during the take-off, the stress is quickly relaxed. The maximum creep deformation is attained to a level nearly half of the plastic strain at the end of taking-off. When material A is substituted with material B, the maximum stress and plastic strain are almost the same, by the same mechanism of plasticity, but the creep accumulation in material B is larger and hence the stress is more relaxed at the end of take-off, as shown Figure 14.9.

The responses of material A and B are also compared for the cruise condition for nearly 2×10^6 sec, as shown in Figure 14.10. It is interesting to see, surprisingly, that creep strain accumulation is much larger in NGV of material A than material B during the cruise, even though the former is supposed to be creep resistant. The phenomenon can be reasoned as follows. The material B that exhibits higher creep rate yields at lower stress upon loading and relaxes faster as the total strain is constrained by the thermal gradient in the component. Therefore, the lower stress profile in the material B-NGV generates less creep strain under the subsequent cruise condition. On the contrary, material A that exhibits lower creep rate yields at high stress upon loading and relaxes slower, so the high stress profile generates a higher creep strain accumulation over time. Also, the creep strain accumulation mechanisms in the NGV are different during take-off and cruise. During

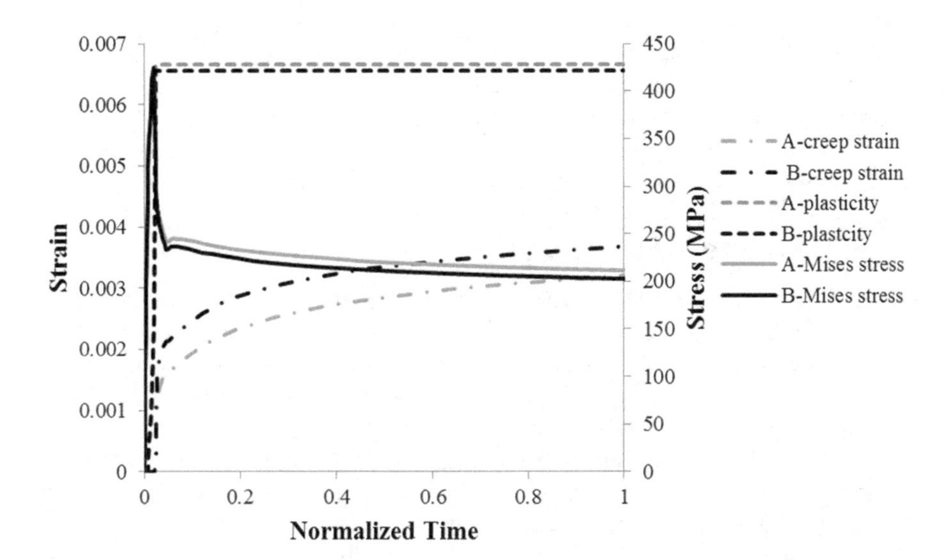

Figure 14.9. Accumulation of plastic and creep strain and relaxation of von Mises stress during take-off.

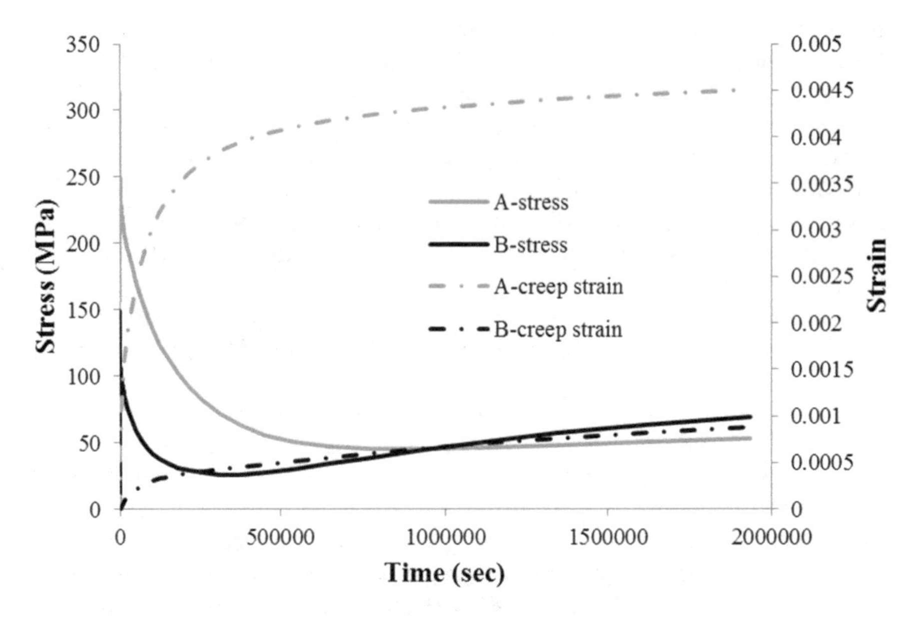

Figure 14.10. Accumulation of plastic and creep strain and relaxation of von Mises stress during cruise.

take-off, the initial stress relaxation occurs mostly by intragranular creep mechanisms; but during most of the cruise time, creep strain is accumulated by GBS, when the stress drops down below 200 MPa at the cruise operating temperature. According to the simulation under the strain-controlled condition as induced by thermal gradients, creep resistance generally results in a high stress profile, which would adversely affect fatigue life. Therefore, component design against creep-fatigue interaction should consider the separate contributions of plasticity and creep strains to evaluate which mechanism has the most detrimental effect on life. The ICFT can describe the component failure mode as induced by the underlying mechanisms.

14.6.2 Failure Mode Analysis

According to the simulation results, take-off creates the maximum plastic strain, which would contribute to fatigue; while creep strain accumulation contributes to internally distributed damage, even though it is moderate (<1%) after long cruise. Figure 14.11 shows a service-induced crack emanating from the corner of gas outlets, which corroborates with the critical fracture location identified in the simulation. Figure 14.12 shows the fracture mode of a service induced crack. It is seen that the fracture surface is predominantly transgranular in nature, covered with cobalt oxides. In addition, a few grain boundary facets appear, indicating that creep also plays a secondary role. For engineering purpose, the crack initiation size can be estimated by the distribution size of the life-limiting strain amount. Crack growth analysis is not performed because NGV is not a safety critical component, but from the strain distribution pattern shown in Figures 14.7 and 14.8, it can be envisaged that crack growth would proceed along the maximum strain controlled paths, which actually occurred as shown in Figure 14.12.

To further understand the relative contribution of each deformation/damage mechanism, here we construct the LCF mechanism map, as shown in Figure 14.13, where the life fractions (N/N_f for fatigue, and Nh/a_c for oxidation, and the total strain is indicated for each curve) are plotted against the homologous temperature (T/T_m, where T_m is the melting temperature in Kelvin). The effect of material's internal damage via creep ($1/D$) is represented by the solid curve. At any temperature, the sum of mechanical fatigue fraction and oxidation fraction is equal to $1/D$, according to Eq. (4.18). At low temperatures, the internal damage by creep is minimal, and the material's life is almost exclusively limited by fatigue due to rate-independent plasticity alone. As temperature increases, the internal damage grows by creep, and at the same time the oxidation life fraction, Nh/a_c, rises up to dominance.

Material selection for high temperature applications usually follow the "maximum strength" principle, that is, the material must have high tensile strength, high creep strength; and good oxidation resistance. Given the example as shown above, the intricate play of the above factors at the component level has to be evaluated by FEM analysis incorporating ICFT. Particularly, since the stress in a component that can be "shaken down" with accumulation of creep strains, resulting in stress redistribution. It seems that a creep compliant material with good oxidation resistance and minimal cavity accumulation seems to be a good choice to counter thermal fatigue. Actually, replacement of alloy B with alloy A did not result in twice service

Figure 14.11. A service-exposed NGV (alloy B) containing a crack.

Figure 14.12. Fracture mode of service-exposed NGV (alloy B).

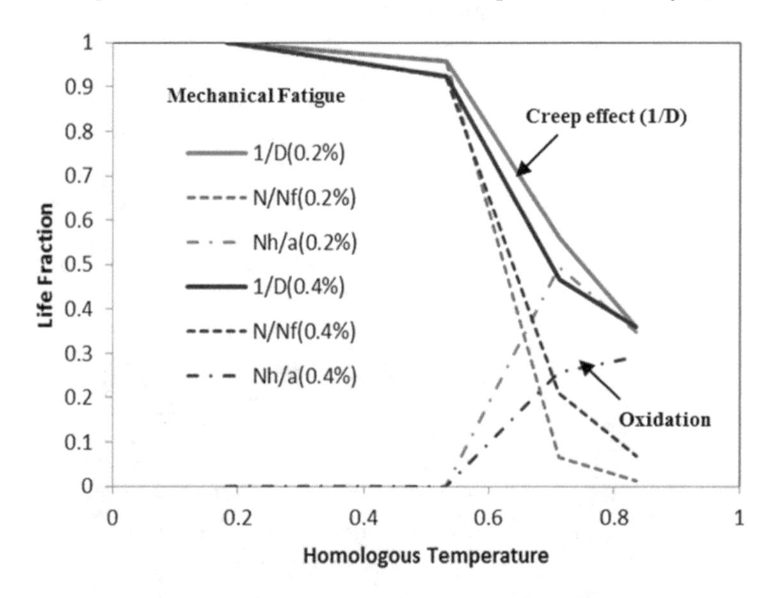

Figure 14.13. LCF mechanism map for Mar-M 509.

life, as its creep property would seem to promise. This further stresses on the importance of mechanism-delineation for the component life prediction.

14.7 Component Damage Tolerance Analysis

In either DTA or HOSIP, analysis of crack growth in 3D components is an important part of life prediction (the former starts with an NDI identified crack size, and the latter starts with a microstructurally-defined crack size). The process of crack growth in a 3D component is going through a 3D stress field induced by the service loading on that component, the characterization of which requires a robust and expedient method to calculate the crack growth driving force. In addition to the methods mentioned in section 14.4, here we propose a 3D weight function method.

14.7.1 The 3D Weight Function Method

The weight function method (WFM) is based on the reciprocal theorem of elasticity, by which the stress intensity factor of a crack can be obtained through integration of the stress distribution $\sigma(x, y)$ in the uncracked solid, multiplied by the weight function, as:

$$K_{P'} = \iint \sigma(P(x,y))m(P,P')dA_P \tag{14.1}$$

A 3D weight function $m(P(x, y), P'(x, y))$ has been proposed (Zhang and Wu 2011), modifying the formulation for embedded elliptical crack (Wang et al. 1998) with surface and boundary correction factors, as:

$$m(P,P') = F_s \frac{\sqrt{2s}}{\pi^{3/2}l^2}\sqrt{1-\frac{s}{8\rho_1}-\frac{s}{8\rho_2}-\frac{s}{8\rho_3}-\frac{s}{8\rho_4}}\left[\sqrt{\sec\left(\frac{\pi\lambda}{2L_1}\right)}\right]^{1.25-\frac{a}{c}} \tag{14.2a}$$

where

$$F_s = \begin{cases} 1 & \text{for an embedded crack} \\ 1.12 & \text{for a surface crack} \\ (1.12)^2 & \text{for a corner crack} \end{cases} \tag{14.2b}$$

The parameters of Eq. (14.2) are defined schematically in Figure 14.14, where l is the distance from P (the integration point) to P′ (the crack front point), s is the shortest distance from P to the eclipse; ρ_i (i = 1, 2, 3, 4) are the distances of the ellipse to the intersection point C on the major axis; λ is the normal line distance of the crack front point P' to the major axis C', L_1 is the distance between C' and the component boundary along this normal line extension, c and a are the major and minor axes of the ellipse, respectively.

The stress distribution on the presumed crack plane can be extracted from the FEM results for the uncracked component. Then, the integration, Eq. (14.1)-(14.2), is carried out over the entire elliptical crack domain to evaluate the stress intensity factor for the current crack profile, with the appropriate crack boundary correction factor, F_s. Crack advance is then evaluated based on the material's crack growth law and the crack profile is updated as a new elliptical crack (defined by crack increment along the major and minor axes). The above calculations are repeated until a dysfunction crack size is reached.

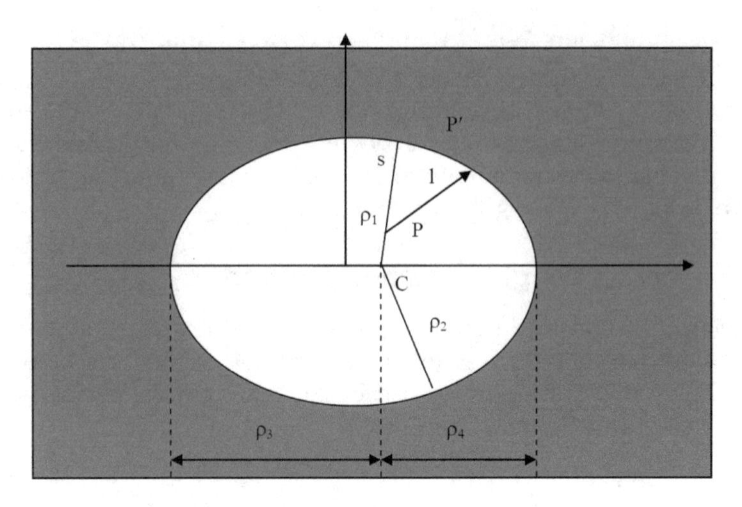

Figure 14.14. Weight function for an elliptical crack.

A computer code has been written in Patran Command Language to implement the WFM in MSC.Patran as a post-processing module such that damage tolerance analysis can be performed for a component after the FEM stress analysis (Zhang and Wu 2014). For the convenience of description, this code is called *3DCrackpro*.

As WFM requires, FEM stress analysis of the component with no crack must have been performed first. MSC.Patran can post-process the stress result files generated using most popular commercial FEM solvers such as MSC. Nastran, MSC.Marc, Abaqus and ANSIS. Hence, integration of 3DCrackpro with MSC.Patran allows the versatility to work with those FEM solvers to handle challenging DTA cases by user's choice. The details are described by Zhang and Wu (2014). Several case studies are shown in the sections below.

14.7.2 Validation with FEM for Single Notch Specimen

The first case study is performed on a rectangular bar, length 32 mm and width 10.68 mm, with a circular notch of radius 1.24 mm. This geometry represents a single edge notch specimen to study short fatigue crack growth. The stress intensity factors for this type of specimens have been evaluated by (Newman 1992), using FEM. This case is used to validate Eq. (14.1-14.2) and the integration algorithm. An FEM was created for the specimen (without crack) subjected to uniform tension of 500 MPa. The stress results are loaded into the Crack module, as shown in Figure 11.15. The WFM evaluation of the stress intensity factor along the front of a centered semi-circular crack is shown in Figure 14.16, in comparison with the Newman's FEM results and the finite element analysis. The differences between these results are below 7% over the entire angular range (Φ: $0 \rightarrow \pi/2$), which is quite satisfactory for engineering evaluation. Validation against other specimen-crack geometries are documented in the Ref. (Zhang and Wu 2011, 2014).

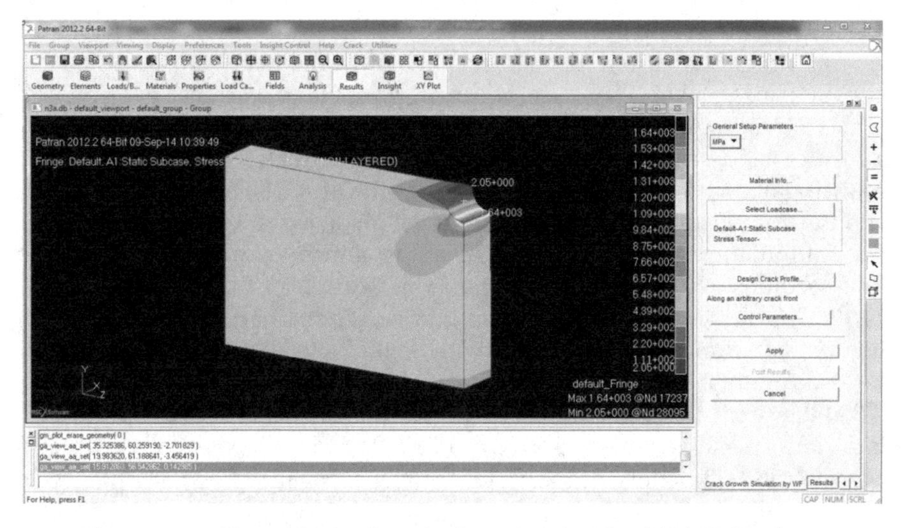

Figure 14.15. The analysis of notched specimen with added 3DCrackpro post-processing features in MSC.Patran.

Color version at the end of the book

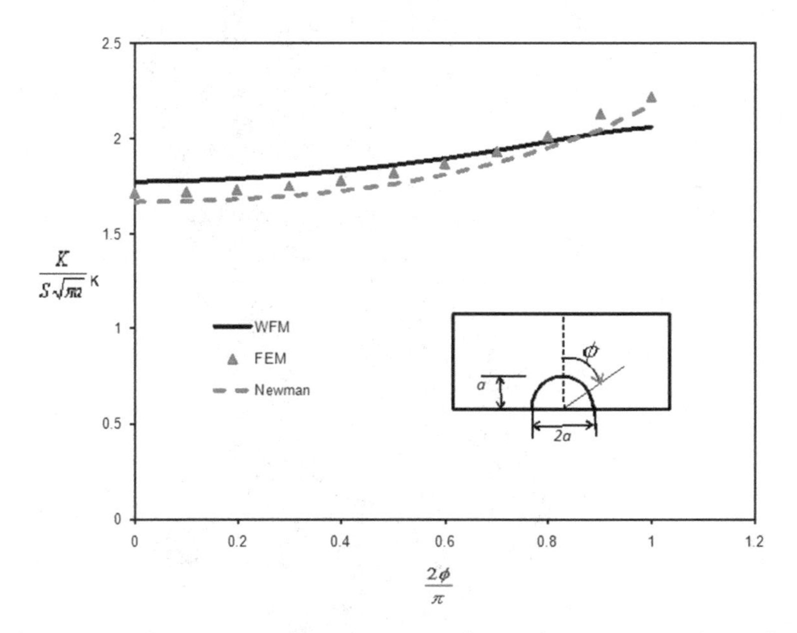

Figure 14.16. Comparison of stress intensity factors for a semi-circular surface crack in single notched specimen.

14.7.3 Fracture Analysis of Turbine Blade

During engine operation, turbine blade serrated roots usually experience the highest stress due to the centrifugal force induced at high rotation speed, and crack may be first initiated by fretting wear. If fractured, the loose blade may destroy other blades and also affect down-stream components, causing the engine to lose power. Therefore, failure of rotor blade is critical, even though it may be contained within the casing. A finite element model is created for a turbine blade and stress analysis is conducted with the consideration of centrifugal force and the blade contact with the disc. The finite element mesh and the von Mises stress distribution in the blade is shown in Figure 14.17.

An initial crack of semi-elliptical shape is inserted in the trough of the first serration on the pressure side, as indicated by the arrow in Figure 14.17. The principal stress over the first serration root plane is shown in Figure 14.18. Simulation of crack growth is conducted using the above weight function method. A parallel simulation is also conducted using ZENCRACK/ABAQUS with the same initial crack profile. The ZENCRACK block and local FEM mesh is shown in Figure 14.19. The simulated crack depth and aspect ratio as functions of the cycle number (normalized by its failure cycle

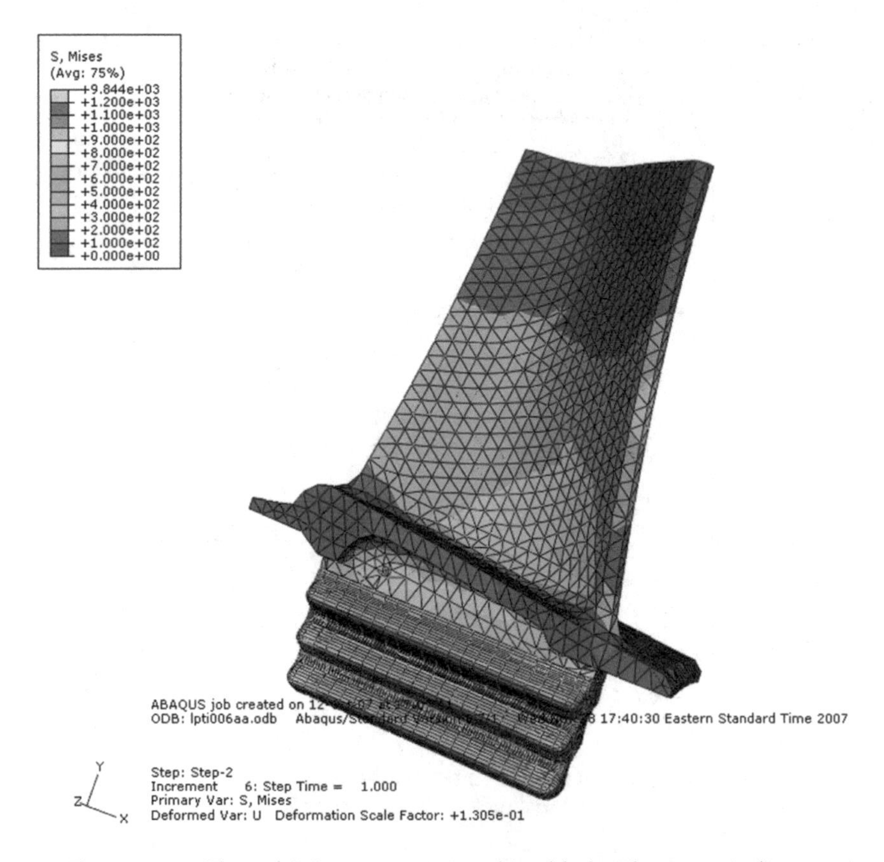

Figure 14.17. The von Mises stress in a turbine blade. The arrow indicates where a crack would form.

Color version at the end of the book

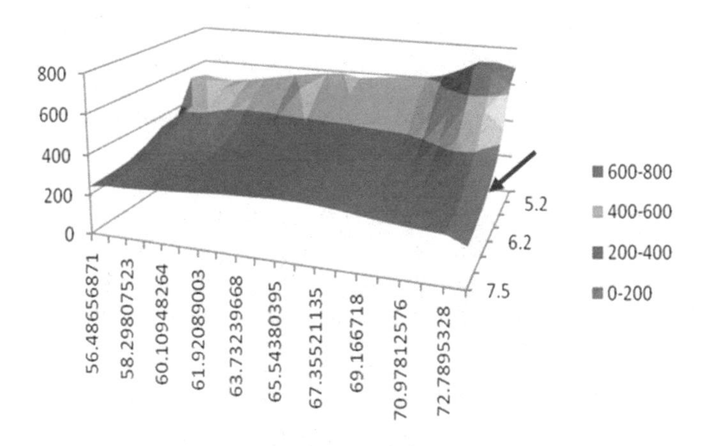

Figure 14.18. Stress distribution (in unit of MPa) over the first serration root plane (in unit of mm). The arrow indicates where the initial crack existed.

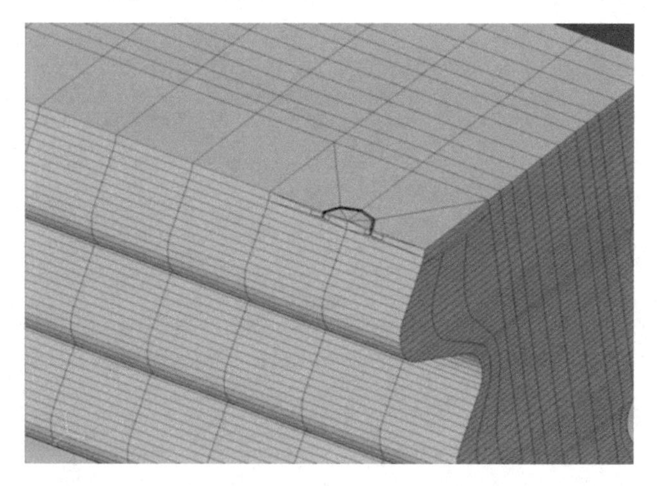

Figure 14.19. ZENCRACK mesh for the blade root crack.

number) are shown in Figure 14.20. As the crack depth increases, the crack aspect ratio a/c increases to a value of 0.93, and then slightly decreases. The aspect ratio change reflects the elliptical crack shape change under the effect of stress distribution on the cracking plane in the component. When the crack was small, the stress over the crack was almost constant, the maximum stress intensity factor of the semi-elliptical crack occurred at the deepest crack front, which drove the crack toward a "circular" shape. However, when the crack became large, the stress distribution would have an effect. Since the stresses were relatively high at the notch root surface, then the stress intensity factor at the surface point began to accelerate, which then drove the crack to grow faster in the surface length direction. The WFM can easily simulate the entire crack growth process. The initial crack propagation profile by ZENCRACK simulation agrees with the weight function profile very well, but it only ran to a depth of 1.4 mm, due to excessive distortion of the mesh, as the crack approach the boundary of the inserted crack block. Also, ZENCRACK took a much longer time to complete the simulation. The WFM evaluated stress intensity factors at both the surface and the deepest points are shown in Figure 14.21.

Post-mortem examination of the fracture surface revealed: a) the initial and b) the final crack profiles, as shown in Figure 14.22. The weight-function-method crack growth simulation agree with the fractographic observations. Indeed, the initial semi-elliptical crack grew into a nearly semi-circular shape on the actual component.

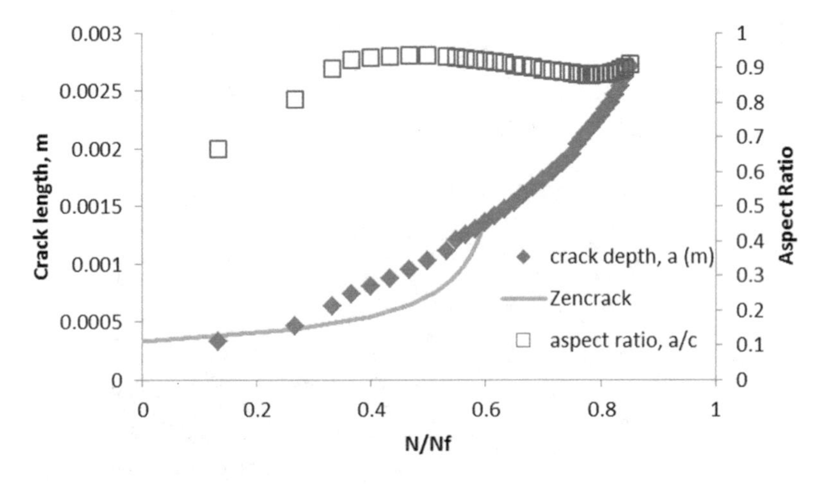

Figure 14.20. Crack depth and aspect ratio as functions of the number of cycles.

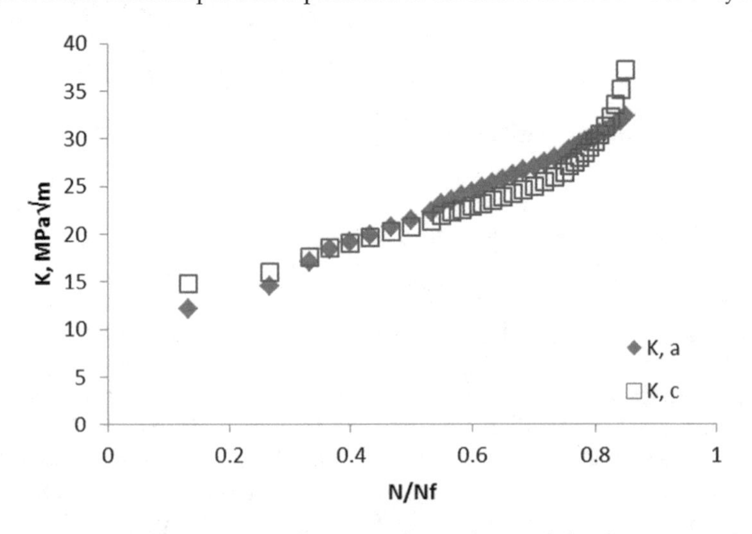

Figure 14.21. Stress intensity factors at the surface and the deepest points.

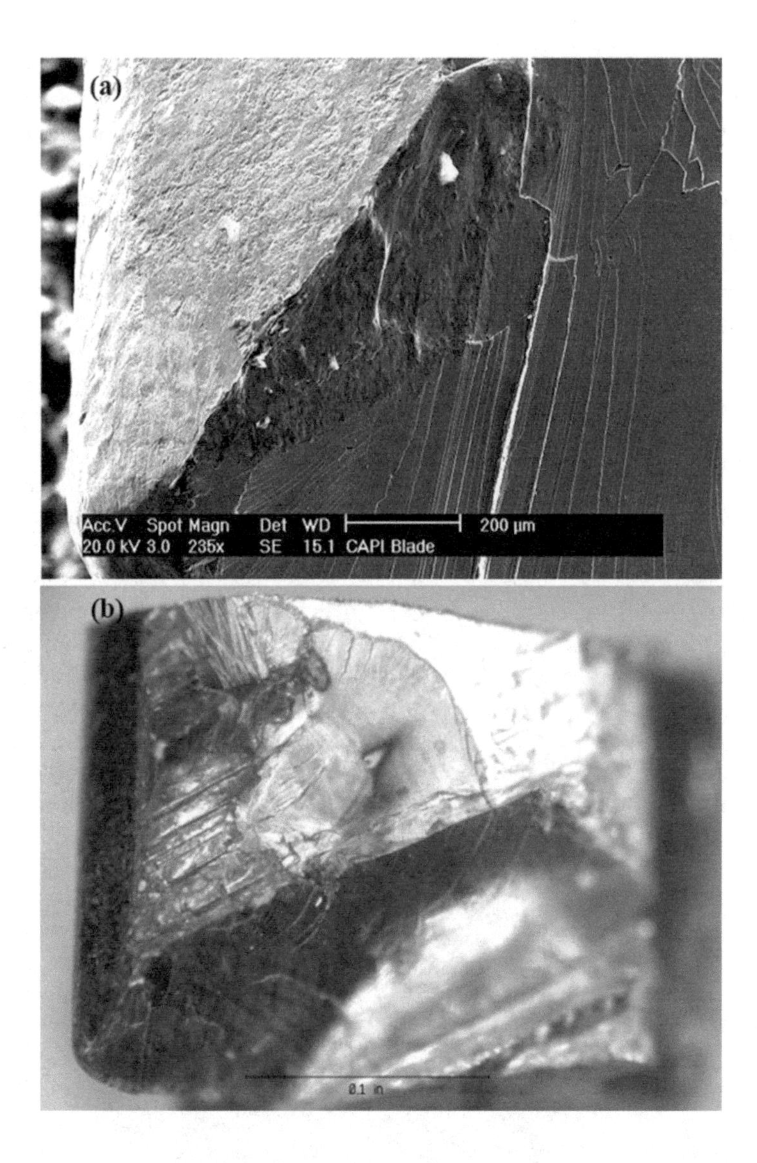

Figure 14.22. (a) The initial crack profile, and (b) the final crack profile.

14.7.4 Fracture Analysis of Spacer

A spacer life validation program was conducted using a spin rig at the National Research Council Canada (Beres et al. 2004). The spin rig test was stopped after 10,206 cycles, and NDI was performed using eddy current technique. Fifteen (15) cracks were found in a service-exposed spacer with an estimated 2,100 cycles of service history. The largest crack of 0.38 inch was found to be located at the aft bore corner. The cracked region was then cut open as shown in Figure 14.23. The discoloration region (the dark region) is indicative of oxidation of the fracture surface during testing in a low vacuum environment of 0.35 torr. Therefore, it can be inferred that the darkest area should correspond to the area of crack nucleation. Deeper into the material, at the flange, the crack surface became progressively less discolored, which indicates that the main crack surface was created during the later stage of crack growth.

Scanning electron microscopy (SEM) fractographic examinations of the most discolored area (near the bore/chamfer corner) showed that a mixed mode of faceted and striated fatigue fracture occurred on the crack surface, Figure 14.24, where the local crack growth directions are marked by arrows. Areas A, B, C, D and E appear to be crack nucleation sites. Particularly, areas C and D consist of two large twin-boundary facets. There are no striations or fatigue marks on those flat facets, which indicate that they formed in a brittle manner. Therefore, it can be envisaged that crack nucleation occurred by twin boundary fracture. The tearing at the bore/chamfer surface suggests that initially the facet cracks nucleated in the subsurface region and broke into the surface, initiating a surface crack. Within the high stress region enclosed by the line connecting the points 15, 16, 14, 13, 25, 27, 30 and 31, there appears to be multiple crack nucleation sites, and the initial crack growth direction points to all directions. Beyond the crack front as along the line that connects, fatigue striation directions appear to indicate crack propagation towards the inside of the spacer. The subsequent crack growth is dominated by the propagation of the main surface crack. In the crack propagation region, large flat fracture facets also appeared, but isolated. The transgranular/

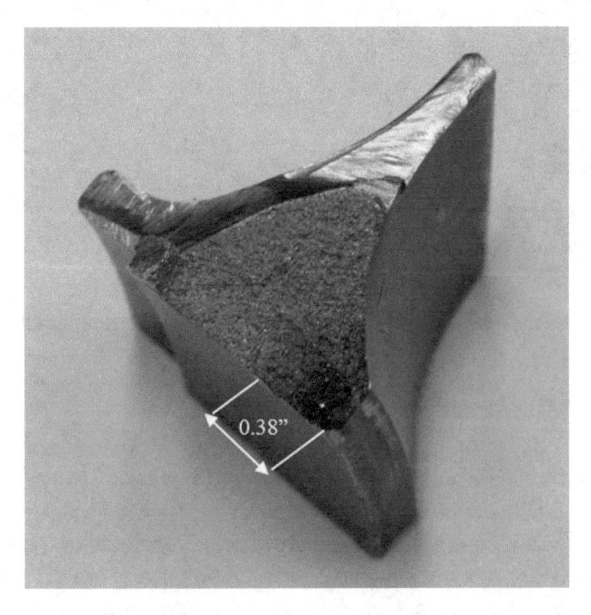

Figure 14.23. Fracture surface of the largest crack found on the spacer.

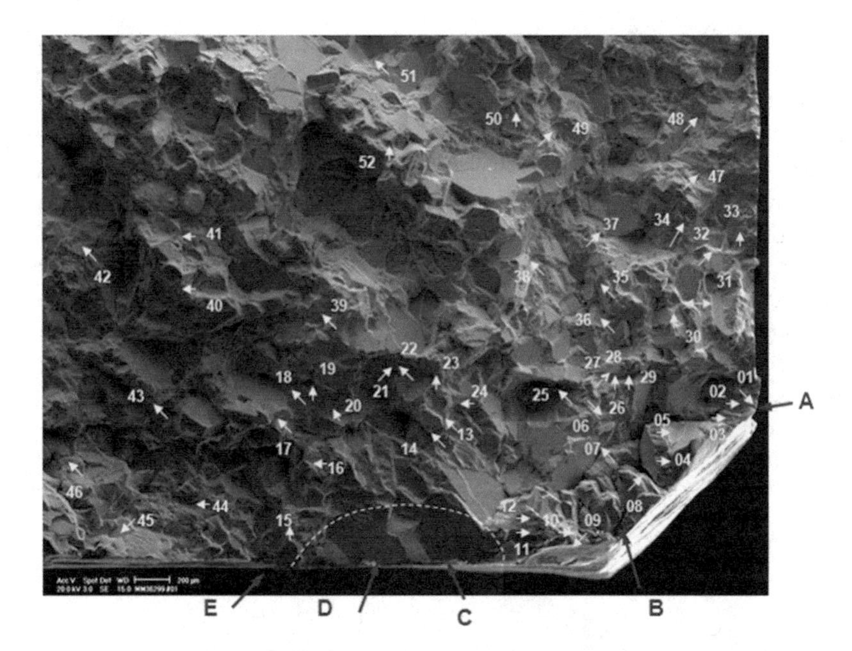

Figure 14.24. SEM micrograph showing the fracture mode with arrows to indicate the local crack propagation direction.

twin boundary facture mode seems to be an intrinsic characteristic of crack propagation in this service-exposed material.

A FEM stress analysis was performed for the spacer under the spin rig test conditions (Beres 2002). Taking advantage of the circular symmetry, a 15° segment of the spacer was meshed, as shown in Figure 14.25 (a). The stress distribution on the nominal crack plane based on the FEM analysis is shown in Figure 14.25 (b). This is the plane of maximum principal stress, on which the crack propagated. The stress distribution on this plane without the presence of crack was then taken as the stress function in the weight function integration. A ZENCRACK analysis was also performed for the spacer (Beres and Murzionak 2009), which assumed an initial semi-circular crack size of 0.01 inch formed at the bore/chamfer corner (corresponding to B). The crack propagation life of that crack was calculated to be ~ 40000 cycles, which is 4 times longer than the test cycles.

From the post-mortem fractographic examination, we could see that a semi-elliptical surface crack had nucleated, corresponding to the areas C and D at the bore surface (marked by the white dashed line in Figure 14.24) with a size of $2c = 0.054$ inch. Then, crack growth simulation was performed using the weight function method, to the point when the stress intensity factor reached the fracture toughness of the material. The simulated crack growth in terms of the major axis, c, and minor axis, a, are shown in Figure 14.26, and the crack profiles at selected cycles are plotted in Figure 14.27.

The WFM-calculated crack growth life for the spacer is ~16000 cycles with the crack growth law provided by the original equipment manufacturer (OEM). The total actual (service + spin rig testing) life of the spacer is 12,306 cycles (not fractured yet). In this simulation, crack nucleation at the chamfer corner, A and B, were neglected. The coalescence of A and B with C and D would result in a reduced fatigue crack growth life. Nonetheless, the WFM simulation is very close to the actual observation both in terms of life cycles and crack propagation profiles. This analysis emphasizes that subsurface crack nucleation is important to the overall structural integrity of gas turbine components. The damage tolerance of this component would be counted from either the NDI limit or the maximum grain size (considering twin boundary breaking).

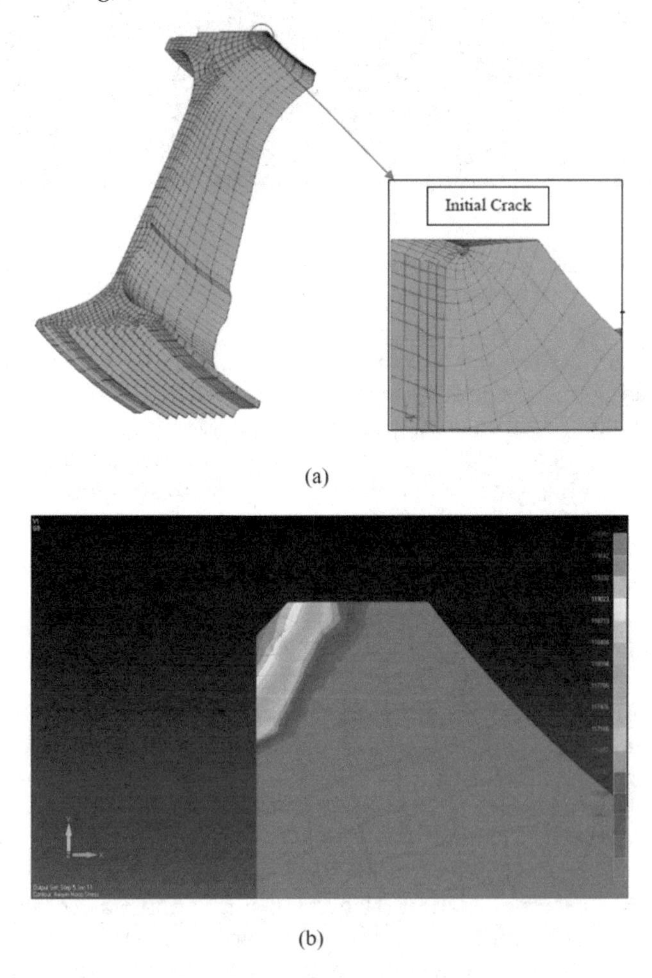

Figure 14.25. FEM mesh of the spacer model (a) and stress distribution in the vicinity of aft bore corner (Beres 2002).

Color version at the end of the book

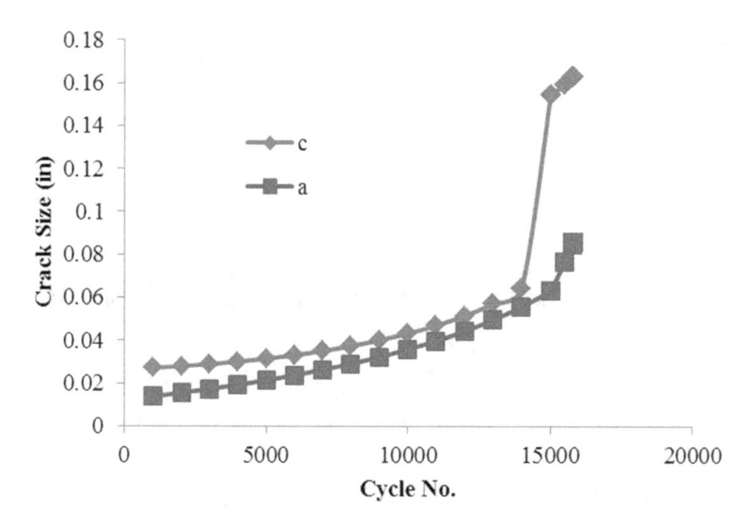

Figure 14.26. WFM simulated crack growth curves, initially starting as a semi-elliptical crack at the bore surface (Note that the jump represents the crack profile changed from semi-elliptical to corner crack).

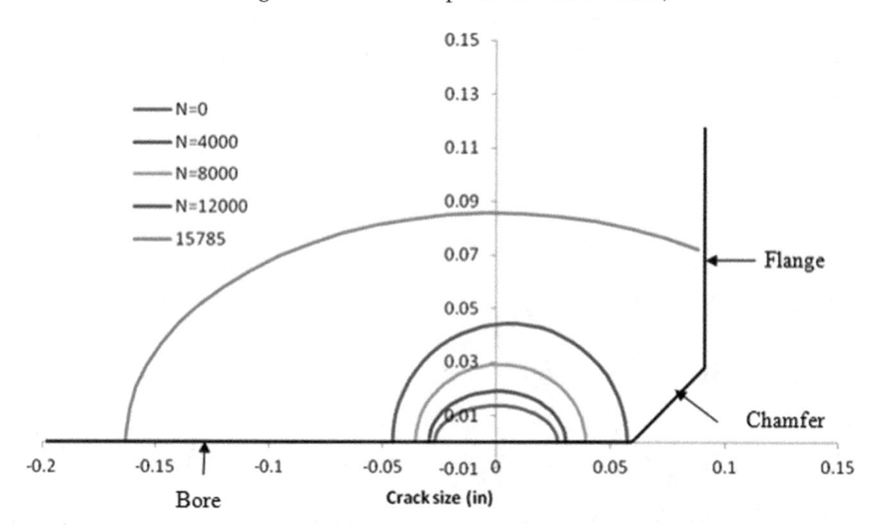

Figure 14.27. Simulated crack profiles at selected cycles.

Appendix A

Solving Dislocation Distributions for a ZSK Crack

The governing equations for dislocation distributions of a ZSK crack are multi-length integrals with couplings as induced by the effect of anisotropic elasticity, Eq. (9.14). For the given order of mode dominance, $a < c_3 < c_1 < c_2$, the solution can be obtained stepwise as shown in the following.

Taking advantage of the condition, $B_i(x_1) = 0$ when $|x_1| > c_i$, the integrations limits on the left side of Eq. (9.14) can be extended to c_2, as:

$$\int_{-c_2}^{c_2} \frac{F_{ij} B_j(\xi) d\xi}{2\pi(x_1 - \xi)} = t_i(x_1), \qquad |x_1| < c_i \tag{A.1}$$

where

$$t_i = \begin{cases} t_i^F & a < x_1 < c_i \\ 0 & -a < x_1 < a \\ -t_i^F & -c_i < x_1 < -a \end{cases} \tag{A.2}$$

Then, by inverse matrix operation,

$$\int_{-c_2}^{c_2} \frac{B_i(\xi) d\xi}{2\pi(x_1 - \xi)} = F_{ij}^{-1} t_j(x_1), \tag{A.3}$$

Because of the condition $a < c_3 < c_1 < c_2$,the solution can be first sought for dislocations with the smallest distribution size, i.e., $B_3(x_1)$, as given by

$$\int_{-c_3}^{c_3} \frac{B_3(\xi) d\xi}{2\pi(x_1 - \xi)} = F_{3j}^{-1} t_j(x_1) \tag{A.4}$$

The solution of this integral equation, which is bounded at $x_1 = \pm c_3$, can be obtained, directly using Muskhelishvili's method, in the following form:

$$B_3(x_1) = -\frac{2\sqrt{c_3^2 - x_1^2}}{\pi} \int_{-c_3}^{c_3} \frac{F_{3j}^{-1} t_j(\xi) d\xi}{(x_1 - \xi)\sqrt{c_3^2 - \xi^2}} \tag{A.5}$$

and the following condition should be satisfied

$$\int_{-c_3}^{c_3} \frac{x_1 F_{3j}^{-1} t_j(x_1) dx_1}{\sqrt{c_3^2 - x_1^2}} = \frac{b_T^{(3)}}{2} \tag{A.6}$$

where $b_T^{(3)}$ is the total burgers vector of screw dislocations along the x_3 coordinate.

Eq. (A.5) leads to

$$B_3(x_1) = \frac{2F_{3j}^{-1} t_j^F}{\pi} \psi(x_1, c_3) \qquad (j = 1, 2, 3) \quad |x_1| < c_3 \tag{A.7}$$

where

$$\psi(x, c) = \ln \left| \frac{\sqrt{c^2 - a^2} + \sqrt{c^2 - x^2}}{\sqrt{c^2 - a^2} - \sqrt{c^2 - x^2}} \right| \tag{A.8}$$

while Eq. (A.6) results in

$$\sqrt{c_3^2 - a^2} = \frac{b_T^{(3)}}{4F_{3j}^{-1} t_j^F} \tag{A.9}$$

With $B_3(x_1)$ solved, the governing equation for $B_1(x_1)$ can be rearranged into

$$\int_{-c_1}^{c_1} \frac{B_1(\xi) d\xi}{2\pi(x_1 - \xi)} = \omega_1(x_1) = M_{1J}^{-1} \left(t_J(x_1) - \int_{-c_3}^{c_3} \frac{F_{J3} B_3(\xi) d\xi}{2\pi(x_1 - \xi)} \right)$$

$$(J = 1, 2) \quad |x_1| < c_1 \tag{A.10}$$

where matrix $\mathbf{M}^{-1}$ is the inverse of a principal submatrix of **F**, as defined by

$$M^{-1} = \begin{bmatrix} F_{11} & F_{12} \\ F_{21} & F_{22} \end{bmatrix}^{-1} \tag{A.11}$$

Noting that **F** is real symmetric and positive definite, so is the matrix **M**. Then, substituting Eq. (A.7) into Eq. (A.10), and also using Muskhelishvili's method, we obtain

$$B_1(x_1) = \frac{2M_{1J}^{-1} t_J^F}{\pi} \psi(x_1, c_1) - M_{1J}^{-1} F_{J3} B_3(x_1) \qquad |x_1| < c_1 \tag{A.12}$$

while c_1 satisfies the condition

$$\int_{-c_1}^{c_1} \frac{x_1 \omega_1(x_1) dx_1}{\sqrt{c_1^2 - x_1^2}} = \frac{b_T^{(1)}}{2} \tag{A.13}$$

Substituting Eq. (A.10) into Eq. (A.13) and noting that

$$\int_{-c_1}^{c_1} \frac{xdx}{\sqrt{c_1^2 - x_1^2}} \int_{-c_3}^{c_3} \frac{F_{J3}B_3(\xi)d\xi}{2\pi(x_1 - \xi)} = -\int_{-c_1}^{c_1} F_{J3}B_3(\xi)d\xi \int_{-c_3}^{c_3} \frac{xdx}{2\pi(\xi - x_1)\sqrt{c_1^2 - x_1^2}} = -\frac{F_{J3}b_T^{(3)}}{2} \tag{A.14}$$

we find

$$\sqrt{c_1^2 - a^2} = \frac{b_T^{(1)} + F_{J3}b_T^{(3)}}{4M_{1J}^{-1}t_J^F} \qquad (J = 1, 2) \tag{A.15}$$

Finally, for $B_2(x_1)$, the governing equation can be written as

$$\int_{-c_2}^{c_2} \frac{B_2(\xi)d\xi}{2\pi(x_1 - \xi)} = \omega_2(x_1) \qquad |x_1| < c_2 \tag{A.16}$$

where

$$\omega_2(x_1) = F_{22}^{-1}\left(t_2(x_1) - \int_{-c_1}^{c_1} \frac{F_{21}B_1(\xi)d\xi}{2\pi(x_1 - \xi)} - \int_{-c_3}^{c_3} \frac{F_{23}B_3(\xi)d\xi}{2\pi(x_1 - \xi)} \right) \tag{A.17}$$

The solution of Eqs (A.16) can be obtained as

$$B_2(x_1) = \frac{2F_{22}^{-1}t_2^F}{\pi}\psi(x_1, c_2) - F_{22}^{-1}F_{21}B_1(x_1) - F_{22}^{-1}F_{23}B_3(x_1) \qquad |x_1| < c_2 \tag{A.18}$$

with the condition

$$\int_{-c_2}^{c_2} \frac{x_1\omega_2(x_1)dx_1}{\sqrt{c_2^2 - x_1^2}} = \frac{b_T^{(2)}}{2} \tag{A.19}$$

which leads to

$$\sqrt{c_2^2 - a^2} = \frac{F_{2j}b_T^{(j)}}{4t_2^F} \tag{A.20}$$

Appendix B

Solving Dislocation Distributions for a BCS Crack

For a BCS crack with $a < c_3 \le c_1 \le c_2$, the equilibrium condition, Eq. (9.35), can be generalized as:

$$\int_{-c_2}^{c_2} \frac{F_{ij} B_j(\xi) d\xi}{2\pi(x_1 - \xi)} = t_i(x_1), \qquad |x_1| < c_i \tag{B.1}$$

where

$$t_i(x_1) = \begin{cases} -t_i^0 & |x_1| < a \\ -t_i^0 + t_i^F & a < |x_1| < c_i \end{cases} \tag{B.2}$$

Again, given the mode dominance order of $a < c_3 < c_1 < c_2$, the mode III with the smallest plastic zone size can be solved first. The governing equation for the dislocation density function, $B_3(x_1)$, can be obtained by inversion of the matrix relation, Eq. (B.1), and taking into consideration of the fact that $B_3(x_1) = 0$ when $|x_1| > c_3$, as

$$\int_{-c_3}^{c_3} \frac{B_3(\xi) d\xi}{2\pi(x_1 - \xi)} = \omega_3(x_1) = \begin{cases} -F_{3j}^{-1} t_j^0 & |x_1| < a \\ F_{3j}^{-1}(t_j^F - t_j^0) & a < |x_1| < c_3 \end{cases} \tag{B.3}$$

The solution to Eq. (B.3), according to Muskhelishvili (1953), is given by

$$B_3(x_1) = -\frac{2\sqrt{c_1^2 - x_1^2}}{\pi} \int_{-c_3}^{c_3} \frac{\omega_3(\xi) d\xi}{(x_1 - \xi)\sqrt{c_1^2 - \xi^2}} = \frac{2F_{3j}^{-1} t_j^F}{\pi} \varphi(x_1, c_1) \tag{B.4}$$

where

$$\varphi(x, y) = \ln \left| \frac{a\sqrt{y^2 - x^2} + x\sqrt{y^2 - a^2}}{a\sqrt{y^2 - x^2} - x\sqrt{y^2 - a^2}} \right| \tag{B.5}$$

The condition for the bounded mode II solution is given by

$$\int_{-c_3}^{c_3} \frac{\omega_3(x_1)dx_1}{\sqrt{c_1^2 - x_1^2}} = 0 \tag{B.6}$$

which leads to

$$\frac{a}{c_3} = \cos\frac{\pi F_{3j}^{-1} t_j^0}{2F_{3j}^{-1} t_j^F} \tag{B.7}$$

Similarly, once $B_3(x_1)$ is known, the governing equation for $B_1(x_1)$ can be rearranged from Eq. (B.1) as

$$\int_{-c_2}^{c_2} \frac{F_{\alpha\beta}B_\beta(\xi)d\xi}{2\pi(x_1 - \xi)} = \left(t_\alpha(x_1) - \int_{-c_3}^{c_3} \frac{F_{\alpha 3}B_3(\xi)d\xi}{2\pi(x_1 - \xi)} \right) \quad |x_1| < c_\alpha \quad (\alpha, \beta = 1, 2) \tag{B.8}$$

Upon inversion,

$$\int_{-c_1}^{c_1} \frac{B_1(\xi)d\xi}{2\pi(x_1 - \xi)} = \omega_1(x_1) = M_{1\alpha}^{-1} \left(t_\alpha(x_1) - \int_{-c_3}^{c_3} \frac{F_{\alpha 3}B_3(\xi)d\xi}{2\pi(x_1 - \xi)} \right) \quad (\alpha = 1, 2)$$
$$|x_1| < c_1 \tag{B.9}$$

where matrix $\mathbf{M}^{-1}$ is the inverse of a principal submatrix of $\mathbf{F}$, as defined by Eq. (A.11).

Then, $B_1(x_1)$ can be obtained as

$$B_1(x_1) = \frac{2M_{1J}^{-1} t_J^F}{\pi} \varphi(x_1, c_1) - M_{1\alpha}^{-1} F_{\alpha 3} B_3(x_1) \qquad |x_1| < c_1 \tag{B.10}$$

The condition for the bounded mode III solution is given by

$$\int_{-c_1}^{c_1} \frac{\omega_1(x_1)dx_1}{\sqrt{c_1^2 - x_1^2}} = 0 \tag{B.11}$$

which leads to

$$\frac{a}{c_1} = \cos\frac{\pi M_{1\alpha}^{-1} t_\alpha^0}{2M_{1\alpha}^{-1} t_\alpha^F} \qquad (\alpha = 2, 3) \tag{B.12}$$

Finally, for $B_2(x_1)$ (mode I),

$$\int_{-c_2}^{c_2} \frac{B_2(\xi)d\xi}{2\pi(x_1 - \xi)} = \omega_2(x_1) \qquad |x_1| < c_2 \tag{B.13}$$

where

$$\omega_2(x_1) = F_{22}^{-1}\left(t_2(x_1) - \int_{-c_1}^{c_1} \frac{F_{21}B_1(\xi)d\xi}{2\pi(x_1-\xi)} - \int_{-c_3}^{c_3} \frac{F_{23}B_3(\xi)d\xi}{2\pi(x_1-\xi)} \right) \tag{B.14}$$

The condition for the bounded mode-I solution is given by

$$\int_{-c_2}^{c_2} \frac{\omega_2(x_1)dx_1}{\sqrt{c_2^2 - x_1^2}} = 0 \tag{B.15}$$

which leads to

$$\frac{a}{c_2} = \cos\frac{\pi t_2^0}{2t_2^F} \tag{B.16}$$

Then the dislocation density function can be obtained as

$$B_2(x_1) = \frac{2F_{22}^{-1}t_2^F}{\pi}\varphi(x_1, c_2) - F_{22}^{-1}[F_{21}B_1(x_1) + F_{23}B_3(x_1)] \qquad |x_1| < c_2 \tag{B.17}$$

References

Alain, R., P. Violan and J. Mendez. 1997. Low cycle fatigue behaviour in vacuum of a type 316 austenitic stainless steel between 20 and 600 °C. Part I: Fatigue resistance and cyclic behaviour. Mater. Sci. & Eng. A 229: 87-94.

Allen, D. and S. Garwood. 2007. Energy materials-strategic research agenda. Q2. Materials Energy 414 Review, IoM3, London.

Andrade, E.N.D. 1910. On the viscous flow in metals and allied phenomena. Proc. Roy. Soc. A 84: 1-12.

Andrade, E.N.D. 1914. The flow in metals under large stresses. Proc. Roy. Soc. A 90: 329-342.

Antolovich, B., S. Antolovich and A. Saxena. 1997. A mechanistic based analysis of fatigue crack propagation in single crystal nickel base superalloys. pp. 1348-1358. *In*: Proc. 13th International Symposium on Air Breathing Engines. September 7-12, 1997. Chattanooga, Tennessee.

Arakere, N.K. and E. Orozco. 2001. Analysis of low cycle fatigue properties single crystal nickel-base turbine blade superalloys. High Temperature Materials and Processes 20(5-6): 403-419.

Ardell, A.J. and J.C. Huang. 1988. Antiphase boundary energies and transition from shearing to looping in alloys strengthened by ordered precipitates. Phil. Mag. Letters 58: 189-197.

Argon, A.S. and S. Yip. 2006. The strongest size. Phil. Mag. Lett. 86: 713-720.

Arsenlis, A. and D.M. Parks. 2002. Modeling the evolution of crystallographic dislocation density in crystal plasticity. J. Mech. and Phys. Solids 50: 1979-2009.

Ashby, M.F. 1972. A first report on deformation mechanism map. Acta Metall. 20: 887-897.

Ashby, M.F. 1983. Mechanisms of deformation and fracture. Advances in Applied Mechanics 23: 117-177.

Ashby, M.F and B.F. Dyson. 1984. Creep damage mechanics and micromechanisms. pp. 3-30. *In*: Proc. 6th International Conference on Fracture. New Delhi, India. 4-10 December 1984.

ASME International. 2016. Power Piping. ASME Code for Pressure Piping, B31.

ASTM standard—E2368, 2010. Standard practice for strain controlled thermomechanical fatigue testing. American Society for Testing of Materials. West Conshohocken, PA.

Asundi, A. and W. Deng. 1995. Rigid inclusions on the interface between two bonded anisotropic media. J. Mech. Phys. Solids 43: 1045-1058.

Bache, M.R. 2003. A review of dwell sensitivity fatigue in titanium alloys: the role of microstructure, texture and operating conditions. Int. J. Fatigue 25: 1079-1087.

Bache, M.R., M. Cope, H.M. Davies, W.J. Evans and G. Harrison. 1997. Dwell sensitive fatigue in a near alpha titanium alloy at ambient temperature. Int. J. Fatigue 19: S83-S88.

Bailey, R.W. 1935. The utilization of creep test data in engineering design. Proc. I. Mech. E. 131: 209-284.

Bain, K.R., M.L. Gambone, J.H. Hyzak and M.C. Thomas. 1988. Development of damage tolerant microstructure in Udimet 720. pp. 13-22. *In*: Superalloys 1988. The Metallurgical Society. Warrendale, PA.

Bano, N., A.K. Koul and M. Nganbe. 2014. A Deformation mechanism map for the 1.23Cr-1.2Mo-0.26V rotor steel and its verification using neural networks. Metall. Mater. Trans. A 45A: 1928-1936.

Barnett, D.M. and R.J. Asaro. 1972. The fracture mechanics of slit-like cracks in anisotropic elastic media. J. Mech. Phys. 20: 353-366.

Barnby, J.T. 1975. Crack propagation during steady state creep. Eng. Fract. Mech. 7: 299-304.

Basquin, O.H. 1910. The exponential law of endurance tests. Proceedings of American Society for Testing and Materials 10: 625-630.

Bauschinger, J. 1886. On the change of the position of the elastic limit of iron and steel under cyclic variations of stress. Mitt. Mech. Tech. Lab. 13(1): 1-115. Munich.

Bennett, A. 1986. Properties of thermal barrier coatings. Mater. Sci. Tech. 2: 257-261.

Benson, J.P. and D.V. Edmonds. 1978. Effects of microstructure on fatigue in threshold region in low-alloy steels. Metal Sci. 12: 223-232.

Beres, W. 2002. Structural analysis of spin rig test of T56 series III engine turbine spacer 1-2. LTR-SMPL-2002-0240. National Research Council Canada.

Beres, W., S. Robertson and M. Brothers. 2004. Spin rig test of the spacer 1-2 for the T56 series III gas turbine engine. LTR-SMPL-2004-0171. National Research Council Canada.

Beres, W., D. Fread, L. Harris, P. Haupt, J. Kappas, R. Olson et al. 2008. Critical components life update for gas turbine engines – case study of an international collaboration (GT2008-50655). *In*: Proceedings of ASME Turbo Expo 2008: Power for Land, Sea and Air. Berlin, Germany.

Beres, W. and A. Murzionak. 2009. Simulation of crack growth at the bore of the 1-2 stage turbine spacer for T56 series III gas turbine engine. LTR-SMPL-2009-0223. National Research Council Canada.

Berger, C., J. Granacher and A. Thoma. 2001. Creep rupture behaviour of Ni base superalloys for 700 °C-steam turbines. pp. 489-499. *In*: E.A. Loria (ed.). Superalloys 718, 625, 706 and Various Derivatives. TMS. Warrendale, PA.

Bilby, B.A., A.H. Cottrell and K.H. Swinden. 1963. The spread of plastic yield from a notch. Proc. R. Soc. Lond. A 272: 304-314.

Brandl, W., H.J. Grabke, D. Toma and J. Krüger. 1996. The oxidation behaviour of sprayed MCrAlY coatings. Surface and Coatings Technology 86-87: 41-47.

Brown, M.W. and K.J. Miller. 1973. A theory for fatigue under multiaxial stress-strain conditions. Proc. Inst. Mech. Engrs 187: 745-755.

Bruckner-Foit, A. and X. Huang. 2008. On the determination of material parameters in crack initiation laws. Fat. Fract. Eng. Mat. Struct. 31: 980-988.

Budiansky, B., J.W. Hutchinson and A.G. Evans. 1986. Matrix fracture in fiber-reinforced ceramics. J Mech Phys Solids 34: 167-189.

Bueno, L.O. 2008. Effect of oxidation on creep data. Part 1: comparison between some constant load creep results in air and vacuum on 2.25Cr-1Mo steel from 600 to 700 °C. Materials at High Temperatures 25: 213-221.

Busso, E., J. Lin and S. Sakurai. 2001. A mechanistic study of oxidation-induced degradation in a plasma sprayed thermal barrier coating system. Part II: Life prediction model. Acta Mater. 49: 1529-1536.

Carry, C. and J.L. Strudel. 1977. Apparent and effective creep parameters in single crystals of a nickel base superalloy—I. Incubation period. Acta Metall. 25: 767-777.

Carry, C. and J.L. Strudel. 1978. Apparent and effective creep parameters in single crystals of a nickel base superalloy—II. Secondary creep. Acta Metall. 26: 859-870.

Castillo, R., A.K. Koul and E.H. Toscano. 1987. Lifetime prediction under constant load creep conditions for a cast Ni-base superalloy. J. Engng. for Gas Turbine and Powers, 109: 99-106.

Castillo, R., A.K. Koul and J.-P. Immarigeon. 1988. The effect of service exposure on the creep properties of cast IN-738LC subjected to low stress high temperature creep conditions. pp. 805-813. *In*: S. Reichmen, D.N. Duhl, G. Maurer, S. Antolovich and C. Lund (eds.). Superalloys 1988. The Metallurgical Society. Warrendale, PA.

Chaboche, J.L. 1989. Constitutive equations for cyclic plasticity and cyclic viscoplasticity. Int. J. Plasticity 5: 247-302.

Chaboche, J.L. 2008. A review of some plasticity and viscoplasticity constitutive theories. Int. J. Plasticity 24: 1642-1693.

Chan, K.S., J.E. Hack and G.R. Leverant. 1986. Fatigue crack propagation in Ni-base superalloy single crystals under multiaxial cyclic loads. Metall. Trans. 17A: 1739-1750.

Chan, K.S., N.S. Cheruvu and G.R. Leverant. 1998. Coating life prediction under cyclic oxidation conditions. Transactions of the ASME. Journal of Engineering for Gas Turbines and Power 120: 609-614.

Chan, K.S., N.S. Cheruvu and G.R. Leverant. 1999. Coating life prediction for combustion turbine blades. Transactions of the ASME. Journal of Engineering for Gas Turbines and Power 121: 484-488.

Chan, K.S. and N.S. Cheruvu. 2004. Degradation mechanism characterization and remaining life prediction for NiCoCrAlY coatings, GT2004-53383. Proceedings of ASME Turbo Expo 2004. Power for Land, Sea, and Air. Vienna, Austria.

Chang, M., A.K. Koul, P. Au and T. Terada. 1994. Damage tolerance of wrought alloy 718 Ni- Fe-base superalloy. J. Mater. Engineering & Performance 3: 356-366.

Chang, M., A.K. Koul and C. Cooper. 1996. Damage tolerance of P/M turbine disc materials. *In*: Superalloy 1996. TMS. Warrendale, PA.

Che, C., G.Q. Wu, H.Y. Qi, Z. Huang and X.G. Yang. 2009. Depletion model of aluminum in bond coat for plasma-sprayed thermal barrier coatings. Advanced Materials Research 75L: 31-35.

Chen, W.R., X.J. Wu, B. Marple and P. Patnaik. 2005. Oxidation and crack nucleation/growth in an air-plasma-sprayed thermal barrier coating with NiCrAlY bond coat. Surf. Coat. Technol. 197: 109-115.

Chen, W.R., X.J. Wu, B. Marple and P. Patnaik. 2006. The growth and influence of thermally grown oxide in a thermal barrier coating. Surf. Coat. Technol. 201: 1074-1079.

Chen, W.R., X.J. Wu, B. Marple, R. Lima and P. Patnaik. 2008. Pre-oxidation and TGO growth behaviour of an air-plasma-sprayed thermal barrier coating. Surf. Coat. Technol. 202: 3787-3796.

Chen, W.R., X.J. Wu, B. Marple, D. Nagy and P. Patnaik. 2008. TGO growth behaviour in TBCs with APS and HVOF bond coats. Surf. Coat. Technol. 202: 2677-2683.

Chen, W.R., K. Chen and X.J. Wu. 2012a. Oxidation kinetics and diffusivity of HVOF-produced CoNiCrAlY coating. LTR-SMPL-2012-0043. National Research Council Canada.

Chen, W.R., X.J. Wu and D. Dudzinski. 2012b. Influence of thermal cycle frequency on the TGO growth and cracking behaviours of an APS-TBC. Journal of Thermal Spray Technology 21(6): 1294-1299.

Chen, W.R. 2014. Degradation of a TBC with HVOF-CoNiCrAlY bond coat. Journal of Thermal Spray Technology 23: 876-884.

Chermant, J.L., G. Boitier, S. Darzens, G. Farizy, J. Vicens and J.C. Sangleboeuf. 2002. The creep mechanism of ceramic matrix composites at low temperature and stress by a material science approach. Journal of the European Ceramic Society 22: 2443-2460.

Cocks, A.C.F. and M.F. Ashby. 1982. The growth of a dominant crack in a creeping material. Scripta Metall. 16: 109-114.

Coffin, L.F. 1954. A study of the effects of cyclic thermal stresses on a ductile metal. Transactions of the American Society of Mechanical Engineers 76: 931-950.

Cottrell, A.H. 1953. A note on the Portevin-Le Chatelier effect. Philos. Mag. 44: 829-832.

Dalby, S. and J. Tong. 2005. Crack growth in a new nickel-based superalloy at elevated temperature. Part I: Effects of loading wave form and frequency on crack growth. J. Mater. Sci. 40: 1217-1228.

Danflou, H. ., M. Marty and A. Walder. 1992. Formation of serrated grain boundaries and their effect on the mechanical properties in a P/M nickel base superalloy. pp. 63-72. *In*: S.D. Antolovich, R.W. Stusrud, R.A. Mackay, D.L. Anton, T. Khan, R.D. Kissinger and D.L. Klarstrom (eds.). Superalloy 1992. The Metallurgical Society. Warrendale, PA.

Danielewski, M. 1992. Phenomenology of the diffusion controlled reactions in solids, high temperature oxidation of metals. Solid State Phenomena 21&22: 103-134.

Dassault Systèmes. 2008. Abaqus Analysis User's Manual. Vol. II: 11.26.1-3. Dassault Systèmes, France.

Davidson, D.L. and J. Lankford. 1986. High resolution techniques for the study of small cracks. pp. 455-70. *In*: R.O. Ritchie and J. Lankford (eds.). Small Fatigue Cracks. The Metallurgical Society of the American Institute of Mining, Metallurgical and Petroleum Engineers. Warrendale, PA.

DeMasi, J., S.L. Manning, M. Ortiz and K. Sheffler. 1987. Thermal barrier coating life prediction model development. pp. 385-399. *In*: NASA Technical Report 89N17333. NASA Lewis Research Center. Cleveland, Ohio.

Department of Defense. 2002. Engine Structural Integrity Program (ENSIP). MIL HDBK 1783B. Department of Defense, USA.

Diboine, A. and A. Pineau. 1987. Creep crack initiation and growth in inconel 718 alloy at 650 °C. Fatigue Fract. Eng. Mater. Struct. 10: 141-151.

Dimelfi, R.J. and W.D. Nix. 1977.The stress dependence of the crack growth rate during creep. Int. J. Fract. 13: 341-348.

Drucker, D.C. 1950. Some implications of work hardening and ideal plasticity. Quarterly of Applied Mechanics 7: 411-418.

Dugdale, D.S. 1960. Yielding of steel sheets containing slits. Journal of the Mechanics and Physics of Solids 8: 100-108.

Dyson, B.F. 1983. Continuous cavity nucleation and creep fracture. Scripta Metall. 17(1): 31-37.

Dyson, B.F. and T.B. Gibbons. 1987. Tertiary creep in nickel-base superalloys: analysis of experimental data and theoretical synthesis. Acta Metall. 35: 2355-2369.

Dyson, B.F. and M. McLean. 1983. Particle-coarsening, σ_0 and tertiary creep. Acta Metall. 31: 17-27.

Dyson, B.F. and M. McLean. 2000. Modelling the effects of damage and microstructural evolution on the creep behaviour of engineering alloys. J. Eng. Mater. Technol. 122: 273-278.

Dyson, B.F. and S. Osgerby. 1987. Modelling creep-corrosion interactions in nickel-base superalloys. Materials Science and Technology 3: 545-553.

Eggeler, G., J.C. Earthman, N. Nilsvang and B. Ilschner. 1989. Microstructural study of creep rupture in a 12% chromium ferritic steel. Acta Metall. 37(1): 49-60.

Elber, W. 1970. Fatigue crack closure under cyclic tension. Eng. Fract. Mech. 2: 37-45.

Elber, W. 1971. The significance of fatigue crack closure. pp. 230-242. *In*: Damage Tolerance in Aircraft Structures. ASTM STP 486. American Society for Testing and Materials. Philadelphia, PA.

Eshelby, J.D., W.T. Read and W. Shockley. 1953. Anisotropic elasticity with applications to dislocation theory. Acta Metall. 1: 252-259.

Essmann, U., U. Gösele and H. Mughrabi. 1981.A model of extrusions and intrusions in fatigued metals I. Point-defect production and the growth of extrusions. Phil. Mag. 44: 405-426.

Evans, A.G. 1990. Perspective on the development of high-toughness ceramics. J Am Ceram Soc 73: 187-206.

Evans, A.G., J.-M. Domergue and E. Vagagini. 1994. Methodology for relating the tensile constitutive behaviour of ceramic-matrix composites to constituent properties. J Am Ceram Soc 77: 1425-1435.

Evans, A.G., M.Y. He and J.W. Hutchinson. 2001. Mechanics-based scaling laws for the durability of thermal barrier coatings. Progress in Materials Science 46: 249-271.

Evans, R.W. and B. Wilshire. 1985. Creep of metals and alloys. The Institute of Metals. London, UK.

Evans, W.J. 1987. Dwell sensitive fatigue in a near alpha-titanium alloy. J. Mater. Sci. Letters 6: 571-574.

Fang, B., Y.S. Hong and Y.L. Bai. 1995. Experimental and theoretical study on numerical density evolution of short fatigue cracks. Acta Mechanica Sinica (English Edition) 11: 144-152.

Farkas, D., S. Van Petegem, P.M. Derlet and H. Van Swygenhoven. 2005. Dislocation activity and nano-void formation near crack tips in nanocrystalline Ni. Acta Mater. 53: 3115-3123.

Fatemi, A. and D. Socie. 1988. A critical plane approach to multiaxial fatigue damage including out-of-phase loading. Fat. Fract. Eng. Mat. Struct. 11(3): 149-165.

Floreen, S. 1975. The creep fracture of wrought Ni-base alloys by a fracture mechanics approach. Metall. Trans. A 6: 1741-1749.

Floreen, S. and R.H. Kane. 1976. A critical strain model for the creep fracture of nickel-base superalloys. Metall. Trans. A 7A: 1157-1160.

Fong, C. and D. Thomans. 1988. Stage I: corrosion fatigue crack crystallography in austenitic stainless steel (316L). Metall. Trans. A 19A: 2753-2764.

Forsyth, P.J.E. 1963. Fatigue damage and crack growth in aluminium alloys. Acta Metall. 11: 703-715.

Frost, H. and M.F. Ashby. 1982. Deformation Mechanism Maps. Pergamon Press. Elmsford, NY.
Fuller, E.R. Jr., R.J. Fields, T.-J. Chuang and S. Singhal. 1984. Characterization of creep damage in metals using small angle neutron scattering. Journal of Research of the National Bureau of Standards 89(1): 35-45.
Furrillo, F.T., J.M. Davidson, J.K. Tien and L.A. Jackman. 1979. The effects of grain boundary carbides on the creep and back stress of a nickel-base superalloy. Mater. Sci. Eng. 39: 267-273.
Furillo, F.T., S. Purushothaman and J.K. Tien. 1978. Further discussion on "understanding the Larson-Miller parameter". Script. Metall. 12(4): 331-332.
Garofalo, F. 1963. An empirical relation defining the stress dependence of minimum creep rate in metals. Trans AIME 227: 351-356.
Garofalo, F. 1965. Fundamentals of Creep and Creep-Rupture in Metals. Macmillan Publishing, NY.
Gedwill, M.A. 1980. Improved Bond Coat Coatings for Use with Thermal Barrier Coatings. NASA TM-81567.
Ghosh, R.N., R.V. Curtis and M. McLean. 1990. Creep deformation of single crystal superalloys – modeling the crystallographic anisotropy. Acta Met. Mater. 38: 1977-1992.
Gilman, J.J. 1940. Mechanical Behaviour of Materials at Elevated Temperatures. J. Dorn (ed.). McGraw-Hill, pp. 17-44.
Glieter, H. and E. Hornbogen. 1967/68. Precipitation hardening by coherent particles. Mater. Sci. Eng. 2: 285-302.
Gouws, J.J., R.M. Morris and J.A. Visser 2008. Modelling of a gas turbine combustor using a network solver. South African Journal of Science 102. November/ December.
Gooch, D.J. 1984. The effect of cold work on low temperature ($0.35T_m$) creep crack growth in CrMn steels. Mater. Sci. Eng. 64: 183-196.
Goodman, J. 1919. Mechanics Applied to Engineering. Longmans, Green and Co. London, pp. 631-636.
Graham, A. and K. Walles. 1955. Relationships between long and short time creep and tensile properties of a commercial alloy. J. Iron Steel Inst. 179: 104-121.
Greenwood, G.W. 1969. Cavity nucleation in the early stages of creep. Phil. Mag. 19: 423-427.
Griffith, A.A. 1921. The phenomena of rupture and flow in solids. Phil. Trans. Roy. Soc. London A 221: 163-198.
Grujicic, M., R. Galgalikar, S. Ramaswami and J.S. Snipes. 2016. Derivation, parameterization and validation of a creep deformation/rupture material constitutive model for SiC/SiC ceramic matrix composites (CMCs). AIMS Materials Science 3(2): 591-619.
Hack, J.E. and G.R. Leverant. 1982. The influence of microstructure on the susceptibility of titanium alloys to internal hydrogen embrittlement. Metall. Trans. 13A: 1729-1738.
Halford, G.R. 1991. Evolution of creep-fatigue life prediction models. pp. 43-57. *In*: G.K. Haritos and O.O. Ochoa (eds.). Creep-Fatigue Interaction at High Temperature. American Society of Mechanical Engineers. New York.
Halford, G.R., J.F. Saltsman and M.H. Hirschberg. 1977. Ductility-normalized strain range partitioning life relations for creep-fatigue life predictions. NASA TM-73737.

Han, S., S. Li. and D.J. Smith. 2001. Comparison of phenomenological and crystallographic models for single crystal nickel-base superalloys—I: analytical and identification. Mech. Mater. 33: 251-266.
Hart, E.W. 1970. A phenomenological theory for plastic deformation of polycrystalline metals. Acta Metall. 18: 599-610.
Hardt, S., H.J. Maier and H.-J. Christ. 1998. The effect of creep pre-deformation on high temperature fatigue behaviour of the titanium alloy IMI 834. pp. 9-14. *In*: K.T. Rie and P.D. Portella (eds.). Low Cycle Fatigue and Elasto-Plastic Behaviour of Materials. Vol. 4. Elsevier Appl. Sci. London.
Hardt, S., H.J. Maier and H.-J. Christ. 1999. High-temperature fatigue damage mechanisms in near-a titanium alloy IMI 834. Int. J. Fatigue 21: 779-789.
Hayhurst, D.R. 2005. CDM mechanism-based modelling of tertiary creep: ability to predict the life of engineering components. Arch. Mech. 57: 103-132.
Haynes, J.A., M.K. Ferber, W.D. Porter and E.D. Rigney. 1999. Characterization of alumina scales formed during isothermal and cyclic oxidation of plasma-sprayed TBC systems at 1150 °C. Oxidation of Metals 52: 31-76.
Haynes, J.A., E.D. Rigney, M.K. Ferber and W.D. Porter. 1996. Oxidation and degradation of a plasma-sprayed thermal barrier coating system. Surface and Coatings Technology 86-87: 102-108.
Henderson, M.B. and J.W. Martin. 1992. The Influence of orientation, temperature and frequency on fatigue crack growth in a nickel-based superalloy single crystal. pp. 707-716. *In*: Superalloy 1992, TMS.
Hertzberg, R.W. 1996. Deformation and Fracture Mechanics of Engineering Materials. John Wiley & Sons, Inc.
Hill, R. 1998. The Mathematical Theory of Plasticity. Oxford University Press.
Hindam, H. and D.P. Whittle. 1982. Microstructure, adhesion and growth kinetics of protective scales on metals and alloys. Oxidation of Metals 18: 245-284.
Hirth, J.P. and J. Lothe. 1992. Theory of Dislocation. Krieger Publishing Company, New York.
Hoeppner, D.W. 1986. Damage tolerance concepts to critical engine components. pp. 4-1–4-16. *In:* AGARD Conference Proceedings No. 393. AGARD, NATO.
Holmes, J.W. and J.L. Chermant. 1994. High Temperature Ceramic Matrix Composites. R. Naslain, J. Lamon and D. Doumeingts (eds.). Cambridge: Woodhead Publishing, pp. 633-647.
Hong, S.-G. and S.-B. Lee. 2005. Mechanism of dynamic strain aging and characterization of its effect on the low-cycle fatigue behaviour in type 316L stainless steel. J. Nuclear Materials 340: 307-314.
Hong, S.-G., S.-B. Lee and T.-S. Byun. 2007. Temperature effect on the low-cycle fatigue behaviour of type 316L stainless steel: cyclic non-stabilization and an invariable fatigue parameter. Mater. Sci. & Eng. A 457: 139-147.
Hull, D. and D.E. Rimmer. 1959. The growth of grain-boundary voids under stress. Phil. Mag. A 4: 673-687.
Hurst, J. 2018. NASA transformational tools and technologies project: 2700 °F CMC/EBC technology challenge. *In*: Proceedings of ASME Turbo Expo 2018 Turbomachinery Technical Conference and Exposition. GT2018-77282. June 11-15, 2018. Oslo, Norway.
Hutchings, I.M. 1983. The behaviour of metals under ballistic impact at sub-ordnance velocities. pp. 161-196. *In*: J. Mescall and V. Weiss (eds.). Materials Behaviour under High Stress and Ultra High Loading Rates. Plenum Press, New York.

Hutchinson, J.W. 1968. Singular behaviour at the end of a tensile crack in a hardening material. J. Mech. Phys. Solids 16: 13-31.

Inglis, G.E. 1913. Stresses in a plate due to the presence of cracks and sharp corners. Transactions of the Institution of Naval Architects 55: 219-230.

Immarigeon, J.-P., W. Beres, P. Au, A. Fahr, W. Wallace, A.K., Koul et al. 2001. Life Cycle Management Strategies for Aging Engines, RTO-MP-079(II)-SM 17/1-15.

Inoue T., M. Okazaki, T. Igari, M. Sakane and S. Kishi. 1991. Evaluation of fatigue-creep life prediction methods in multiaxial stress state—the second report of the benchmark project (B) by the Subcommittee on Inelastic Analysis and Life Prediction of High Temperature Materials, JSMS. Nuclear Engineering and Design 126: 13-21.

Irwin, G.R. and J.E. Kies. 1954. Critical energy rate analysis of fracture strength. Welding Journal 33: 193-198.

Isik, M.I., A. Kostka and G. Eggeler. 2014. On the nucleation of Laves phase particles during high-temperature exposure and creep of tempered martensite ferritic steels. Acta Mater. 81: 230-240.

Jennions, I.K. (ed.). 2011. Integrated Vehicle Health Management: Perspectives on an Emerging. SAE International.

Jezernik, N., J. Kramberger, T. Lassen and S. Glodež. 2010. Numerical modelling of fatigue crack initiation and growth of martensitic steels. Fat. Fract. Eng. Mat. Struct. 33: 714-723.

Johnston, G.W. 1962. Yield points and delay times in single crystals. J. Appl. Phys. 33: 2716-2730.

Joyce, M.R., X.J. Wu and P.A.S. Reed. 2003. The effect of environment and orientation on fatigue crack growth behaviour of CMSX-4 nickel base single crystal at 650 °C. Materials Letters 58: 99-103.

Kachanov, L.M. 1958. On creep rupture time. Proc. Acad. Sci. USSR. Div. Eng. Sci. 8: 26-31.

Kandil, F.A., M.W. Brown and K.J. Miller. 1982. Biaxial low cycle fatigue of 316 stainless steel at elevated temperatures. pp. 203-210. *In*: Mechanical Behaviour and Nuclear Applications of Stainless Steel at Elevated Temperatures. Vol. 14. Metals Soc. London.

Kassner, M.E. and T.A. Hayes. 2003. Creep cavitation in metals. Int. J. Plasticity 19: 1715-1748.

Kanninen, M.F. and C.H. Popelar. 1987. Advanced Fracture Mechanics. Oxford, UK.

Kim, W.H and C. Laird. 1978. Crack nucleation and state I propagation in high strain fatigue II mechanism. Acta Metall. 26: 789-799.

Kimura, K. and Y. Takahashi. 2012. Evaluation of long-term creep strength of ASME Grades 91, 92, and 122 type steels. ASME 2012 Press. Vessel. Pip. Conf. 1–8. doi:10.1115/PVP2012-78323.

Kimura, K., M. Tabuchi, Y. Takahashi, K. Yoshida and K. Yagi. 2011. Long-term creep strength and strength reduction factor for welded joints of ASME Grades 91, 92 and 122 type steels. Int. J. Microstruct. Mater. Prop. 6: 72.

Klueh, R.L. 2005. Elevated temperature ferritic and martensitic steels and their application to future nuclear reactors. Inter. Mater. Rev. 50(5): 287-310.

Kobayashi, T., K. Nishino, Y. Kimoto, Y. Awano, Y. Hibino and H. Ueno. 1998. 673K embrittlement of ferritic spheroidal graphite cast iron by magnesium. Casting Engineering 70: 273-278.

Kocks, U.F., A.S. Argon and M.F. Ashby. 1975. Thermodynamics and kinetics of slip. Prog. Mater. Sci. 19: 1-271.

Koehler, J.S. 1952. The production of large stresses by dislocations. Phy. Rev. 85: 480-481.

Komorowski, J.P. 2003. New tools for aircraft maintenance. Aircraft Engng Aerospace Technol 75(5): 453-460.

Komorowski, J.P. 2016. From Science to Engineering Practice – Evolving a Structural Integrity Framework (ASTM Jo Dean Morrow Fatigue Lecture). ASTM E08 Executive Committee. May 3, 2016. San Antonio, TX, USA. http://www.astm.org/COMMIT/ASTM%202016%20Fatigue%20Lecture%20Komorowski.pdf

Kramberger, J., N. Jezernik, P. Goncz and S. Glodež. 2010. Extension of the Tanaka-Mura model for fatigue crack initiation in thermally cut martensitic steels. Eng. Fract. Mech. 77: 2040-2050.

Koul, A.K. and G.H. Gessinger. 1983. On the mechanism of serrated grain boundary formation in Ni-based superalloys. Acta Metall. 31: 1061-1069.

Koul, A.K., P. Au, N. Bellinger, R. Thamburaj, W. Wallace and J.-P. Immarigeon. 1988. Development of a damage tolerant microstructure for Inconel 718 turbine disc alloy. pp. 3-12. *In*: D.N. Duhl, G. Maurer, S. Antolovich, C. Lund and S. Reichman (eds.). Superalloys 1988, TMS. Warrendale, PA.

Koul, A.K. and R. Castillo. 1988. Assessment of service induced microstructural damage and its rejuvenation in turbine blades. Metall. Trans. A 19A: 2049-2066.

Krausz, A.S. 1968. A rate theory of dislocation mobility. Acta. Met. 16: 897-902.

Krausz, A.S. and H. Eyring. 1975. Deformation Kinetics. John Wiley & Sons. New York.

Kuznetsov, V. (1993). Aluminum-Nickel-Oxygen. Vol. 7. pp. 434-440. *In*: G. Petzow and G. Effenberg (eds.). Ternary Alloys. VCH Publishers. New York, NY.

Laird, C. and G.C. Smith. 1982.Crack propagation in high stress fatigue. Phil. Mag. 7: 847-857.

Landes, J.D. and J.A. Begley. 1976.A fracture mechanics approach to creep crack growth. pp. 128-148. *In*: Mechanics of Crack Growth. ASTM STP 590. American Society for Testing and Materials, Philadelphia, PA.

Langdon, T.G. 1970. Grain boundary sliding as a deformation mechanism during creep. Phil. Mag. 22: 689-700.

Lankford, J. 1985. The growth of small fatigue cracks in 7075-T6 aluminum. Fatigue of Engineering Materials and Structures 5(3): 233-248.

Larson, F.R. and J. Miller. 1952. Time-temperature relationships for rupture and creep stresses. Trans. ASME 74: 765-771.

Larson, J.M. and S. Floreen. 1977. Metallurgical factors affecting the crack growth resistance of a superalloy. Metall. Trans. A 8: 51-55.

Larsen, J.M., T. Nicholas, A.W. Thompson and J.C. Williams. 1986. Small crack growth in titanium-aluminum alloys. pp. 499-512. *In*: R.O. Ritchie and J. Lankford (eds.). Small Fatigue Cracks. The Metallurgical Society of the American Institute of Mining, Metallurgical and Petroleum Engineers. Warrendale, PA.

Lee, J.J. and W.N. Sharpe. 1986. Short fatigue cracks in notched aluminum specimens. pp. 323-339. *In*: R.O. Ritchie and J. Lankford (eds.). Small Fatigue Cracks. The Metallurgical Society of the American Institute of Mining, Metallurgical and Petroleum Engineers. Warrendale, PA.

Lee, E.Y., D.M. Chartier, R.R. Biederman and R.D. Sisson Jr. 1987. Modelling the microstructural evolution and degradation of the M–Cr–Al–Y coatings during high temperature oxidation. Surf. Coat. Technol. 32: 19-39.

Legros, M., B.R. Elliott, M.N. Rittner, J. Weertman and K.J. Hemker. 2000. Microsample tensile testing of nanocrystalline metals. Phil. Mag. A 80: 1017-1026.

Lemaitre, J. and J.L. Chaboche. 1999. Mécanique des Matériaux Solides. Dunod, Bordas, Paris.
Lerch, B.A. and N. Jayaraman. 1984. A study of fatigue damage mechanisms in Waspaloy from 25 to 800 °C. Mater. Sci. & Eng. 66: 151-166.
Li, S.X. and D.J. Smith. 1998. Development of an anisotropic constitutive model for single crystal superalloy for combined fatigue and creep loading. Int. J. Mech. Sci. 40(10): 937-948.
Liao, M. 2009. Probabilistic modeling of fatigue related microstructural parameters in aluminum alloys. Eng. Fract. Mech. 76: 668-680.
Liao, M. 2010. Dislocation theory based short crack model and its application for aircraft aluminum alloys. Eng. Fract. Mech. 77: 22-36.
Lii, M.-J., X.-F. Chen, Y. Katz and W.W. Gerberich. 1990. Dislocation modeling and acoustic emission observation of alternating ductile/brittle events in Fe-3wt%Si crystals. Acta Metall.38: 2345-2353.
Lin, I.-H. and R. Thomson. 1986. Cleavage, dislocation emission, and shielding for cracks under general loading. Acta Metall. 34: 187-206.
Liu, C.D., Y.F. Han, M.G. Yan and M.C. Chaturvedi. 1991. Creep crack growth behavior of Alloy 718. pp. 537-548. *In*: Proc. Superalloys 718, 625, and Various Derivatives. TMS. Warrendale, PA.
Liu, Y.J. and T.R. Hsu. 1985. A general treatment of creep crack growth. Eng. Fract. Mech. 21: 437-452.
Liu, X., G. Quan, X.J. Wu and Z. Zhang. 2015. Simulation of Thermomechanical Fatigue of Ductile Cast Iron and Lifetime Calculation. SAE Technical Paper 2015-01-0552, doi:10.4271/2015-01-0552.
Lloyd, G.J. and J. Wareing. 1981. Life-prediction methods for combined creep-fatigue endurance. Metals Technology (August): 297-305.
Louat, N., K. Sadanada, M. Duesbery and A.K. Vasudevan. 1993. A theoretical evaluation of crack closure. Metall. Trans. A 24: 2225-2232.
Lüthy, H., R.A. White and O.D. Sherby. 1979. Grain boundary sliding and deformation mechanism maps. Mater. Sci. Eng. 39: 211-216.
Ma, B.-T. and C. Laird. 1989. Overview of fatigue behaviour in copper single crystals - II. Population, size, distribution and growth kinetics of stage I cracks for tests at constant strain amplitude. Acta Metall. 37: 337-348.
Maccagno, T., A.K. Koul, J.-P. Immarigeon, L. Cutler and G. L'Esperance. 1990. Rejuvenation of Alloy 713 C turbine blades. Metall. Trans. 21A: 3115-3125.
MacLachlan, D.W., L.W. Wright, S. Gunturi and D.M. Knowles. 2001. Constitutive modeling of creep deformation in single crystal blade alloys SRR99 and CMSX-4. Int. J. Plasticity 17: 441-467.
Malakondiah, G., N. Prasad, G. Sundararajan and P. Rama Rao. 1988. An analysis of the transient stage in low stress viscous creep. Acta Metall. 36: 2167-2181.
Manson, S.S. 1954. Behaviour of materials under conditions of thermal stresses. National Advisory Commission on Aeronautics Report 1170. Lewis Flight Propulsion Laboratory. Cleveland, OH.
Manson, S.S. 1965. Fatigue: a complex subject—some simple approximations. Experimental Mechanics 5(4): 193-226.
Manson, S.S. and G.R. Halfords. 2006. Fatigue and Durability of Structural Materials. ASM International. Materials Park, OH.
Manson, S.S., G.R. Halfords and M.H. Hirschberg. 1971. Creep fatigue analysis by strain range partitioning. Proc. First Symposium on Design for Elevated Temperature Environment, pp. 12-24.

Marinis, A., S.A. Aquilino, P.S. Lund, D.G. Gratton, C.M. Stanford, A.M. Diaz-Arnold and F. Qian. 2013. Fracture toughness of yttria-stabilized zirconia sintered in conventional and microwave ovens. The Journal of Prosthetic Dentistry 109: 165-171.

Marshall, D.B. and A.G. Evans. 1985. Failure mechanisms in ceramic-fiber/ceramic-matrix composites. J Am Ceram Soc 68: 225-231.

Marshall, D.B., B.N. Cox and A.G. Evans. 1985. The mechanics of matrix cracking in brittle-matrix fiber composites. Acta Metall 33: 2013-2021.

Maruyama, K., K. Sawada and J. Koike. 2001. Advances in physical metallurgy and processing of steels. Strengthening mechanisms of creep resistant tempered martensitic steel. ISIJ Int. 41: 641-653.

Marx, M., H. Vehoff, W. Schäf and M. Welsch.2006. The mechanisms of the interaction of microcracks with grain boundaries: in-situ investigations in SEM by FIB, EBSD and electron channelling contrast. Proc. 16th European Conference on Fatigue (CD-ROM). July 4-7, 2006. Alexandropoulis, Greece.

Masing, G. 1926. Eigenspannungen und verfestigungbeim messing. *In*: Proc. Second Congress for Appl. Mech., Zurich.

Masumura, R.A., P.M. Hazzledine and C.S. Pande. 1998. Yield stress of fine grained materials. Acta Mater. 46: 4527-4534.

Maruyama, K., K. Sawada and J.I. Koike. 2001. Strengthening mechanisms of creep resistant tempered martensitic steel. ISIJ Inter. 41(6): 641-653.

Masatsugu, Y., N. Kaoru and N. Sosuke. 2016. Re-evaluation of long-term creep strength of welded joint of ASME grade 91 type steel. pp. 1-7. *In*: Proceedings of ASME 2016 Pressure Vessels and Piping Conference (PVP 2016).

McClintock, F.A. 1963.On the plasticity of the growth of fatigue cracks. pp. 65-102. *In*: D.C. Drucker and J.J. Gilman (eds.). Fracture of Solids. John Wiley & Sons. New York.

McClintock, F.A. 1971. Plasticity aspects of fracture. pp. 47-225. *In*: H. Liebowitz (ed.). Fracture, Vol. 3. Academic Press, New York.

Meier, G.H., F.S. Pettit and J.L. Smialek. 1995. The effects of reactive element additions and sulfur removal on the adherence of alumina to Ni- and Fe-base alloys. Materials and Corrosion 46: 232-240.

Merati, A. and H. Awatta. 2005. Microstructural characterization of 2024-T3 aluminum alloy plate. LM-SMPL-2005-0126. National Research Council Canada.

Merati, A., W. Beres and X.J. Wu. 2011. Effect of foreign object damage (FOD) on fatigue performance of fan blade Ti-6Al-4V material. Canadian Forces – DRDC International Defence Applications of Materials Meeting. Halifax, NS. June 7-8, 2011.

Meric, L. and G. Cailletaud. 1991. Single crystal modelling for structural applications, part 1 – model presentation. J. Eng. Mater. Tech. 113: 162-170.

Meric, L. and G. Cailletaud. 1991. Single crystal modelling for structural applications, part 2 – finite element implementation. J. Eng. Mater. Tech. 113: 171-182.

Miller, A.K. (ed.). 1987. Constitutive Equations for Creep and Plasticity. Elsvier Applied Science.

Miller, D.A. and R. Pilkington. 1980. Diffusion and deformation controlled creep crack growth. Metall. Trans. A 11A: 177-180.

Miller, K.J. 1982. The short crack problem. Fat. Eng. Mater. Struct. 5: 223-232.

Miller, K.J. 1989. Short fatigue cracks. pp. 3-22. *In*: Advances in Fatigue Science and Technology, Vol. 159. NATO ASI Series.

Miller, M.P., D.L. McDowell, R.L.T. Oehmke and S.D. Antolovich. 1993. A life prediction model for thermomechanical fatigue based on microcrack propagation. ASTM STP 1186: 35-49.

Miller, M.D., P.A.S. Reed, M.R. Joyce, M.B. Henderson, J.W. Brooks, I. Wilcock and X. Wu. 2007. Effect of environment on notch fatigue behaviour in CMSX4. Mater. Sci. Tech. 23: 1439-1445.

Miller, R.A. 1984. Oxidation-based model for thermal barrier coating life. J. Am. Ceram. Soc. 67: 517-521.

Miller, R.J. 2000. Design approaches for high temperature composite aeroengine components. Comprehensive Composite Materials 6: 181-207, Elsevier.

Milne, I., R.A. Ainsworth, A.R. Dowling and A.T. Stewart. 1988. Assessment of the integrity of structures containing defects. Int. J. Pressure Vessels and Piping 32: 3-104.

Miner, M.A. 1945. Cumulative damage in fatigue. Trans ASME J. Appl. Mech. 67: A159-A167.

Miura, S., S. Hashimoto and T. Fujii. 1988. Effect of the triple junction on grain boundary sliding in aluminum tricrystals. Journal de Physique Colloques 49(C5): C5-599-604.

Mizuno, M., S. Zhu, Y. Kagawa and H. Kaya. 1998. Stress, strain and elastic modulus behaviour of SiC/SiC composites during creep and cyclic fatigue. Journal of the European Ceramic Society 18: 1869-1878.

Mohamed, F.A. and M. Chauhan. 2006. Interpretation of the creep behaviour of nanocrystalline Ni in terms of dislocation accommodated boundary sliding. Metall. Mater. Trans. A 37: 3555-3567.

Mohamed, F.A. and H. Yang. 2010. Deformation mechanisms in nanocrystalline materials. Metall. Mater. Trans A 41A: 823-837.

Molski, K. and G. Glinka. 1981. A method of elastic-plastic stress and strain calculation at a notch root. Materials Science and Engineering 50: 93-100.

Monkman, F.C. and N.J. Grant. 1956. An empirical relationship between rupture life and minimum creep rate in creep-rupture tests. Proc. ASTM 56: 593-620.

Morrow, J. 1965. Cyclic plastic strain energy and fatigue of metals. Internal Friction, Damping and Cyclic Plasticity. ASTM STP 378: 45-84.

Mughrabi, H. 1980. Microscopic mechanisms of metal fatigue. pp. 1615-1639. *In*: P. Haasen, V. Gerold and G. Kostorz (eds.). The Strength of Metals and Alloys, Vol. 3. Oxford: Pergamon Press.

Mughrabi, H. 2009. Cyclic slip irreversibilities and the evolution of fatigue damage. Metall. Mater. Trans. A 40A: 1257-1279.

Mughrabi, H. and H.W. Höppel. 2010. Cyclic deformation and fatigue properties of very fine-grained metals and alloys. Int J Fatigue 32: 1413-1427.

Mukherjee, A.K., J.E. Bird and J.E. Dorn. 1969. Experimental correlations for high-temperature creep. Trans. ASM 62: 155.

Mura, T. and C.T. Lin. 1974. Theory of fatigue crack growth for work hardening materials. Int. J. Fracture 10: 284-287.

Murayama, M., J.M. Howe, H. Hidaka and S. Takaki. 2002. Atomic-level observation of disclination dipoles in mechanically milled nanocrystalline Fe. Science 295: 2433-2435.

Muskhelishvili, N.I. 1953. Some Basic Problems of the Mathematical Theory of Elasticity. Noordhoff Leyden.

Nabarro, F.R.N. 1967. Theory of Crystal Dislocations. Oxford University Press, London.

Navarro, A. and E.R. de los Rios. 1988. Short and long fatigue crack growth: a unified model. Phil. Mag. A 57: 15-36.
Needleman, A. and J.R. Rice. 1980. Plastic creep flow effects in the diffusive cavitation of grain boundaries. Acta Met. 28: 1315-1332.
Neu, R. and H. Sehitoglu. 1989. Thermo-mechanical fatigue, oxidation and creep: part 2 – life prediction. Metall. Trans. A. 20A: 1769-1783.
Neuber, H. 1961. Theory of stress concentration for shear-strained prismatic bodies with arbitrary non-linear stress-strain law. Trans. of the ASME, J. Appl. Mech. 27: 544-551.
Neumann, P. 1974. New experiments concerning the slip processes at propagating fatigue cracks – I. Acta Metall. 22: 1155-1178.
Newman, J.C. Jr. 1977. Finite-element analysis of crack growth under monotonic and cyclic loading. ASTM STP 637: 56-80.
Newman, J.C. Jr. and P.R. Edward. 1988. Short-crack growth behaviour in an aluminum alloy: an AGARD cooperative test programme. AGARD-R-732, NATO. Advisory Group for Aerospace Research and Development.
Newman, J.C. Jr. 1992. Fracture mechanics parameters for small fatigue cracks, small crack, small crack test methods. pp. 6-28. *In*: J. Allison and J. Larsen (eds.). ASTM STP 1149. American Society for Testing and Materials, Philadelphia, PA.
Nikbin, K.M., G.A. Webster and C.E. Turner. 1977. A comparison of methods of correlating creep crack growth. p. 627. *In*: D.M.R. Taplin (ed.). Fracture 1977, Vol. 2. ICF-4. Waterloo, Pergamon, NY.
NIMS. 2014. Creep Data Sheet No. 43A. National Institute for Materials Science. Japan.
North Atlantic Treaty Organization. 1978. Advisory Group for Aerospace Research and Development—Structure. AGARD Conference Proceedings, CP-243.
Norton, F.H. 1929. The Creep of Steel at High Temperatures. McGraw-Hill. London, UK.
Nouailhas, D. and J. Chaboche. 1991. Anisotropic constitutive modeling for single crystal superalloys using a continuum phenomenological approach. pp. 213-218. *In*: C.S. Desai (ed.). Proc. 3rd International Conference on Constitutive Las for Engineering Materials. Tucson, AZ. ASME Press. 2.
Norton, F.H. 1929. The Creep of Steel at High Temperatures. McGraw-Hill. London, UK.
Nusier, S.Q., G.M. Newaz and Z.A. Chaudhury. 2000. Experimental and analytical evaluation of damage processes in thermal barrier coatings. Int. J. Solids Struct. 37: 2495-2506.
Nye, J.F. 1957. Physical Properties of Crystals. Oxford University Press.
Orowan, E. 1934. Zur Kristallplastizität. III. Zeitschrift für Physik. 89(9-10): 634-659.
Orowan, E. 1940. Problems of plastic gliding. Proc. Phys. Soc. 52: 8-22.
Orowan, E. 1948. Discussion on internal stresses. pp. 451-453. *In*: Symp. Internal Stresses in Metals and Alloys. The Institute of Metals. London.
Ostergren, W.J. 1976. Correlation of hold-time effects in elevated temperature low cycle fatigue using a frequency modified damage function. Proc. ASME-MPC Symposium on Creep-Fatigue Interaction. MPC-3: 179-202.
Ovid'ko, I.A. 2002. Deformation of nanostructures. Science 295: 2386.
Palmgren, A. 1924. Die Lebensdauer von Kugellagern, Z.V.D.I. 68(14): 339-341.
Paris, P.C. and F. Erdogan. 1963. A critical analysis of crack propagation laws. Journal of Basic Engineering 85: 528-534.
Pearson, S. 1975. Initiation of fatigue cracks in commercial aluminum alloys and the subsequent propagation of very short cracks. Eng. Fract. Mech. 7: 235-247.

Perry, A.J. 1974. Cavitation in creep. J. Mater. Sci. 9: 1016-1039.
Peters, J.O., O. Roder, B.L. Boyce, A.W. Thompson and R.O. Ritchie. 2000. Role of foreign-object damage on thresholds for high-cycle fatigue in Ti-6Al-4V. Metall. Mater. Trans. A 31A: 1571-1583.
Petrenec, M., K. Obrtlik and J. Polák. 2005. Inhomogeneous dislocation structure in fatigued INCONEL 713 LC superalloy at room and elevated temperatures. Mater. Sci. Eng. A 400-401: 485-488.
Pettit, F.S. and G.H. Meier. 1984. Oxidation and hot corrosion of superalloys. pp. 651-687. *In*: Superalloys, TMS. Warrendale, PA.
Pollock, T.M. and A.S. Argon. 1992. Creep resistance of CMSX-3 nickel base superalloy single crystals. Acta Metall. Mater. 40: 1-30.
Plumtree, A. and H.A. Abdel-Raouf. 2001. Cyclic stress-strain response and structure. Int. J. Fatigue 23: 799-805.
Polák, J., V. Mazánová, M. Heczko et al. 2017. The role of extrusions and intrusions in fatigue crack initiation. Eng Fract Mech. 85: 46-60.
Polanyi, M. 1934. Über eine Art Gitterstörung, die einen Kristall plastisch machen könnte. Z. Physik 89: 660-664.
Prager, W. 1958. Problemès de PlasticitiéTheoriqué. Dunod, Paris.
Prandtl, L. 1924. Spannungverteilung in plastischen korpen. p. 43. *In*: Proceedings of the First International Conference on Applied Mechanics. Delt, The Netherlands.
Pugh, P. 2001. The magic of a name, the Rolls-Royce story. Part 2: the power behind the jets 1945–1987. UK/USA: Icon Books/Totem Book.
Qiu, Y., J.C. Pang, C.L. Zou, M.X. Zhang, S.X. Li and Z.F. Zhang. 2018. Fatigue strength model based on microstructures and damage mechanism of compacted graphite iron. Materials Science & Engineering A 724: 324-329.
Rabiei, A. and A.G. Evans. 2000. Failure mechanisms associated with the thermally grown oxide in plasma-sprayed thermal barrier coatings. Acta Materialia 48: 3963-3976.
Rae, C.F.M. and R.C. Reed. 2007. Primary creep in single crystal superalloys: origins, mechanisms and effects. Acta Mater. 55: 1067-1081.
Raj, R. 1975. Transient behaviour of diffusion-induced creep and creep rupture. Metall. Trans. A 6A: 1499-1509.
Raj, R. and M.F. Ashby. 1971. On grain boundary sliding and diffusional creep. Metall. Trans. 2A: 1113-1125.
Raj, R. and M.F. Ashby. 1975. Intergranular fracture at elevated temperature. Acta Metall. 23: 653-666.
Ramberg, W. and W.R. Osgood. 1943. Description of stress-strain curves by three parameters. Technical Note No. 902. National Advisory Committee for Aeronautics. Washington DC.
Reed, W.T. 1953. Dislocations in Crystals. McGraw-Hill Book Co. New York.
Reed, P.A.S., X.D. Wu and I. Sinclair. 2000. Fatigue crack path prediction in Udimet 720 nickel based alloy single crystals. Met. Mater. Trans. A 31A: 109-123.
Reed, R.C. 2006. The Superalloys—Fundamentals and Applications. Cambridge University Press. Cambridge, UK.
Rees, D.W.A. 1987. Life prediction techniques for combined creep and fatigue. Progress in Nuclear Energy 19(3): 211-239.
Renusch, D., M. Schorr and M. Schütze. 2008. The role that bond coat depletion of aluminum has on the lifetime of APS-TBC under oxidizing conditions. Mater. Corr. 59(7): 547-555.

Reuchet, J. and L. Remy. 1979. High temperature fatigue behaviour of a cast cobalt base superalloy. Fat. Fract. Eng. Mater. Struct. 2: 51-62.
Reuchet, J. and L. Remy. 1983. High temperature low cycle fatigue of MAR-M 509 superalloy II: the influence of oxidation at high temperatures. Mater. Sci. Eng. 58: 33-42.
Reuss, A. 1930. Mathematik und Mechanik 10: 266.
Rice, J.R. 1967. Mechanics of crack tip deformation and extension by fatigue. pp. 247-311. *In*: Fatigue Crack Propagation. ASTM STP 415. American Society for Testing and Materials. Philadelphia, PA.
Rice, J.R. 1968. A path independent integral and the approximate analysis of strain concentration by notches and cracks. J. Appl. Mech. 35: 379-386.
Rice, J.R. and G.F. Rosenger. 1968. Plane strain deformation near a crack tip in a power-law hardening material. J. Mech. Phys. Solids 16: 1-12.
Riedel, H. and J.R. Rice. 1980. Tensile cracks in creeping solids. pp. 112-130. *In*: Fracture Mechanics. ASTM STP 700. American Society for Testing and Materials. Philadelphia, PA.
Riedel, H. 1989. Creep crack growth. pp. 101-126. *In*: Fracture Mechanics: Perspectives and Directions (Twentieth Symposium). ASTM STP 1020. American Society for Testing and Materials. Philadelphia, PA.
Riedel, H. 1990. Creep crack growth under small-scale creep conditions. Int. J. Fract. 42: 173-188.
Ritchie, R.O. 1979. Near-threshold fatigue-crack propagation in steels. International Metals Reviews 20: 205-230.
Robinson, E.I. 1952. Effect of temperature variation on the long time rupture strength of steels. Trans ASME 74(5): 777-780.
Robotnov, Y.N. 1969. Creep Problems in Structural Members. North Holland.
Romannoski, G.R. Jr. 1982. Mechanisms of Deformation and Fracture in High Temperature Low-Cycle Fatigue of Rene 80 and IN 100. NASA Contractor Report 16549.
Rösler, J. and E. Arzt. 1988a. The kinetics of dislocation climb over hard particles - I. Climb without attractive particle-dislocation interaction. Acta Metall. 36: 1043-1051.
Rösler, J. and E. Arzt. 1988b. The kinetics of dislocation climb over hard particles - II. Effects of an attractive particle-dislocation interaction. Acta Metall. 36: 1053-1060.
Rospars, C., J.L. Chermant and P. Ladevèze. 1998. On a first creep model for a 2D SiCf–SiC composite. Materials Science and Engineering A 250: 264-269.
Sadanada, K. 1978. A theoretical model for creep crack growth. Metall. Trans. A 9A: 635-641.
Sadananda, K. and P. Shahinian. 1977. Creep crack growth in alloy 718. Metall. Trans. A 8A: 439-449.
Saeidi, S., K.T. Voisey and D.G. McCartney. 2011. Mechanical properties and microstructure of VPS and HVOF CoNiCrAlY coatings. J. Therm. Spray Technol. 20: 1231-1243.
Saxena, A., R.S. Williams and T.T. Shih. 1981. A model for representing and predicting the influence of hold times on fatigue crack growth behaviour at elevated temperatures. pp. 86-99. *In*: Fracture Mechanics, Thirteenth Volume. ASTM STP 743. American Society for Testing and Materials. Philadelphia, PA.
Saxena, A. 1986. Creep crack growth under non-steady-state conditions. pp. 185-201. *In*: Fracture Mechanics, Seventeenth Volume. ASTM STP 905. American Society for Testing and Materials. Philadelphia, PA.

Sehitoglu, H. 1992. Thermo-mechanical fatigue life prediction methods. pp. 47-76. *In*: Advances in Fatigue Lifetime Predictive Techniques. ASTM STP 1122. American Society for Testing and Materials. Philadelphia, PA.
Seifert, T. and H. Riedel. 2010. Mechanism-based thermomechanical fatigue life prediction of cast iron. Part I: Models. Int. J. Fatigue 32: 1358-1367.
Seifert, T., G. Maier, A. Uihlein, K.-H. Lang and H. Riedel. 2010a. Mechanism-based thermomechanical fatigue life prediction of cast iron. Part II: Comparison of model predictions with experiments. Int. J. Fatigue 32: 1368-1377.
Seifert, T., C. Schweizer, M. Schlesinger, M. Moser and M. Eibl. 2010b. Thermomechanical fatigue of 1.4849 cast steel—experiment and life prediction using a fracture mechanics approach. Int. J. Mat. Res. 101(8): 942-950.
Seifert, T., R. Hazime and S. Dropps. 2014. TMF life prediction of high temperature components made of cast iron HiSiMo. Part II: Multiaxial implementation and component assessment. SAE Int. J. Mater. Manf. 7(2): doi:10.4271/2014-01-0905.
Sessions, M.L., C.J. McMahon and J.L. Walker. 1977. Further observations on the effect of environment on the creep/rupture behavior of a nickel-base high temperature alloy: grain size effects. Material Sci. & Eng. 27: 17-24.
Schmunk, R.E. and G.E. Korth. 1981. Tensile and low-cycle fatigue measurements on cross-rolled tungsten. J Nuclear Mater 103&104: 943-948.
Shrestha, T., M. Basirat, I. Charit, G.P. Potirniche and K.K. Rink. 2013. Creep rupture behavior of Grade 91 steel. Mater. Sci. Eng. A 565: 382-391.
Sieborger, D., H. Knake and U. Glatzel. 2001. Temperature dependence of the elastic moduli of the nickel-base superalloy CMSX-4 and its isolated phases. Mater. Sci. Eng. A 298: 26-33.
Sih, G.C. 1973. Handbook of Stress-intensity Factors. Lehigh University.
Singh, C.V., A.J. Mateos and D.H. Warner. 2011. Atomistic simulations of dislocation–precipitate interactions emphasize importance of cross-slip. Scripta Materialia 64: 398-401.
Skelton, R.P. 2013. The energy density exhaustion method for assessing the creep-fatigue lives of specimens and components. Materials at High Temperatures 30(3): 183-201.
Slavik, D. and H. Sehitoglu. 1986. Constitutive models suitable for thermal loading. ASME Journal of Engineering Materials and Technology 108(4): 303-312.
Smialek, J.L., J.A. Nesbitt, C.A. Barrett and C.E. Lowell. 2000. Cyclic oxidation testing and modelling: a NASA Lewis perspective. NASA TM-1999-209769.
Smith, R.A. and J.F. Cooper. 1989. A finite element model for the shape development of irregular planar cracks. Int. J. Pressure Vessels and Piping 36: 315-326.
Smith, K.N., P. Watson and T.H. Topper. 1970. A stress-strain function for the fatigue of materials. J. Mater. 5: 767-778.
Smith, K.N., P. Watson and T.H. Topper. 1970. A stress-strain function for the fatigue of metals. J. Mater. 5(4): 767-778.
Smith, W.F. 2004. Foundations of Materials Science and Engineering 3rd ed. McGraw-Hill.
Snowden, K.U.1963. Dislocation arrangements during cyclic hardening and softening in A1 crystals. Acta Metall. 11: 675-684.
Socie, D.F., P. Kurath and J. Koch. 1985. A multiaxial fatigue damage parameter. The Second International Symposium on Multiaxial Fatigue. Sheffield, U.K.
Song, Z. and D.W. Hoeppner. 1988. Dwell time effects on the fatigue behaviour of titanium alloys. Int. J. Fatigue 10: 211-218.

Speidel, M.O. 1973. Modulus of elasticity and fatigue crack growth. pp. 212-221. *In*: High Temperature Materials in Gas Turbine. Proceedings of the Symposium on High Temperature Materials in Gas Turbines. Brow, Boveri & Company Limited. Baden, Switzerland.

Spindler, M.W. 2007. An improved method for calculation of creep damage during creep–fatigue cycling. Materials Science and Technology 23: 1461-1470.

Stephens, Ralph I. and Henry O. Fuchs. 2001. Metal Fatigue in Engineering (Second edition). John Wiley & Sons, Inc. ISBN 0-471-51059-9.

Stroh, A.N. 1954. The formation of cracks as a result of plastic flow. Proc. Roy. Soc. London 223: 404.

Stroh, A.N. 1957. A theory of the fracture of metals. Advances in Physics 6: 418-465.

Stroh, A.N. 1958. Dislocations and cracks in anisotropic elasticity. Phil. Mag. 3: 625-646.

Strong, A. and D.J. Gooch. 1997. Microstructural Development and Stability in High Chromium Ferritic Power Plant Steels. Institute of Materials.

Suresh, S. 1998. Fatigue of Materials. Cambridge University Press, New York.

Suresh, S. and R.O. Ritchie. 1984. Near-threshold fatigue crack propagation: a perspective on the role of crack closure. pp. 227-261. *In*: D. Davidson and S. Suresh (eds.). Fatigue Crack Growth Threshold Concepts. The Metallurgical Society of AIME. Warrendale, PA.

Taira, S., K. Tanaka and Y. Nakai. 1978. A model of crack-tip slip band blocked by grain boundary. Mech. Res. Commun 5: 375-381.

Taira, S., R. Ohtani and T. Kitamura. 1979. Application of J-integral to high-temperature crack propagation. Part I: Creep crack propagation. J. Eng. Mater. Tech. 101: 154-161.

Tanaka, K. and T. Mura. 1981. A dislocation model for fatigue crack initiation. J. App. Mech 48: 97-103.

Tanaka, K. and Y. Nakai 1983. Propagation and non-propagation of short fatigue cracks at a sharp notch. Fatigue of Engineering Materials and Structures 6: 315-327.

Tanaka, K., Y. Akiniwa and R.P. Wei. 1986. Modelling of small fatigue crack growth interacting with grain boundary. Engng Fract Mech. 24: 803-819.

Tang, F. and J.M. Schoenung. 2006. Evolution of Young's modulus of air plasma sprayed yttria-stabilized zirconia in thermally cycled thermal barrier coatings. Scripta Materialia 54: 1587-1592.

Taylor, G.I. 1934. The mechanism of plastic deformation of crystals. Part I: Theoretical. Proc. Roy. Soc. London A 145: 362-387.

Telesman, J., T.P. Gabb, A. Garg, P. Bonacuse and J. Gayda. 2008. Effect of microstructure on time dependent fatigue crack growth behaviour in a P/M turbine disc alloy. pp. 807-816. *In*: Superalloys 2008. The Metallurgical Society. Warrendale, PA.

Thompson, N., N.J. Wadsworth and N. Louat. 1956. The origin of fatigue fracture in copper. Phil. Mag. 1: 113-126.

Timoshenko, S. and J.M. Gere. 1961. Theory of Elastic Stability, 2nd ed. McGraw Hill, New York.

Ting, T.C.T. 1996. Anisotropic Elasticity: Theory and Applications. Oxford University Press, Oxford.

Tomkins, B. 1968. Fatigue crack propagation—an analysis. Phil. Mag. 18: 1041-1066.

Toshimitsu, A. and Yokobori Jr. 1999. Difference in the creep and creep crack growth behaviour between creep ductile and brittle materials. Engineering Fracture Mechanics 62: 61-78.

Tsang, J., R.M. Kearsey, P. Au, S. Oppenheimer and E. McDevitt. 2011. Microstructural study of fatigue and dwell fatigue crack growth behaviour of ATI 718Plus alloy. Can. Metall. Quart. 50(3): 222-231.

Tryon, R.G. and T.A. Cruse. 1998. A reliability-based model to predict scatter in fatigue crack nucleation life. Fatigue & Fracture of Engineering Materials & Structures 21: 257-267.

Tyson, W.R. and W.A. Miller. 1977. Surface free energies of solid metals: estimation from liquid surface tension measurements. Surface Science 62: 267-276.

Vaßen, R., G. Kerkhoff and D. Stöver. 2001. Development of a micromechanical life prediction model for plasma sprayed thermal barrier coatings. Mater. Sci. Eng. A 303: 100-109.

Venkateswara Rao, K.T., W. Yu and R.O. Ritchie. 1988. Fatigue crack propagation in aluminum-lithium alloy 2090. Part II: Small crack behaviour. Metall. Trans. A 19A: 563-568.

Vitek, V. 1978. A theory of diffusion controlled intergranular creep crack growth. Acta Metall. 26: 1345-1356.

vonMises, R. 1913. Mechanik der Festen Korperimplastischdeformablen Zustand. Göttin. Nachr. Math. Phys. 1: 582-592.

Wadsworth, N.J. 1963. Work hardening of copper crystals under cyclic straining. Acta Metall. 11: 663-673.

Wadsworth, J., O.A. Ruano and O.D. Sherby. 2002. Denuded zones, diffusional creep, and grain boundary sliding. Metall. Mater. Trans. A 33A: 219-229.

Walker, K. 1970. The effect of stress ratio during crack propagation and fatigue for 2024-t3 and 7075-t6 aluminum, effects of environment and complex load history on fatigue life. pp. 1-14. *In*: ASTM STP 462. American Society for Testing and Materials. West Conshohocken, PA.

Wallace, W., J.C. Beddoes and M.C. de Malherbe. 1981. A new approach to the problem of stress corrosion cracking in 7075-T6 aluminum. Can. Aero & Space J. 27(3): 221-232.

Wallwork, G.R. and A.Z. Hed. 1971. Some limiting factors in the use of alloys at high temperatures. Oxidation of Metals 3: 171-184.

Wang, X., S.B. Lambert and G. Glinka. 1998. Approximate weight functions for embedded elliptical cracks. Engineering Fracture Mechanics 59(3): 381-392.

Wareing, J.H. and G. Vaughan. 1979. Influence of surface finish on low-cycle fatigue characteristics of Type 316 stainless steel at 400 °C. Metal Science January: 1-8.

Warren, R. 1992. Ceramic-matrix Composites. Blackie and Son, Glasgow.

Weaver, C.W. 1959-60. Intergranular cavitation, structure, and creep of Nimonic 89A-type alloy. J Inst Met. 88: 296-300.

Weertman, J. 1955. Theory of steady-state creep based on dislocation climb. J. Appl. Phys. 26: 1213-1217.

Weetman, J. 1966. Rate of growth of fatigue cracks calculated from the theory of infinitesimal dislocations distributed on a crack plane. Int. J. Fract. Mech. 2: 460-467.

Weertman, J. 1981. Fatigue crack growth in ductile metals. pp. 11-19. *In*: T. Mura (ed.). Mechanics of Fatigue, ADM. Vol. 47, ASME Publication.

Weetman, J. 1996. Dislocation Based Fracture Mechanics. World Scientific.

Wei, R.P. and G. Shim. 1984. Fracture mechanics and corrosion fatigue. *In*: Corrosion Fatigue: Mechanics, Metallurgy, Electrochemistry, and Engineering. ASTM-STP 801: 5-25.

Westergaard, H.M. 1939. Bearing pressures and cracks. J. Appl. Mech. 61: A49-A53.
Williams, M.L. 1957. On the stress distribution at the base of a stationary crack. Journal of Applied Mechanics 24: 109-114.
Wilshire, B. 2002. Observations, theories, and predictions of high-temperature creep behaviour. Metall. Mater. Trans. A 33A: 241-248.
Wilshire, B. and P.J. Scharning. 2009. Theoretical and practical approaches to creep of Waspaloy. Mat. Sci. Tech. 25: 243-248.
Wohler, A. 1867. Wohler's experiments on the strength of metals. Engineering 4: 160-161.
Wu, X.J., A.K. Koul and A.S. Krausz. 1993. A transgranular fatigue crack growth model based on restricted slip reversibility. Metall. Trans. A 24A: 1373-1380.
Wu, X.J. and A.K. Krausz. 1994. A kinetics formulation for low-temperature plasticity. J. Mater. Eng. Performance 3: 169-177.
Wu, X.J. and W. Wallace. 1994. On low-temperature environment-assisted fatigue crack growth. Metall. Trans. A 25A: 658-659.
Wu, X.J., W. Wallace, M.D. Raizenne and A.K. Koul. 1994. The orientation dependence of fatigue crack growth rate in 8090 aluminum-lithium plate. Metall. Trans. A 25A: 575-588.
Wu, X.J. 1995. An energy approach to crack closure. Int. J. Fract. 73: 263-272.
Wu, X.J. and A.K. Koul. 1995. Grain boundary sliding in the presence of grain boundary precipitates during transient creep. Metall. Trans. A 26A: 905-913.
Wu, X.J., W. Wallace and A.K. Koul. 1995a.A new approach to fatigue threshold. Fat. & Fract. Eng. Struct. & Mat. 18: 833-845.
Wu, X.J., W. Wallace, A.K. Koul and M.D. Raizenne. 1995b. Near-threshold fatigue crack growth in 8090 Al-Li alloy. Metall. Trans. A 26A: 2973-2982.
Wu, X.J. and A.K. Koul. 1996. Modelling creep in complex engineering alloys. pp. 3-19. *In*: H. Merchant (ed.). Creep and Stress Relaxation in Miniature Structures and Components. TMS. The Metallurgical Society. Warrendale, PA.
Wu, X.J. and A.K. Koul. 1997. Grain boundary sliding at serrated grain boundaries. Advanced Performance Materials 4: 409-420.
Wu, X.J., W. Deng, A.K. Koul and J.-P. Immarigeon. 2001. A continuously distributed dislocation model for fatigue cracks in anisotropic crystalline materials. Int J. Fatigue 23: S201-S206.
Wu, X.J., M.D. Raizenne, R.T. Holt and W. Wallace. 2001. Thirty years of retrogression and re-aging (RRA). Can. Aero. Space J. 47: 131-137.
Wu, X.J., S. Yandt, P. Au and J.-P. Immarigeon. 2002. Modeling thermo-mechanical fatigue by evolution of its activation energy. pp. 3-14. *In*: M.A. McGaw, S. Kalluri, J. Bressers and S.D. Peteves (eds.). Thermomechanical Fatigue Behaviour of Materials, 4th Volume. ASTM STP 1428. American Society for Testing and Materials. West Conshohocken, PA.
Wu, X.J., P.C. Patnaik, M. Liao and W.R. Chen. 2003. A statistical assessment of the damage state in plasma-sprayed thermal barrier coating, GT-38790. *In*: Proceedings of ASME TURBO EXPO 2003. Turbo Expo Power: Land, Sea & Air. Atlanta, Georgia, USA.
Wu, X.J. 2005. A continuously distributed dislocation model of Zener-Stroh-Koehler cracks in anisotropic materials. Int. J. Solids Struct 42: 1909-1921.
Wu, X.J., C. Poon and D. Raizenne. 2005. Method and System for Prediction of Precipitation Kinetics in Precipitation-Hardenable Aluminum Alloys, US 6925352.
Wu, X.J. and P. Au. 2007. Deformation kinetics during dwell fatigue. Materials Science and Technology 23: 1446-1449.

Wu, X.J., J. Bird and P.C. Patnaik. 2007. A Framework of Prognosis and Health Management – The NRC Approach. Proc. ASME TURBO EXPO 2007. Montreal, QC.

Wu, X.J. 2009. A model of nonlinear fatigue-creep (dwell) interactions. ASME J. Gas Turbine Powers 131: 032101-1-6.

Wu, X.J. 2010. Life prediction of gas turbine materials. pp. 215-282. *In*: G. Injeti (ed.). Gas Turbines. InTech, Rejika, Croatia.

Wu, X.J., P. Au, D. Seo, L. Lafleur and O. Lupandina. 2010. Uniaxial Creep Lifing Methodology—III: Modeling the Effect of Pre-Creep on Fatigue Lives of Waspaloy and Udimet 720 Li. LTR-SMPL-2010-0219. National Research Council Canada.

Wu, X.J., S. Williams and D. Gong. 2012. A true-stress creep model based on deformation mechanisms for polycrystalline materials. J. Mater. Eng. Perform. 21(11): 2255-2262.

Wu, X.J. and Z. Zhang. 2013. T-56 1st Stage NGV Material Behaviours and Modeling. LTR-SMPL-2013-0078. National Research Council Canada.

Wu, X.J., G. Quan, R. MacNeil, Z. Zhang and C. Sloss. 2014. Failure mechanisms and damage model of ductile cast iron under low-cycle fatigue conditions. Metall. and Mater. Trans. A 45A: 5088-5097.

Wu, X.J. 2015. An integrated creep-fatigue theory for material damage modeling. Key Engineering Materials 627: 341-344.

Wu, X.J., G. Quan, R. MacNeil, Z. Zhang and C. Sloss. 2015. Thermomechanical fatigue of ductile cast iron and its life prediction. Metall. and Mater. Trans. A 46A: 2530-2542.

Wu, X.J. and Z. Zhang. 2016. A mechanism-based approach from low cycle fatigue to thermomechanical fatigue life prediction. Journal of Engineering for Gas Turbines and Power 138: 072503-1-7.

Wu, X.J., G. Quan and C. Sloss. 2016. Low cycle fatigue of cast austenitic steel. *In*: Z. Wei, K. Nikbin, P. McKeighan and D. Harlow (eds.). Fatigue and Fracture Test Planning, Test Data Acquisitions and Analysis. ASTM STP 1598.

Wu, X.J., Z. Zhang, L. Jiang and P. Patnaik. 2017. Material selection issues for a nozzle guide vane against service induced failure. Journal of Engineering for Gas Turbines and Power 139: 052101.

Wu, X.J. 2017. Comments on: A physical model and constitutive equations for complete characterization of S-N fatigue behavior of metals. Scripta Materialia 133: 113.

Wu, X.J. 2018a. On Tanaka-Mura's fatigue crack nucleation model and validation. Fat. Fract. Eng. Mat. & Struct. 41(4): 894-899.

Wu, X.J. 2018b. A fatigue crack nucleation model for anisotropic materials. Fatigue Fract. Eng. Mater. Struct. 2018: 1-7. https://doi.org/10.1111/ffe.12907.

Wu, X.J. 2019. The crack number density theory on air-plasma-sprayed thermal barrier coating. Surface & Coatings Technology 358: 347–352.

Wu, X.J., D. Seo, M. Head and S. Chan. 2018. Effect of pre-creep strain on high cycle fatigue life of TI 834, GT2018-75259. Journal of Engineering for Gas Turbines and Power 141 (2019): 052101-1-6.

Xu, S., X.J. Wu, A.K. Koul and J.I. Dickson. 1999. An intergranular creep crack growth model based on grain boundary sliding. Metall. Trans. A 30A: 1039-1045.

Yandt, S., D. Seo, P. Au and N. Tsuno. 2011. The effect of compressive dwell on the low cycle fatigue behaviour of a nickel-base single crystal alloy at 1100 °C. Can. Metall. Quart. 50: 303-310.

Yandt, S., X.J. Wu, N. Tsuno and A. Sato. 2012. Cyclic-dwell fatigue behaviour of single crystal Ni-base superalloys with/without rhenium. pp. 501-508. *In*: E.S. Huron et al. (eds.). Superalloys 2012. The Minerals Metals and Materials Society (TMS). Warrendale PA.

Yang, Y., X.M. Zhang, Z.H. Li and Q.Y. Li. 1996. Adiabatic shear band on the titanium side in the Ti/mild steel explosive cladding interface. Acta Mater. 44(2): 561-565.

Yao, Z., M. Zhang and J. Dong. 2013. Stress rupture fracture model and microstructure evolution for Waspaloy. Metall. Mater. Trans. A 44A: 3084-3089.

Yokobori, T. and M. Yoshida. 1974. Kinetic theory approach to fatigue crack propagation in terms of dislocation dynamics. Int. J. Fracture 10: 467-470.

Yokobori, T., A.T. Yokobori Jr. and A. Kamei. 1975. Dislocation dynamics theory for fatigue crack growth. Int. J. Fracture 11: 781-788.

Yurechko, M., C. Schroer, O. Wedemeyer, A. Skrypnik and J. Konys. 2011. Creep-to-rupture of 9%Cr steel T91 in air and oxygen-controlled lead at 650 °C. Journal of Nuclear Materials 419: 320-328.

Zener, C. 1948. Fracturing of Metals, Cleveland. American Society for Metals, pp. 3-31.

Zhao, L.G. and J. Tong. 2008. A viscoplastic study of crack-tip deformation and crack growth in a nickel-based superalloy at elevated temperature. J. Mech. Phys. Solids 56: 3363-3378.

Zhang, X.Z., X.J. Wu, R. Liu, J. Liu and M.X. Yao. 2017a. Deformation-mechanism-based modeling of creep behaviour of modified 9Cr-1Mo steel. Mater. Sci. & Eng. A 689: 345-352.

Zhang, X.Z., X.J. Wu, R. Liu, J. Liu and M.X. Yao. 2017b. Influence of Laves phase on creep strength of modified 9Cr-1Mo steel. Mater. Sci. & Eng. A 706: 279-286.

Zhang, X.Z., X.J. Wu, R. Liu and M.X. Yao. 2019. Effects of oxidation-resistant coating on creep behavior of modified 9Cr-1Mo steels. Mater. Sci. & Eng. A 743: 418-424.

Zhang, Z. and X.J. Wu. 2011. The Weight Function Method for Damage Tolerance Analysis of Gas Turbine Components. LTR-SMPL-2011-0073. National Research Council Canada.

Zhang, Z. and X.J. Wu. 2014. A 3D Stress Weight Function-Based Software for Automated Damage Tolerance Analysis. LTR-SMM-2014-0222. National Research Council Canada.

Zhang, Z.F., H.C. Gu and X.L. Tan. 1998. Low-cycle fatigue behaviours of commercial-purity titanium. Mater. Sci. & Eng. A 252: 85-92.

Zheng, Z., D.S. Balint and F.P.E. Dunne. 2016. Dwell fatigue in two Ti alloys: an integrated crystal plasticity and discrete dislocation study. J. Mech. and Phy. Solids 96: 411-427.

Zhu, D., S.R. Choi and R.A. Miller. 2004. Development and thermal fatigue testing of ceramic thermal barrier coatings. Surface and Coatings Technology 188-189: 146-152.

Zhu, S., M. Mizuno, Y. Kagawa, J. Cao, Y. Nagano and H. Kaya. 1997. Creep and fatigue behaviour of SiC fiber reinforced SiC composite at high temperatures. Materials Science and Engineering A 225: 69-77.

Zhu, S., M. Mizuno, Y. Kagawa and Y. Mutoh. 1999. Monotonic tension, fatigue and creep behaviour of SiC-fiber-reinforced SiC-matrix composites: a review. Composites Science and Technology 59: 833-851.

Zou, C.L., J.C. Pang, M.X. Zhang, Y. Qiu, S.X. Li, L.J. Chen, J.P. Li, Z. Yang and Z.F. Zhang. 2018. The high cycle fatigue, deformation and fracture of compacted graphite iron: influence of temperature. Materials Science & Engineering A 724: 606-615.

Index

Color Plate Section

Chapter 8

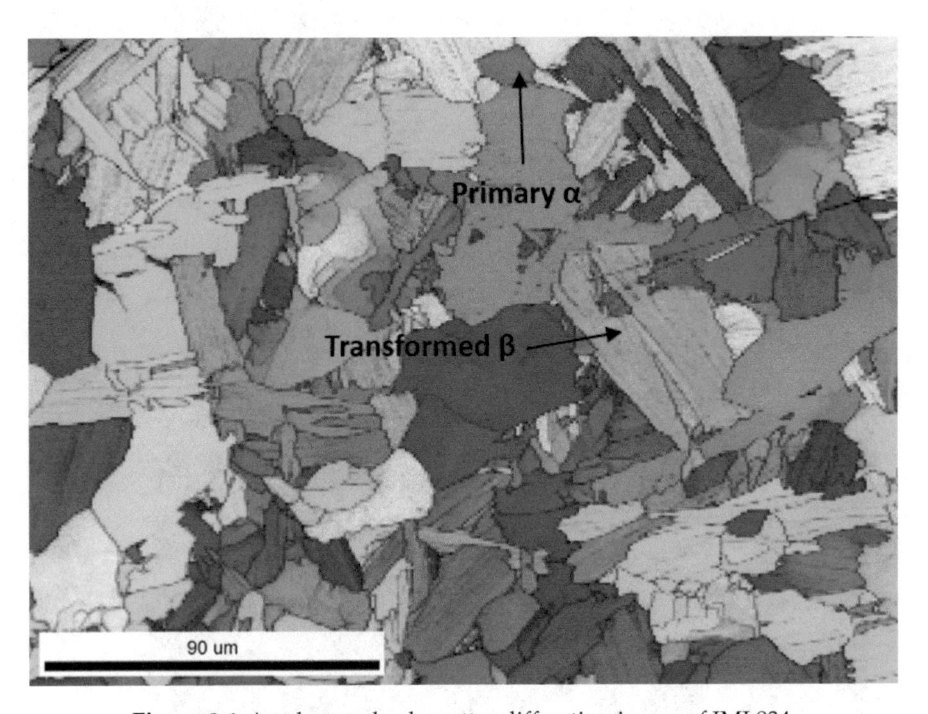

Figure 8.6. An electron back-scatter diffraction image of IMI 834.

Chapter 14

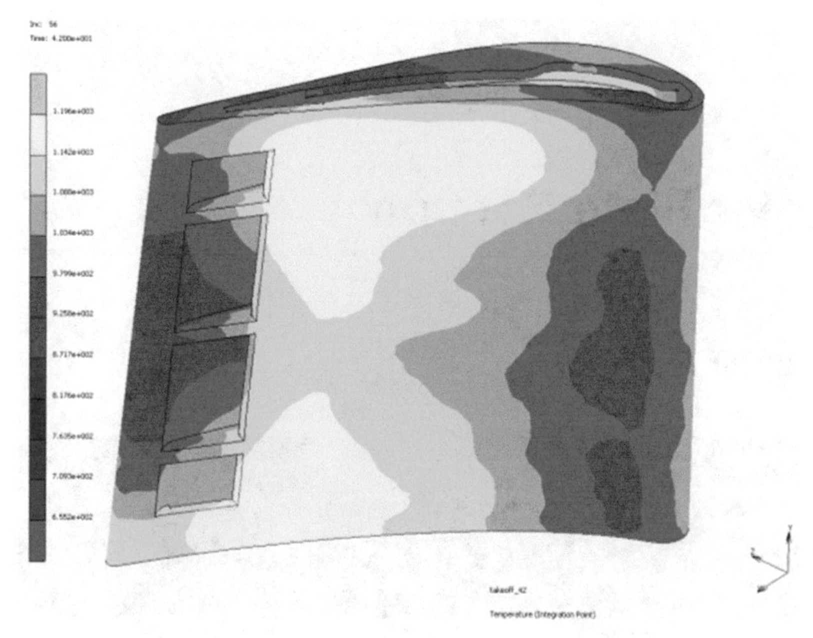

Figure 14.5. Temperature distribution in the NGV.

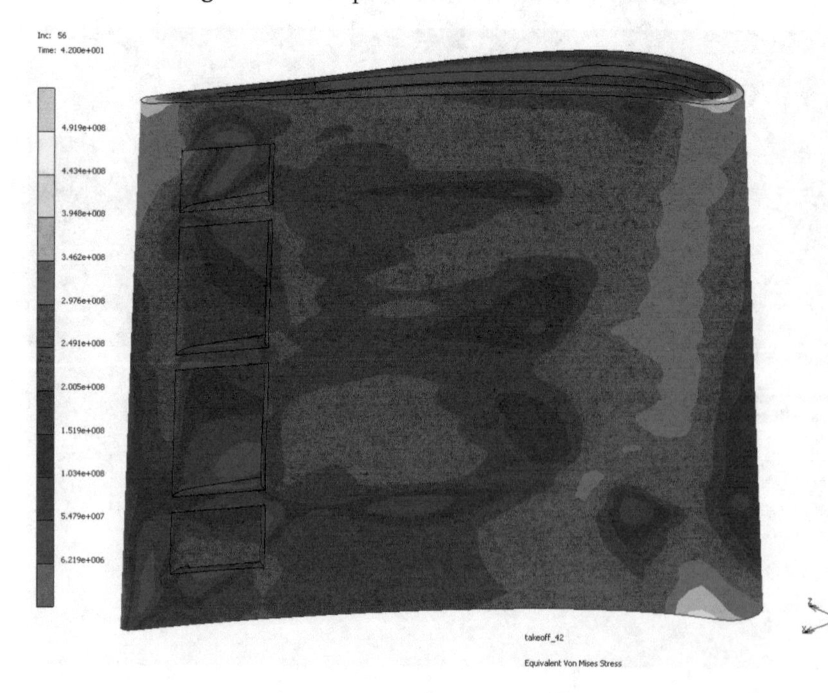

Figure 14.6. Stress distribution in the NGV (material A).

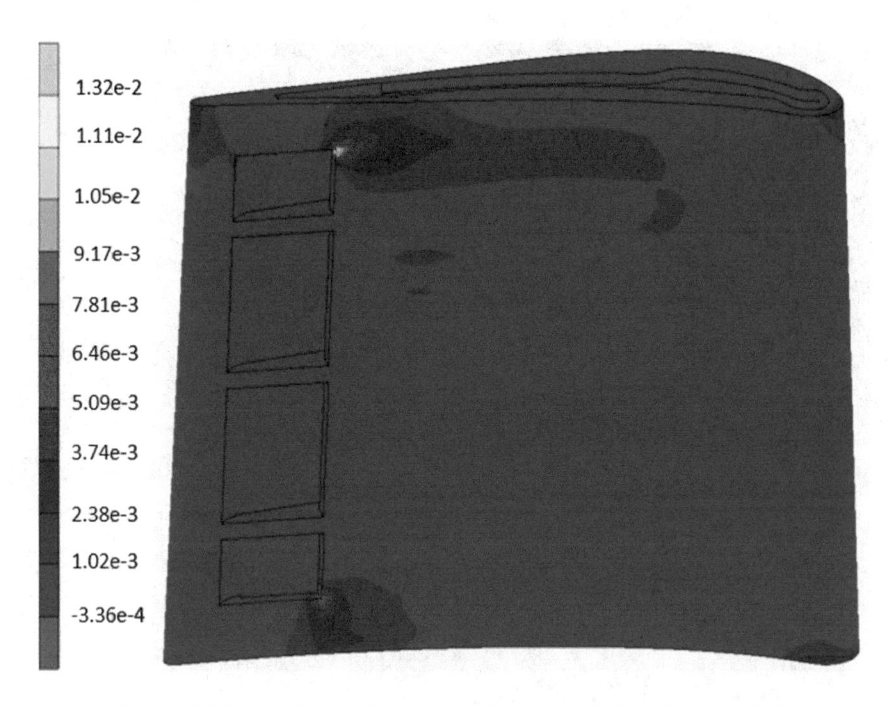

Figure 14.7. Locations of plastic strain accumulation (material A).

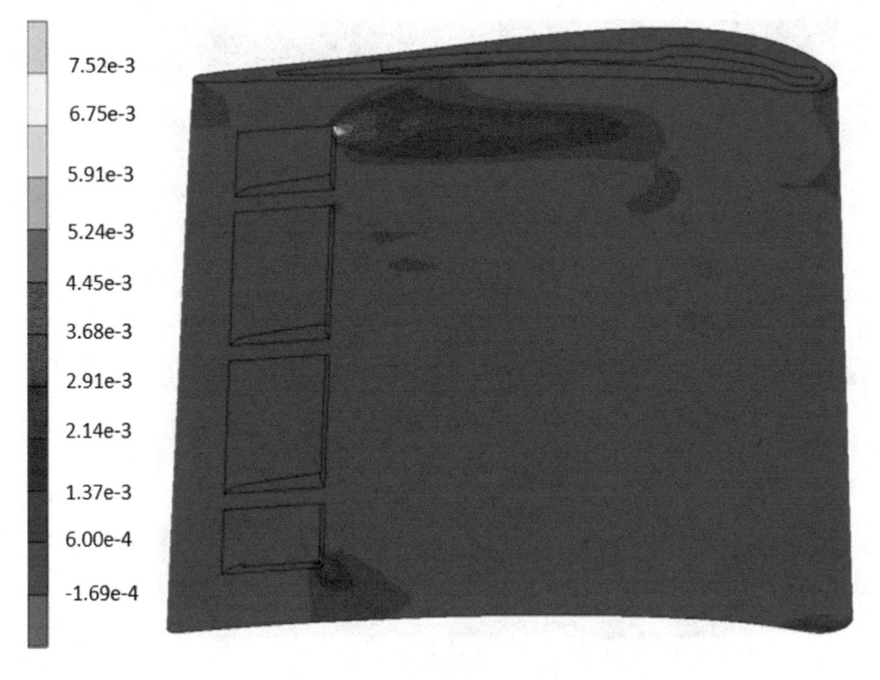

Figure 14.8. Locations of creep strain accumulation (material A).

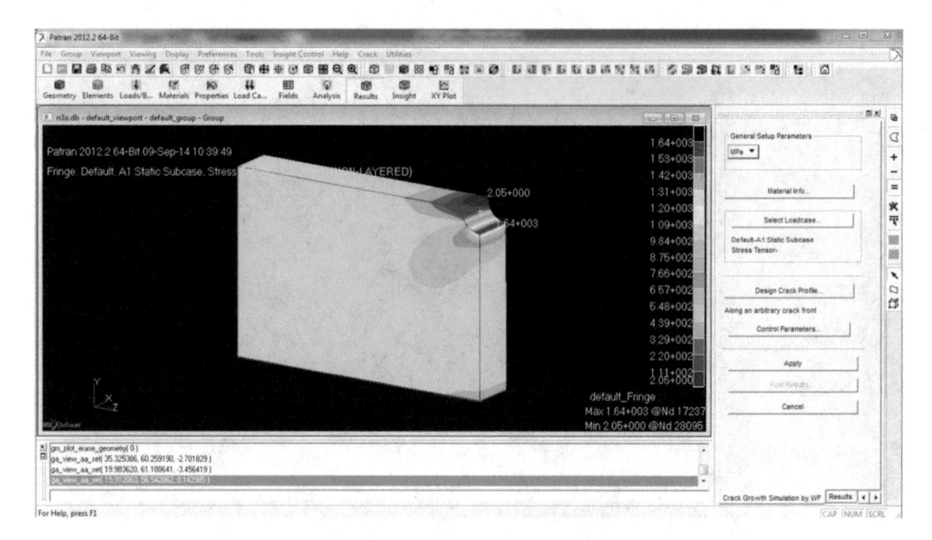

Figure 14.15. The analysis of notched specimen with added 3DCrackpro post-processing features in MSC.Patran.

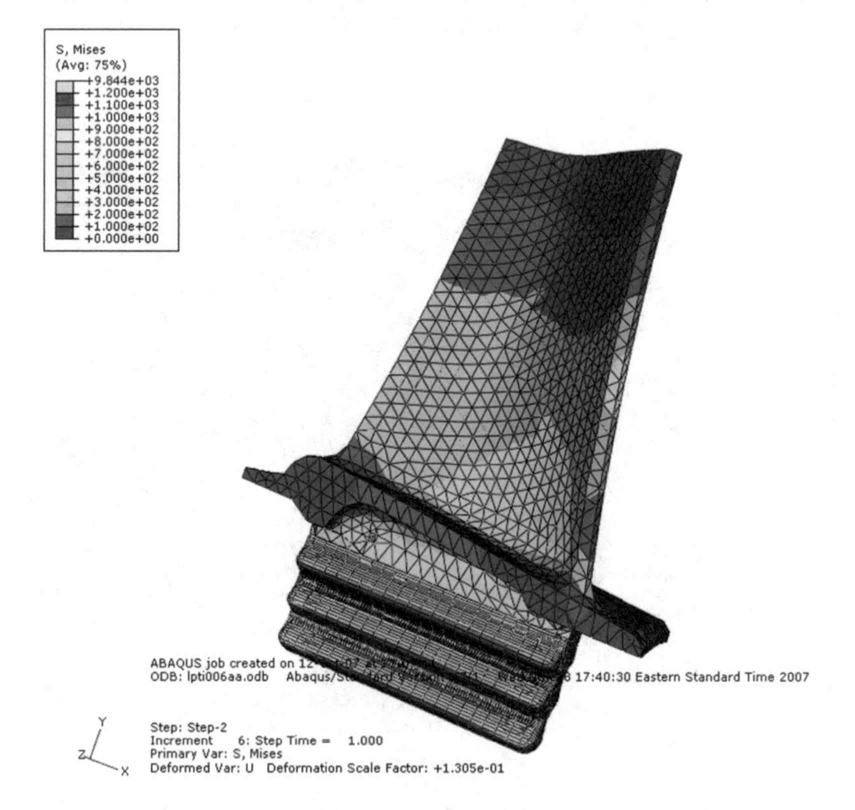

Figure 14.17. The von Mises stress in a turbine blade. The arrow indicates where a crack would form.

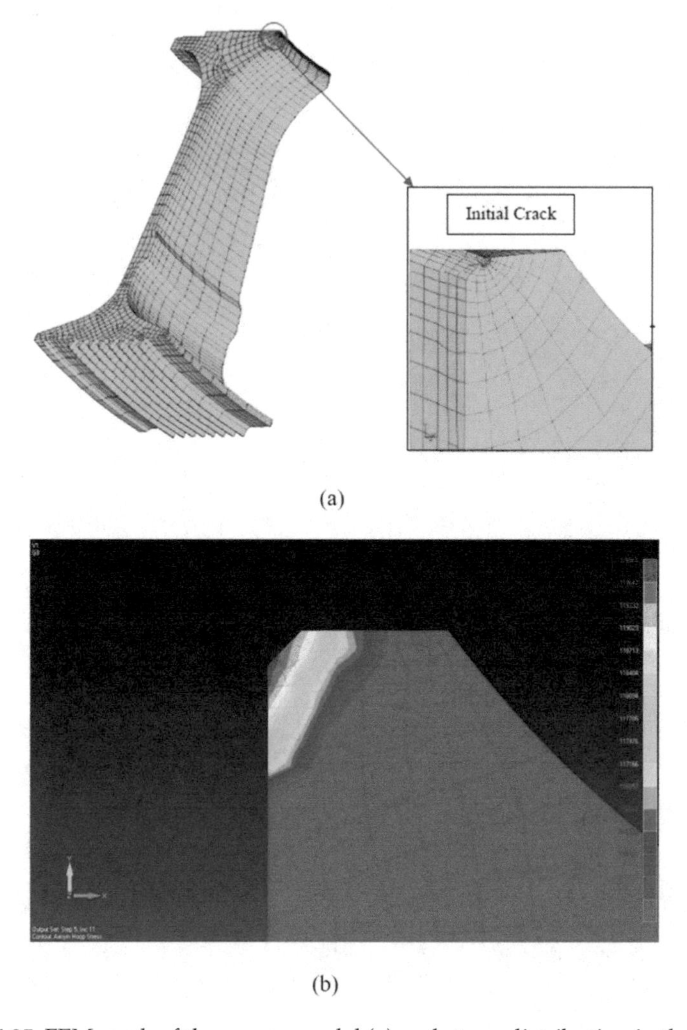

(a)

(b)

Figure 14.25. FEM mesh of the spacer model (a) and stress distribution in the vicinity of aft bore corner (Beres 2002).